Progress in Mathematics 1

Edited by
J. Coates and
S. Helgason

Herbert Gross

# Quadratic Forms in Infinite Dimensional Vector Spaces

Birkhäuser
Boston, Basel, Stuttgart

Author

Herbert Gross
Mathematisches Institut
Universität Zürich
Freiestrasse 36
CH-8032 Zürich
Switzerland



**Library of Congress Cataloging in Publication Data**

Gross, Herbert, 1936-
Quadratic forms in infinite-dimensional vector spaces.

(Progress in mathematics ; 1)
Includes bibliographical references and indexes.
1. Forms, Quadratic. 2. Vector spaces. I. Title.
II. Series: Progress in mathematics (Boston) ; 1. QA243.G76 512′.33 79-16626
ISBN 3-7643-1111-8

**CIP—Kurztitelaufnahme der Deutschen Bibliothek**

*Gross, Herbert:* Gross, Herbert:
Quadratic forms in infinite dimensional vector spaces / Herbert Gross. — Boston, Basel, Stuttgart : Birkhäuser, 1979.
(Progress in mathematics ; 1)
ISBN 3-7643-1111-8



ISBN 3-7643-1111-8

Cover design: Albert Gomm SWb/asg, Basel

Printed in USA

To Esther

# Preface

For about a decade I have made an effort to study quadratic forms in infinite dimensional vector spaces over arbitrary division rings. Here we present in a systematic fashion half of the results found during this period, to wit, the results on denumerably infinite spaces ("$\aleph_0$- forms"). Certain among the results included here had of course been published at the time when they were found, others appear for the first time (the case, for example, in Chapters IX , X , XII where I include results contained in the Ph.D.theses by my students W. Allenspach, L. Brand, U. Schneider, M. Studer).

If one wants to give an introduction to the geometric algebra of infinite dimensional quadratic spaces, a discussion of $\aleph_0$- dimensional spaces ideally serves the purpose. First, these spaces show a large number of phenomena typical of infinite dimensional spaces. Second, most proofs can be done by recursion which resembles the familiar procedure by induction in the finite dimensional situation. Third, the student acquires a good feeling for the linear algebra in infinite dimensions because it is impossible to camouflage problems by topological expedients (in dimension $\aleph_0$ it is easy to see, in a given case, whether topological language is appropriate or not).

Two more remarks are in order. Since classical Hilbert spaces have either finite or uncountable dimensions there will be no overlapping with Hilbert space theory here. And, finally, we wish to point out that we have made no steps to generalize away from vector spaces even in cases where such a possibility was in view.

The manuscripts for the book have been critically read and reread by Dr. Werner Bäni. He has eliminated a large number of errors. Yet of greatest importance to me has been his acute mathematical judgement on disputable matters in the texts. I express my warmest thanks to him.

Zurich, March 1979 Herbert Gross

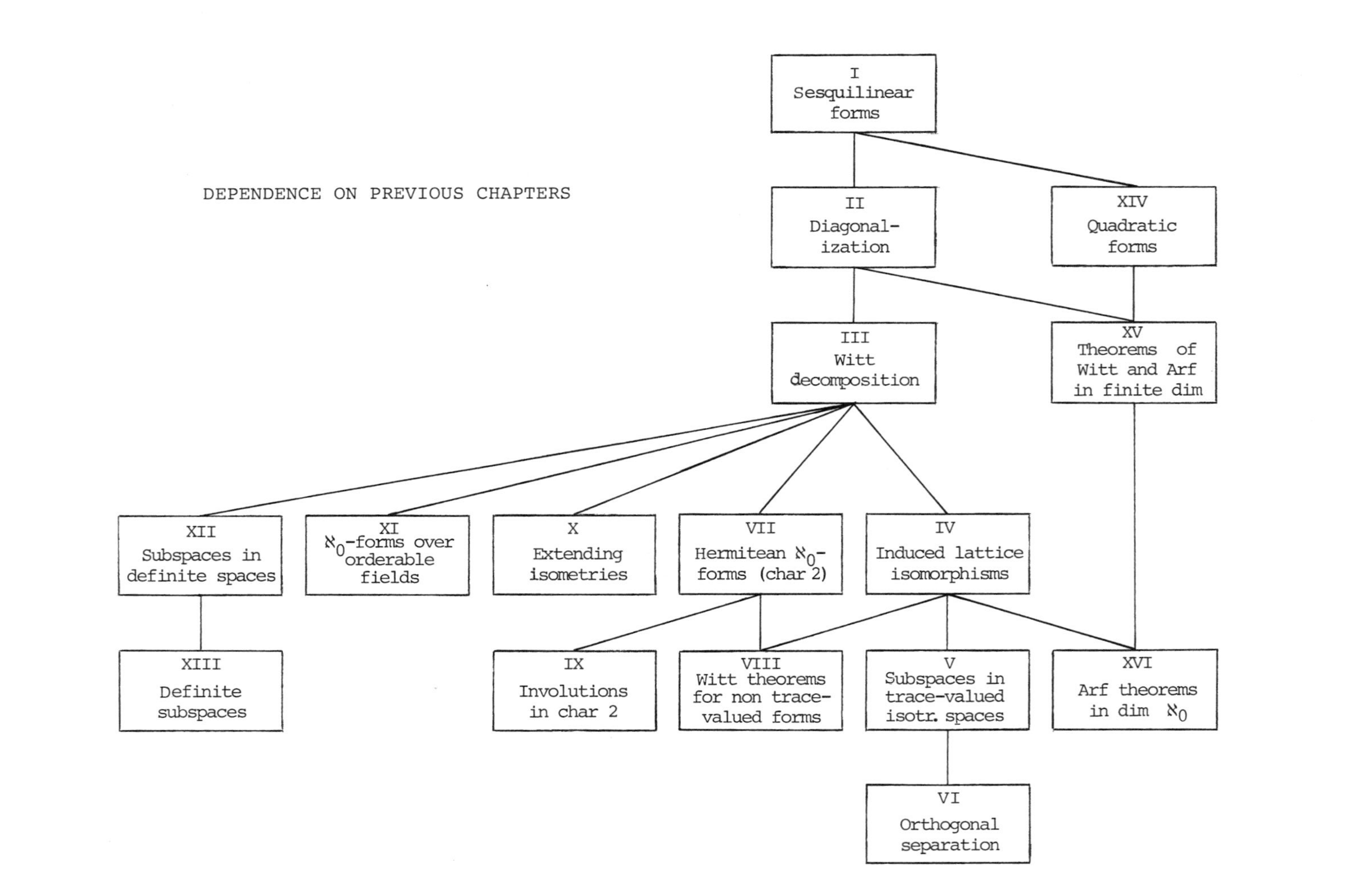
DEPENDENCE ON PREVIOUS CHAPTERS
I
Sesquilinear forms
II
Diagonal-ization
XIV
Quadratic forms
III
Witt decomposition
XV
Theorems of Witt and Arf in finite dim
XII
Subspaces in definite spaces
XI
$\aleph_0$-forms over orderable fields
X
Extending isometries
VII
Hermitean $\aleph_0$-forms (char 2)
IV
Induced lattice isomorphisms
XIII
Definite subspaces
IX
Involutions in char 2
VIII
Witt theorems for non trace-valued forms
V
Subspaces in trace-valued isotr. spaces
XVI
Arf theorems in dim $\aleph_0$
VI
Orthogonal separation

# Contents

## III. WITT DECOMPOSITIONS FOR HERMITEAN $\aleph_0$- FORMS



## IV. ISOMORPHISMS BETWEEN LATTICES OF LINEAR SUBSPACES WHICH ARE INDUCED BY ISOMETRIES

V. SUBSPACES IN TRACE - VALUED SPACES WITH MANY ISOTROPIC VECTORS



VI. ORTHOGONAL AND SYMPLECTIC SEPARATION



VII. CLASSIFICATION OF HERMITEAN FORMS IN CHARACTERISTIC TWO



VIII. SUBSPACES IN NON - TRACE - VALUED SPACES

## IX. INVOLUTIONS IN HERMITEAN SPACES IN CHARACTERISTIC TWO



## X. EXTENSION OF ISOMETRIES



## XI. CLASSIFICATION OF FORMS OVER ORDERED FIELDS

XII. CLASSIFICATION OF SUBSPACES IN SPACES WITH DEFINITE FORMS



XIII. CLASSIFICATION OF ⊥-DENSE SUBSPACES WITH DEFINITE FORMS

## XIV. QUADRATIC FORMS



## XV. WITTs THEOREM IN FINITE DIMENSIONS



## XVI. ARFs THEOREM IN DIMENSION $\aleph_0$

# INTRODUCTION

No one would assert that finite dimensionality is an intrinsic feature of the concept of quadratic form. Yet, apart from a very small number of results (see References to Chapter XI ) there has been, as far as we know, only Kaplansky's 1950 paper on infinite dimensional spaces pointing our way, namely in the direction of a purely algebraic theory of quadratic forms on infinite dimensional vector spaces over "arbitrary" division rings. Such a theory would, naturally, leave aside the highly developed theory of Hilbert spaces and its relatives, Krein spaces, Pontrjagin spaces (see [2] for an orientation on these topics). Furthermore, when we speak of infinite dimensional geometric algebra we do not, in this book, mean discussion of the ramifications into geometry of hypotheses belonging to set theory nor, reversely, the study of axioms forced upon set theory by geometry. We simply mean that the (algebraic) dimension of the quadratic spaces is allowed to be infinite. Many problems of the finite dimensional setting remain perfectly meaningful and invite an investigation when the finiteness condition is removed. Our results show that it is possible to generalize, without rarefying, classical results from finite dimensional orthogonal geometry.

There are also rather specific problems which call for a bit of general theory. As an illustration consider the classification problem of hilbertean spaces in the sense of [8] ( A nondegenerate quadratic space $E$ is called hilbertean if for all subspaces $X$ of $E$ that are closed, $X = X^{\perp\perp}$ (biorthogonal), we have $E = X \oplus X^{\perp}$ ; loc. cit. p. 177 .) . A complete classification would be of some interest to quantum logic. Nobody has been able to produce an infinite dimensional hilbertean space other than classical Hilbert space (real or complex or quaternionic). Under certain additional restrictions it has, in fact, been possible to prove that there are none (cf. [3], [5], [9] in Appendix I to Chapter I ).

These proofs resemble "proofs by circumstantial evidence": a number of exotic possibilities has to be ruled out in order to get at what looks like the probable issue. The elimination requires knowledge about "arbitrary" forms. (Incidentally, either outcome of the classification would be exciting: the existence of an infinite dimensional hilbertean space other than Hilbert space or else the characterization of Hilbert space by the property ⟨hilbertean⟩ .)

Apart from a few glimpses into the uncountable (e.g. in the introduction to Chapter II or in Section 8 of Chapter X ) infinite dimension in this volume means dimension $\aleph_0$ . Our text does not presuppose any knowledge about proofs in finite dimensional orthogonal geometry; however, the book is intended for readers who are acquainted with some of the classical results. Where should motivation for our endeavor come from if not from finite dimensional geometry! For the novice we mention that [1] [3] [4] [5] [6] [7] are a few of the excellent texts that treat finite dimensional orthogonal geometry. On the other hand, we do assume thorough familiarity with linear algebra. On a very few occasions we do assume a superficial acquaintance with topology. Besides the characterization of isometry classes of quadratic spaces we have focused our main effort in this book onto the characterization of subspaces in quadratic spaces (modulo the action of the orthogonal group associated to the space). Problems of this and related kinds are often referred to by using the adjective "Witt" (adjectives which cannot be negated should be spelled with capital initials). By using the table of contents it is easy to locate where what is being proved.

Needless to say that there is a number of pretty and unsolved problems in connection with $\aleph_0$- forms (let alone the higher dimensions) . Perhaps we shall write up a list some day. Here is a sample which I found several years ago and which has already puzzled some specialists:

Is there any commutative field which admits no anisotropic $\aleph_0$- form but which has infinite u-invariant, i.e. admits, for each $n \in \mathbb{N}$, some anisotropic form in n variables?

*


[1] E. Artin, Geometric Algebra. Interscience Publ. New York 1957.

[2] J. Bognár, Indefinite inner product spaces. Ergebnisse Band 78, Springer, Berlin Heidelberg New York 1974.

[3] J.W.S. Cassels, Rational Quadratic Forms. L.M.S. Monographs vol. 13 Academic Press, London New York 1978.

[4] M. Eichler, Quadratische Formen und orthogonale Gruppen. Grundlehren Band 63, 2. Aufl., Springer, Berlin Heidelberg New York 1974.

[5] I. Kaplansky, Linear Algebra and Geometry. Allyn and Bacon, Boston 1969.

[6] J. Milnor and D. Husemoller, Symmetric Bilinear Forms. Ergebnisse Band 73, Springer, Berlin Heidelberg New York 1973.

[7] O.T. O'Meara, Introduction to quadratic forms. Grundlehren Band 177 Springer, Berlin Heidelberg New York 1974.

[8] V.S. Varadarajan, Geometry of Quantum Theory, vol. 1. van Nostrand Princeton 1968.


<u>Postscript</u>. A fortnight after the above introduction had been written Hans A. Keller sent me a detailed description of an infinite dimensional hilbertean space different from the classical Hilbert spaces. (Ein nicht-klassischer Hilbertraum, pp. 1-16, letter to the author dated March 20 1979.) Refer to a forthcoming publication.

# CHAPTER ONE

# FUNDAMENTALS ON SESQUILINEAR FORMS

## Introduction

Chapter I contains some of the basic concepts and facts upon which subsequent chapters are built. The reader will find the terminology and notations that are used throughout the text. A number of fundamental definitions have been inserted in later chapters; whenever it had been possible to introduce a concept right where it is needed without interrupting the flow of ideas we have postponed its introduction.

On a very few occasions in this book we have made hints about questions relating to spaces of nondenumerable dimension (and made references to the literature; see e.g. the introduction of Chapter II). A number of references to work on sesquilinear spaces of uncountable dimension, upon which we shall not touch, is given in the bibliography at the end of this chapter; these are [3] , [10] , [11] , [14] , [16] , [20] , [24] , [28] , [37] , [42] .

## 1. Orthosymmetric sesquilinear forms

1.1 The underlying division rings. Let $k$ be a division ring. A bijection $\nu : k \to k$ is called antiautomorphism if $\nu(\alpha+\beta) = \nu(\alpha)+\nu(\beta)$ and $\nu(\alpha\beta) = \nu(\beta)\nu(\alpha)$ for all $\alpha , \beta \in k$ . If $k$ is commutative then the identity map is an antiautomorphism (it may be the only one as is witnessed by $k = \mathbb{Q} , \mathbb{R}$ ); if $k$ is skew then there may be none (see Appendix 1). All division rings in this book are assumed to admit antiautomorphisms. Antiautomorphisms permit the convertion of left vector spaces into right vector spaces and vice versa: if $E$ is a k-left vector space and $x \in E$ we set $x\lambda := \nu^{-1}(\lambda)x$ and verify all axioms of a k-right vector space. For example, the set $E^* = \mathrm{Hom}_k(E,k)$ of all k-linear maps $f : E \to k$ naturally is a k-right vector space under pointwise addition and multiplication by scalars from $k$ ; it can be made into a k-left vector space by defining $\lambda f = f\nu(\lambda)$ , i.e.

$$(\lambda f)(x) = f(x)\cdot\nu(\lambda) \quad \text{for all} \quad x \in E \,.$$

Notation: Sometimes we write $\lambda^{\nu}$ instead of $\nu(\lambda)$ , especially so if the antiautomorphism is written as $\lambda \mapsto \lambda^{*}$ or $\lambda \mapsto \bar{\lambda}$ .

1.2 The concept of sesquilinear form. Let $\nu : k \to k$ be an anti-automorphism of the division ring $k$ . With the k-left vector space $E$ we associate its dual $E^{*}$ , converted into a left vector space via $\nu$ as explained above. Both spaces being k-left vector spaces it is possible to consider linear maps

$$\varphi : E \longrightarrow E^{*} \,.$$

Each $\varphi$ gives raise to a function $\Phi : E \times E \to k$ by setting

$$\Phi(x,y) = \varphi y(x) \,. \tag{1}$$

The map $\Phi$ is additive in both arguments; it is linear in the first and $\nu$ - semilinear in the second argument:

$$\begin{aligned}
&\text{(i)} && \Phi(x+y,z) = \Phi(x,z) + \Phi(y,z) && \\
&\text{(ii)} && \Phi(x,y+z) = \Phi(x,y) + \Phi(x,z) && (x,y,z \in E)\\
&\text{(iii)} && \Phi(\lambda x,y) = \lambda\Phi(x,y) && \\
&\text{(iv)} && \Phi(x,\lambda y) = \Phi(x,y)\lambda^{\nu} && (x,y \in E\,;\ \lambda \in k)
\end{aligned}$$

Any map $\Phi : E \times E \to k$ which satisfies the four conditions (i) through (iv) is called a sesquilinear form with respect to $\nu$ ( ses-qui∿ = $1\frac{1}{2}$∿ ; from the Latin ) . For fixed $(k,\nu)$ we let $\mathrm{Sesq}_{\nu}(E)$ be the additive group of all such forms. If $\Phi \in \mathrm{Sesq}_{\nu}(E)$ and $y \in E$ are kept fixed then $x \mapsto \Phi(x,y)$ is a linear map $\Phi_y : E \to k$ by (i) and (iii). By (ii) and (iv) we see that $y \mapsto \Phi_y$ (from $E$ into $E^{*}$ )

is linear if $E^*$ is considered as a left vector space as in 1.1. Thus $\Phi$ is obtained from $\varphi : y \mapsto \Phi_y$ via (1). We see that (1) establishes a bijection between the set of linear maps $\varphi : E \to E^*$ and the set $\mathrm{Sesq}_\nu(E)$.

If $\nu$ is the identity, which forces $k$ to be commutative, then we obtain the concept of bilinear form.

1.3 Orthosymmetric forms are $\varepsilon$-hermitean. Let $\Phi \in \mathrm{Sesq}_\nu(E)$. We say that the vector $x \in E$ is perpendicular - or orthogonal - to the vector $y \in E$ if and only if $\Phi(x,y) = 0$, and then we write $x \perp y$. In this book we are interested only in forms $\Phi$ for which $\perp$ turns out a symmetric relation,

(2) $\qquad x \perp y \iff y \perp x \qquad$ (axiom of orthosymmetry)

Forms with this property are termed orthosymmetric ($\perp$- symmetric for short). If $\Phi$ is orthosymmetric and $Y \subset E$ we define $Y^\perp$, as usual, to be $\{x \in E \mid \Phi(x,y) = 0 \text{ for all } y \in Y\}$; if $y \in E$ then $y^\perp$ is $\{y\}^\perp$. If $Y$ is a subspace we call $Y^\perp$ the orthogonal complement of $Y$ in $E$, or simply the orthogonal of $Y$ in $E$. We have for all $X, Y \subset E$

(3) $\qquad X \subset Y \implies Y^\perp \subset X^\perp$

(4) $\qquad X \subset (X^\perp)^\perp$

from which we conclude

(5) $\qquad X^\perp = ((X^\perp)^\perp)^\perp$.

Indeed, (3) applied to (4) yields $\supset$ in (5); application of (4) to $X^\perp$ gives the converse inclusion in (5). This establishes (5). In the following we shall save on brackets, $X^{\perp\perp} = (X^\perp)^\perp$ etc. From (3) and (4) it follows furthermore that

(6) $\qquad (X+Y)^\perp = X^\perp \cap Y^\perp$.

<u>Proof</u>. Since $X^{\perp} \cap Y^{\perp} \subset X^{\perp}$ we obtain $X \subset X^{\perp\perp} \subset (X^{\perp} \cap Y^{\perp})^{\perp}$ by (3) and (4) and, symmetrically, $Y \subset (X^{\perp} \cap Y^{\perp})^{\perp}$ so that $X + Y \subset (X^{\perp} \cap Y^{\perp})^{\perp}$. Therefore, by (3) and (4), $X^{\perp} \cap Y^{\perp} \subset (X^{\perp} \cap Y^{\perp})^{\perp\perp} \subset (X+Y)^{\perp}$. On the other hand, $X \subset X + Y$ so $(X+Y)^{\perp} \subset X^{\perp}$ by (3) and, of course, $(X+Y)^{\perp} \subset Y^{\perp}$ by symmetry. Thus $(X+Y)^{\perp} \subset X^{\perp} \cap Y^{\perp}$.

Notice that we <u>cannot</u> prove the dual property "$(X \cap Y)^{\perp} = X^{\perp} + Y^{\perp}$" in this style: Interchanging $+$ and $\cap$ corresponds to switching from $\subset$ to $\supset$ ; but axiom (4) is not immune to turning around $\subset$ . (In fact, the attractiveness of the infinite dimensional theory of forms derives, to a large extent, from the fact that this property does <u>not</u> carry over from finite dimensional geometry.) We now turn to the assertion in the caption.

<u>Theorem 1</u>. Let $\Phi \in \mathrm{Sesq}_{\nu}(E)$ be orthosymmetric and assume that $\dim E / E^{\perp} > 1$ . Then the square $\nu^2$ is an inner automorphism of $k$ ,

$$\nu^2(\lambda) = \varepsilon^{-1} \lambda \varepsilon \qquad \text{(for all } \lambda \in k \text{)} , \text{ further} \tag{7}$$

$$\nu(\varepsilon) \cdot \varepsilon = \varepsilon \cdot \nu(\varepsilon) = 1 , \tag{8}$$

and $\Phi$ satisfies

$$\Phi(y, x) = \varepsilon\, \Phi(x, y)^{\nu} \qquad \text{(for all } x, y \in E \text{)} . \tag{9}$$

<u>Proof</u>. Let $E_o$ be a supplement of $E^{\perp}$ in $E$ ; $\dim E_o \geq 2$ by assumption of the theorem. For fixed $y \in E_o$ we consider the linear maps $x \mapsto \Phi(x, y)$ and $x \mapsto \nu^{-1}(\Phi(y, x))$ from $E$ into $k$ . By orthosymmetry their kernels coincide so the two functionals are proportional: there is nonzero $\mu(y) \in k$ such that $\Phi(x,y) = [\nu^{-1}(\Phi(y,x))]\mu(y)$ . How does $\mu$ depend on $y$ ? For $y_1, y_2 \in E_o$ we have $\Phi(x, y_1 + y_2) = [\nu^{-1}(\Phi(y_1 + y_2, x))]\mu(y_1 + y_2)$ ; on the other hand $\Phi(x,y_1) + \Phi(x,y_2) = [\nu^{-1}(\Phi(y_1,x))]\mu(y_1) + [\nu^{-1}(\Phi(y_2,x))]\mu(y_2)$ . Assume first that $y_1, y_2 \in E_o$ are linearly independent. We assert that $\mu(y_1) = \mu(y_2)$ .

Indeed, we cannot have $y_1^{\perp} \subset y_2^{\perp}$ or $y_2^{\perp} \subset y_1^{\perp}$ for otherwise there would be a nonzero linear combination $\eta_1 y_1 + \eta_2 y_2 \perp E$ which is impossible since $E_o \cap E^{\perp} = (0)$ . Hence we can pick $x \in E$ perpendicular to one of the y's and not perpendicular to the other. Substitutions of such $x$ lead to $\mu(y_1) = \mu(y_1 + y_2) = \mu(y_2)$ . If, on the other hand $y_1$ and $y_2$ are dependent, then we pick $z \in E_o$ not on the line $(y_1)$ and conclude $\mu(y_1) = \mu(z) = \mu(y_2)$ by the former result. Thus $\Phi(x,y) = [\nu^{-1}(\Phi(y,x))]\mu$ for all $x \in E$, $y \in E_o$ . Since this equality holds trivially when $y \in E^{\perp}$ we conclude that it holds for all $y \in E$ . Setting $\varepsilon := \nu(\mu^{-1})$ we obtain (9). Applying (9) twice yields $\Phi(y,x) = \varepsilon\Phi(x,y)^{\nu} = \varepsilon(\varepsilon\Phi(y,x)^{\nu})^{\nu} = \varepsilon\nu^2(\Phi(y,x))\nu(\varepsilon)$ for all $x , y \in E$ . For arbitrary $\lambda \in k$ there exist $x , y \in E$ with $\Phi(y,x) = \lambda$ . Choosing $\lambda = 1$ yields $1 = \varepsilon \cdot \nu(\varepsilon)$ , hence (8) and $\lambda = \varepsilon\nu^2(\lambda)\varepsilon^{-1}$ for all $\lambda \in k$ (as asserted by (7) ) . This finishes the proof of Theorem 1.

We call <u>$\varepsilon$ - hermitean</u> any sesquilinear form $\Phi \in \mathrm{Sesq}_{\nu}(E)$ which satisfies (9); such forms are orthosymmetric, obviously. Furthermore, (9) implies (7) and (8). Therefore, instead of speaking about orthosymmetric forms we may just as well consider $\varepsilon$ - hermitean forms attached to a structure $(k, \nu, \varepsilon)$ that satisfies (7) and (8). In particular, with each orthosymmetric form (which has $\dim E / E^{\perp} > 1$) is associated, via Theorem 1, the additive subgroup $S$ of $k$ consisting of the <u>symmetric elements</u>,

(10) $$S := \{ \xi \in k \mid \xi = \varepsilon \xi^{\nu} \}$$

and the subgroup $T \subset S$ of <u>traces</u>,

(11) $$T := \{ \xi + \varepsilon \xi^{\nu} \mid \xi \in k \} .$$

If the characteristic of $k$ is not 2 then $T = S$ because each symmetric $\xi$ is of the form $\xi = (\tfrac{1}{2}\xi) + \varepsilon(\tfrac{1}{2}\xi)^{\nu}$ ; if the characteristic is 2 then the quotient $S/T$ is a crucial object (Chapters VIII, IX, XIV, XVI). We still may have $S = T$ in characteristic 2; e.g. whenever there exists an element $\gamma$ in the center with $\gamma \neq \gamma^{\nu}$ (if $\xi \in S$ then $\xi = \eta + \varepsilon\eta^{\nu}$ with $\eta := \gamma(\gamma + \gamma^{\nu})^{-1}\xi$).

Remark 1. Bourbaki ([6], p. 49), in his definition of $\varepsilon$-hermitean form, stipulates that $\varepsilon$ belong to the center of $k$. The antiautomorphism $\nu$ is then an involution by (7), $\nu^2 = \mathbb{1}$. We have no need to make such an assumption.

Definition 1. A sesquilinear space is a pair $(E, \Phi)$ where $E$ is a k-left vector space and $\Phi : E \times E \to k$ an $\varepsilon$-hermitean form in the sense of (9).

If there is no risk of confusion we shall often speak about the sesquilinear space $E$ and omit mention of the form; we shall sometimes use the terms "space" and "form" interchangeably, e.g. we shall say that the form is $\aleph_0$-dimensional when we mean that the space has dimension $\aleph_0$; etc.

In euclidean 3-space $(\mathbb{R}, \Phi)$, where $\Phi$ is the inner product ("dot product") $\Phi((x_i),(y_i)) = \sum_{i=1}^{3} x_i y_i$, the number $\Phi(x,x)$ is the square of the "length" of the vector $x$ ($\Phi(x,x)$ is sometimes called the norm of the vector $x = (x_i)$). Most of the division rings admitted in the theory of forms are such that $\Phi(x,x)$ will not be the square of anything in $k$ so that there is no analogue of the classical length. Because "norm" has too many meanings already we simply call length of $x$ the element $\|x\| := \Phi(x,x)$. A vector of length 1 is called a unit vector. By $\|E\|$ (or $\|\Phi\|$) we mean the set $\{\|x\| \mid x \in E \setminus \{0\}\}$ of lengths of all nonzero vectors, i.e. the set of all elements in $k$ that are nontrivially represented by $\Phi$. Notice that $\|\Phi\| \subset S$ as $\Phi$ is $\varepsilon$-hermitean. Similarly, for $F \subset E$ a subspace, $\|F\| := \{\|f\| \mid f \in F \setminus \{0\}\}$.

1.4 Zoology of forms. Let $\Phi : E \times E \longrightarrow k$ be an $\varepsilon$ - hermitean form with respect to the antiautomorphism $\nu$ . If $\varepsilon = 1$ then the form is called hermitean. If $\varepsilon = 1$ and $\nu = 1\!\!1$ (possible only when $k$ is commutative) then $\Phi$ is called symmetric. If $\varepsilon = -1$ the form is termed skew-hermitean; if $\varepsilon = -1$ and $\nu = 1\!\!1$ the form is termed skew-symmetric. A form $\Phi$ which has $\Phi(x,x) = 0$ for all $x \in E$ is termed alternate; we have:

(12) If the restriciton of $\Phi$ to the subspace $X \subset E$ is alternate but not identically zero then $\Phi$ is skew-symmetric on the entire space $E$ ; in particular, $k$ is commutative.

Indeed, pick $x , y \in X$ with $\Phi(x,y) = 1$ . Then, for all $\lambda \in k$ , $0 = \|\lambda x + y\| = 0 + 0 + \lambda + \varepsilon \lambda^{\nu}$ . Hence $\varepsilon = -1$ (substitute $\lambda = 1$ ) and therefore $\lambda^{\nu} = \lambda$ , i.e. $\nu = 1\!\!1$ .

Alternate forms are skew-symmetric; if char $k \neq 2$ then the converse holds true too. As a corollary of Theorem 1 we note

(13) Let $\Phi \in \mathrm{Sesq}_{1\!\!1}(E)$ be orthosymmetric and assume that $\dim E / E^{\perp} > 1$ . Then $\Phi$ is symmetric or alternate.

Indeed, if $\Phi$ is not alternate, then for some nonzero $\Phi(x,x)$ we have $\Phi(x,x) = \varepsilon \Phi(x,x)$ by (9), hence $\varepsilon = 1$ .

1.5 Scaling of forms. If $\Phi \in \mathrm{Sesq}_{\nu}(E)$ and $\mu \in k \setminus \{0\}$ then the right multiple $\Phi_1 := \Phi\mu$ belongs to $\mathrm{Sesq}_{\nu_1}(E)$ where

(14) $$\nu_1(\lambda) = \mu^{-1} \nu(\lambda) \mu$$

If, in addition, $\Phi$ is $\varepsilon$ - hermitean then $\Phi_1$ is $\varepsilon_1$ - hermitean with $\varepsilon_1 = \varepsilon \nu(\mu^{-1}) \mu$ . We assert:

(15) Let $\Phi$ be $\varepsilon$-hermitean. If $\Phi$ is not skew-symmetric then there are $\mu_1 \in k \setminus \{0\}$ such that $\Phi_1 = \Phi\mu_1$ is hermitean; if $\Phi_1$ is not symmetric then there are $\mu_2 \in k \setminus \{0\}$ such that $\Phi_2 = \Phi_1\mu_2$ is skew-hermitean (relative to antiautomorphisms which differ among themselves by factors which are inner).

Proof. If $\Phi$ is not skew-symmetric then there are $a, b \in E$ with $\Phi(a,b) \neq -\Phi(b,a)$ so $\rho := \Phi(a,b) + \varepsilon\Phi(a,b)^\nu \neq 0$. We have $\varepsilon\rho^\nu = \rho$ and $\lambda^\tau := (\rho\lambda\rho^{-1})^\nu$ defines an involution $\tau : k \to k$ relative to which $\Phi_1 := \Phi\mu_1$, $\mu_1 = \rho^\nu$, turns out hermitean. Assume then that $\Phi_1$ (or any hermitean form) is not symmetric: there are $\alpha \in k$ with $\alpha^\tau \neq \alpha$ so $\beta := \alpha^\tau - \alpha \neq 0$. Setting $\lambda^\sigma := \beta^{-1}\lambda^\tau\beta$ defines another involution of $k$ with respect to which $\Phi_2 = \Phi_1\beta$ turns out to be skew-hermitean.

The transition from $\Phi \in Sesq_\nu(E)$ to $\Phi_1 = \Phi\mu \in Sesq_{\nu_1}(E)$, with $\nu_1$ defined by (14), is called <u>scaling</u>. Certain properties are immune to scaling, e.g. "$\Phi$ is orthosymmetric"; others are not, e.g. "$\Phi(x,x) = 1$ has a solution". For the behaviour of $S$, $T$ in (10), (11) under scaling see Appendix I to Chapter XIV.

1.6 <u>Existence of $\varepsilon$-hermitean forms</u>. Let $\Phi \in Sesq_*(E)$ and $(e_\iota)_{\iota \in I}$ be some fixed basis of $E$. Incidentally, all bases in this book are bases in the sense of linear algebra (so-called <u>Hamel bases</u>): each $x \in E$ is a <u>finite</u> linear combination of some $e_\iota$. Thus, if another basis $(f_\iota)_{\iota \in I}$ is introduced, $f_\iota = \Sigma\alpha_{\iota\kappa}e_\kappa$, then the "matrix" $A = (\alpha_{\iota\kappa})_{\iota,\kappa \in I}$ is <u>row-finite</u>: in each "row" all but a finite number of entries from $k$ are zero (meaning, as usual, that for each $\iota \in I$ there are finitely many $\kappa_1, \ldots, \kappa_n$ such that $\alpha_{\iota\kappa} = 0$ for all $\kappa \in I \setminus \{\kappa_1, \ldots, \kappa_n\}$). Clearly, such a substitution matrix $A$ need not be column-finite. We abbreviate $\gamma_{\iota\kappa} := \Phi(e_\iota, e_\kappa)$ and call $C := (\gamma_{\iota\kappa})_{\iota,\kappa \in I}$ the matrix of the form $\Phi$ relative to the basis $(e_\iota)_{\iota \in I}$ (the hybrid quantities of old. See [1], § 10, pp.51). For $x = \Sigma\, \xi_\iota e_\iota$, $y = \Sigma\, \eta_\iota e_\iota$ two typical elements of E we find, by sesquilinearity,

$$\Phi(x,y) = \sum_{\iota,\kappa \in I} \xi_\iota \gamma_{\iota\kappa} \eta^*_\kappa \quad . \tag{16}$$

As there are only finitely many nonzeros among $\xi_\iota$, $\eta_\kappa$ the double sum in (16) makes sense even if C has infinite rows and columns. If $\Phi$ is $\varepsilon$-hermitean then ${}^{tr}C = \varepsilon C^*$,

$$\gamma_{\iota\kappa} = \varepsilon \gamma^*_{\kappa\iota} \quad . \tag{17}$$

Conversely, we can use (16), (17) in order to define $\varepsilon$-hermitean forms on any vector space E over a field $(k, *, \varepsilon)$ where (7) and (8) hold, i.e. $(\lambda^*)^* = \varepsilon^{-1} \lambda \varepsilon$ and $\varepsilon^* \varepsilon = \varepsilon \varepsilon^* = 1$. In this connection the following Theorem by Albert is relevant ([1], Thm. 19, Chap. X):

(18) Let the division ring k be finite dimensional over its center. If k admits an antiautomorphism leaving every element of the center fixed then k admits an involution of the first kind (i.e. leaving every element of the center fixed).

For proofs of this theorem see also [21], [38]. Thus, over such k we can define $\varepsilon$-hermitean forms, and then quite a few. (Cf. Appendix I)

Definition 2. A basis $\mathcal{B} = (e_\iota)_{\iota \in I}$ of a sesquilinear space $(E, \Phi)$ is called orthogonal basis if $e_\iota \perp e_\kappa$ for all $\iota \neq \kappa$ in I, i.e. if the matrix of the form $\Phi$ with respect to $\mathcal{B}$ is diagonal. If card I ( = dim E ) $\leq \aleph_0$ we sometimes write $\Phi = \langle \alpha_1, \ldots, \alpha_n \rangle$ or $\Phi = \langle \alpha_1, \alpha_2, \ldots \rangle$ where $\alpha_\iota = \Phi(e_\iota, e_\iota)$ and $(e_\iota)_{\iota \in I}$ is an orthogonal basis for $\Phi$. $\mathcal{B}$ is called orthonormal if it is an orthogonal basis consisting of unit vectors, i.e. $\Phi(e_\iota, e_\kappa) = \delta_{\iota\kappa}$ (Kronecker).

Definition 3. Let $(E,\Phi)$ be a sesquilinear space and $X \subset E$ a subspace. The radical rad $X$ of $X$ is the subspace $X \cap X^{\perp}$; $X$ is called nondegenerate iff rad $X = (0)$ .

In particular, $E$ is nondegenerate if and only if $E^{\perp} = (0)$ , i.e. the zero vector is the only vector perpendicular to all of $E$ . If $E$ is nondegenerate we call $\Phi$ nondegenerate (since subspaces are always equipped with the restricted forms we say that $\Phi$ is nondegenerate on $X$ iff $X$ is nondegenerate etc.). If $C$ is the matrix of $\Phi$ with respect to some basis $\mathcal{B}$ then we see that $E$ is nondegenerate if and only if $C$ has full rank for any choice of $\mathcal{B}$ : if finitely many rows, say $(\gamma_{\iota\kappa})_{\kappa \in I}$ for $\iota \in J = \{\iota_1, \ldots, \iota_n\}$ , were linearly dependent, i.e. $\sum_J \lambda_\iota \gamma_{\iota\kappa} = 0$ for all $\kappa \in I$ and for suitable $\lambda_\iota$ $(\iota \in J)$ , not all zero, then $\sum \lambda_\iota e_\iota \perp E$ . The argument may be reversed. Since $C$ is an $\varepsilon$ - hermitean matrix we may furthermore confound left row - rank and right column - rank.

## 2. Trace - valued forms and hyperbolic planes

Let $\Phi : E \times E \longrightarrow k$ be $\varepsilon$ - hermitean with respect to some division ring $(k,*,\varepsilon)$ . A vector $x \in E$ is called isotropic iff $\Phi(x,x) = 0$ ; the space $E$ is called isotropic if it contains a nonzero isotropic vector. A subspace $X \subset E$ is called totally isotropic if $X \subset X^{\perp}$; i.e. if $\Phi$ vanishes on $X \times X$ .

Definition 4. A nondegenerate plane is called hyperbolic if it is spanned by two isotropic vectors.

Thus a hyperbolic plane $P$ always admits "canonical" bases such that the matrix of the form takes the shape

$$\begin{pmatrix} 0 & 1 \\ \varepsilon & 0 \end{pmatrix} ;$$

we see that $\|P\| = T$ where $T$ is as in (11).

Lemma 1. Let $(E, \Phi)$ be a nondegenerate isotropic space over $(k, *, \varepsilon)$. The following are equivalent: (i) $E$ admits a basis consisting of isotropic vectors and each nondegenerate isotropic plane in $E$ is hyperbolic; (ii) $\|E\| \subset T$ ($T$ as in (11)).

Definition 5. ([9], § 10). An $\varepsilon$-hermitean form $\Phi$ which satisfies (ii) in Lemma 1 is called trace-valued ("for each $x$ there is $\xi$ such that $\Phi(x,x) = \xi + \varepsilon\xi^*$).

Proof (of Lemma 1). Assume (ii). Let $\Phi(x,x) = 0$ and $\Phi(x,y) \neq 0$, say $\Phi(x,y) = 1$ for given $x,y \in E$. The equation $\|y - \xi x\| = \Phi(y,y) - (\xi + \varepsilon\xi^*) = 0$ can be solved for $\xi$ by (ii) so the plane $k(x,y)$ is hyperbolic. In particular, $E$ contains two isotropic vectors $x,x'$ with $\Phi(x,x') = 1$. If $z$ is any vector we set $x'' = z - \zeta x - \zeta' x'$. It is very easy to verify that we can solve (by (ii)) the equation $\|x''\| = 0$ for $\zeta$ and $\zeta'$. Hence $z = \zeta x + \zeta' x' + x''$ is a sum of three isotropic vectors. Since the set of isotropic vectors in $E$ is thus proved to be a set of generators we can select a basis of the required sort. This establishes (ii) $\Rightarrow$ (i). Assume (i). We have to show that $\Phi(x,x)$ is a trace for arbitrarily fixed $x \in E$. This is obvious for $x = 0$. If $x \neq 0$ then any basis of $E$ will contain a member $e$ with $\Phi(x,e) \neq 0$ since $E$ is nondegenerate. By (i) we may assume $e$ isotropic so $k(x,e)$ is a nondegenerate isotropic plane and hence hyperbolic: there is an isotropic vector $\alpha x + \beta e$ not on the line $(e)$, i.e. with $\alpha \neq 0$. This means that $0 = \|x + \alpha^{-1}\beta e\|$ from which we conclude that the equation $\Phi(x,x) = \xi + \varepsilon\xi^*$ has the solution $\xi = -\alpha^{-1}\beta\Phi(e,x)$. This completes the proof of the lemma.

We shall see that trace-valued forms have a theory which is considerably different from that of non-trace-valued forms. By an earlier remark (made just after (11)) forms are invariably trace-valued when the characteristic is not 2 or when the center is not left pointwise fixed (this takes place when $k$ is commutative and $*$ is not $\mathbb{1}$). If the characteristic is 2 and $\Phi$ is symmetric then $\Phi$ is trace-valued if and only if $\Phi$ is alternate (because zero is the only trace in this case).

Definition 6. A symplectic basis of a sesquilinear space $(E,\Phi)$ is a basis $\{r_\iota, r'_\iota \mid \iota \in I\}$ with $\Phi(r_\iota, r'_\iota) = 1$ for all $\iota \in I$ and all other products between members of the family equal to zero (symplectic = intertwined; from the Greek).

Clearly, if $E$ admits a symplectic basis $\{r_\iota, r'_\iota \mid \iota \in I\}$ then $E$ is the orthogonal sum of the hyperbolic planes $P_\iota$ spanned by $\{r_\iota, r'_\iota\}$, $E = \oplus P_\iota$. Conversely, if $E$ is such a sum and each $P_\iota$ is spanned by the isotropic basis $\{r_\iota, r'_\iota\}$ with $\Phi(r_\iota, r'_\iota) = 1$ then the family $\{r_\iota, r'_\iota \mid \iota \in I\}$ is a symplectic basis.

The two halves of a symplectic basis of $E$ (if there is such a basis) span totally isotropic subspaces, $R = k(r_\iota)_{\iota \in I}$, $R' = k(r'_\iota)_{\iota \in I}$, $E = R \oplus R'$. Notice that by nondegeneracy it follows that $R^\perp = R$ and $R'^\perp = R'$.

Notation: It is convenient to write $A \overset{\circ}{\oplus} B$ if in the sum $A \oplus B$ we have $A \subset A^\perp$ and $B \subset B^\perp$; this is in analogy to $A \overset{\perp}{\oplus} B$ where $A \subset B^\perp$ and $B \subset A^\perp$.

Remark 2 ([41], Satz 15, p.39). Let $*$ be an involution on $k$ and $\operatorname{card} k > 5$. Let $a_1 = \mu + \mu^*$, $a_2 = \sigma + \sigma^*$ be nonzero elements in $k$. Pick $\nu$ in the prime field of $k$ such that $\nu \neq 0$, $\nu^2\mu + \sigma \neq 0$, $\nu^2\mu^* - \sigma \neq 0$. For $\xi_1 \in k \setminus \{0\}$ a given element determine $\xi_2$ from $\xi_2(\nu^2\mu + \sigma) + \xi_1 a_1 \nu = 0$. Then we have

$$\xi_1 a_1 \xi_1^* + 0 a_2 0^* = (\xi_1 + \nu \xi_2) a_1 (\xi_1 + \nu \xi_2)^* + \xi_2 a_2 \xi_2^*$$

with $\xi_1 + \nu \xi_2 \neq 0$ , $\xi_2 \neq 0$ . Iteration of this procedure (and Sec. 1.5) shows: If $e_1, \ldots, e_n$ is an orthogonal basis of the trace-valued $\varepsilon$-hermitean form $\Phi = \langle a_1, \ldots, a_n \rangle$ and if there exists a point $\neq 0$ on the quadric $\Phi(x,x) = c$ then there exists a point $x = \sum_1^n \xi_i e_i$ on the same quadric with $\xi_i \neq 0$ for all $i = 1, \ldots, n$ .

## 3. Positive forms

If the underlying division ring has the additional feature of admitting an order structure then we can single out "positive" forms.

Definition 7. Let $\Phi : E \times E \longrightarrow k$ be $\varepsilon$-hermitean over $(k,*,\varepsilon)$. Assume that $k$ contains a (Hilbert-) ordered subdivision ring $(k_o,<)$ such that $\|E\| \subset k_o$ . We say that $\Phi$ is definite [semidefinite] iff

(19) $\qquad \|x\| \cdot \|y\| > 0 \quad [\geq 0] \quad$ for all $x, y \in E \setminus \{0\}$ .

If $\Phi$ is semidefinite and then, for all $x \in E$ , $\|x\|$ is either invariably $> 0$ , or $\geq 0$ , or $< 0$ , or $\leq 0$ then $\Phi$ is called (accordingly) positive definite, positive semidefinite, negative definite, negative semidefinite.

Remark 3. A Hilbert ordering on a skew field $k$ (= division ring) is defined just as on a commutative field: $\dot{k} = k \setminus \{0\}$ is assumed to contain a subset $P$ - whose elements are termed positive - such that the following "axioms" hold: for all $\alpha, \beta \in P$ and all $\gamma \in \dot{k}$ either $\gamma \in P$ or $-\gamma \in P$ , $\alpha + \beta \in P$ , $\alpha\beta^{-1} \in P$ . In other words, $P$ is an additively closed multiplicative subgroup of $\dot{k}$ and of index 2 in $\dot{k}$ . If $\alpha - \beta \in P$ then one writes $\alpha > \beta$ or $\beta < \alpha$ and the usual laws on inequalities can be verified. The reader who is interested in the existence of such orderings may jump to Section 4 in Chapter XI .

<u>Example</u>. Let $(k_o,<)$ be some ordered commutative field and $\alpha,\beta$ negative elements of $k_o$. The 4-dimensional $k_o$-algebra $(\frac{\alpha,\beta}{k_o})$ of so-called (generalized) <u>quaternions</u> $q = \xi_o + \xi_1 e_1 + \xi_2 e_2 + \xi_3 e_3$ $(\xi_i \in k_o)$ has the multiplication table $e_1 e_2 = -e_2 e_1 = e_3$, $e_1^2 = \alpha$, $e_2^2 = \beta$. The assignment $q \mapsto \bar{q} := \xi_o - \xi_1 e_1 - \xi_2 e_2 - \xi_3 e_3$ is an antiautomorphism (called "conjugation"). One finds that $N(q) := q\bar{q} = \bar{q}q = \xi_o^2 - \alpha\xi_1^2 - \beta\xi_2^2 + \alpha\beta\xi_3^2$. Thus $N(q) \leqq 0$ if and only if $q = 0$ because $\alpha, \beta < 0$. Therefore each nonzero $q$ has an inverse $q^{-1} = N(q)^{-1}\bar{q}$ and $k$ is a division ring. We now define hermitean forms $\Phi : E \times E \longrightarrow k$ over $(k, \bar{\ }, 1)$. By choosing diagonal the matrix $(\Phi(e_\iota, e_\kappa))$ of $\Phi$ (with respect to some basis $(e_\iota)_I$) and with positive diagonal elements $\Phi(e_\iota, e_\iota) \in k_o = S$ the form turns out positive definite. If $x = \Sigma\, q_\iota e_\iota$ is a typical vector we have $\|x\| = \Sigma\, \Phi(e_\iota, e_\iota) N(q) \leqq 0$ iff $x = 0$.

By scaling the hermitean form in the example we get new ($\varepsilon$-hermitean) definite forms. There are no other skew examples by the following

<u>Theorem 2</u>. Let $\Phi$ be hermitean over the division ring $k$ with involution $*$. Assume that $k$ contains a (Hilbert-) ordered division subring $(k_o,<)$ with $\|\Phi\| \subset k_o$. If $k$ is noncommutative and $\Phi$ definite on at least one line then $k = (\frac{\alpha,\beta}{k_o})$ with suitable negative $\alpha, \beta \in k_o$ and $*$ is the usual conjugation. If $*$ is not $\mathbb{1}$ but $k$ commutative and $\Phi$ is definite on at least one line then $k = k_o(\sqrt{\gamma})$ for some negative $\gamma \in k_o$ and $(a + b\sqrt{\gamma})^* = a - b\sqrt{\gamma}$ for all $a, b \in k_o$.

It is a very easy exercise to prove the assertion in the commutative case; in the skew situation it follows from a result by Dieudonné ([7], Lemma 1, p. 367 and [8], Sec. 4, p.677) as we shall show.

Lemma 2 (Dieudonné). Let $(k,*)$ be a noncommutative involutorial division ring with center $C$. Let $[S]$ be the ring generated in $k$ by the subset $S$ of $*$-symmetric elements. Then either $[S] = k$ or else char $k \neq 2$ and $k = (\frac{\alpha, \beta}{C})$ for suitable $\alpha, \beta \in C$ and $*$ is conjugation.

Remark 4. Notice that in the case of a quaternion algebra $k$ in characteristic 2 (Appendix I to Chapter XVI) we have $\dim_C S = 3$ and $[S] = k$.

Proof of Theorem 2 in the noncommutative case. There exists $\gamma \in \|\Phi\| \setminus \{0\}$. For any $\alpha \in S$ we have
$\alpha = (\alpha\gamma^{-1} + \frac{1}{2})\gamma(\alpha\gamma^{-1} + \frac{1}{2})* - (\alpha\gamma^{-1})\gamma(\alpha\gamma^{-1})* - \frac{1}{2}\gamma\frac{1}{2}* \in \|\Phi\| - \|\Phi\| - \|\Phi\| \subset k_o$.
Thus $S \subset k_o$ and therefore $[S] \subset k_o$. We first show that $[S] = k$ is impossible. Indeed, as $*$ is not $\mathbb{1}$ there exists $\lambda \neq 0$ with $\lambda* = -\lambda$ so $\|\lambda x\| \cdot \|x\| = -(\lambda \|x\|)^2$. If we had $k = k_o$ then $\|\lambda x\| \cdot \|x\|$ would be negative for all $x \in E$ thereby contradicting the assumption on definiteness. Hence $k$ is a quaternion algebra $(\frac{\alpha, \beta}{C})$ and $*$ is conjugation by Dieudonné's Lemma. Again, since $\Phi$ is definite on some line $(x_o)$ the term $\|q x_o\| = N(q)\|x_o\|$ may not change sign as $q$ ranges in $k$; therefore $\alpha, \beta$ must be negative elements of $C \cap k_o$. Since $k_o$ contains $N(q)$ for all $q \in k$ ($k_o$ contains $N(q)\|x_o\|$ and is a division ring) it contains in particular $C = C^2 - C^2$. On the other hand, $k_o$ cannot properly contain $C$: We have seen that $k \neq k_o$ so assume that $[k_o : C] = 2$, say $k_o = C(q)$. The discriminant of the quadratic equation $X^2 - (q + \bar{q})X + N(q) = 0$ (over $C$) for the element $q$ is the square of the "pure" quaternion $q - \bar{q}$, hence equal to an invariably negative element $\theta = \alpha\xi_1^2 + \beta\xi_2^2 - \alpha\beta\xi_3^2$. Hence $k_o = C(\sqrt{\theta})$ cannot be ordered. Thus we have proved that $k_o = C$. This finishes our proof.

Let $(k,*,\varepsilon)$ be as in Definition 7 and $\Phi : E \times E \longrightarrow k$ an $\varepsilon$-hermitean form. If there is an orthogonal decomposition

$$(20) \qquad E = E_+ \overset{\perp}{\oplus} E_-$$

such that the restriction of $\Phi$ to $E_+ \times E_+$ is positive definite and the restriction to $E_- \times E_-$ is negative definite then the pair $(\dim E_+ , \dim E_-)$ is uniquely determined. Indeed, if $E = F_+ \overset{\perp}{\oplus} F_-$ is a second decomposition of the same kind and, say, $\dim F_+ \neq \dim E_+$, e.g. $\dim F_+ > \dim E_+$, then $F_+ \cap E_- \neq (0)$ which is absurd. This uniqueness is known as Sylvester's law of inertia. The spaces $E_+ , E_-$ in (20) are by no means unique.

Remark 5. The concept of positive form in the sense of Definition 7 has its legitimation by the positive symmetric forms. In Appendix 1 we present a different kind of order structure which seems particularly appropriate for the study of forms over noncommutative fields.

## 4. Dense subspaces

In this section $(E,\Phi)$ is an $\varepsilon$-hermitean space. The map $\bar{\ }$ which assigns to each linear subspace $X \subset E$ its biorthogonal, $X \mapsto \bar{X} := X^{\perp\perp}$, satisfies all the axioms of a so-called closure operation: (i) $X \subset \bar{X}$, (ii) $\bar{X} = \bar{\bar{X}}$, (iii) if $X \subset Y$ then $\bar{X} \subset \bar{Y}$. Hence

Definition 8. A subspace $X$ of a sesquilinear space $E$ is called $\perp$-closed ("orthogonally closed") iff $X = X^{\perp\perp}$; a subspace $Y$ is called $\perp$-dense iff its $\perp$-closure is all of $E$, $Y^{\perp\perp} = E$ (in particular if $E$ is nondegenerate then $Y$ is $\perp$-dense iff $Y^{\perp} = (0)$ ).

In Section 8 below we shall define a topology $\sigma(\Phi)$ that can be introduced on any sesquilinear space $(E,\Phi)$ and such that $X^{\perp\perp}$ is the closure, in this topology, of the linear subspace $X$. Here we show that each infinite dimensional $(E,\Phi)$ contains proper subspaces that are $\perp$-dense.

Lemma 3. Assume that $(E,\Phi)$ is nondegenerate and of infinite dimension. Then there are $\perp$-dense subspaces $Y \subset E$ with $\dim E/Y = \dim E$.

Proof. Assume that we are given an infinite dimensional subspace $F \subset E$ with the property that $F \not\subset H^{\perp\perp}$ for all hyperplanes $H$ of $F$. It follows that the map which assigns to each $x \in E$ the linear map $f \mapsto \Phi(f,x)$, defined on $F$, is an epimorphism of $E$ onto $F^* := \mathrm{Hom}_k(F,k)$. The kernel is $F^\perp$, thus we obtain $\dim E/F^\perp = \dim F^*$; a fortiori $\dim E \geq \dim F^* = (\mathrm{card}\, k)^{\dim F} > \dim F$. Therefore, if we pick a subspace $F \subset E$ with $\dim E = \dim F$ then $F$ must contain a hyperplane $H$ such that $H \subset F \subset H^{\perp\perp}$; in particular $H^\perp = F^\perp$. Now we can describe how to find $\perp$-dense $Y \subset E$ of large codimension. Since $\dim E$ is infinite, we may in many ways decompose $E$ into a direct sum $E = \oplus\{F_\iota \mid \iota \in I\}$ such that $\mathrm{card}\, I = \dim E = \dim F_\iota$ for all $\iota \in I$. There are hyperplanes $H_\iota \subset F_\iota$ with $H_\iota^\perp = F_\iota^\perp$. Set $Y := \oplus\{H_\iota \mid \iota \in I\}$. We find $\dim E/Y = \sum_I 1 = \dim E$ and $Y^\perp = \bigcap_I H_\iota^\perp = \bigcap_I F_\iota^\perp = (\bigoplus_I F_\iota)^\perp = E^\perp = (0)$.

## 5. Finite dimensional subspaces

In this section $(E,\Phi)$ is a nondegenerate $\varepsilon$-hermitean space. We establish a small number of very basic facts used extensively but tacitly throughout the rest of the book. The first observation is

(21) If $X \subset E$ has finite $\dim X$ then $E = X \oplus X^\perp$ if and only if $X$ is nondegenerate.

Indeed, if there is such a decomposition then $\mathrm{rad}\, X \subset \mathrm{rad}\, E = (0)$ by the nondegeneracy of $E$. Assume conversely that $e_1,\ldots,e_n$ is a basis of the nondegenerate $X$. This means that the homogeneous system of $n$

equations $\sum_{i=1}^{n} \xi_i e_i \perp e_j$ $(j = 1,\ldots,n)$ has only the trivial solution $(\xi_1,\ldots,\xi_n) = (0,\ldots,0)$. Hence for fixed $z \in E$ we can always solve the $n$ inhomogeneous equations $z - \sum \xi_i e_i \perp e_j$ $(j = 1,\ldots,n)$ for the unknowns $\xi_1,\ldots,\xi_n$. Therefore $E \subset X + X^{\perp}$.

(22) If $r_1,\ldots,r_n$ is an orthogonal family of linearly independent isotropic vectors in $E$ there exists an orthogonal family $y_1,\ldots,y_n$ in $E$ with $\Phi(r_i,y_j) = \delta_{ij}$ (Kronecker).

Indeed, if $n = 1$ this is obvious by nondegeneracy of $E$. Upon induction assumption we find mutually orthogonal $x_1,\ldots,x_{n-1} \in E$ such that $\Phi(r_i,x_j) = \delta_{ij}$, $1 \leq i,j \leq n-1$. Hence the plane $P_i$ spanned by $r_i, x_i$ is nondegenerate and $P_i \perp P_j$ $(i \neq j)$. Hence $X := P_1 \overset{\perp}{\oplus} \ldots \overset{\perp}{\oplus} P_{n-1}$ qualifies for (21) : $E = X \oplus X^{\perp}$. The vector $r := r_n - \sum_{i=1}^{n-1} \Phi(r_n,x_i)\, r_i$ is in $X^{\perp}$; $r$ is isotropic and $X^{\perp}$ nondegenerate, thus there is $y_n \in X^{\perp}$ with $\Phi(r,y_n) = 1$; therefore $\Phi(r_n,y_n) = 1$ and $r_i \perp y_n$ for $i = 1,\ldots,n-1$. Again by (21) $E = P_n \oplus P_n^{\perp}$ and once more we apply the induction assumption to $r_1,\ldots,r_{n-1} \in P_n^{\perp}$ to find $y_1,\ldots,y_{n-1} \in P_n^{\perp}$. The family $(y_i)_{1 \leq i \leq n}$ has the requisite properties.

Combining (21) and (22) we obtain the

Lemma 4. Let $Z$ be a finite dimensional subspace in the nondegenerate $\varepsilon$-hermitean space $(E,\Phi)$. If $r_1,\ldots,r_m$ is a fixed basis of $\operatorname{rad} Z$ and $Z_o$ a fixed supplement of $\operatorname{rad} Z$ in $Z$ then there exist mutually orthogonal vectors $y_1,\ldots,y_m \in Z_o^{\perp}$ with $\Phi(r_i,y_j) = \delta_{ij}$ (Kronecker), $1 \leq i,j \leq m$. Thus there is a decomposition

$$E = k(r_1,y_1) \overset{\perp}{\oplus} \ldots \overset{\perp}{\oplus} k(r_m,y_m) \overset{\perp}{\oplus} Z_o \overset{\perp}{\oplus} E_o . \tag{23}$$

Indeed, $E = Z_o \oplus Z_o^{\perp}$ by (21) as $Z_o$ is nondegenerate; then we can apply (22) to $r_1,\ldots,r_m$ in $Z_o^{\perp}$.

From (23) we read off:

(24) If $E$ is nondegenerate and $Z \subset E$ is finite dimensional then $\dim E/Z^{\perp} = \dim Z$ (if $E$ is degenerate we have $\dim E/Z^{\perp} \leq \dim Z$ .

(25) Let $Z$ be the span of the linearly independent vectors $z_1, \ldots, z_n \in E$ ( $E$ nondegenerate) and $\alpha_1, \ldots, \alpha_n \in k$ . There exists $z \in E$ with $\Phi(z, z_i) = \alpha_i$ for $1 \leq i \leq n$ .

Proof of (25). It suffices to prove the assertion for arbitrarily fixed basis of $Z$ and arbitrary $\alpha_i$ . Use (23).

Remark 6. If we deal with trace - valued forms then by Lemma 1 in Section I.2 we can choose the vectors $y_i$ in (22) and (23) to be isotropic. Thus a finite dimensional nondegenerate alternate space is an orthogonal sum of hyperbolic planes (choose $n$ maximal in (22) ).

Lemma 5. Let $F$ be an arbitrary subspace in the nondegenerate space $(E,\Phi)$ . There exists a nondegenerate subspace $G \subset E$ with $F \subset G$ and $\dim G/F \leq \dim \operatorname{rad} F$ .

Proof. Let $(r_\iota)_{\iota \in I}$ be a basis of $R := \operatorname{rad} F$ . If $I$ is finite let $G = F \oplus k(y_i)_{i \in I}$ where the $y_i$ are as in (23). If $I$ is infinite let $J$ be the set of all finite sets of vectors $r_\iota$ ; card $J$ = card $I = \dim R$ . Let $W$ be a supplement of $R^{\perp}$ in $E$ . By (22) we find for each $S = \{r_{\iota_1}, \ldots, r_{\iota_m}\} \in J$ a set $T(S) = \{y_{\iota_1}, \ldots, y_{\iota_m}\} \subset W$ such that $\Phi(r_{\iota_i}, y_{\iota_j}) = \delta_{ij}$ . Let $H$ be the span of the set $\bigcup \{T(S) \mid S \in J\}$ . We have $H \subset W$ and $\dim H \leq \dim R$ . Because $W \cap R^{\perp} = (0)$ it is easy to verify that $F \oplus H$ is nondegenerate.

## 6. Closed subspaces

In this section $(E,\Phi)$ is an $\varepsilon$-hermitean space. By making use of the isomorphism theorem $A+B/A \cong A/A\cap B$ for subspaces $A, B \subset E$ we obtain some basic facts about $\perp$-closed spaces.

*

Let $U \subset V \subset E$ be subspaces with $\dim V/U < \infty$. Pick a supplement $X$ of $U$ in $V$. We have $U^{\perp}/V^{\perp} = U^{\perp}/(U+X)^{\perp} = U^{\perp}/U^{\perp}\cap X^{\perp} \cong U^{\perp}+X^{\perp}/X^{\perp} \subset E/X^{\perp}$. In particular $\dim U^{\perp}/V^{\perp} \leqq \dim E/X^{\perp}$. Hence by (24) $\dim U^{\perp}/V^{\perp} \leqq \dim X = \dim V/U < \infty$ and the argument may be repeated with $V^{\perp} \subset U^{\perp} \subset E$. Threefold repetition yields

$$\dim (V+U^{\perp\perp})/U^{\perp\perp} \leqq \dim V^{\perp\perp}/U^{\perp\perp} = \dim U^{\perp}/V^{\perp} \leqq \dim V/U . \tag{26}$$

<u>Lemma 6</u>. If $U \subset V \subset E$ ( $E$ degenerate or not) and $\dim V/U < \infty$ then (26) holds. In particular, if $U$ is $\perp$-closed then $V$ is $\perp$-closed and $\dim U^{\perp}/V^{\perp} = \dim V/U$.

Indeed, if $U = U^{\perp\perp}$ then we have equality throughout in (26). If we assume $E$ nondegenerate then $U = (0)$ is $\perp$-closed, hence

<u>Corollary 1</u>. All finite dimensional subspaces in a nondegenerate space are $\perp$-closed.

<u>Corollary 2</u>. If $(E,\Phi)$ is nondegenerate then the following are equivalent: (i) $\dim E$ is finite, (ii) $X = X^{\perp\perp}$ for all $X \subset E$, (iii) if $X \subset E$ & $X \neq E$ then $X^{\perp} \neq (0)$.

<u>Proof</u>. (i) => (ii) by Corollary 1 ; (ii) => (iii) is trivial ; (iii) => (i) by Lemma 3.

<u>Corollary 3</u>. If $(E,\Phi)$ is degenerate and $F$ is finite dimensional then $F^{\perp\perp} = F + \operatorname{rad} E$.

Proof. $E^{\perp} = \text{rad } E$ is closed, so $F + \text{rad } E$ is closed by Lemma 6 and thus contains $F^{\perp\perp}$. The converse inclusion is obvious.

Lemma 7. Let $(E,\Phi)$ be an $\varepsilon$-hermitean space (degenerate or not) with the property that for all subspaces $X$ we have

(27) $$\dim E/X^{\perp} \leqq \dim X$$

(this always takes place if $E$ can be decomposed into a direct orthogonal sum of finite dimensional subspaces). Then for all subspaces $U \subset V \subset E$ we have (26). Thus, if $E$ is nondegenerate we see (choose $U = (0)$) that $\dim V = \dim V^{\perp\perp}$ for all $V \subset E$; a fortiori $\dim W = \dim E$ for all $\perp$-dense subspaces.

Example. We shall give an illustration concerning (27). Let $k$ be an uncountable commutative field and $a$, $b$ cardinals with $\aleph_0 \leqq a < b \leqq \text{card } k$. Let $V, W$ be $k$-vector spaces, $\dim V = a$, $\dim W = b$. Select subsets $I, J \subset k$ such that $\text{card } I = a$, $\text{card } J = b$, $\iota + \kappa \neq 0$ for all $\iota \in I$, $\kappa \in J$. Let then $(v_\iota)_{\iota \in I}$, $(w_\kappa)_{\kappa \in J}$ be bases of $V$ and $W$ respectively. We now consider symmetric bilinear forms $\Phi$ on $V \oplus W$ which have $\Phi(v_\iota, w_\kappa) = \Phi(w_\kappa, v_\iota) = (\iota + \kappa)^{-1}$. If $v = \Sigma \lambda_\iota v_\iota \in V$ has precisely $m$ nonzero coefficients $\lambda_\iota$ then, if $v \perp w_\kappa$ for at least $m$ different $\kappa \in J$, it follows that $v = 0$. This simply obtains from the fact that any $n \times n$ determinant

$$\det\left(\frac{1}{\iota_i + \kappa_j}\right) = \prod_{i<j} (\iota_i - \iota_j) \prod_{i<j} (\kappa_i - \kappa_j) \prod_{i,j} (\iota_i + \kappa_j)^{-1}$$

is nonzero for different $\iota_1, \ldots, \iota_n$ and different $\kappa_1, \ldots, \kappa_n$. In particular, $V \cap W^{\perp} = (0)$ and, symmetrically, $W \cap V^{\perp} = (0)$. If we assume that at least one of the bases $(v_\iota)_I$, $(w_\kappa)_J$ is orthogonal, then we can verify that $(V \oplus W, \Phi)$ is nondegenerate. Thus we may in particular choose $V$ and $W$ totally isotropic for $\Phi$, $E = V \overset{\circ}{\oplus} W$.

We then have $V^{\perp} = V$, $W^{\perp} = W$. Therefore $\dim E/V^{\perp} > \dim V$, $\dim E/W^{\perp} < \dim W$. This illustrates that we can find in the same nondegenerate space $(E,\Phi)$ all three cases $\dim E/X^{\perp} \gtreqless \dim X$ (on the other hand, (27) may actually hold in spaces which are very far from admitting orthogonal bases; we refer to Example 2 in the next chapter). For a further example see the space $(F \overset{\circ}{\oplus} F^*, \Phi)$ defined in Remark 9 of Section 8 below.

(28) If $V \subset E$ is $\perp$-dense and $F \subset E$ finite dimensional then $(V \cap F^{\perp})^{\perp} = F$.

Indeed, since $\dim V / V \cap F^{\perp} = \dim V + F^{\perp}/F^{\perp} \leqq \dim E / F^{\perp} = \dim F < \infty$ we may quote Lemma 6 and obtain $\dim (V \cap F^{\perp})^{\perp}/V^{\perp} \leqq \dim F$, i.e. $\dim (V \cap F^{\perp})^{\perp} \leqq \dim F$ since $V^{\perp} = (0)$. Since $F \subset (V \cap F^{\perp})^{\perp}$ we have equality as asserted.

We now turn to a lemma of eminent utility for geometric constructions. The following situation often occurs. Given $n$ linearly independent vectors $f_1, \ldots, f_n$ in a space $E$ and scalars $\alpha_1, \ldots, \alpha_n \in k$ one should find a vector $x$ inside some prescribed subspace $V \subset E$ such that we have

$$(29) \qquad \Phi(x, f_i) = \alpha_i \quad (i = 1, \ldots, n).$$

By (25) we know that there always exist $x \in E$ which satisfy (29); the problem is to pick $x$ in $V$. We have

Lemma 8 ([22], Lemma 5, p. 12). Let $F$ be the span of linearly independent vectors $f_1, \ldots, f_n$ in the nondegenerate $\varepsilon$-hermitean space $(E,\Phi)$. Let $V$ be a subspace of $E$. In order that for arbitrarily prescribed $\alpha_1, \ldots, \alpha_n \in k$ there exists $x \in V$ with (29) it is necessary and sufficient that $V^{\perp} \cap F = (0)$.

Proof. Suppose (29) holds for $x \in V$ and randomly given $\alpha_1, \dots, \alpha_n$. A nonzero vector $d = \Sigma \beta_i f_i \in V^{\perp} \cap F$ would yield a nontrivial relation $\Phi(x,d) = 0 = \Sigma \alpha_i \beta_i^*$. Hence we must have $V^{\perp} \cap F = (0)$. If this is assumed then $V + F^{\perp}$ is $\perp$-dense; but $V + F^{\perp}$ modulo the closed $F^{\perp}$ is finite dimensional by (24), hence $V + F^{\perp}$ is also $\perp$-closed by Lemma 6. Therefore $V + F^{\perp} = E$. Pick some $x$ in $E$ which has (29) and decompose, $f = v + f'$ with $v \in V$, $f' \in F^{\perp}$. The vector $v$ responds to the problem.

We finish this section with a lemma on $\perp$-closed supplements. Its proof uses a technique which is of independent interest. We first formulate our assertion.

Lemma 9 ( [32] ). Let $(\bar{E},\bar{\Phi})$ be an $\varepsilon$-hermitean space which admits an orthogonal basis, $E$ a subspace and both $\bar{\Phi}$ and $\Phi := \bar{\Phi}|_{E \times E}$ nondegenerate forms. Let $E = F \oplus G$ be a given decomposition of $E$. Then there exists a decomposition $E = F_o \oplus G$ with $F_o \subset F^{\perp\perp}$ (the biorthogonal in $(E,\Phi)$ ) and $F_o^{\perp\bar{E}\perp\bar{E}} \cap E = F_o$.

Corollary. Let $(E,\Phi)$ be a nondegenerate $\varepsilon$-hermitean space which possesses an orthogonal basis. If the subspace $G \subset E$ admits a totally isotropic supplement it admits a totally isotropic $\perp$-closed supplement.

Indeed, if $F$ is totally isotropic in Lemma 9 then so is $F^{\perp\perp}$ and hence $F_o$ as well.

In order to prove Lemma 9 we first establish another lemma.

Lemma 10 ( [32] ). Let $H$ be a linear subspace of $E$ and $(e_\iota)_{\iota \in J}$ a fixed basis of $E$. With $x \in E$ associate the finite set $M(x) = \{\iota \in J \mid \xi_\iota \neq 0$ in the representation $x = \sum_J \xi_\iota e_\iota \}$. Then $H$ possesses a basis $(h_\kappa)_{\kappa \in I}$ such that for all $\kappa \in I$ we have $M(h_\kappa) \not\subset \cup \{ M(h_\iota) \mid \iota \in I \setminus \{\kappa\} \}$.

<u>Proof</u> (Lemma 10). Adjoin a new index $\circ$ to the index set $J$, $J_o := J \cup \{\circ\}$ and wellorder $J_o$ such that $\circ$ is the first element. Define a map $\mu : E \to J_o$ as follows: with each nonzero $x \in H$ associate the largest index in $M(x)$ ; furthermore $\mu(0) = \circ$ and, if $x \notin H$ and $M(x) \cap \mu(H) \neq \emptyset$ , then let $\mu(x)$ be the largest index in $M(x) \cap \mu(H)$ , otherwise (i.e. if $M(x) \cap \mu(H) = \emptyset$ ) let $\mu(x) = \circ$ .

We first observe that for each $x \in E$ there is $x' \in E$ with $x' \equiv x \pmod H$ and $\mu(x') = \circ$ . If $\mu(x) = \circ$ we may of course choose $x' = x$ ; if $\mu(x) \neq \circ$ then there is $y_1 \in H$ with $\mu(x) = \mu(y_1)$ . We can determine a scalar $\lambda_1$ such that $\mu(x) \notin M(x - \lambda_1 y_1)$ ; i.e. we shall have $\mu(x - \lambda_1 y_1) < \mu(x)$ . The step may be repeated. Since there are no infinite descending sequences in $J_o$ we arrive at $\mu(x - \lambda_1 y_1 - \ldots - \lambda_n y_n) = \circ$ after a finite number of steps.

Next we show: for each $\kappa \in \mu(H)$ and $\kappa \neq \circ$ there is $z \in H$ with $\mu(z) = \kappa$ and $M(z) \cap \{\iota \in \mu(H) \mid \iota < \kappa\} = \emptyset$ . Indeed, $H_\kappa = \{y \in H \mid \mu(y) < \kappa\}$ is a linear subspace of $H$ . Let $z' \in H$ with $\mu(z') = \kappa$ . By what we have just proved there is $z \in H$ with $z \equiv z' \pmod{H_\kappa}$ and $M(z) \cap \mu(H_\kappa) = \emptyset$ , and we have $\mu(z) = \mu(z') = \kappa$ . For each $\kappa \in I := \mu(H) \setminus \{\circ\}$ define $A_\kappa := \{y \in H \mid \mu(y) = \kappa, \iota \notin M(y)$ for all $\iota \in \mu(H)$ and $\iota < \kappa\}$ . $A_\kappa \neq \emptyset$ by what we have shown. Pick one $h_\kappa$ from each $A_\kappa$ to obtain the family $(h_\kappa)_{\kappa \in I}$ . There remains to show that $(h_\kappa)_{\kappa \in I}$ is a basis of $H$ .

It is obvious that $(h_\kappa)_{\kappa \in I}$ is linearly independent since each $h_\kappa$ has some $e_{\iota_o}$ in its representation $h_\kappa = \Sigma\, \xi_{\kappa\iota} e_\iota$ which does not show up in the representations of all the other $h_\iota$ . Let then $H_o$ be the span of $(h_\kappa)_{\kappa \in I}$ , $H_o \subset H$ . For $x \in H$ there exists $x' \in H$ with $x' \equiv x \pmod{H_o}$ and $M(x') \cap \mu(H_o) = \emptyset$ . If we had $x' \neq 0$ then $\mu(x') \in M(x')$ and thus $M(x') \cap \mu(H_o) = M(x') \cap \mu(H) \neq \emptyset$ , contradiction. Thus $x' = 0$ , i.e. $x \in H_o$ . This establishes Lemma 10.

Proof (Lemma 9). Here we let $(e_\iota)_{\iota \in J}$ be an orthogonal basis of $\bar{E}$, $(f_\kappa)_{\kappa \in K}$ a basis of $F$ and $H := F^{\perp\perp} \cap G$. We choose a basis $(h_\iota)_{\iota \in I}$ of the kind constructed in the previous proof: $\mu(h_\iota) = \iota$, $\mu(H \setminus \{0\}) = I \subset J$, $\nu \notin M(h_\iota)$ for all $\nu \in I$ and $\nu \neq \iota$. We now set $f_\kappa' := f_\kappa - \Sigma \lambda_{\iota\kappa} h_\iota$ with $\lambda_{\iota\kappa}$ determined such that $\bar{\Phi}(e_\nu, f_\kappa') = 0$ for all $\nu \in I$ (this is possible since $\Phi(e_\nu, f_\kappa) \neq 0$ for finitely many $\nu$ only). Let $F_o$ be the span of the $f_\kappa'$. We have $E = F_o \oplus G$, $F_o \subset F^{\perp\perp}$. There remains the assertion on $F_o^{\perp\bar{E}\perp\bar{E}} \cap E$.

We have $F_o \subset F^{\perp\bar{E}\perp\bar{E}} \cap E$, $F_o^{\perp} \subset F_o^{\perp\bar{E}}$, so $F_o^{\perp\bar{E}\perp\bar{E}} \subset F_o^{\perp\perp\bar{E}}$, $F_o^{\perp\bar{E}\perp\bar{E}} \cap E \subset F_o^{\perp\perp\bar{E}} \cap E = F_o^{\perp\perp} \subset F^{\perp\perp}$. Let $x \in F_o^{\perp\bar{E}\perp\bar{E}} \cap E$ but assume $x \notin F_o$. Decompose $x = y + z$, $y \in F_o$, $z \in G$, $z \neq 0$. As $z \in H \setminus \{0\}$ we have $\mu(z) = \iota_o \in I$ (for some $\iota_o$), i.e. $\bar{\Phi}(e_{\iota_o}, z) \neq 0$. On the other hand, $\bar{\Phi}(e_{\iota_o}, f_\kappa') = 0$ for all $\kappa \in K$ (by construction of the $f_\kappa'$). Thus $z \notin F_o^{\perp\bar{E}\perp\bar{E}}$. A fortiori $z \notin F_o^{\perp\bar{E}\perp\bar{E}} \cap E$ contradicting the choice of $x$. Therefore $z = 0$, i.e. $F_o^{\perp\bar{E}\perp\bar{E}} \cap E = F_o$ as asserted.

## 7. Isometries between sesquilinear spaces

Let $\Phi : E \times E \longrightarrow k$, $\bar{\Phi} : \bar{E} \times \bar{E} \longrightarrow k$ be $\varepsilon$-hermitean forms over $(k, *, \varepsilon)$. The spaces $(E, \Phi)$ and $(\bar{E}, \bar{\Phi})$ are called isometric iff there is a $k$-linear bijection $\varphi : E \longrightarrow \bar{E}$ satisfying $\bar{\Phi}(\varphi x, \varphi y) = \Phi(x,y)$ for all $x, y \in E$; $\varphi$ is called an isometry. If it is clear from the context what the forms are we may simply say that $E$ and $\bar{E}$ are isometric (notation $E \cong \bar{E}$). If $E \cong \bar{E}$ we also say that the two spaces are of the same isometry type or that they belong to the same isometry class.

Consider a family $\mathcal{F} = (f_\iota)_{\iota \in I}$ of linearly independent vectors in $E$ and let $\bar{\mathcal{F}} = (\bar{f}_\iota)_{\iota \in I}$ be an analogous object in $\bar{E}$. On says that $\mathcal{F}$ and $\bar{\mathcal{F}}$ are congruent if

$$\bar{\Phi}(\bar{f}_\iota, \bar{f}_\kappa) = \Phi(f_\iota, f_\kappa) \qquad (\iota, \kappa \in I). \tag{30}$$

If we let $F$ and $\bar{F}$ be the linear spans of $\mathcal{F}$ and $\bar{\mathcal{F}}$ , $\Phi_o$ , $\bar{\Phi}_o$ the restrictions to $F \times F$ and $\bar{F} \times \bar{F}$ of $\Phi$ and $\bar{\Phi}$ respectively, then it is obvious that the assignment $f_\iota \mapsto \bar{f}_\iota$ induces an isometry of the spaces $(F,\Phi_o)$ , $(\bar{F},\bar{\Phi}_o)$ . Conversely, each isometry between sesquilinear spaces maps the bases of one space onto congruent bases of the other.

The isometries $\varphi : E \longrightarrow E$ of $(E,\Phi)$ onto itself form a group under composition, called the <u>orthogonal group</u> of the sesquilinear space $(E,\Phi)$ . A collection of objects that characterizes the orbit of a subspace $F$ (in the set of subspaces of $E$ ) under the the action of the orthogonal group is called a complete set of <u>orthogonal invariants</u> of $F$ ; these invariants determine the position of $F$ inside $E$ up to isometric automorphisms of $E$ .

<u>Remark 7</u>. There is an enormous literature on the orthogonal groups in the finite dimensional case. As a first orientation the reader may consult [9] , [33] . There is a zoology of groups that runs parallel to that of forms (symplectic groups, unitary groups, ... ); for lack of results in the infinite dimensional case we do not need it here. (Investigations into the infinite dimensional case are e.g. [18],[19],[31], [36] ; the matter will not be pursued in this book.)

The most fundamental theorem in the theory of finite dimensional sesquilinear forms is Witt's theorem. It will frequently be used in subsequent chapters. In order to state it we introduce

<u>Definition 9</u>. Let $(E,\Phi)$ be an $\varepsilon$ - hermitean form over $(k,*,\varepsilon)$ and $T = \{ \xi + \varepsilon \xi^* \mid \xi \in k \}$ the additive subgroup of traces in $k$ . The linear subspace $E_* := \{ x \in E \mid \Phi(x,x) \in T \}$ is called the trace - valued part of $E$ .

Lemma 11. Each element $\varphi$ of the orthogonal group of the nondegenerate space $(E,\Phi)$ leaves $E_*^{\perp}$ pointwise fixed.

Proof. $\varphi$ maps $E_*$ onto $E_*$ hence $E_*^{\perp}$ onto $E_*^{\perp}$. Let $z \in E_*^{\perp}$ and $x \in E$. As $\varphi x - x$ invariably belongs to $E_*$ we have $0 = \Phi(\varphi z, \varphi x - x) = \Phi(z,x) - \Phi(\varphi z, x) = \Phi(z - \varphi z, x)$, i.e. $z - \varphi z \in \operatorname{rad} E = (0)$.

Theorem 3 ("Witt"). Let $(E,\Phi)$ be a nondegenerate $\varepsilon$-hermitean space. An isometry $\varphi_o : F \to \bar{F}$ between finite dimensional subspaces $F, \bar{F} \subset E$ can be extended to an element of the orthogonal group of $(E,\Phi)$ if and only if the following condition is satisfied

(31) $F \cap E_*^{\perp} = \bar{F} \cap E_*^{\perp}$ and $\varphi_o : F \to \bar{F}$ leaves $F \cap E_*^{\perp}$ pointwise fixed.

Corollary ("Cancellation Theorem"). Let $(E,\Phi)$ be a nondegenerate $\varepsilon$-hermitean space. If $E$ is decomposed, $E = F \oplus F^{\perp} = G \oplus G^{\perp}$ with $F, G$ finite dimensional isometric subspaces of $E_*$, then $F^{\perp} \cong G^{\perp}$.

The proof of Theorem 3 is given in Chapter XV where the topic is treated in a broader context. Here we shall make a few comments.

First, if forms are assumed trace-valued, i.e. $E = E_*$, then (31) is vacuous and the assertion of the theorem is classical. (Generalizing away from finite dimensional $E$ as long as $\dim F$ is kept finite is a triviality in the trace-valued situation.)

Second, if forms are not trace-valued the result appears in [35]. The special case where $\dim E/E_* = 1$ is assumed had been treated independently in [34] and by the author. For further details, e.g. when $F \subset E_*$ is assumed, see Chapter XV.

Third, in the case of trace-valued spaces Theorem 3 and its corlary are equivalent statements. In fact, Lemma 4 in Section 5 immediately reduces the extension problem to the cancellation problem. This is not so in the general case.

Forth, the above corollary on cancellation is identical with Kaplansky's Lemma 2 in [22] ; his (purely computational) proof is that fast that we believe every student of the field should know it. It runs as follows (cf. [23] pp. 34-38 ): Choose congruent bases in $F$ and $G$ and join them with bases of $F^{\perp}$ and $G^{\perp}$ in order to obtain two bases of $E$ . We get two matrices $M, N$ of $\Phi$ which split, $M = \begin{pmatrix} A & 0 \\ 0 & B \end{pmatrix}$ , $N = \begin{pmatrix} A & 0 \\ 0 & C \end{pmatrix}$ and which are congruent under some matrix $P$ , $P^{\circ} M P = N$ , where $P^{\circ} := {}^{tr}(M^{*})$ . Split $P$ into the same size blocks, $P = \begin{pmatrix} W & X \\ Y & Z \end{pmatrix}$ and expand $P^{\circ} M P$ to get $W^{\circ} A W + Y^{\circ} B Y = A$ , $W^{\circ} A X + Y^{\circ} B Z = 0$ , $X^{\circ} A W + Z^{\circ} B Y = 0$ , $X^{\circ} A X + Z^{\circ} B Z = C$ . For arbitrary $U$ (same size as $A$ ) one then checks the identity:

$$(Z^{\circ} + X^{\circ} U^{\circ} Y^{\circ})\, B\, (Z + Y U X) \;=\; C + X^{\circ} R X$$

$$\text{where}\quad R := U^{\circ} A U - (U^{\circ} W^{\circ} + \mathbb{1})\, A\, (W U + \mathbb{1}) .$$

The problem is to find $U$ such that $R = 0$ (for then $B$ and $C$ will be congruent, i.e. $F^{\perp} \cong G^{\perp}$ ). Now it is obvious (by an induction argument) that the corollary has to be proved only when $F$ cannot be further (orthogonally) decomposed into proper subspaces. In other words, $A$ is $1$ by $1$ or else $A$ is $\begin{pmatrix} 0 & 1 \\ 1 & 0 \end{pmatrix}$ and $(k,*,\varepsilon) = (k,\mathbb{1},-1)$ and thus $k$ commutative. If $A$ is $1$ by $1$ and not $W = 1 = -1$ then one of the equations $W U + \mathbb{1} = \pm U$ can be solved for $U$ and $R$ is zero; if, on the other hand, $W = 1 = -1$ then $R = U^{\circ} A + A U + A$ and this can be made zero by the very assumption on trace-valuedness of the subspace $F$. In the $2$ by $2$ case left we may first dismiss the possibility of alternate $E$ , for then $F^{\perp} \cong G^{\perp}$ simply by equality of dimensions (Remark 6 in Section 5). Thus we are left with char $k = 2$ . We first note that $L^{\circ} A L = A$ for any $2$ by $2$ matrix $L$ with determinant $1$ . We can get

$R = 0$ by solving $LU = WU + 1$ for $U$ ; and solution is impossible only when $W + L$ is singular for every such $L$ . But there is always $L$ among $\begin{pmatrix}1&0\\0&1\end{pmatrix}, \begin{pmatrix}1&1\\0&1\end{pmatrix}, \begin{pmatrix}1&0\\1&1\end{pmatrix}, \begin{pmatrix}0&1\\1&0\end{pmatrix}, \begin{pmatrix}1&1\\1&0\end{pmatrix}, \begin{pmatrix}0&1\\1&1\end{pmatrix}$ to make $W + L$ nonsingular.

Fifth, there do exist infinite dimensional trace - valued spaces such that the cancellation theorem holds for arbitrary dimension of $F$ ; i.e. cancellation is possible (unconditionally) just as in the finite dimensional case. These are the "generic" spaces in [18] ; the cancellation property is verified in [40] .

Example. Let $k$ be commutative and of characteristic 2 and $\Phi = \langle 1\rangle \oplus \langle 1\rangle \oplus \langle 1\rangle$ . Thus, if $e_1, e_2, e_3$ is an orthonormal basis we can introduce the new basis $e_1 + e_2 + e_3, e_1 + e_2, e_1 + e_3$ to get an isometry $\langle 1\rangle \overset{\perp}{\oplus} \langle 1\rangle \overset{\perp}{\oplus} \langle 1\rangle \cong \langle 1\rangle \overset{\perp}{\oplus} H$ , $H$ a hyperbolic plane. The assignment $\varphi_o : e_1 \mapsto e_1 + e_2 + e_3$ violates (31) because $E_*^{\perp}$ contains $e_1 + e_2 + e_3$ but not $e_1$ . Hence there is no extension of $\varphi_o$ and, in particular, no cancellation.

We finish this section with a concept closely related to that of isometry:

Definition 10. Let $(E,\Phi)$ and $(\bar{E},\bar{\Phi})$ be sesquilinear spaces over division rings $k$ and $\bar{k}$ . The two spaces are called similar iff there exists an isomorphism $\kappa : k \to \bar{k}$ of division rings and a $\kappa$ - semilinear bijection $\varphi : E \to \bar{E}$ and some fixed nonzero $\alpha \in \bar{k}$ ("multiplyer for $\varphi$ ") such that for all $x,y \in E$ we have $\bar{\Phi}(\varphi x,\varphi y) = \Phi(x,y)^{\kappa}\cdot\alpha$ ; $\varphi$ is called similitude.

If $(E,\Phi)$ and $(\bar{E},\bar{\Phi})$ are similar, then each similitude $\varphi : E \to \bar{E}$ induces a bijection $\hat{\varphi}$ between the sets $L(E)$ , $L(\bar{E})$ of linear subspaces in $E$ and $\bar{E}$ respectively; in fact, $\hat{\varphi}$ is a lattice isomorphism because it obviously respects the operations sum and intersection. Furthermore, $\hat{\varphi}$ respects the orthogonality relations $\perp$ given on $L(E)$, $L(\bar{E})$ by $\Phi$ and $\bar{\Phi}$. However, interest is directed in the opposite direction: If, for any nondegenerate $(E,\Phi)$, $(\bar{E},\bar{\Phi})$ of dimensions at least 3 over division rings $k$ and $\bar{k}$, there exists an "ortho - isomorphism" $\hat{\varphi}$ between the lattices $L(E)$, $L(\bar{E})$ then $\hat{\varphi}$ must be induced by a similitude $\varphi$, in particular, $(E,\Phi)$ and $(\bar{E},\bar{\Phi})$ are similar. This follows easily from the First Fundamental Theorem of Projective Geometry ( [2] , p. 44 ). For the discussion of a related consequence of the Fundamental Theorem see Theorem 1 in [15] .

Remark 8. It is not difficult to arrive at the appropriate version of Theorem 3 for similitudes. Let $(E,\Phi)$ and $(\bar{E},\bar{\Phi})$ be nondegenerate sesquilinear spaces over $k$ and $\bar{k}$ respectively. If $\varphi_o : F \to \bar{F}$ is a similitude relative to a fixed isomorphism $\kappa : k \to \bar{k}$ and with multiplyer $\alpha$ between finite dimensional subspaces $F \subset E$, $\bar{F} \subset \bar{E}$ then there exists an extension $\varphi : E \to \bar{E}$ of $\varphi_o$ that is a similitude relative to $\kappa$ and with multiplyer $\alpha$ if and only if the following are satisfied: There exists at least one similitude $\psi : E \to \bar{E}$ relative to $\kappa$ and with multiplyer $\alpha$ and $F \cap E^{*\perp} = (\psi^{-1}\bar{F}) \cap E^{*\perp}$ and $\varphi_o$ coincides with $\psi$ on $F \cap E^{*\perp}$. (Apply Theorem 3 to the isometry $\psi^{-1} \circ \varphi_o : F \to (\psi^{-1}\bar{F})$ .)

## 8. The weak linear topology $\sigma(\Phi)$ on $(E,\Phi)$

Let $K$ be a topological division ring and $E$ a $K$-vector space. A topology $\tau$ on $E$ is called a vector space topology if the two composition laws $(x,y) \mapsto x+y$, $(\lambda,x) \mapsto \lambda x$ from $E \times E$ and $K \times E$ into $E$ are continuous ( $E \times E$ and $K \times E$ carrying the product topologies). Since then, for fixed $b \in E$, the map $x \mapsto b+x$ is a homeomorphism of $E$ into itself we obtain a neighbourhood basis for $b$

simply by translating a neighbourhood basis $\mathfrak{B}(0)$ of the origin to $b + \mathfrak{B}(0) = \{ b + U \mid U \in \mathfrak{B}(0) \}$ .

The only vector space topologies which we shall consider here are linear vector space topologies in the sense of Lefschetz ([27], Chap.II, § 25). This means the following additional features:

1. The division ring carries the discrete topology.
2. There is a neighbourhood filter consisting of linear subspaces.

We shall not by definition require that linear topologies be hausdorff. Thus, if $\mathfrak{B} = \{ U_\alpha \mid \alpha \in A \}$ is any filter basis consisting of linear subspaces in $E$ we gain a linear topology $\tau$ on $E$ by declaring $b + \mathfrak{B}$ to be a neighbourhood basis $\mathfrak{B}(b)$ of $b \in E$ . This topology is hausdorff iff $\cap U_\alpha = (0)$ . Each neighbourhood $b + U$ of $b \in E$ is both open and closed; hence if $\tau$ is hausdorff then $E$ is totally disconnected. Finite dimensional $(E,\tau)$ are discrete if hausdorff; thus the concept is of interest only in the infinite dimensional case. A linear vector space topology $\tau$ is induced - as is every topology of a topological group - by a uniform structure on $E$ , a basis $\{N_\alpha \mid \alpha \in A\}$ for a uniformity being given by $N_\alpha = \{ (x,y) \in E \times E \mid x - y \in U_\alpha \}$ . Thus it will make sense to talk about Cauchy filters, completions etc.

Let us look for a linear topology $\tau$ on sesquilinear spaces $(E,\Phi)$ that makes $\Phi$ separately continuous, i.e. makes continuous, for all $y \in E$ , the maps $x \mapsto \Phi(x,y)$ , $x \mapsto \Phi(y,x)$ . Let $y$ be fixed. Since $\Phi(0,y) = 0$ and $\{0\}$ is a 0-neighbourhood of $0 \in k$ there must be a 0-neighbourhood $U$ of $0 \in E$ with $\Phi(U,y) \subset \{0\}$ , i.e. $U \subset y^\perp$ . Hence we see that the neighbourhood filter $\mathfrak{B}(0)$ for $\tau$ must contain at least all orthogonals $y^\perp$ ( $y \in E$ ) if $\tau$ is to render $\Phi$ separately continuous. But $\tau$ is separately continuous if we let

$\{y^{\perp} \mid y \in E\}$ be a subbasis for $\mathfrak{B}(0)$. Hence, the $\tau$ thus defined is the coarsest such topology.

Definition 11. Let $(E,\Phi)$ be an $\varepsilon$-hermitean space (degenerate or not). The linear topology on $E$ which has all linear subspaces $y^{\perp}$ ($y \in E$) as a subbasis of a neighbourhood filter of $0 \in E$ (equivalently: which has all orthogonals $F^{\perp}$, where $F$ runs through the finite dimensional subspaces of $E$, as a $0$-neighbourhood basis) is called the weak linear topology $\sigma(\Phi)$ associated with $(E,\Phi)$. Thus, $\sigma(\Phi)$ is hausdorff if and only if $\Phi$ is nondegenerate.

The first observation of interest is

(32) The $\sigma(\Phi)$-closure of a linear subspace $X \subset E$ is $X^{\perp\perp}$.

Proof. Let $\bar{X}$ be the $\sigma(\Phi)$-closure of $X$. We have $\bar{X} \perp X^{\perp}$ by separate continuity. Conversely, to see that $X^{\perp\perp} \subset \bar{X}$ let $z \in X^{\perp\perp}$ and $F^{\perp}$ be a typical $\sigma(\Phi)$-neighbourhood. For $f_1, \dots, f_n$ a basis of a supplement $F_o$ of $X^{\perp} \cap F$ in $F$ we let $\alpha_i = \Phi(z,f_i)$, $1 \leqq i \leqq n$. Trivially $X^{\perp} \cap F_o = (0)$ so, by Lemma 8, there is $x_1 \in X$ with $\Phi(x_1,f_i) = \alpha_i$. In other words, $z - x_1 \in F_o^{\perp}$. Because $z \perp X^{\perp}$ and $x_1 \perp X^{\perp}$ we have in fact that $z - x_1 \in F^{\perp}$. Therefore $(z + F^{\perp}) \cap X \neq \emptyset$; $z$ is an accumulation point of $F$.

If $Y \subset E$ is a subspace and $f : Y \longrightarrow k$ a $\sigma(\Phi)$-continuous linear function into (the discrete) $k$ then the kernel $X = \ker f$ is $\sigma$-closed. Assume that $f$ is not identically zero, thus $Y = X \oplus (y)$, $f(y) = 1$. There exists $z \in X^{\perp} \setminus y^{\perp}$, say $\Phi(y,z) = 1$. We see that the map $x \longmapsto \Phi(x,z)$ is $\sigma(\Phi)$-continuous on $E$ and extends $f$ to all of $E$. What we have seen can also be put as follows: if $X$ is a $\perp$-closed subspace of $E$ and $y \notin X$ then there exists a $\perp$-closed hyperplane $H$ with $X \subset H$ and $y \notin H$ (namely $z^{\perp}$). Finally, this can also be expressed as follows

(33) If $F = F^{\perp\perp} \subset E$ then $F = \bigcap \{ H \mid H^{\perp\perp} = H \supset F \ \&\ \dim E/H = 1 \}$

We see that the injection $E \to E^* = \mathrm{Hom}_k(E,k)$ which sends a vector $y$ into the linear map $x \mapsto \Phi(x,y)$ is onto the supspace $E' \subset E^*$ consisting of all $\sigma(\Phi)$ - continuous functionals on $E$.

The next observation of interest is

(34) If $\Phi$ is nondegenerate and $(E,\sigma(\Phi))$ complete then $\dim E < \infty$.

Proof. Endow the algebraic dual $E^* = \mathrm{Hom}_k(E,k) = \prod_I k$ ( card $I = \dim E$ ) with the product topology $\pi$ of the discrete topology on $k$. Consider the mapping $\varphi : E \to E^*$ defined by $x \mapsto \Phi(.,x) = (\Phi(e_\iota, x))_{\iota \in I}$ where $(e_\iota)_{\iota \in I}$ is some fixed basis of $E$. $\varphi$ is injective because $\Phi$ is nondegenerate. From the definition of the topologies it is immediate that $\varphi$ is homeomorphic onto the image; $\varphi$ is in fact a dense embedding (for, density is here tantamount to Lemma 8). Hence $E \cong E^*$ if $E$ is complete.

Remark 9. It may very well happen that $(E,\Phi)$ contains an infinite dimensional subspace which is complete under the topology induced by $\sigma(\Phi)$. Let us give an example. The algebraic dual $F^*$ of a k-left vector space may be turned into a left k - space by means of an involution $\nu : k \to k$ of the underlying division ring $k$ (cf. Sec. 1.1). Hence we may form the direct sum $E = F \oplus F^*$ and define a hermitean form $\Phi$ on $E$ by

$$\Phi(f + f', g + g') := g'(f) + \nu(f'(g))$$

for all $f, g \in F$ and $f', g' \in F^*$. Here the subspace $F^*$ is complete under the topology $\sigma(\Phi)$ restricted to $F^*$ (cf. [26], § 10.10); this is not difficult to verify.

We always have $\sigma(\Phi|_{H\times H}) \leqq \sigma(\Phi)|_H$ for $H$ a subspace of $(E,\Phi)$ . (Strict inequality is witnessed by $H = F^*$ in the above example (Remark 9).) Equality holds only when trivially so:

(35) $$\sigma(\Phi|_{H\times H}) = \sigma(\Phi)|_H \text{ if and only if } H + H^{\perp} = E .$$

Proof. We begin by noting that the assertion can be reduced to the case of nondegenerate $\Phi$ . Hence $\sigma(\Phi)$ is hausdorff and so is $\sigma(\Phi)|_H$ . If we assume equality of the two topologies on $H$ then $\sigma(\Phi|_{H\times H})$ is hausdorff and thus $\Phi|_{H\times H}$ nondegenerate, $H \cap H^{\perp} = (0)$ . Let $E = H \oplus G$ for some supplement $G$ with $H^{\perp} \subset G$ . We show that $G \subset H^{\perp}$ . Let $g \in G$ and consider $g^{\perp} \cap H$ which is, of course, $\sigma(\Phi)|_H$-closed and thus $\sigma(\Phi|_{H\times H})$-closed: there must be $h \in H$ with $h^{\perp} \cap H = g^{\perp} \cap H$ . Hence $H^{\perp}$ contains a suitable linear combination $\alpha h + g$ . If $g$ were not in $H^{\perp}$ then $\alpha \neq 0$ and $h \in H^{\perp} + G = G$ , thus $h \in G \cap H = (0)$, contradiction. Hence $G \subset H^{\perp}$ and $E = H + H^{\perp}$ .

Lemma 12. Let $V, \bar{V}$ be subspaces in the nondegenerate sesquilinear space $(E,\Phi)$ . If there is an isometry $\varphi : V \cong \bar{V}$ which is a homeomorphism with respect to the topologies $\sigma(\Phi)|_V$ and $\sigma(\Phi)|_{\bar{V}}$ then we have $\dim\,(\text{rad}\,V)^{\perp}/(V + V^{\perp}) = \dim\,(\text{rad}\,\bar{V})^{\perp}/(\bar{V} + \bar{V}^{\perp})$.

Proof. We first show that $\dim E/V + V^{\perp} = \dim E/\bar{V} + \bar{V}^{\perp}$ . Let $(v_\iota)_{\iota \in I}$ be a basis of $V$ and $\bar{v}_\iota := \varphi v_\iota$ . A minimal family $(e_\kappa)_{\kappa \in K}$ of vectors in $E \setminus (V + V^{\perp})$ such that the $v_\iota^{\perp} \cap V$ and the $e_\kappa^{\perp} \cap V$ $(\iota \in I, \kappa \in K)$ add up to a subbasis of the zero-neighbourhood filter of $\sigma(\Phi)|_V$ is a basis of a supplement of $V + V^{\perp}$ in $E$ . $\varphi$ maps all $e_\kappa^{\perp} \cap V$ into a system $S$ of $\sigma(\Phi)|_{\bar{V}}$-open neighbourhoods which, together with all $\bar{v}_\iota^{\perp} \cap \bar{V}$ form a subbasis for the zero-neighbourhood filter of $\sigma(\Phi)|_{\bar{V}}$ . Since the elements of $S$ are $\sigma(\Phi)|_{\bar{V}}$-closed hyperplanes of $\bar{V}$ they must be of the form $x_\iota^{\perp} \cap \bar{V}$ .

As $S$ is minimal all these vectors $x_\iota$ span a supplement of $\bar{V} + \bar{V}^\perp$ in $E$. Hence $\dim E/(V + V^\perp) = \dim E/(\bar{V} + \bar{V}^\perp)$. Now we can arrange it so that certain among the $e_\kappa$ span a supplement of $V + V^\perp$ in $(\mathrm{rad}\ V)^\perp$. Since $\varphi$ is an isometry we shall have $\mathrm{rad}\ V \subset e_\iota^\perp \cap V$ if and only if $\mathrm{rad}\ \bar{V} \subset x_\iota^\perp \cap \bar{V}$ i.e. $x_\iota \in (\mathrm{rad}\ \bar{V})^\perp$. Thus $e_\iota \mapsto x_\iota$ maps a supplement of $V + V^\perp$ in $(\mathrm{rad}\ V)^\perp$ into a supplement of $\bar{V} + \bar{V}^\perp$ in $(\mathrm{rad}\ \bar{V})^\perp$. Hence the assertion.

Corollary. The existence of a $\varphi$ as in Lemma 12 implies that for closed $V$, the dimensions of quotient spaces of neighbouring elements in the lattice generated by $V$ under the operations $+, \cap, \perp$ (taking the orthogonal)

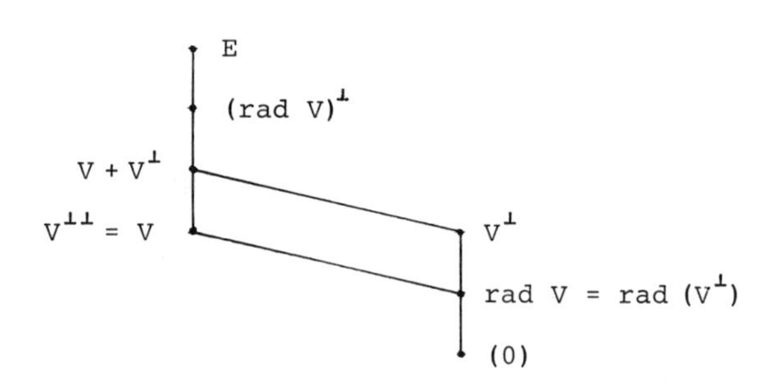


coincide with the corresponding cardinals for $\bar{V}$ except, of course, for $\dim V^\perp/\mathrm{rad}\ (V^\perp)$.

Lemma 12 is the natural background for Allenspach's Theorem 4 in Section 5 of Chapter X.

Remark 10. It is a little hard for the beginner to visualize the linear topology $\sigma(\Phi)$. Nevertheless, since topology is such a suggestive language, it sometimes helps to find and to motivate proofs if $\perp \circ \perp$ is remembered as the closure of the topology. Here is a simple example of what we mean. Let $F, G, H$ be $\perp$-closed subspaces in a non-degenerate space $(E,\Phi)$ with $F \subset H$ and such that $G^\perp + H^\perp = E$.

We assert: if $G + H$ is $\perp$-closed then $F + G$ is $\perp$-closed. A proof for this can be figured out but it will hardly be straight forward. Provided one knows that " $G^{\perp} + H^{\perp} = E$ " simply means that $\sigma(\Phi)|_{G+H}$ is the product topology of $\sigma(\Phi)|_{G}$ and $\sigma(\Phi)|_{H}$ (Theorem 2 in Chapter X) then the proof of the assertion becomes natural and easy. Similar remarks could be made on other occasions. Section 7 of Chapter X is a good example of a topologically motivated deduction of algebraic results.

## 9. Orthostable lattices of subspaces

An abstract lattice is a partially ordered set $(V,\leq)$ such that each set $\{a,b\} \subset V$ possesses a least upper bound (supremum) $c$ and a largest lower bound (infimum) $d$ (standard notation $c = a \vee b$, $d = a \wedge b$). Example 1. $V$ is the set $L(E)$ of all linear subspaces of a vector space $E$ and $\leq$ on $V$ is $\subseteq$; it follows that $X \vee Y = X + Y$ and $X \wedge Y = X \cap Y$ for all $X, Y \in L(E)$. Example 2. $V$ is the set $L_{\perp\perp}(E)$ of all $\perp$-closed subspaces of a nondegenerate sesquilinear space $(E,\Phi)$ and $\leq$ on $V$ is again $\subseteq$; it follows that here $X \vee Y = (X+Y)^{\perp\perp}$ and $X \wedge Y = X \cap Y$. Most of the lattices of interest to us are complete which means that arbitrary sets of lattice elements possess suprema and infima (this is the case in the two examples above). A lattice is called modular if it satisfies the

modular law: $a \leq c \implies (a \vee b) \wedge c = a \vee (b \wedge c)$ ;

it is called distributive if it satisfies the stronger

distributive identities:
$$(a \vee b) \wedge (a \vee c) = a \vee (b \wedge c) ,$$
$$(a \wedge b) \vee (a \wedge c) = a \wedge (b \vee c) .$$

(Actually, the two distributive identities are equivalent in any lattice, if one holds for all terns $a, b, c$ then so does the other. This does not mean that they are equivalent for fixed elements $a, b, c$.)

Both of the above lattices $L(E)$ and $L_{\perp\perp}(E)$ fail to be distributive (unless, of course, $\dim E = 1$). $L(E)$ is modular; on the other hand, $L_{\perp\perp}(E)$ is modular if and only if $\dim E$ is finite by the Theorem of H.A. Keller ( [12] , [25] ). In a lattice which fails to be modular the concept of a modular pair is of interest. This topic is taken up in Section 3 of Chapter X .

Definition 12. The lattices $(V,\leq)$ and $(\bar{V},\leq)$ are isomorphic, and the map $\varphi : V \longrightarrow \bar{V}$ is an isomorphism, if $\varphi$ is a bijection such that $x \leq y$ in $V$ iff $\varphi x \leq \varphi y$ in $\bar{V}$ . The lattices $(V,\leq)$ , $(\bar{V},\leq)$ are antiisomorphic (or dually isomorphic), and $\tau : V \longrightarrow \bar{V}$ is an antiisomorphism (or dual isomorphism), if $\tau$ is a bijection such that $x \leq y$ in $V$ iff $\tau y \leq \tau x$ in $\bar{V}$ ; if in particular $(V,\leq) = (\bar{V},\leq)$ and $\tau$ is an antiisomorphism with $\tau^2$ = identity then $\tau$ is called polarity.

The assignment $X \mapsto X^{\perp}$ is a polarity in the lattice $L_{\perp\perp}(E)$ of Example 2 . If the space is anisotropic, then this polarity is a so-called orthocomplementation in the lattice. We shall make no use of this particular kind of polarity.

If $(E,\Phi)$ is a sesquilinear space then $X \mapsto X^{\perp}$ is an antitone mapping $\perp : L(E) \longrightarrow L(E)$ . We shall be interested in the sublattice $V$ of $L(E)$ orthostably generated by a family $(V_\iota)_{\iota \in I}$ of subspaces of $E$ . By this we mean the smallest sublattice $V$ of $L(E)$ that contains $V_\iota$ for all $\iota \in I$ and which contains along with each element $X$ its orthogonal $X^{\perp}$ . Sometimes it is convenient to formally require that the spaces $(0)$ and $E$ are elements of $V$ also.

In the case where $(E,\Phi)$ is finite dimensional, the concept of orthostably generated sublattice of $L(E)$ reduces to the concept of sublattice because of the rules $(X \cap Y)^{\perp} = X^{\perp} + Y^{\perp}$ , $(X + Y)^{\perp} = X^{\perp} \cap Y^{\perp}$ , both of which hold when $\dim E$ is finite. Thus, the lattice $\mathcal{V} = \mathcal{V}(F_\iota)_{\iota\in I}$ orthostably generated by the family $(F_\iota)_{\iota\in I}$ simply is the sublattice in $L(E)$ generated by the <u>two</u> families $(F_\iota)_{\iota\in I}$, $(F_\iota^\perp)_{\iota\in I}$.

<u>Example 3</u>. The lattice $\mathcal{V} = \mathcal{V}(V)$ orthostably generated by one single subspace $V$ of a nondegenerate sesquilinear space $(E,\Phi)$ is finite and distributive. It consists of the following fourteen elements (the proof is straight forward): $(0)$, $V\cap V^\perp$, $(V\cap V^\perp)^{\perp\perp}$, $V^\perp\cap V^{\perp\perp}$, $V^\perp$, $V$, $V+(V\cap V^\perp)^{\perp\perp}$, $V+(V^\perp\cap V^{\perp\perp})$, $V+V^\perp$, $V^{\perp\perp}$, $V^{\perp\perp}+V^\perp$, $(V+V^\perp)^{\perp\perp}$, $(V\cap V^\perp)^\perp$, $E$. A diagram is given in Section 2 of Chapter V below. Since $\mathcal{V}$ is the union of two chains it is easy to see that the distributive identities are satisfied (because there are always two comparable elements among $a$, $b$, $c$ in the identities). It is a good exercise for the beginner to figure out examples where all fourteen spaces in the above list are different. The lattice appears in [22]. If we have <u>two</u> generators, then the lattice orthostably generated will, in general, be infinite and nondistributive. Of this treat the following two examples.

<u>Example 4</u>. In [4], p. 64 it is shown that the "free" modular lattice with four generators is infinite by making use of the fact that the harmonic net generated by a complete quadrangle is infinite. The four generators may be picked as lines $U_1$, $U_2$, $U_3$, $U_4$ in the vector space $E = \mathbb{R}^3$. Let then $E^* := \mathrm{Hom}_k(E,k)$ be the algebraic dual and define a symmetric nondegenerate bilinear form $\Phi$ on $E \oplus E^*$ as explained in Remark 9 of Section 8, namely $\Phi(e+e', f+f') = f'(e) + e'(f)$, where $e, f \in E$ and $e', f' \in E^*$. In $E \oplus E^*$ consider the two subspaces $U_1 \oplus U_4^\circ$, $U_2 \oplus U_3^\circ$ where, for $X \subset E$, we set $X^\circ := X^\perp \cap E^* = \{ f \in E^* \mid f(x) = 0 \text{ for all } x \in X \}$. With respect to $\Phi$ we find $(U_1 \oplus U_4^\circ)^\perp = U_1^\perp \cap (U_4^\circ)^\perp = (E \oplus U_1^\circ) \cap (U_4 \oplus E^*) = U_4 \oplus U_1^\circ$, $(U_2 \oplus U_3^\circ)^\perp = U_3 \oplus U_2^\circ$. Therefore, the <u>orthostable</u> lattice $\mathcal{V}$ generated in the

6-dimensional space $(E \oplus E^*, \Phi)$ by the two subspaces $U_1 \oplus U_4^\circ$, $U_2 \oplus U_3^\circ$ contains $U_3 \oplus U_2^\circ$, $U_4 \oplus U_1^\circ$. Hence, by the previous remark on $U_1, U_2, U_3, U_4$, it will be infinite (watch the "components" in $E$ of the four particular lattice elements $U_1 \oplus U_4^\circ, \ldots$ ). If we allow for infinite dimensions of $(E,\Phi)$ then we can give examples of orthostable lattices $V$ with two generators such that $V$ has infinite chains. Here is an example due to my former student Vinnie Miller.

Example 5. Let $(e_i)_{i \in \mathbb{N}}$ be a basis of the k-vector space $E$, $k$ a commutative field. Define a symmetric form $\Phi$ on $E \times E$ as follows: $\Phi(e_i, e_i) = 1$ for $i = \frac{1}{2}n(n+1)$, $n \in \mathbb{N}$, and $\Phi(e_i, e_i) = 0$ otherwise; furthermore $\Phi(e_{2i}, e_{2j}) = \Phi(e_{2i}, e_{2j+1}) = \Phi(e_{i+j}, e_{i+j})$ and $\Phi(e_{2i+1}, e_{2j+1}) = \Phi(e_{i+j+1}, e_{i+j+1})$ for $i, j \in \mathbb{N}$. Set $F := k(e_{2i} - e_{2i+1})_{i \geq 0}$, $G := k(e_{2i+1} - e_{2i+2})_{i \geq 0}$ (thus $F \cap G = (0)$ and $\dim E/F \oplus G = 1$ ). One verifies that $F^\perp = k(e_{2i})_{i \geq 0}$, $G^\perp = k(e_{2i+1})_{i \geq 0}$ (thus $(F+G)^\perp = F^\perp \cap G^\perp = (0)$ so $\Phi$ must be non-degenerate). Define recursively $A_o := F$ and, for $s \in \mathbb{N}$, $A_{4s+1} = A_{4s} \cap F$, $A_{4s+2} = A_{4s+1} + F^\perp$, $A_{4s+3} = A_{4s+2} \cap G$, $A_{4s+4} = A_{4s+3} + G^\perp$. It is routine to verify that for all $s \in \mathbb{N}$ we have

$$A_{4s+1} = k(e_{2(i+s)} - e_{2(i+s)+1})_{i \geq 0} .$$

Thus we get the infinite and properly descending chain $A_1 \supset A_5 \supset A_9 \supset \ldots$ in the lattice $V(F,G)$ $\perp$-stably generated by $F$ and $G$. Notice that both $F$ and $G$ are $\perp$-closed. Furthermore $F + F^\perp = E$ so $G \cap (F + F^\perp) = G$, whereas $G \cap F + G \cap F^\perp = (0)$. We see that $V(F,G)$ is not distributive.

Definition 13. Let $(V, \leq)$ be a lattice and $\perp : V \longrightarrow V$ an antitone map ( $x \leq y \Rightarrow y^\perp \leq x^\perp$ ) such that

$$x \leq (x^\perp)^\perp \qquad (x \in V) . \tag{36}$$

If $x = (x^{\perp})^{\perp}$ we call $x$ closed; if $x \leq y^{\perp}$ we say that $x$ and $y$ are perpendicular (notice that $x \leq y^{\perp}$ is a symmetric relation). Let $(V_1, \leq_1)$ be a second lattice equipped with an antitone $\perp_1 : V_1 \to V_1$ and satisfying (36). Then $(V, \leq, \perp)$ and $(V_1, \leq_1, \perp_1)$ (or simply $V$ and $V_1$ if there is no risk of confusion) are ortho-isomorphic, and $\varphi : V \to V_1$ is an ortho-isomorphism iff $\varphi$ is a lattice isomorphism with $\perp_1 \circ \varphi = \varphi \circ \perp$. We recall that, from (36), it follows that $x^{\perp\perp\perp} = x^{\perp}$ and $(x \vee y)^{\perp} = x^{\perp} \wedge y^{\perp}$ for all $x, y \in V$ (cf. the proof of (6) in 1.3).

The answer to the following question is overdue by now: Why are we interested in orthostable sublattices of $L(E)$ when $E$ is a sesquilinear space? Suppose that - for some reason - we wish to classify, say, pairs of subspaces $F, G$ in $E$ modulo the operations of the orthogonal group of $E$. In different terms, given a second pair $\bar{F}, \bar{G}$, we are to decide whether there exists an isometry $\varphi : E \to E$ with $\varphi F = \bar{F}$ and $\varphi G = \bar{G}$. Assume that there is such a $\varphi$. Since it respects the form on $E$ it will not only send $F$ in $\bar{F}$, $G$ in $\bar{G}$, $F \cap G$ in $\bar{F} \cap \bar{G}$ etc., it will send $F^{\perp}$ into $\bar{F}^{\perp}$, $F^{\perp} \cap G$ into $\bar{F}^{\perp} \cap \bar{G}$, etc. In short, $\varphi$ will induce a lattice isomorphism $\bar{\varphi} : V(F,G) \to V(\bar{F},\bar{G})$ between the lattices orthostably generated by the pairs. This produces a host of "obvious" invariants of the pair $F, G$ or, more correctly, invariants of the <u>orbit</u> of the pair under the action of the orthogonal group (on the set of pairs of subspaces). In other words, we get obvious (= necessary) conditions for such a $\varphi$ to exist. Besides the lattice $V(F,G)$, which is an invariant attached to the pair $F, G$, let us mention a few other typical invariants as will turn up again and again. Foremost among these are the dimensions of the quotient spaces $X/Y$ defined by neighbouring elements $Y \subset X$ in the lattice; the cardinals $\dim X/Y$ are not changed under $\bar{\varphi}$ (we call them indices).

Further invariants are the isometry classes $\hat{X}$ of the elements $X \in V(F,G)$ . These invariants reveal little about how $X$ is immersed into $E$ . An invariant of $X$ that is sensitive to the "surrounding" space is, e.g., the subset $\cap \{\| X \cap F^{\perp} \| \mid F \subset E \ \& \ \dim F < \infty\}$ of the underlying division ring (such invariants might be called "arithmetical" in contrast to the above "cardinal" invariants and "metric" invariants). There is no limit to dreaming up invariants. However, the utility of a characterization theorem depends entirely on the <u>choice</u> of the invariants. It should be possible to handle the invariants in applications and they should permit insight.

It is plausible that a thorough knowledge of the lattice $V(F,G)$ would shed light on the problem of classifying pairs (classification has been possible, thus far, for special classes of pairs only, say for disjoint, modular and dual modular pairs. Cf. Chapter VI ). Chapters IV , V , VI , VII , VIII , IX and XVI are motivated by the plan to use lattices as the principal guide on our expeditions. We hasten to add that the idea to proceed along this line is quite easy to have. But as easy as it is to make such a plan as difficult it is to actually carry it out.

Whereas the lattice $L_{\perp\perp}(E)$ has - at least in the case of an anisotropic space $E$ - been the object of study (see Chapter II.14 on ortholattices in [4] ), it appears that lattices equipped with just an antitone mapping $\perp$ satisfying (36) have received no attention worthy of note. Our results on the classification of subspaces in a sesquilinear space bear out, however, that the contemplation of $L_{\perp\perp}(E)$ and sublattices of it are of little use, even in very special circumstances. On the other hand it becomes clear, that lattice theoretic results on lattices with antitone $\perp$ would be susceptible of application in view of Chapter IV below.

We terminate with a very interesting result on lattices and forms.

Remarks on a representation theorem. In this short remark we shall freely use the terminology in Maeda's book [30] . The object of discussion is the following

Representation Theorem ("Birkhoff - v. Neumann"). Let $L$ be an irreducible, complete AC-lattice of length $\geq 4$ equipped with a polarity $\perp : L \longrightarrow L$ . Then there exists a division ring k with an antiautomorphism $*$ , a k-vector space E and an orthosymmetric sesquilinear form $\Phi : E \times E \longrightarrow k$ relative to $*$ such that $(L, \perp)$ is ortho-isomorphic to the lattice $L_{\perp\perp}(E)$ of $\perp$-closed subspaces in $(E, \Phi)$ .

The assertion of the theorem coincides with Theorem (34.5) in [30] or Theorem 4.4 in [29] if "polarity" is replaced by "ortho-complementation" (One has to pay heed to certain differences in terminology; cf. Remark (8.20) in [30] ). In order to obtain the above sharpened version one merely needs to generalize, say, Theorem 5.1 in [29] to the case of polarities in lieu of orthocomplementations. This is not difficult (as a matter of fact, an assumption on anisotropy of the form is alien to the issue at hand). A different proof may be modelled on Maeda's account of (34.5) where the required form is defined "locally". (The principal ingredients in the above representation theorem are the result proved in the Appendix of [5] to the effect that polarities of $L(E)$ , for finite dimensional E , are induced by sesquilinear forms (cf. Prop. 1 p.102 in [2]) and, on the other hand, the Representation Theorem for DAC-lattices proved in Chapter VII of [2] (cf. p. 302) ).

By the results proved in [17] it is furthermore possible (see [13]) to formulate, in lattice theoretical terms, conditions on the lattice $L$ in the above Representation Theorem which are necessary and sufficient in order for $(E,\Phi)$ in the theorem to turn out "euclidean" [ to turn out "preeuclidean" ] . Here $(E,\Phi)$ is termed <u>euclidean</u> iff $E$ splits into an orthogonal sum of finite dimensional subspaces; $(E,\Phi)$ is called <u>preeuclidean</u> iff $(E,\Phi)$ is a subspace of some euclidean space.

*

## References to Chapter I

[1] A.A. Albert, Structure of Algebras. ( 3rd print of rev. ed. ) AMS Coll Publ. XXIV, New York 1968.

[2] R. Baer, Linear Algebra and Projective Geometry. Academic Press, New York 1952.

[3] W. Baur and H. Gross, Strange inner product spaces. Comment. Math. Helv. 52 (1977) 491-495.

[4] G. Birkhoff, Lattice Theory. AMS Coll Publ. XXV ( 3rd ed ,2nd print) Providence, R.I. 1973.

[5] - and J. v. Neumann, The logic of quantum mechanics. Ann. of Math. 37 (1936) 823-843.

[6] N. Bourbaki, Formes sesquilinéaires et formes quadratiques. Hermann Paris 1959.

[7] J. Dieudonné, On the structure of unitary groups. Trans. Amer. Math. Soc. 72 (1952) 367-385.

[8] - , On the structure of unitary groups II. Amer. J. Math. 75 (1953) 665-678.

[9] - , La géometrie des groupes classiques, 3ième éd.. Ergebnisse der Mathematik, Heft 5, Springer Berlin, Heidelberg 1971.

[10] H.R. Fischer and H. Gross, Quadratic Forms and Linear Topologies I . Math. Ann. 157 (1964) 296-325.

[11] H.R. Fischer and H. Gross, Tensorprodukte linearer Topologien (Quadratische Formen und lineare Topologien III). Math. Ann. 160 (1965) 1-40.

[12] A. Frapolli, Generalizzazione di un teorema di H.A. Keller sulla modularità del reticolo dei sottospazi ortogonalmente chiusi di uno spazio sesquilineare. Masters Thesis, Univ. of Zurich 1975. (This concerns some technicalities when $\operatorname{char} k = 2$; in [25] it was assumed that $\operatorname{char} k \neq 2$ )

[13] H. Gross, On a representation theorem for AC - lattices. Mimeographed notes 1974.

[14] - , Linearly topologized spaces without continuous bases (Quadratic forms and linear topologies V). Math. Ann. 194 (1971) 313-315.

[15] - and H.A. Keller, On the definition of Hilbert space. Manuscripta math. 23 (1977) 67-90.

[16] - and V.H. Miller, Continuous forms in infinite dimensional spaces (Quadratic forms and linear topologies IV). Comment.Math. Helv. 42 (1967) 132-170.

[17] - and E. Ogg, On completions. Ann. Acad. Sci. Fenn. Ser. A.I 584 (1974) 1-19.

[18] - and - , Quadratic spaces with few isometries. Comment. Math. Helv. 48 (1973) 511-519.

[19] O. Hamara, On the structure of the orthogonal group. Math. Scand. 21 (1967) 219-232.

[20] - , Quadratic forms on linearly topologized vector spaces. Portugal. Math. 27 (1968) 15-30.

[21] I.N. Herstein, On a Theorem of Albert. Scripta Mathematica XXIX (1973) 391-394.

[22] I. Kaplansky, Forms in infinite dimensional spaces. Anais da Academia Brasileira de Ciencias 22 (1950) 1-17.

[23] - , Linear Algebra and Geometry. Allyn and Bacon, Boston 1969.

[24] H.A. Keller, Stetigkeitsfragen bei lineartopologischen Cliffordalgebren. Ph.D. Thesis, University of Zurich 1971.

[25] - , Ueber den Verband der orthogonal abgeschlossenen Teilräume eines hermiteschen Raumes. Letter to the author of Nov. 7 1973 pp. 1-6.

[26] G. Köthe, Topological Vector Spaces I. Grundlehren Band 159, Springer Verlag, Heidelberg New York 1969.

[27] S. Lefschetz, Algebraic Topology. AMS Colloquium Publ. vol. XXVII. Reprinted 1963 by AMS New York.

[28] E.A. Lüssi, Ueber Cliffordalgebren als quadratische Räume. Ph.D. Thesis, University of Zurich 1971.

[29] M.D. Mac Laren, Atomic orthocomplemented lattices. Pacific J. Math. 14 (1964) 597-612.

[30] F. Maeda and S. Maeda, Theory of symmetric lattices. Grundlehren Band 173, Springer, Berlin Heidelberg New York 1970.

[31] G. Maxwell, Infinite symplectic groups over rings. Comment. Math. Helv. 47 (1972) 254-259.

[32] E. Ogg, Ein Satz über orthogonal abgeschlossene Unterräume. Comment. Math. Helv. 44 (1969) 117-119.

[33] O.T. O'Meara, Symplectic groups. Math. Surveys vol. 16, AMS Providence R.I. 1978.

[34] V. Pless, On Witt's theorem for nonalternating symmetric bilinear forms over a field of characteristic 2 . Proc. Amer. Math. Soc. 15 (1964) 979-983.

[35] V. Pless, On the invariants of a vector subspace of a vector space over a field of characteristic two. Proc. Amer. Math. Soc. 16 (1965) 1062-1067.

[36] C.E. Rickart, Isomorphisms of infinite - dimensional analogues of the classical groups. Bull. Amer. Math. Soc. 57 (1951) 435-448.

[37] J. Saranen, Ueber die Verbandcharakterisierung einiger nichtentarteter Formen. Ann. Acad. Sci. Fenn. Ser. A.I vol. 1 (1975) 85-92.

[38] W. Scharlau, Zur Existenz von Involutionen auf einfachen Algebren. Math. Z. 145 (1975) 29-32.

[39] J.A. Schouten, Ricci - Calculus. 2nd ed. Grundlehren vol. 10, Springer, Berlin Heidelberg 1954.

[40] U. Schneider, Ueber Räume mit wenig orthogonalen Zerlegungen. Masters Thesis, University of Zurich 1972.

[41] E. Witt, Theorie der quadratischen Formen in beliebigen Körpern. J. reine angew. Math. 176 (1937) 31-44.

*

[42] W. Meissner, Untersuchungen unendlich dimensionaler quadratischer Räume im Hinblick auf modelltheoretische Uebertragungsprinzipien. This Ph.D. thesis (University of Konstanz) is nearing completion. Refer to forthcoming publications by Meissner.

## APPENDIX I

## A DIVISION RING WHICH ADMITS NO SESQUILINEAR FORMS AND A REMARK ON BAER ORDERABILITY

### Introduction

We reproduce here a classical example of a 9-dimensional division algebra $k$ over $\mathbb{Q}$ ([2] pp. 73-75). Any antiautomorphism of $k$ must leave the center $\mathbb{Q}$ pointwise fixed, hence there are none by Brauer-group theory. However, this can directly be established as we shall see below. By a slight variation of the construction we obtain an example used by Holland in [4] to exhibit some features of Baer orderability. As we believe this concept to be of importance to the theory of quadratic forms we shall draw up a docket on the matter. We could have shortened our description by quoting various standard theorems of Algebra (such as the Skolem-Noether theorem) as we have done in other appendices. However, by keeping computations elementary here we intended to offer to the beginner some examples of skew fields (other than Hamilton's quaternions) which he can actually handle knowing basic facts about field theory only.

*

1. Dickson's Example. Consider in $\mathbb{C}$ the cubic extension $K$ of $\mathbb{Q}$ defined by the polynomial

(1) $$f(X) = X^3 + X^2 - 2X - 1 .$$

It is irreducible over $\mathbb{Q}$ for, if it has a decomposition it has a linear factor and hence a root $r = \frac{n}{m} \in \mathbb{Q}$ , $n$ , $m$ relatively prime. Because the coefficient of $X^3$ is 1 substitution shows that $r$ must be in $\mathbb{Z}$ . Since all coefficients are integral $r$ must divide the constant term. But $f(\pm 1) \neq 0$ .

If $\Theta \neq 1$ is a $7^{th}$ root of unity in $\mathbb{C}$, $\Theta^6 + \Theta^5 + \Theta^4 + \Theta^3 + \Theta^2 + \Theta + 1 = 0$, then we consider the numbers

$$u := \Theta + \Theta^6 \quad , \quad v = \Theta^2 + \Theta^5 \quad , \quad w = \Theta^3 + \Theta^4 \quad . \tag{2}$$

We have $u + v + w = -1$. A short calculation shows that $u^2 = 2 + v$, $v^2 = 2 + w$, $w^2 = 2 + u$; furthermore $uv = u + w$, $vw = u + v$, $uw = v + w$. By adding the last three equations we obtain $uv + vw + wu = -2$. Finally $uvw = (u+w)w = uw + w^2 = (v+w) + (u+2) = 1$. Thus $u$, $v$, $w$ in (2) are the three roots of $f$ in (1).

We define a $\mathbb{Q}$-linear automorphism $\sigma: K \to K$ by setting $\sigma(u) = v$, $\sigma(v) = w$, $\sigma(w) = u$ and we find for $N(\lambda) := \lambda\cdot\sigma(\lambda)\cdot\sigma^2(\lambda)$, where $\lambda = xu + yv + zw$ is a typical element of $K$ $(x,y,z \in \mathbb{Q})$, the expression

$$N = x^3 + y^3 + z^3 - 4(x^2z+y^2x+z^2y) + 3(x^2y+y^2z+z^2x) - xyz \tag{3}$$

Let $k$ be a 3-dimensional K-right vector space with basis $\{a_0,a_1,a_2\}$. We define a multiplication on $K$ by first defining it for basis vectors and then extending it to arbitrary vectors. The $a_i$ are to be multiplied according to the table

| | $a_0$ | $a_1$ | $a_2$ |
|---|---|---|---|
| $a_0$ | $a_0$ | $a_1$ | $a_2$ |
| $a_1$ | $a_1$ | $a_2$ | $2a_0$ |
| $a_2$ | $a_2$ | $2a_0$ | $2a_1$ |

For arbitrary $\lambda \in K$ we set down the further rules $(a_i\lambda)a_o = (a_ia_0)\lambda$, $(a_i\lambda)a_1 = (a_ia_1)\sigma(\lambda)$, $(a_i\lambda)a_2 = (a_ia_2)\sigma^2(\lambda)$ where $\sigma$ is the automorphism operating on $K$ as defined above. Arbitrary vectors in $k$ are now multiplied distributively. One checks that an associative ring is obtained. It has a unit element 1, namely $1 = a_0$; if we write $a$

instead of $a_1$ then $a_2 = a^2$ and $a^3 = 2$ . What we have defined here is the so-called crossed product of the (normal) extension K of $\mathbb{Q}$ with its Galois group $\{1\!\!1,\sigma,\sigma^2\}$ . (Here the Galois group is cyclic; standard notation for k is $(K/\mathbb{Q},\sigma,2)$ .) k is obviously an algebra of dimension 9 over the field $\mathbb{Q}$ .

What is the center of k ? Let $x \in k \setminus \mathbb{Q}$ . The field $\mathbb{Q}(x) \subset k$ is not all of k as k is not commutative. Therefore $[\mathbb{Q}(x):\mathbb{Q}] = 3$ . Hence $Q(x)$ is maximal as a <u>commutative</u> field contained in the $\mathbb{Q}$-algebra k . Hence x cannot commute with all elements of k . Thus the center of k is $\mathbb{Q}$ .

Finally we show that k is a division algebra over $\mathbb{Q}$ . It suffices to show that every nonzero element x in k admits a right inverse. This in turn will follow if we can show that left multiplication $x_L\colon \xi \mapsto x\xi$ in the K-right vector space k is an injective endomorphism. Now it is very easy to compute the matrix of $x_L$ relative to the basis $1$ , $a$ , $a^2$ . Setting $x = \lambda_0 + a\lambda_1 + a^2\lambda_2$ we find

$$(4) \qquad \det(x_L) = \det\begin{pmatrix} \lambda_0 & 2\sigma(\lambda_2) & 2\sigma^2(\lambda_1) \\ \lambda_1 & \sigma(\lambda_0) & 2\sigma^2(\lambda_2) \\ \lambda_2 & \sigma(\lambda_1) & \sigma^2(\lambda_0) \end{pmatrix} = N(\lambda_0) +$$

$$2N(\lambda_1)+4N(\lambda_2)-2(\lambda_0\sigma(\lambda_1)\sigma^2(\lambda_2)+\sigma(\lambda_0)\sigma^2(\lambda_1)\lambda_2+\sigma^2(\lambda_0)\lambda_1\sigma(\lambda_2)) .$$

We can show that $\det(x_L)$ is nonzero by a "parity check" as follows. First we may assume that $\lambda_0$ , $\lambda_1$ , $\lambda_2$ have integral components with respect to the basis $\{u,v,w\}$ of K . Secondly, by multiplying x from the left, if necessary, by a suitable power of $a^{-1}$ we may achieve that $\lambda_0 = xu + yv + zw$ has not all its components x , y , z even. A direct inspection of (3) now shows that $N(\lambda_0)$ is odd in this case. Thus by (4)

(5) $$\det(x_L) \equiv N(\lambda_0) \equiv 1 \pmod 2$$

and so $x_L$ is an automorphism. Q. E. D. We have shown that $k$ is a 9-dimensional division algebra over its center $\mathbb{Q}$ .

2. There is no antiautomorphism. Let $j: K \to k$ be any injective homomorphism; $\tau := j\circ\sigma\circ j^{-1}$ is then an automorphism of the subfield $j(K) \subset k$ . By the Skolem-Noether theorem this automorphism is induced by an inner automorphism of $k$ , i.e. there exists $x \in k$ such that

(6) $$x^{-1}j(\xi)x = j(\sigma(\xi)) \quad \text{for all} \quad \xi \in K .$$

We can establish the existence of such an $x$ directly as follows: Suppose we have two elements $\alpha$ , $\beta \in k$ which have the same irreducible cubic polynomial $h \in \mathbb{Q}[Y]$ . We want to show that there is $\gamma \in k$ such that $\gamma^{-1}\alpha\gamma = \beta$ . Let $h(Y) = Y^3 + rY^2 + sY + t$ . Then $0 = h(\alpha) - h(\beta) = [\alpha(\alpha^2+\alpha\beta+\beta^2)-(\alpha^2+\alpha\beta+\beta^2)\beta] + r[\alpha(\alpha+\beta)-(\alpha+\beta)\beta] + s(\alpha-\beta)$ . Set $\gamma = \alpha^2 + \alpha\beta + \beta^2 + r(\alpha+\beta) + s$ . We have $\alpha\gamma = \gamma\beta$ . If $\gamma \neq 0$ we are done. If $\gamma = 0$ we can interchange the roles of $\alpha$ and $\beta$ to get a $\gamma'$ with $\beta\gamma' = \gamma'\alpha$ . If $\gamma' \neq 0$ we are done. If $\gamma' = 0$ then $\gamma = \gamma'$ which means that $\alpha\beta = \beta\alpha$ ; i.e. $\beta \in \mathbb{Q}(\alpha)$ . In that case we insert an intermediate element $\delta$ which has $h$ as its minimal polynomial over $\mathbb{Q}$ and which is not in $\mathbb{Q}(\alpha)$ . By the foregoing $\alpha$ and $\delta$ are then conjugate and so are $\beta$ and $\delta$ and hence we are done. There are such $\delta$ : Pick some $\gamma \in k$ with $\gamma^{-1}\alpha\gamma \neq \alpha$ . Then for all natural $n$ , $m$ and $n \neq m$ we have $(\gamma+n)^{-1}\alpha(\gamma+n) \neq (\gamma+m)^{-1}\alpha(\gamma+m)$ . Not all of these conjugates of $\alpha$ are in $\mathbb{Q}(\alpha)$ . This establishes (6).

If we particularize $j$ in (6) to the embedding $K \subset k$ then we already know that we can choose $x = a$ in (6) because $\xi a = a\sigma(\xi)$ by the construction of $k$ . By the extension theorem just proved we can also extend an arbitrary $j: K \to k$ to an inner automorphism of $k$ ,

$y^{-1}\xi y = j(\xi)$ for suitable $y \in k$ and all $\xi \in K$. Substitution into (6) shows that $ay$ and $yx$ differ only by a factor $z$ which is in $K$. Thus, if we pass to the determinants of the K-endomorphisms $k \to k$ induced by the left multiplications we find $\det(x_L) = \det(a_L)\cdot\det(z_L) = 2N(z)$ by (4). We have proved:

(7) for each injection $j: K \to k$ and $x \in k$ with (6) we have $\det(x_L) = 2N(\lambda)$ for some $\lambda \in K$.

Direct inspection of (3) shows that ([2], p. 75)

(8) If $0 \neq \lambda \in K$ then $N(\lambda) = 8^r \cdot \frac{s}{t}$ ; $r, s, t \in \mathbb{Z}$ and $s, t$ odd.

The conjunction of (7) and (8) exhibits a quality of non-symmetry of $k$, or rather, of non-antisymmetry. Assume by way of contradiction that $*: k \to k$ were an antiautomorphism. Since $a^{-1}\xi a = \sigma(\xi)$ for all $\xi \in K$ we find $a^*\xi^*a^{*-1} = \sigma(\xi)^*$, i.e. $a^{*-1}$ qualifies for $x$ in (6) if $j$ is the restriction of $*$ to $K$. Hence $\det(x_L)$ is of the shape described in (7) for $x = a^{*-1}$. But $a$ and $a^*$ have the same minimal polynomial over $\mathbb{Q}$ and hence they are conjugate by what we have proved further up; consequently $\det(a^*_L) = \det(a_L) = 2$. This means that $\frac{1}{2}$ $(= \det(a^{*-1}_L))$ is of the shape $2N(\lambda)$ where $N(\lambda)$ is as described in (8), a flagrant nonsense.

Thus we have shown that Dickson's 9-dimensional algebra $k$ over $\mathbb{Q}$ admits no antiautomorphism whatever. Hence we can define no sesquilinear forms over $k$. In the next section we shall see that a small change in the construction allows for antiautomorphisms.

3. Modifying Dickson's example. We keep working with a $7^{th}$ root of unity $\Theta \neq 1$ in $\mathbb{C}$ and let $\omega := \Theta + \Theta^2 + \Theta^4$. We find $\omega^2 + \omega = 2(\Theta^6 + \Theta^5 + \Theta^4 + \Theta^3 + \Theta^2 + \Theta + 1) - 2 = -2$ ; thus $\omega$ is a root of $g(X) = X^2 + X + 2 \in \mathbb{Q}[X]$ . The other root of $g$ is

$$\omega^* = -\omega - 1$$

where $*$ is complex conjugation in $\mathbb{C}$ ; hence $*$ is an automorphism of the quadratic extension

$$Z := \mathbb{Q}(\omega) .$$

The cubic $f$ in (1) remains irreducible over $Z$ . Thus with $u$ as defined in (2) we define

$$K := Z(u) = Z(u,v,w)$$

and we have $[K:Z] = 3$ . Now we define $k$ as a K-right vector space spanned by a basis $a_0 = 1$ , $a_1 = a$ , $a_2 = a^2$ and with a multiplication almost identical with that in Dickson's example. We merely replace "$a^3 = 2$" by

(9) $$a^3 = \gamma \quad \text{where} \quad \gamma := \frac{\omega}{\omega^*} .$$

Here $\sigma: K \to K$ remains just as before, it sends $u$ , $v$ , $w$ into $v$ , $w$ , $u$ respectively. We obtain a 9-dimensional algebra over the center $Z$ .

Is $k$ a division algebra? Again, it suffices to show that $\det(x_L) \neq 0$ if $x_L$ is the endomorphism $\xi \mapsto x\xi$ in the K-right vector space $k$ and $x \neq 0$ is in $k$ . Dickson's argument that followed (4) above can be reproduced by making use of the fact that the ring $R := \mathbb{Z}[\omega] = \{n+m\omega \mid n,m \in \mathbb{Z}\}$ is euclidean and has $Z$ as its field of quotients. Indeed, let for $a = n + m\omega \in R$ $|a|$ be its usual complex norm, $|a|^2 = (n+m\omega)(n+m\omega^*) = n^2 - nm + 2m^2$ . For nonzero $a$ , $b \in R$ compute $ab^{-1} = r + s\omega$ which is in $Z$ . Then pick $t = p + q\omega \in R$ with $|p-r| \le \frac{1}{2}$ , $|q-s| \le \frac{1}{2}$ and, if $(r,s)$ happens to be the center of some square in the tessellation of the plane $\mathbb{C}$ by $\mathbb{Z} \times \mathbb{Z}$ , require furthermore that $(p-r)(q-s) = +\frac{1}{4}$ (only the vertices in the southwest or northeast of the squares are eligible). We then invariably have $|a-tb|^2 = |b|^2|ab^{-1}-t|^2 < |b|^2$ . Thus $R$ is euclidean and has unique factorization into primes in $R$ . E.g.,

$$2 = \omega\omega^*$$

is the decomposition of $2$ into prime numbers in $R$ . Call an element $r = n + m\omega \in R$ even if $\omega$ is a factor in $r$ . Thus if $r$ is even then $\omega|r$ so $\omega|n$ thus $\omega\omega^*|n^2$ (by taking norms in $\mathbb{C}$ ) and thus $2|n$ in $\mathbb{Z}$ . The argumentation can be reversed and we see that

$$\omega|n + m\omega \quad \text{in} \quad R \iff 2|n \quad \text{in} \quad \mathbb{Z} . \tag{10}$$

In other words $R/\omega R \cong \mathbb{Z}/2\mathbb{Z}$ . Hence we are able to repeat the parity check in the formula for $\det(x_L)$ for which, this time, we obtain

$$\begin{aligned} \det(x_L) &= N(\lambda_0) + \gamma N(\lambda_1) + \gamma^2 N(\lambda_2) - \\ &\gamma(\lambda_0\sigma(\lambda_1)\sigma^2(\lambda_2) + \sigma(\lambda_0)\sigma^2(\lambda_1)\lambda_2 + \sigma^2(\lambda_0)\lambda_1\sigma(\lambda_2)) . \end{aligned} \tag{11}$$

After some simple normalizations we may assume that the components of the $\lambda_i$ are all in $R$ and, furthermore, that $x$ , $y$ , $z$ in $\lambda_0 = xu + yv + zw$ are not all even elements in $R$ . Now, if we had that $\det(x_L) = 0$ then we would obtain the following congruence in $R$ ,

$$0 = \omega^{*2}\det(x_L) \equiv \omega^{*2}N(\lambda_0) \pmod{\omega} .$$

Hence $N(\lambda_0) \equiv 0 \pmod{\omega}$ . But from (3) we see that $N(\lambda_0) \equiv x^3 + y^3 + z^3 + x^2y + y^2z + z^2x + xyz \equiv 1 \pmod{\omega}$ since at least one of $x$ , $y$ , $z$ is odd in $R$ .

We have established that $k$ is a 9-dimensional division algebra over its center $Z$. And this time there exists an antiautomorphism! We define

$$(12) \qquad a^* := a^{-1} = a^2 \cdot \frac{1}{\gamma}$$

and extend $*$ by additivity and anticommutativity to all of $k$,

$$(13) \qquad (\lambda_0 + a\lambda_1 + a^2\lambda_2)^* = \lambda_0^* + \lambda_1^* a^* + \lambda_2^* a^{*2}$$

(recall that $\lambda^*$ is the complex conjugate of $\lambda \in K$). It is not difficult to verify that (12) and (13) define an antiautomorphism of $k$. Furthermore $(x^*)^* = x$ for all $x \in k$. Notice that the involution $*$ does not leave the center pointwise fixed (of that kind there still can be none).

We end this section by the following remark ([4]). If $x = \lambda_0 + a\lambda_1 + a^2\lambda_2 \in k$ then an easy computation shows that the coefficient of $1 = a^0$ in $xx^*$ is $\lambda_0\lambda_0^* + \sigma^2(\lambda_1)\sigma^2(\lambda_1)^* + \sigma(\lambda_2)\sigma(\lambda_2)^*$. Therefore, an equation of the type

$$(14) \qquad xx^* + yy^* + \cdots = 0$$

in $k$ yields an equation $\lambda_0\lambda_0^* + \sigma^2(\lambda_1)\sigma^2(\lambda_1)^* + \sigma(\lambda_2)\sigma(\lambda_2)^* + \mu_0\mu_0^* + \sigma^2(\mu_1)\sigma^2(\mu_1)^* + \sigma(\mu_2)\sigma(\mu_2)^* + \cdots = 0$ in $K \subset \mathbb{C}$. Because such an equation in $\mathbb{C}$ entails $\lambda_0 = \lambda_1 = \lambda_2 = \mu_0 = \mu_1 = \mu_2 = \ldots = 0$ we see that (14) implies

$$(15) \qquad x = y = \ldots = 0 .$$

4. Baer ordered *-fields. The crossed product $(K/Z, \sigma, \frac{\omega}{\omega^*})$ constructed in the previous section was presented in [4] to exhibit some salient features of Baer orderability; Holland calls Baer-ordered an involutorial division algebra $(k,*)$ that contains a subset $\Pi$ with

the following properties

(i) $\Pi \subset S := \{\xi \in k \mid \xi^* = \xi\}$

(ii) $1 \in \Pi$ , $0 \notin \Pi$

(iii) $\Pi + \Pi \subset \Pi$

(iv) $\rho \Pi \rho^* \subset \Pi$ for all $\rho \neq 0$

(v) $-\Pi \cup \Pi = S \setminus \{0\}$

If $\lambda$ , $\mu \in S$ then one defines $\lambda > \mu$ if $\lambda - \mu \in \Pi$ etc. The notion was articulated by Baer in [1, Chapter IV, Appendix 1, p. 127-128] .

Orderability in this sense is an adequate concept if one wishes to talk about <u>positive</u> hermitean forms. In the commutative situation the concept had been put to use by Prestel in [7], [8]; for a survey in the commutative case one should consult [6].

A Baer ordered $(k,*)$ is called <u>archimedean</u> when $0 \leq \lambda < \frac{1}{n}$ for all $n = 1, 2, \ldots$ implies $\lambda = 0$ . The following characterization is proved in [4].

<u>Theorem</u> [4]. An archimedean ordered $*$-field is $*$- and order isomorphic to a subfield of the real numbers $\mathbb{R}$ , the complex numbers $\mathbb{C}$ or the real quaternions $\mathbb{H}$ .

Since all semiorderings on algebraic number fields are orderings in the usual sense by a result of Prestel ([7] Korollar 1.5) the theorem has the following

<u>Corollary</u> [4]. Other than a quaternion algebra, no finite dimensional noncommutative $*$-field central over an algebraic number field admits an ordering in the sense of Baer.

From this result it follows directly that the 9-dimensional $*$-field $k$ central over the algebraic number field $Z$ , as detailed in the previous section, admits no ordering in the sense of Baer. On the other hand, we have seen that $k$ is formally real in the sense that $\Sigma x_i x_i^* = 0$ implies $x_i = 0$. Thus formal reality does not, as in the Artin-Schreier Theory, imply orderability. (Orderability does, of course, imply formal reality by properties (ii) and (iv).) See also Section 5 in Appendix 1 to Chapter II where an example of a noncommutative involutorial division ring is given that is formally real and Hilbert ordered.

We refer to another example in [4] of a Baer ordered field $(k,*)$ (loc. cit. pp. 215-219). It is of interest to us because it provides a noncommutative field, other than quaternions, that admits positive hermitean $\aleph_0$-forms and is such that we are able to give a reasonable classification of $\perp$-dense subspaces (along the line of Chapter XII.8 below).

We terminate this section by pointing out that in [5] Baer orderings have been put to use in the problem of classifying the <u>infinite</u> dimensional hermitean spaces $E$ that possess the following property on subspaces $X \subset E$:

(16) $\qquad$ if $X^{\perp\perp} = X$ then $X \oplus X^{\perp} = E$.

This problem has turned out to be surprisingly difficult. Although no orderings or topologies are involved in (16) progress has been made so far only under additional provisos involving orderings or Baer orderings. See [3] and [9] where the same problem is investigated.

See furthermore the postscript added to the introduction of the book on page 3.

*

## References to Appendix I


[1] R. Baer, Linear Algebra and Projective Geometry. Academic Press Inc., New York 1952.

[2] L.E. Dickson, Algebren und ihre Zahlentheorie. Orell Füssli Verlag Zürich (Switzerland) 1927.

[3] H. Gross and H.A. Keller, On the definition of Hilbert space. Manuscripta math. 23 (1977) 67-90.

[4] S.S. Holland, Orderings and Square Roots in *-Fields. J. Algebra 46 (1977) 207-219.

[5] - , Orthomodular forms over ordered *-fields. To appear.

[6] A. Prestel, Lectures on Formally real Fields. Monografias de Mat. 22, Inst. de Mat. Pura e Aplicada, Rio de Janeiro 1975.

[7] - , Quadratische Semiordnungen und quadratische Formen. Math. Z. 133 (1973) 319-342.

[8] - , Euklidische Geometrie ohne das Axiom von Pasch. Abh. Math. Sem. Hamburg 41 (1974) 82-109.

[9] W.J. Wilbur, On characterizing the standard quantum logics. Trans. Amer. Math. Soc. 233 (1977) 265-282.

CHAPTER TWO

# DIAGONALIZATION OF $\aleph_0$-FORMS

## 1. Introduction

In this chapter we shall prove that $\aleph_0$-dimensional sesquilinear spaces are orthogonal sums of lines and planes and we characterize the cases where a decomposition into mutually orthogonal lines is impossible. The problem of "normalizing" bases brings us to stability and the beginner is confronted with the first Ping-Pong style proof with its characteristic back-and-forth argument (Theorem 2). These matters are basic and their knowledge is tacitly assumed in the rest of the book.

Diagonalization of forms in dimension $\aleph_0$ is a simple affair. However, in order to grasp just how exclusive the property of admitting decompositions into orthogonal summands of small dimensions (i.e. smaller than that of the entire space) really is, the presentation of some examples from the uncountable is enlightening. Here they are.

Example 1. Let $k$ be an uncountable field of any characteristic and $(e_\iota)_{\iota \in I}$ the basis of a $\aleph_1$-dimensional k-vector space. We define a symmetric form by setting $\Phi(e_\iota, e_\kappa) = X_{\{\iota,\kappa\}}$ where the family $X_{\{\iota,\kappa\}}$ is algebraically independent over the prime field in $k$. $(E,\Phi)$ turns out nondegenerate and in [8] we established that it enjoys the following outlandish property on subspaces $X \subset E$:

$$(1) \qquad \dim X \geq \aleph_0 \Rightarrow \dim X^\perp \leq \aleph_0 .$$

In particular, we see from (1) that $(E,\Phi)$ possesses very few orthogonal splittings,

$$(2) \qquad E = X \oplus X^\perp \Rightarrow \dim X < \infty \quad \text{or} \quad \dim X^\perp < \infty .$$

Needless to say that there can be no orthogonal decomposition of $(E,\Phi)$ into finite dimensional subspaces. Incidentally, by using this construction one can exhibit $2^{\aleph_1}$ nondegenerate symmetric $\aleph_1$-forms over the field $\mathbb{C}$ which are <u>not</u> isometric, surprising since there is but 1 isometry class in each dimension $\leq \aleph_0$ . (Remark: One can construct spaces with (1) over <u>finite</u> and <u>countable</u> fields also; however, a little more imagination is needed being that short of scalars. See [1].)

<u>Example 2</u>. Let $J$ be a well-ordered set of cardinality $\aleph_1$ , $I := J \times \mathbb{Z}$ , $(e_\iota)_{\iota\in I}$ the basis of a vector space $E$ . Order $I$ lexicographically, $(\mu,m) \leq (\nu,n)$ iff $\mu < \nu$ or else $\mu = \nu$ & $m \leq n$ . We equip $E$ with the symmetric bilinear form $\Phi$ defined by

$$\Phi(e_{(\mu,m)},e_{(\nu,n)}) = \begin{cases} m & \text{if} \quad (\mu,m) \leq (\nu,n) \\ n & \text{if} \quad (\mu,m) \geq (\nu,n) \ . \end{cases}$$

Set

$$h_{(\mu,m)} := -e_{(\mu,m-1)} + 2e_{(\mu,m)} - e_{(\mu,m+1)} \ .$$

We see that we have secured <u>dual systems</u>: $\Phi(e_\iota,h_\kappa) = \delta_{\iota\kappa}$ (Kronecker) for all $\iota , \kappa \in I$ . Observe that the $h_\kappa$ span a proper subspace of $E$ . The existence of such a dual system provides for many properties shared by spaces with orthogonal bases. Thus, we see that here orthogonal decompositions abound; e.g.,

$$\text{(3)} \qquad \text{for all} \quad \iota \in I : \quad E = \bigoplus_{\kappa\leq\iota} (e_\kappa) \overset{\perp}{\oplus} \bigoplus_{\kappa>\iota} (e_\kappa - e_\iota) \ .$$

In sharp contrast to Example 1 we see furthermore equations such as $\dim X = \dim X^{\perp\perp}$ , $\dim E/X^\perp = \dim X$ to hold for arbitrary subspaces $X$ (use (3)). Yet $E$ admits no orthogonal basis ([2], pp. 36-37).

<u>Example 3</u>. Let $(\bar{E},\Phi)$ be a nondegenerate space which <u>is</u> spanned by an uncountable orthogonal basis $(e_\iota)_{\iota\in I}$ . Let $E$ be the hyperplane

spanned by all $e_\iota - e_{\iota_0}$, $\iota \in I$. This hyperplane $(E,\Phi)$ admits no orthogonal decomposition into finite dimensional subspaces, in spite of the fact that it appears as a subspace in a space with such a decomposition ([7], Satz 1 p. 105). In this respect it radically differs from the first two examples which both cannot appear as subspaces in orthogonal sums of finite dimensional spaces.

Remark. The three examples invite for an investigation into the existence of orthogonal bases when dimensions are uncountable or of embeddability into spaces admitting such bases. This is beyond the scope of this book, and we refer to [2, 3, 7, 9, 10, 14]. My discovering Example 3 gave the motivation to look for a theory. E. Ogg - then a student in my algebra class - after listening to the topological setting of the density theorem for irreducible modules, came up with the startling idea to use the countable analogue of the weak linear topology of the sesquilinear form in order to attack the existence problem on orthogonal bases. It was a marvellous idea [14]. For a survey on the emerging theory one should consult [3].

*

## 2. Diagonalization

Our starting point is the following fundamental

Theorem 1. Each sesquilinear space of dimension at most $\aleph_0$ is a direct orthogonal sum of lines and planes.

Proof. Let $E_0$ be a supplement of $E^\perp$ in the sesquilinear space $E$ $(\dim E \le \aleph_0)$. We shall construct a decomposition of $E_0$ into mutually orthogonal lines and planes and join it with any direct decomposition of $E^\perp$ into lines in order to get a decomposition of $E$ of the required sort. The proof is modelled on the well-known Gram-Schmidt orthogonalization process ([5], [15]).

For the construction we need a countable family $(e_i)_{i \in I}$ of generators of $E_0$ (e.g. a basis of $E_0$ ). Choose the index set $I$ to either be $\mathbb{N}$ or a finite initial segment of $\mathbb{N}$ . We show how to get a family $(F_j)_{j \in J}$ , for some $J \subset I$ , of nondegenerate lines and planes $F_j$ in $E_0$ such that, for all $m \in I$ , we have

(4) $\qquad$ if $i < m$ then $e_i \in F_0 \overset{\perp}{\oplus} \cdots \overset{\perp}{\oplus} F_m$ .

For any such family the subspace $\bigoplus_{j \in J} F_j$ contains all generators $e_i$ and therefore is the entire space $E_0$ ; thus $E_0 = \bigoplus_{j \in J} F_j$ then is a decomposition of $E_0$ of the required sort.

How shall we define a family of such $F_j$ ? If the first generator $e_0$ is not isotropic we let $F_0$ be the line $(e_0)$ ; if $e_0$ happens to be isotropic then there is $x \in E_0$ not perpendicular to $e_0$ (because $E_0$ is nondegenerate) and we let $F_0$ be the span of $e_0$ and $x$ (it is nondegenerate). Assume that for some $m \in I$ (e.g. $m = 0$ ) we have defined mutually orthogonal nondegenerate lines and planes $F_0 , F_1 , \ldots , F_m$ such that (4) holds. The sum $S := F_0 \oplus \cdots \oplus F_m$ is nondegenerate and of finite dimension so $E_0 = S \overset{\perp}{\oplus} S'$ . If $S' = (0)$ we let $J = \{0,\ldots,m\}$ and we are through. Assume that $S' \neq (0)$ whence $m+1 \in I$ and $e_{m+1} = s + s'$ for some $s \in S$ , $s' \in S'$ . If $s' = 0$ we let $F_{m+1}$ be any anisotropic line or hyperbolic plane in $S'$ ; if $s' \neq 0$ but isotropic then there is $x \in S'$ not perpendicular to $s'$ and we let $F_{m+1}$ be the span of $s'$ and $x$ ; finally if $s'$ is not isotropic then we let $F_{m+1}$ be the line $(s')$ . In sum, we see that we can extend the sequence $F_0 , \ldots , F_m$ by a further nondegenerate (line or plane) $F_{m+1}$ , perpendicular to $F_0 , \ldots , F_m$ , such that (4) holds again with $m+1$ in lieu of $m$ . In other words, the step can be repeated, and at most card $I$ times so if $I$ is finite; if $I = \mathbb{N}$ then we can procure in this fashion, say by Zorn's lemma, a denumerable sequence $(F_i)_{i\in\mathbb{N}}$ such that for each $m \in \mathbb{N}$ condition (4) is satisfied. This terminates the proof of Theorem 1.

The following corollaries to the theorem give precise information as to when a space is an orthogonal sum of lines, i.e. admits an orthogonal basis.

<u>Corollary 1</u>. If $(E,\Phi)$ is nondegenerate and finite dimensional then the following are equivalent: (i) there is no orthogonal basis in $E$, (ii) $\Phi$ is alternate.

<u>Proof</u>. The implication (ii) $\Rightarrow$ (i) is obvious. Assume that $\Phi$ is not alternate. By Theorem 1 $E = \overset{n}{\underset{i=1}{\oplus^{\perp}}} F_i$, $\dim F_i \le 2$. Planes which are not alternate split off a nonisotropic line so that we may assume all planes $F_i$ to be alternate. If there is none we are done. Otherwise the field is commutative and $\Phi$ skew-symmetric on $E$. As $\Phi$ is not alternate it follows that the characteristic is 2. Let $F_1 = (y_1)$ be one of the anisotropic lines and $\{y_2, y_3\}$ a basis of an alternate plane $F_n$, $\Phi(y_2,y_3) = 1$. Set $\alpha = \Phi(y_1,y_1)$. The basis $\{e_1,e_2,e_3\}$ of $F_1 \overset{\perp}{\oplus} F_n$ defined by $e_1 = y_1 + \alpha y_3$, $e_2 = y_1 + y_2 + \alpha y_3$, $e_3 = y_1 + y_2$ is orthogonal. By repeating the procedure we introduce an orthogonal basis in $E$. We have thus shown that not(ii) $\Rightarrow$ not(i).

<u>Corollary 2</u>. Let $(E,\Phi)$ be nondegenerate and $\aleph_0$-dimensional. There is no orthogonal basis for $E$ if and only if we are in one of the following two situations (1) $\Phi$ is alternate, (2) $\Phi$ is skew-symmetric but not alternate (hence the field commutative and of characteristic 2) and $E^*$ is $\perp$-closed and $\dim E/E^*$ is finite (here $E^*$ is the subspace of all isotropic vectors).

Proof. A) We begin by showing that there can be no orthogonal basis if we are in one of the two situations. This is obvious if $\Phi$ is alternate. Assume therefore that we are in case (2). As $E$ is nondegenerate and $\dim E/E^* < \infty$ (by assumption) we must have $\dim E^{*\perp} < \infty$. A fortiori $R := \operatorname{rad} E^*$ is of finite dimension and thus there is a metabolic decomposition $E = (R \oplus R') \overset{\perp}{\oplus} E_0$. Since $E^* \subset R^\perp = R \oplus E_0$ we see that $E^* = R \oplus (E_0)^*$. We read off $E^{*\perp\perp} = R \oplus (E_0)^{*\perp 0 \perp 0}$ where we abbreviated $X^\perp \cap E_0$ as $X^{\perp 0}$. Since $E^* = E^{*\perp\perp}$ in case (2) we see that $(E_0)^{*\perp 0 \perp 0} = (E_0)^*$, i.e., $(E_0)^*$ is an orthogonally closed subspace of $(E_0, \Phi_0)$, $\Phi_0 := \Phi|_{E_0 \times E_0}$. Since $\operatorname{rad}(E_0)^* = (0)$ and $\dim(E_0)^{*\perp 0} \leq \dim E^{*\perp} < \infty$ we see that $(E_0)^* \oplus (E_0)^{*\perp 0}$ is closed and dense in $E_0$, i.e.

$$E_0 = (E_0)^* \oplus (E_0)^{*\perp 0} .$$

Suppose by way of contradiction that $E$ admits an orthogonal basis $(e_i)_{i \in \mathbb{N}}$. Each element of some arbitrary fixed basis of $(R \oplus R') \oplus (E_0)^{*\perp 0}$ is a finite linear combination of some $e_i$. Hence there is $N \in \mathbb{N}$ such that we have for all $i > N$

$$e_i \in (R \oplus R' \oplus (E_0)^{*\perp 0})^\perp = (E_0)^* .$$

In particular, $e_i$ is isotropic for $i > N$. But this is impossible for members of an orthogonal basis of a nondegenerate space. Thus there is no orthogonal basis in the second case.

B) Assume conversely that there is no orthogonal basis and that $E$ is not alternate. By Theorem 1 the space $E$ is an orthogonal sum of lines and planes. Nonalternate planes admit orthogonal bases; hence there must be alternate planes in the decomposition. Therefore $\Phi$ is skew-symmetric on $E$. As $\Phi$ is not alternate (by assumption) we must have characteristic 2. There is a decomposition $E = E_1 \overset{\perp}{\oplus} E_2$ where

$E_1$ admits an orthogonal basis $(e_i)_{i \in I_1}$ and $E_2$ is an orthogonal sum of alternate planes $P_j$. If $I_1$ were infinite then there would be enough $e_i$ to delegate a different $e_i$ to each of the $P_i$. In $e_i \overset{\perp}{\oplus} P_i$ we could introduce orthogonal bases - as we had done in the proof of Corollary 1 - and thus procure an orthogonal basis for $E$. Contradiction! Therefore $\dim E_1 < \infty$. Because $E_2 \subset E^*$ this means $\dim E^*/E_2 \leq \dim E/E_2 = \dim E_1 < \infty$; hence $E^*$ is $\perp$-closed because $E_2$ is closed. Furthermore, $\dim E/E^* \leq \dim E/E_2 = \dim E_1 < \infty$. This shows that we are in case (2) of Corollary 2. Q. E. D.

<u>Corollary 3</u>. In a trace-valued sesquilinear space of dimension $\leq \aleph_0$ which is not alternate every subspace admits an orthogonal basis.

<u>Proof</u>. If $(E,\Phi)$ possesses a subspace without orthogonal basis then (by Corollaries 1 and 2) $\Phi$ must be skew-symmetric on all of $E$ and hence alternate if $\Phi$ is assumed trace-valued.

Since nondegenerate alternate planes are hyperbolic we have the following immediate consequence of Theorem 1:

<u>Corollary 4</u>. A nondegenerate alternate space of dimension $\leq \aleph_0$ is an orthogonal sum of hyperbolic planes.

## 3. Stability (Definition)

We turn to the existence of <u>normalized</u> orthogonal bases in non-degenerate $\aleph_0$-dimensional spaces $(E,\Phi)$. We assume in the first place that the forms are symmetric and that the characteristic of the underlying field $k$ is not $2$. $(E,\Phi)$ will then admit orthogonal bases.

*

If $k$ possesses square roots $\lambda_i$ for each length $\Phi(e_i,e_i)$ where $(e_i)_{i\in I}$ is an orthogonal basis then we can, of course, introduce the <u>orthonormal</u> basis $f_i := \lambda_i^{-1}e_i$, $\Phi(f_i,f_j) = \delta_{ij}$ (Kronecker). If $k$ contains an element $\alpha$ which is <u>not</u> a square, then we can define symmetric forms of arbitrary <u>finite</u> dimension admitting no orthonormal bases: Simply let the matrix $M$ of $\Phi$ with respect to some basis be diagonal and, say, $M = \text{diag}[\alpha,1,1,\ldots,1]$. If $M'$ is the matrix with respect to some other basis then $\det M' = \alpha(\det A)^2$ where $A$ is the substitution matrix; thus $M' = \text{diag}[1,1,1,\ldots,1]$ is impossible. Contrary to what one would expect we shall see that in dimension $\aleph_0$ the following does occur: there are fields $k$ which do not have square roots for all elements and yet <u>each</u> $\aleph_0$-dimensional (nondegenerate) symmetric form $\Phi$ admits an <u>orthonormal</u> basis. As an illustration of how such a thing can come about we establish the following closely related fact ([4] Thm. 8.1 p. 567):

(5) Each positive definite $\mathbb{Q}$-space of dimension $\aleph_0$ admits an orthonormal basis.

The proof will rest on the classical fact that each indefinite $\mathbb{Q}$-form in $5$ variables has a nontrivial zero ([13]; [16] Thm. 22, p. 41). In particular, if $\Psi$ is any positive definite form in 4 variables and $\alpha \in \mathbb{Q}$ is positive then $\Psi$ represents $\alpha$ because $\frac{1}{\alpha}\Psi - x_5^2$ is indefinite (specifically, each positive $\alpha$ is a sum of 4 squares). From this it is obvious that each $\aleph_0$-dimensional positive definite $\mathbb{Q}$-space has $\|E\| = \mathbb{Q}^+$ (the positive rationals) and thus enjoys the following property (cf. 1.4 in [6] 146-147)

(6) For each $\alpha \in \|E\|$ and each finite dimensional subspace $F \subset E$ there exists nonzero $f \in F^{\perp}$ with $\Phi(f,f) = \alpha$.

Proof of (5). Assume that we have determined an orthonormal system $f_0, \ldots, f_m$ in the positive definite $\mathbb{Q}$-space $(E,\Phi)$ $(\dim E = \aleph_0)$. Let $e_r$ be the first member of some (previously fixed) countable set of generators of $E$ not contained in the subspace $S := (f_0) \overset{\perp}{\oplus} \cdots \overset{\perp}{\oplus} (f_m)$. Thus, if we decompose $e_r = x + y$, $x \in S$, $y \in S^{\perp}$ then $y \neq 0$. We shall show how to find a 4-dimensional space $F \subset S^{\perp}$ spanned by an orthonormal basis $f_{m+1}, f_{m+2}, f_{m+3}, f_{m+4}$ and such that $y \in F$. We shall then have $e_r \in \bigoplus_{1 \leq i \leq m+4} (f_i)$. In this fashion we can obtain a denumerable sequence $(f_i)_{\mathbb{N}}$ such that $\Phi(f_i, f_j) = \delta_{ij}$ and the subspace $\bigoplus_{\mathbb{N}}(f_i)$ contains a set of generators for $E$, i.e. is all of $E$; $(f_i)_{\mathbb{N}}$ is then an orthonormal basis for $E$.

In order to find $F$ we first use the fact that each 4-dimensional $\mathbb{Q}$-space $H$ spanned by an orthonormal basis contains a vector $h_1$ of prescribed positive length $\alpha_1 := \Phi(y,y)$. Complete $h_1$ to some orthogonal basis $h_1, h_2, h_3, h_4$ of $H$ and set $\alpha_i := \Phi(h_i, h_i)$, $i = 2,3,4$. Obviously, if in the subspace $S^{\perp} \cap y^{\perp}$ of $E$ we can find any three mutually orthogonal vectors $g_2, g_3, g_4$ with $\Phi(g_i, g_i) = \alpha_i$ $(i = 2,3,4)$ then the subspace $F$ of $S^{\perp}$ spanned by $y, g_2, g_3, g_4$ will admit an orthonormal basis. Are there such $g_i$? Why, this is obvious by (6) because $\alpha_i \in \mathbb{Q}^+ = \|E\|$ : just pick $g_2, g_3, g_4$ of the requisite lengths in turn in $(S+(y))^{\perp}$, $(S+(y)+(g_2))^{\perp}$, $(S+(y)+(g_2)+(g_3))^{\perp}$ respectively. This establish (5). The decisive property (6) used in the proof warrants the following

Definition 1. A nondegenerate $\aleph_0$-dimensional sesquilinear space $(E,\Phi)$ is called stable in itself, or just stable for short, if it satisfies (6). Observe that the defining property (6) can be rendered equivalently as

(7) $$\|E\| = \bigcap \{\|F^{\perp}\| \mid F \subset E \ \& \ \dim F < \dim E\}.$$

Cf. the definition in VII.4.

Examples. 1) Over any given division ring $(k,\varepsilon,*)$ we can define stable $\varepsilon$-hermitean forms. For $\beta \in S := \{\xi \in k \mid \varepsilon\xi^* = \xi\}$ we let $\langle\beta,\ldots\rangle$ be the orthogonal sum of $\aleph_0$ copies of the $\varepsilon$-hermitean line $\langle\beta\rangle$. Then every orthogonal sum $\underset{\beta\in I}{\oplus^{\perp}}\langle\beta,\ldots\rangle$, where $\operatorname{card} I \leq \aleph_0$, is a stable space.

2) Each nondegenerate trace-valued $\varepsilon$-hermitean $\aleph_0$-form $\Phi$ that possesses infinite dimensional totally isotropic subspaces is stable with

$$(8) \qquad \|E\| = T = \bigcap\{\|F^{\perp}\| \mid \dim F < \infty\} .$$

Here $T = \{\xi + \varepsilon\xi^* \mid \xi \in k\}$ is the additive subgroup of traces in $k$. Indeed, let $W \subset W^{\perp} \subset E$ and $\dim W = \aleph_0$. If $F \subset E$ has finite $\dim F$ then $\dim(F_1 \cap W)$ is infinite for $F_1$ any supplement in $F^{\perp}$ of $\operatorname{rad}(F^{\perp})$ Hence $F^{\perp}$ contains hyperbolic planes and therefore $\|F^{\perp}\| = T$. In Section 1 of Appendix 1 we have listed large classes of commutative fields $k$ such that each nondegenerate symmetric $\aleph_0$-form over $k$ is stable.

3) We have just proved that definite forms over $\mathbb{Q}$ in dimension $\aleph_0$ are stable.

## 4. A stable form is determined by the elements it represents

The following theorem brings out a salient feature of stability. The structure $(k,\varepsilon,*)$ is kept fixed and we discuss the stable $\varepsilon$-hermitean $\aleph_0$-forms admitted by it.

Theorem 2 ([12], Thm. 1.1). A stable $\varepsilon$-hermitean space $(E,\Phi)$ is determined up to isometry by the subset $\|E\|$ of $k$.

In order to prove Theorem 2 we shall use a strategy different from that in the proof of (5). Instead of trying to construct, in each space, an orthogonal basis which exhibits a special arithmetic feature, we shall directly set up an isometry between any two stable spaces $E$ , $\bar{E}$ with $\|E\| = \|\bar{E}\|$ . We shall give the proof at a leisurely pace because the arguments will be used in condensed style over and over again in this book.

Proof. Let $E$ , $\bar{E}$ be stable with $\|E\| = \|\bar{E}\|$ . Let $(e_i)_{i\in\mathbb{N}}$ , $(\bar{e}_i)_{i\in\mathbb{N}}$ be any countable sets of generators for $E$ and $\bar{E}$ respectively. Our objective is to construct nested sequences $F_0 \subset F_1 \subset F_2 \subset \ldots$ and $\bar{F}_0 \subset \bar{F}_1 \subset \bar{F}_2 \subset \ldots$ of finite dimensional nondegenerate subspaces and a sequence of isometries $\varphi_i\colon F_i \to \bar{F}_i$ such that the following holds: (i) $\bigcup F_i = E, \bigcup \bar{F}_i = \bar{E}$, (ii) $\varphi_{i+1}$ extends $\varphi_i$ , $\varphi_{i+1}|_{F_i} = \varphi_i$. Then we can define an isometry $\varphi\colon E \to \bar{E}$ simply by defining $\varphi x$ as $\varphi_j x$ where $j$ is any natural number with $x \in F_j$ .

We may start with $F_0 = (0) = G_0$ and $\varphi_0\colon 0 \mapsto 0$ . Assume that we have constructed $F_i$ , $\bar{F}_i$ , $\varphi_i$ for $0 \le i \le m$ . Let $e_r$ be the first generator not in $F_m$ . We shall try to pick finite dimensional nondegenerate isometric spaces $X \subset F_m^\perp$ , $\bar{X} \subset \bar{F}_m^\perp$ such that $e_r \in F_m \overset{\perp}{\oplus} X$ . We then set $F_{m+1} := F_m \oplus X$ and extend $\varphi_m$ to $F_{m+1}$ by joining $\varphi_m$ with any isometry $\psi\colon X \to \bar{X}$ , $\varphi_{m+1}(f+x) = \varphi_m f + \psi x$ $(f \in F_m, x \in X)$ . This will be the $(m+1)^{st}$ construction step. In the $(m+2)^{nd}$ step we pick the first generator $\bar{e}_s$ not contained in $\bar{F}_{m+1}$ and we repeat the procedure with reversed sides: we try to pick isometric nondegenerate finite dimensional spaces $\bar{Y} \subset \bar{F}_{m+1}^\perp$ , $Y \subset F_{m+1}^\perp$ such that $\bar{e}_s \in \bar{F}_{m+1} \overset{\perp}{\oplus} \bar{Y}$; we then set $F_{m+2} := F_{m+1} \oplus Y$ , $\bar{F}_{m+2} := \bar{F}_{m+1} \oplus \bar{Y}$ and extend $\varphi_{m+1}$ to $F_{m+2}$ by joining $\varphi_{m+1}$ with any isometry $Y \to \bar{Y}$ . This bouncing back and forth between $E$ and $\bar{E}$ is necessary to make sure that both unions $\bigcup F_i$ , $\bigcup \bar{F}_i$ will exhaust the entire spaces.

It is sufficient to carry out the $(m+1)^{st}$ step. Since $E = F_m \oplus F_m^\perp$ we decompose $e_r = x + y$ $(x \in F_m, y \in F_m^\perp)$. We have $y \neq 0$. If $\alpha := \Phi(y,y)$ is nonzero we let $X = (y)$. Because $\alpha \in \|\bar{E}\| = \bigcap \|\bar{F}^\perp\|$ where $\bar{F}$ runs through all finite dimensional $\bar{F} \subset \bar{E}$ there is $\bar{y}$ in $\bar{F}_m^\perp$ with $\Phi(\bar{y},\bar{y}) = \alpha$. Set $\bar{X} = (\bar{y})$. Then $y \mapsto \bar{y}$ defines an isometry $\psi: X \to \bar{X}$ and we are done in this case. Assume then that $\alpha = 0$. If $E$ should be alternate, $\|E\| = \{0\} = \|\bar{E}\|$, then $\bar{E}$ is alternate too. It is clear that $y$ is contained in a hyperbolic plane $X \subset F_m^\perp$ and we may let $\bar{X}$ be any hyperbolic plane in the alternate space $\bar{F}_m^\perp$; $X \cong \bar{X}$ obviously. If $E$ is not alternate then, by stability, $F_m^\perp$ is not alternate and $y$ is contained in a nondegenerate plane $X$ that admits an orthogonal basis $x_1$, $x_2$. By stability and $\|E\| = \|\bar{E}\|$ there is $\bar{x}_1 \in \bar{F}_m^\perp$ with $\bar{\Phi}(\bar{x}_1,\bar{x}_1) = \Phi(x_1,x_1)$ and $\bar{x}_2 \in (\bar{F}_m \oplus (\bar{x}_1))^\perp$ with $\bar{\Phi}(\bar{x}_2,\bar{x}_2) = \Phi(x_2,x_2)$. Hence the plane $\bar{X} := k(\bar{x}_1,\bar{x}_2)$ is isometric to $X$. This terminates the proof of Theorem 2.

<u>Remarks</u>. (j) We know that $\|E\| = \mathbb{Q}^+ = \|\bar{E}\|$ for arbitrary positive definite $\aleph_0$-dimensional $\mathbb{Q}$-spaces $E$, $\bar{E}$. Thus it follows by Thm. 2 that any such $E$ is isometric to a space $\bar{E}$ spanned by an orthonormal basis. This establishes anew - but in a different vein - property (5).

(jj) Equality (7) remains perfectly meaningful if we let $\dim E$ be uncountable. Theorem 2 continues to hold for uncountable $\dim E$ and with stability interpreted via (7), <u>provided</u> $E$ is assumed to split into an orthogonal sum of finite dimensional subspaces. Partial isometries between any two such stable spaces $E$, $\bar{E}$, with $\|E\| = \|\bar{E}\|$, are defined only on <u>orthogonal summands</u> of dimensions smaller than $\dim E = \dim \bar{E}$. They can be extended by adding $\aleph_0$ suitable dimensions at a time in such a fashion that the domain of the extended isometries again turn out to be orthogonal summands in $E$ and $\bar{E}$ respectively (we satisfy some kind of union-of-chains condition). It is clear that by picking up all $F_\iota$ , $\bar{F}_\kappa$ of some previously fixed orthogonal splittings, $E = \oplus^\perp_\iota F_\iota$ , $\bar{E} = \oplus^\perp_\kappa \bar{F}_\kappa$ (dim $F_\iota$ and dim $\bar{F}_\kappa$ finite) we can sweep out $E$ and $\bar{E}$ by a nested sequence of partial isometries and thus procure an isomerty $E \cong \bar{E}$ .

<u>Corollary</u>. Let $E$ be of dimension $\aleph_0$ , nondegenerate, trace-valued. If $E$ contains an infinite dimensional totally isotropic subspace then it is an orthogonal sum of hyperbolic planes.

For certain investigations we need rather special stable spaces, which we define now.

<u>Definition 2</u>. A nondegenerate $\aleph_0$-dimensional sesquilinear space $(E,\Phi)$ is called <u>strongly (weakly) universal</u> if for all nondegenerate infinite dimensional subspaces $F$ we have $\|F\| = \|E\|$ ( $\|E\| \subset \|F\| \cup - \|F\|$ ) .

A weakly universal space is always trace-valued because it must be alternate when the characteristic is 2 by the following

<u>Theorem 3</u>. a) If $(E,\Phi)$ is strongly universal and not alternate then it is anisotropic and an orthogonal sum of lines $\langle\alpha\rangle$ for $\alpha$ any fixed element of $\|E\|$ . b) If $(E,\Phi)$ is weakly universal and $\alpha \in \|E\| - \{0\}$ then there is an orthogonal $\pm\alpha$- basis $(e_i)$ , i.e. $\Phi(e_i,e_i) \in \{\alpha,-\alpha\}$ for all $i$ .

A proof can easily be devised along the line of that of assertion (5); it will not be written out.

## 5. Quasistability

This important concept is due to Maxwell [12].

Definition 3. A nondegenerate $\aleph_0$-dimensional sesquilinear space $E$ is quasistable if it is the orthogonal sum of a finite dimensional space and a space that is stable (in itself).

Example. All (nondegenerate) symmetric $\aleph_0$-forms over $\mathbb{Q}$ are quasistable. Indeed, since any such $\Phi\colon E \times E \to \mathbb{Q}$ admits an orthogonal basis $E$ splits, $E = E_+ \overset{\perp}{\oplus} E_-$, where $\Phi$ is positive definite on $E_+$ and negative definite on $E_-$. If one of the indices $n_+ := \dim E_+$, $n_- := \dim E_-$ is zero, then the form is definite and stable by (5). If both dimensions are infinite then $\Phi$ is stable by the corollary to Theorem 2. Hence we are left with infinite $n_+$ and finite nonzero $n_-$ - or the other way round - and then $\Phi$ is quasistable. This proves the assertion. The important thing is that we can still introduce canonical bases. Consider, as an example, the quasistable $\mathbb{Q}$-space

$$\langle -2 \rangle \overset{\perp}{\oplus} \langle 1, \ldots \rangle .$$

We chop off a 4-dimensional orthogonal summand $\langle 1,1,1,1 \rangle$ from the stable summand $E_+ = \langle 1, \ldots \rangle$ fixed in the above decomposition, $E_+ = \langle 1,1,1,1 \rangle \overset{\perp}{\oplus} E_1$. The indefinite form $\langle -2 \rangle \overset{\perp}{\oplus} \langle 1,1,1,1 \rangle$ represents all of $\mathbb{Q}$ by Meyer's result mentioned in the remark following (5). Thus there is a vector of length $-1$ or, for that matter, any previously fixed negative rational, which can be completed to an orthogonal basis,

$$\langle -2 \rangle \overset{\perp}{\oplus} \langle 1,1,1,1 \rangle = \langle -1 \rangle \overset{\perp}{\oplus} \langle \alpha_2, \alpha_3, \alpha_4, \alpha_5 \rangle .$$

Because the "indices" $n_+$, $n_-$ are invariants all $\alpha_i$ must be positive; thus, if we join $\langle \alpha_2, \alpha_3, \alpha_4, \alpha_5 \rangle$ with $E_1$ we do get something positive definite, i.e. something which again admits an orthonormal basis, $\langle\alpha_2,\alpha_3,\alpha_4,\alpha_5\rangle \overset{\perp}{\oplus} E_1 \cong \langle 1,\ldots\rangle$ . The result of the manoeuvre is the isometry

$$\langle -2\rangle \overset{\perp}{\oplus} \langle 1,\ldots\rangle \cong \langle -1\rangle \overset{\perp}{\oplus} \langle 1,\ldots\rangle$$

over $\mathbb{Q}$ . If $n_-$ is greater than 1 then the procedure can be repeated $n_-$ times. Thus we have established the following companion to (5) (cf. [11] Thm. 4 p. 6) :

(9) Each nondegenerate $\mathbb{Q}$- space of dimension $\aleph_0$ admits an orthogonal $\pm 1$- basis.

Remark. The example illustrates the utility of the concept of quasistability. For "suitable" fields, such as $\mathbb{Q}$ , one can prove the existence of canonical bases for quasistable forms even when these forms fail to be stable. The first steps in this direction are found in [11]; the clear distillate of a number of investigations into the classification of $\aleph_0$-forms by various authors is contained in [12]. This and further results that have emerged are the topics of Chapters VII and XI below. Notice that the case of symmetric quasistable forms leaves really only two options for $k$ , to wit, $k$ either formally real or else of characteristic 2, by the following observation:

(10) If $k$ is a commutative nonformally real field of characteristic not 2 then each quasistable symmetric space over $k$ is an orthogonal sum of hyperbolic planes (and thus, in particular, a stable space).

Indeed, this follows immediately from the corollary to Theorem 2.

## 6. Weak stability

As our results in Chapters VII and XI show the following weakening of quasistability is justified (cf. the definition in VII.4).

Definition 4. A nondegenerate $\aleph_0$-dimensional sesquilinear space $(E,\Phi)$ is called weakly stable in itself, or just weakly stable for short, if $\bigcap\{\|F^{\perp}\| \mid F\subset E \;\&\; \dim F<\dim E\} =: \|E\|_{\infty}$ is nonempty.

Lemma. Assume that $(E,\Phi)$ has $\dim E = \aleph_0$. If the characteristic is not two then the following are equivalent: (i) the space $(E,\Phi)$ is weakly stable, (ii) $(E,\Phi)$ admits an orthogonal summand that is stable in itself.

In characteristic two weak stability of $E$ is easily seen to be equivalent with $\{0\}\subset\|E\|_{\infty}$, i.e. with the existence of an infinite dimensional totally isotropic subspace in $E$. More can be said when $\{0\}\subsetneq\|E\|_{\infty}$; cf. Chapter VII.

The proofs are left to the reader; there are no snags.

## 7. A lemma on supplements

Let $(E,\Phi)$ be a nondegenerate sesquilinear space of dimension $\aleph_0$ and $X\subset Y\subset E$ infinite dimensional subspaces, $G$ is a supplement of $X$ in $Y$, $H$ is a supplement of $Y$ in $E$, $\alpha$ and $\beta$ are arbitrary in $\|E\|\setminus\{0\}$ if such there are. The following lemma mentions various possibilities as to how $G$ and $H$ may be chosen. (For a typical application see the proof of the theorem in XII.9.)

Lemma. (i) If $\dim X/(X^{\perp}\cap X)$ is infinite then $G$ and $H$ may be chosen with $G\perp H$.

(ii) If $X^{\perp}=(0)$ and $\Phi$ is not skew symmetric then $G$ and $H$ may be chosen nondegenerate and with $G\perp H$.

(iii) If $\Phi$ is anisotropic and strongly universal then we may choose $G \perp H$ and $G$, $H$ spanned by orthogonal bases $(g_i)_I$, $(h_j)_J$ respectively with $\Phi(g_i, g_i) = \alpha$, $\Phi(h_j, h_j) = \beta$ $(i \in I, j \in J)$.

(iv) If $X^\perp = (0)$ and $\Phi$ is strongly universal then we have the same conclusion as in (iii).

(v) If $\Phi$ is trace-valued and $X$ contains an infinite dimensional totally isotropic subspace $W$ with $W \cap (X \cap X^\perp) = (0)$ then we may choose $G \oplus H$ totally isotropic.

Proof. (i) Assume that we have found finite dimensional spaces $G_n$, $H_n \subset E$ satisfying the following induction assumption:
$G_n \subset Y$, $G_n \cap X = (0)$, $H_n \cap Y = (0)$, $G_n \perp H_n$, $G_n \cap Y^\perp = (0)$, $H_n \cap X^\perp = (0)$.
We may start with $G_n = H_n = (0)$. Let $z$ be a prescribed vector with

$$z \in Y \quad , \quad z \notin X \oplus G_n \ .$$

There exists $x \in X$ with $x + z \in H_n^\perp$. Set $G_n' = G_n \oplus (x+z)$. We have to deal with the possibility that $G_n' \cap Y^\perp \neq (0)$. In that case $x + z \in G_n + Y^\perp$. We first claim that

$$(G_n' + H_n)^\perp \cap X \not\subset G_n + Y^\perp \quad (\subset G_n + X^\perp) \ .$$

Indeed, an inclusion would mean that $\dim X/(X \cap X^\perp) < \infty$, contradicting assumption (i). Hence we may pick $t \in (G_n' + H_n)^\perp \cap X$ with $t \notin G_n + Y^\perp$. Set $G_{n+1} = G_n \oplus (x+z+t)$. $G_{n+1} \cap Y^\perp = (0)$. If, on the other hand, $x + z \notin G_n + Y^\perp$ we set $G_{n+1} = G_n \oplus (x+z)$. We then proceed to construct $H_{n+1} \supset H_n$ such that $Y \oplus H_{n+1}$ contains a prescribed vector $z' \notin Y \oplus H_n$. As before, there is $y' \in Y$ such that the space $H_n' = H_n \oplus (y'+z')$ is $\perp$ to $G_{n+1}$ (we have just proved that $G_{n+1} \cap Y^\perp = (0)$). If $H_n' \cap X^\perp \neq (0)$ we proceed as before: there is $t' \in (G_{n+1}+H_n')^\perp \cap Y$, $t' \notin H_n + X^\perp$ and we set $H_{n+1} = H_n \oplus (y'+z'+t')$; otherwise we set

$H_{n+1} = H'_n$ . We have now constructed $G_{n+1}$ , $H_{n+1}$ satisfying the induction assumption and such that $X \oplus G_{n+1}$ , $Y \oplus H_{n+1}$ contain the prescribed vectors $z$ , $z'$ . In this manner we find sequences $(G_n)$ , $(H_n)$ such that $X + G$ and $Y + H$ , where $G := \bigcup G_n$ , $H := \bigcup H_n$ , contain previously fixed bases of supplements of $X$ in $Y$ and $Y$ in $E$ respectively. Furthermore $G \perp H$ . This proves (i).

(iii) If $\dim E/Y = \dim Y/X$ is infinite then $G$ , $H$ as constructed in the previous proof admit bases of the requisite shape by strong universality.

Assume then that, say, $\dim Y/X$ is finite. $Y$ is nondegenerate and of dimension $\aleph_0$ so it has orthogonal basis $(g_i)$ with $\Phi(g_i,g_i) = \alpha$ . Hence there is a finite dimensional supplement $G_0$ of $X$ in $Y$ spanned by some of the $g_i$ . If $\dim E/Y$ is infinite we may proceed just as in the proof of (i) and construct $G = G_0$ , $H = \bigcup H_n$ . This time we start the recursive construction with the pair $G_0$ , $H_0 = (0)$ . $H$ then admits a basis of the required sort. We are left with the situation where $\dim E/Y$ is finite as well. Assume then that we have found some finite dimensional $H_n$ spanned by an orthogonal basis $(h_j)$ with $\Phi(h_j,h_j) = \beta$ , $H_n \cap Y = (0)$ , $H_n \perp G_0$ . We show that if $Y \oplus H_n \neq E$ we can yet find another vector $y \notin Y \oplus H_n$ with $\Phi(y,y) = \beta$ and $y \in (G_0 \oplus H_n)^{\perp}$ . (Repetition of the argument $\dim E/Y$ times will provide the requisite basis, $H = \bigcup H_n$ .) Now $(G_0 \oplus H_n)^{\perp}$ admits an orthogonal basis whose members $y$ satisfy $\Phi(y,y) = \beta$ . Hence we are stuck only when

$$(G_0 \oplus H_n)^{\perp} \subset Y \oplus H_n .$$

If this takes place then $Y \oplus H_n$ is $\perp$-closed. Furthermore $(Y \oplus H_n)^{\perp} \subset G_0 \oplus H_n \subset Y \oplus H_n$ so $(Y \oplus H_n)^{\perp} = 0$ as $\Phi$ is anisotropic. Hence

$Y \oplus H_n = E$ . This proves (iii).

(iv) Let us keep the notations in the proof of (i). We shall indicate the modifications needed here. Suppose we have found finite dimensional $G_n$ , $H_n \subset E$ satisfying besides the former induction assumptions the further requirement that

$$G_n , H_n \text{ are nondegenerate.}$$

We had shown that we can find vectors $u \in X$ , $v \in Y$ such that $G_n' = G_n \oplus (z+u)$ , $H_n' = H_n \oplus (z'+v)$ satisfy the old induction assumptions in lieu of $G_n$ , $H_n$ .

If now $G_n'$ should be degenerate then its radical $R$ is one-dimensional. There exists a vector $l \in Y \cap H_n'^{\perp}$ with $l \notin R^{\perp}$ (for otherwise $R \subset H_n'$ and so $H_n' \cap Y \neq (0)$ , contradiction). If $\dim Y/X$ is infinite we can escape from $X$ : if $(G_n' \oplus (l)) \cap X \neq 0$ we switch from $l$ to $l'$ where $l - l' \in (G_n' \oplus H_n' \oplus (l))^{\perp} \cap Y$ and $l - l' \notin X \oplus G_n'$ . We have shown: if $\dim Y/X$ is infinite we can find $G_{n+1} \supset G_n$ which is nondegenerate and satisfies $G_{n+1} \subset Y$ , $G_{n+1} \cap X = (0)$ , $G_{n+1} \perp H_n'$ . Provided that $\dim E/Y$ is infinite as well then a nondegenerate space $H_{n+1} \supset H_n'$ can be found such that $G_{n+1}$ and $H_{n+1}$ satisfy the induction assumptions. $G := \bigcup_n G_n$ , $H := \bigcup_n H_n$ are infinite dimensional and nondegenerate and admit bases of the required form by strong universality. If, say, $\dim Y/X$ is finite we find a $G_0$ just as in the proof of (iii). If $\dim E/Y$ happens to be infinite we construct a nondegenerate $H := \bigcup_n H_n$ as we have just explained in detail. We are left with the case where $\dim E/Y$ is finite as well. We proceed as in the proof of (iii). This terminates the proof of (iv).

(ii) The proof of this assertion is actually "contained" in the reasoning given in the proof of (iv). We leave it to the reader to write out details.

(v) Again we consider ascending sequences $G_n$, $H_n$. This time we stipulate that they satisfy besides the old induction assumptions (spelled out at the beginning of the proof of (i)) the additional requirement

$$(G_n + H_n) \cap X^{\perp} = (0) \; ; \quad G_n \subset G_n^{\perp} \, , \quad H_n \subset H_n^{\perp} \, .$$

First we find (just as in (i)) a vector $u \in X$ such that the spaces $G_n' := G_n + (z+u)$, $H_n$ satisfy besides the old conditions the further stipulation

$$z + u \perp G_n + H_n \, .$$

We then consider an auxiliary supplement $X_1$ of $X \cap X^{\perp}$ in $X$ such that $W \subset X_1$. Now $X_1 \cap (G_n' \oplus H_n)^{\perp}$ contains a nondegenerate subspace $X_0$ with finite dim $X_1/X_0$. Hence $X_0 \cap W \neq (0)$ and so, by trace-valuedness, $X_0$ contains hyperbolic planes. Thus there exists $t \in X \cap (G_n' + H_n)^{\perp}$ with $\phi(t,t) = -\phi(z+u,z+u)$. $G_n'' := G_n \oplus (z+u+t)$ is totally isotropic and perpendicular to $H_n$. Some further manipulations are needed when $(G_n'' \oplus H_n) \cap X^{\perp}$ is not $(0)$. Assume therefore that we had $z + u + t \in G_n + H_n + X^{\perp}$. Because the inclusion " $(G_n''+H_n)^{\perp} \cap W \subset G_n + H_n + X^{\perp}$ " contradicts the fact that $W \cap (X \cap X^{\perp}) = (0)$ we may pick $w \in (G_n''+H_n)^{\perp} \cap W$, $w \notin G_n + H_n + X^{\perp}$. Switching from $G_n''$ to $G_n''' := G_n \oplus (z+u+t+w)$ gives $G_n''' \subset (G_n''')^{\perp}$ and $(G_n''' + H_n) \cap X^{\perp} = (0)$. In sum, we have shown that we can find $G_{n+1}$ such that $G_{n+1}$, $H_n$ satisfy all our induction assumptions again and furthermore $z \in X \oplus G_{n+1}$. Similarly we find a $H_{n+1}$. In this fashion we secure a sequence of totally isotropic spaces $G_n \oplus H_n$ such that the spaces $G := \bigcup G_n$, $H := \bigcup H_n$ are supplements of $X$ and $Y$ in $Y$ and $E$ respectively. This terminates the proof of (v).

The proof of the lemma is thus complete.

References to Chapter II

[1] W. Baur and H. Gross, Strange inner product spaces. Comment. Math. Helv. 52 (1977) 491-495.

[2] W. Bäni, Sesquilineare Formen und lineare Topologien. Ph.D.Thesis, University of Zurich (1975).

[3] - , Linear Topologies and Sesquilinear Forms. Comm. in Algebra, 14 (1977) 1561-1587. (Neither of [2] or [3] is a subset of the other.)

[4] C. Everett and H. Ryser, Rational vector spaces. Duke Math. J. 16 (1949) 553-570.

[5] J.P. Gram, Ueber die Entwicklung reeller Funktionen in Reihen mittelst der Methode der kleinsten Quadrate. J. reine angew. Math. 94 (1883) 41-73.

[6] H. Gross, On Witt's theorem in the denumerably infinite case. Math. Annalen 170 (1967) 145-165.

[7] - , Der euklidische Defekt bei quadratischen Räumen. Math. Ann. 180 (1969) 95-137.

[8] H. Gross and E. Ogg, Quadratic Spaces with Few Isometries. Comment. Math. Helv. 48 (1973) 511-519.

[9] - and - , Quadratic forms and linear topologies. On completions. Ann. Acad. Sci. Fenn. A.I. 584 (1975) 1-19.

[10] P. Hafner, Zur Berechnung endlicher euklidischer Defekte in quadratischen Räumen. Comment. Math. Helv. 45 (1970) 135-151.

[11] I. Kaplansky, Forms in infinite-dimensional spaces. Anais da Academia Brasileira de Ciencias, vol. 22 (1950) 1-17.

[12] G. Maxwell, Classification of countably infinite hermitean forms over skewfields. Amer. J. Math. 96 (1974) 145-155.

[13] A. Meyer, Ueber die Auflösung der Gleichung $ax^2+by^2+cz^2+du^2+ev^2=0$ in ganzen Zahlen. Vierteljahrs. Naturforsch. Gesellsch. Zürich 29 (1884) 220-222.

[14] E. Ogg, Die abzählbare Topologie und die Existenz von Orthogonalbasen in unendlichdimensionalen Räumen. Math. Ann. 188 (1970) 233-250.

[15] E. Schmidt, Zur Theorie der linearen und nichtlinearen Integralgleichungen. Math. Ann. 63 (1907) 433-476.

[16] G.L. Watson, Integral quadratic forms. Cambridge Univ. Press (1960).

## APPENDIX I

## A FEW EXAMPLES OF "SUITABLE" FIELDS

### Introduction

Beginners are usually at a loss when they should produce examples of division rings which exhibit certain arithmetic features. For their convenience we have assembled here a number of examples. In the first three sections we stay commutative. The fields in the first four sections are such that all nondegenerate symmetric (hermitean) $\aleph_0$-forms turn out stable or quasistable. Since so much in this book depends on stability of some sort or another it is important to ascertain that examples abound. Everywhere enough hints and references are given so that the student can find his own way into the literature. Only in Section 5 did we give proofs. There we describe a Hilbert ordered noncommutative involutorial division ring which allows for anisotropic hermitean forms of arbitrary (finite or infinite) dimension.

### 1. Commutative nonformally real fields (characteristic $\neq 2$)

In this section we shall list some commutative fields $k$ which enjoy the following property ([6] pp. 5, 6):

(0) There is a natural number $m(k)$, depending on the field $k$ only, such that each symmetric form over $k$ in $m+1$ variables has a nontrivial zero.

We assume throughout that the characteristic is not 2 unless explicitely stated otherwise.

1.1 <u>Nonformally real fields $k$ with finite multiplicative group $\dot{k}/\dot{k}^2$</u> ( = the nonzero elements modulo square factors ). These fields are called Kneser fields (notice that char $k \neq 2$ is assumed). Here we have $m(k) = |\dot{k}/\dot{k}^2|$ ; for an elegant proof due to M. Kneser see [7]. Examples are: algebraically closed fields ($|\dot{k}/\dot{k}^2| = 1$) ; finite fields ($|\dot{k}/\dot{k}^2| = 2$) ; if $k$ is any example and $K := k((X))$ is the field of all formal Laurent series $a_r X^r + a_{r+1} X^{r+1} + \cdots$ ($r \in \mathbb{Z}$) under the usual series addition and series multiplication and with coefficients from $k$, then $K$ is another example and $|\dot{K}/\dot{K}^2| = 2|\dot{k}/\dot{k}^2|$ ; this shows that there are examples with arbitrarily prescribe order $|\dot{k}/\dot{k}^2|$ (notice that this order is necessarily a power of 2 because each element of the group has order 2).

Further examples are the fields $k$ which are complete under a discrete rank 1 valuation $\omega$ and with finite residue class field $K$ (of arbitrary characteristic). Here $|\dot{k}/\dot{k}^2| = 4(\text{card } K)^{\omega(2)}$. For a proof see [12, Thm. 63:9, p. 163] . In particular, $|\dot{k}/\dot{k}^2| = 4$ for the p-adic completions $k = \Omega_p$ when $p \neq 2$ and $|\dot{k}/\dot{k}^2| = 8$ for $k = \Omega_2$ .

Additional examples are provided by maximal algebraic extensions of any algebraic number field (= finite extension of $\mathbb{Q}$ ) in its $p$-adic completions for $p$ any finite spot. These are instances of so-called Hilbert fields investigated in [3].

<u>Remarks</u>. (i) The behaviour of fields with finite $\dot{k}/\dot{k}^2$ under field extensions has been investigated in [5 pp. 298-307]; for more comprehensive results consult [8 pp. 202-203; see in particular the proof of Thm. 3.4, p. 202, for a well motivated proof of our Lemma 2 in [5 p. 298]].

(ii) The following question is natural: Let $K$ be a finite extension of $k$. If $\dot{k}/\dot{k}^2$ is finite, is $\dot{K}/\dot{K}^2$ necessarily finite? In [5] this is answered in the affirmative when $[K:k] = 2$. In a letter to the author Pfister [15] proved that $\dot{K}/\dot{K}^2$ is infinite when $K = \bar{\mathbb{Q}}(\sqrt[3]{2})$ and $\bar{\mathbb{Q}}$ is the quadratic closure of $\mathbb{Q}$. For generalizations of this result see [8 Thm. 2 and Cor. 3 p. 219].

1.2 The $C_i$-fields. The study of these fields was started in [9]. Improvements are contained in [11]. For further developments and references see [17] and [10] (not listed in [17]).

A field $k$ of arbitrary characteristic is said to have property $C_0$ if every $f \in k[X_1,\ldots,X_n]$ which is nonzero and homogeneous of degree $d$, with $n = d > 1$ has a nontrivial zero. For natural $i > 0$ we say that $k$ has property $C_i$ if for each pair $(d,n)$ of natural numbers $\geq 1$ with $n > d^i$ every $f \in k[X_1,\ldots,X_n]$ which is nonzero and homogeneous of degree $d$ has a nontrivial zero in $k$. Theorem: Let $k$ have property $C_i$; if $K$ is an algebraic overfield of $k$ then $K$ is $C_i$; if $K$ is a transcendental extension of $k$ of transcendence degree $r$ then $K$ is $C_{i+r}$ (Thm. 2a, p. 238 in [11]).

We are, of course, interested in the case with $d = 2$. As an illustration we mention the following corollary of the above theorem. Corollary: Let $k$ be any function field in $r$ variables over a finite constant field. Since $k$ is $C_{r+1}$ every symmetric form in more than $2^{r+1}$ variables has a nontrivial zero. (For $r = 1$ this is a classical result of Hasse theory; cf. Thm. 66:2, p. 188 in [12].)

In analogy to Lang's theory Pfister defines a field $K$ to be $C_i^q$ if any system of $r$ quadratic forms over $K$ in $n$ common variables has a nontrivial simultaneous zero in $K$ provided that $n > r\cdot 2^i$.

He proves the following Theorem: Let $p$ be a prime, let $K$ be a p-field, let $f_1,\dots,f_r$ be forms of degrees $d_1,\dots,d_r$ over $K$ in $n$ common variables. Suppose $n > r$ and $p \nmid d_1,\dots,d_r$. Then the system $f_1 = \dots = f_r = 0$ has a nontrivial solution in $K$. (Here $K$ is called a p-field if all finite extensions of $K$ have p-power degree.)
Corollary: A p-field with $p \neq 2$ is a $C_0^q$ -field ([16]).

We terminate this cursory enumeration by referring to Remark 2 in Appendix 1 to Chapter XVI where an entirely different sort of fields with property (0) is mentioned.

*

## 2. Commutative formally real fields

If a field $k$ satisfies (0) in the previous section then each nondegenerate symmetric form in $m(k)$ variables represents $1$. There can be no instance of a formally real field $k$ with this property because $k$ admits negative definite forms. Thus, a natural modification of (0) for formally real fields is as follows ([6] p. 6).

(1) There is a natural number $m'(k)$, depending on the field $k$ only, such that each nondegenerate symmetric form over $k$ in $m'$ variables represents $1$ or $-1$ (or both).

By Hasse-Minkowsky theory we know that $\mathbb{Q}$ satisfies (1) with $m'(\mathbb{Q})=4$. By Hasse-Minkowsky theory it follows further that e.g. each irreducible polynomial $f \in \mathbb{Q}[X]$ of odd degree and with only one real root $\delta \in \mathbb{R}$ yields a formally real field $\mathbb{Q}(\delta)$ with property (1) and $m' = 4$. (See e.g. [12] Thm. 66:1.)

Nondegenerate symmetric $\aleph_0$ - forms over fields with (1) turn out quasistable and are therefore easy to classify. Each nondegenerate symmetric form $\Psi$ in, say, m' variables will represent $\alpha$ or $-\alpha$ (consider the form $\frac{1}{\alpha}\Psi$ and apply (1)). In particular a (nondegenerate) $\aleph_0$- form will represent $\alpha$ or $-\alpha$ (or both) for $\alpha$ arbitrary in $k \setminus \{0\}$ .

Formally real fields with (1) admit, nota bene, at most one ordering, because for each $\alpha$ either $\alpha$ or $-\alpha$ is a sum of m' squares. This is not, however, crucial for quasistability: our lemma in XI.8 gives a handy criterion for fields with more than one ordering to be such that all nondegenerate $\aleph_0$- forms turn out quasistable. For further examples of real fields of interest in this connection refer to XI.3 .

*

## 3. Commutative fields in characteristic 2

Any commutative field $k$ with char $k = 2$ has a known classification of the symmetric $\aleph_0$- forms over $k$ when the degree $[k:k^2]$ of $k$ over its subfield of squares is finite (Lemma 4 (v) in VII.8 ) . It is not difficult to produce examples of such fields. If $k$ itself is finite or if $k$ is algebraically closed then $[k:k^2] = 1$ , obviously. If $X_1,\ldots,X_n$ are algebraically independent over $k$ and $\bar{k}$ is the rational field $k(X_1,\ldots,X_n)$ then $[\bar{k}:\bar{k}^2] = [k:k^2]\cdot 2^n$ (because a basis for $\bar{k}$ over $\bar{k}^2$ is provided by the elements $\alpha\cdot X_1^{\varepsilon_1}\cdot \ldots \cdot X_n^{\varepsilon_n}$ where $\varepsilon_j = 0,1$ and where $\alpha$ runs through a $k^2$- basis of $k$ ). Thus we have examples with $[k:k^2]$ any prescribed power of 2 (notice that $[k:k^2]$ invariably is a power of 2 if finite because all elements of $k$ are quadratic over the subfield $k^2$ ). On the other hand, finite algebraic extensions will not alter this degree. Let $\bar{k}$ be a finite algebraic extension of $k$. We can express $[\bar{k}:k^2]$ as $[\bar{k}:\bar{k}^2][\bar{k}^2:k^2]$ and as $[\bar{k}:k][k:k^2]$; because $[\bar{k}^2:k^2] = [\bar{k}:k]$ we find that $[\bar{k}:\bar{k}^2] = [k:k^2]$. Hence any algebraic function field $k$ in $n$ variables over a finite (or algebraically closed) constant field has $[k:k^2] = 2^n$. Observe that $[\bar{k}:\bar{k}^2] \leq [k:k^2]$ for <u>arbitrary</u> algebraic extensions $\bar{k}$ of $k$ by what we have proved; $<$ is witnessed by the transition to the algebraic closure of $k$.

*

## 4. Involutorial division rings suitable for isotropic hermitean forms

One can contemplate division rings $(k,*)$ which satisfy a property entirely analogous to (0); we merely replace "symmetric" in (0) by the adjective "hermitean". One possibility to construct such $(k,*)$ is to start out with a suitable commutative field $k_0$ (such as described in Section 1) and then pass to a quaternion algebra $k$ over $k_0$ if such there is. $k$ may then be equipped with the usual conjugation $\xi \mapsto \bar{\xi}$ or with an involution $*$ of the kind

$$\xi^* = \alpha \cdot \bar{\xi} \cdot \alpha^{-1}$$

for some fixed quaternion $\alpha \in k$. (These make up <u>all</u> possible involutions that leave the center fixed; in fact, one can always choose $\alpha$ such that $\bar{\alpha} = -\alpha$. See [1], Thm. 11, p. 154. Notice that the "norm" $N: \xi \mapsto \xi\xi^*$ derived from $*$ is not multiplicative if $*$ is not conjugation.) As an illustration we shall give one example; we leave all proofs as exercises.

<u>Example</u>. Let $k_0$ be a commutative field of characteristic not 2 in which $-1$ is a square and where the multiplicative group of nonzero elements modulo square factors has finite order $n$. Assume that

$k_0$ admits a quaternion division ring $k$ over $k_0$. (Take e.g. $k_0 = \Omega_p$ where $p \equiv 1 \pmod 4$ ; cf. Sec. 1.1.) Let $*: k \to k$ be an arbitrary involution that leaves the center fixed. Then each hermitean form over $(k,*)$ in $n^2 + 1$ variables has a nontrivial zero.

Remark. One can modify the previous example so that involutions are obtained which do not leave the center fixed (cf. Thm. 21, p. 161 in [1]). More generally, the following result can be established.

Theorem. Let $(k,*)$ be any involutorial division ring (of arbitrary characteristic) of finite dimension over its center $C$. Let $S = \{\xi \in k \mid \xi^* = \xi\}$ be the set of symmetric elements and $C_0 := C \cap S$. If there is some fixed $m \in \mathbb{N}$ such that each symmetric form in $m + 1$ variables over $C_0$ is isotropic then there exists $m' \in \mathbb{N}$ such that each (trace-valued) hermitean form in $m' + 1$ variables over $k'$ is isotropic.

A crude estimate of $m'$ in the theorem is $m' = (m+1)^n$ where $n = [T:C_0]$; we hasten to add that this is not the most economic choice (compare with the above example).

*

## 5. A formally real involutorial division ring

If $*: k \to k$ is an involutory antiautomorphism of the division ring $k$ then $(k,*)$ is called formally real if an equality $\Sigma x_i x_i^* = 0$ implies $x_i = 0$ for all $i$. If $*$ is the identity, and hence $k$ commutative, then we get the usual concept of a formally real field. Owing to the formal reality of $\mathbb{R}$ we thus see that Hamilton's real quaternions $\mathbb{H}$ are a formally real field with respect to the usual conjugation. We point out that $(k,*)$ may very well be formally real but $(k,^\circ)$, where $\alpha^\circ = \mu\alpha^*\mu^{-1}$ for some fixed nonzero $\mu$, need not be formally real.

In this section we present with some detail a formally real division ring $(k,*)$ which admits an ordering in the sense of Hilbert (see XI.4), i.e., the positive elements of $k$ form an additively closed subgroup of index 2 in the multiplicative group of $k$. Our example is a subfield of Hilbert's famous ordered skew field. However, we do not know whether the involution can be extended to Hilbert's field.

5.1 Ore rings. An integral domain $R$ is called right Ore ring if $aR \cap bR \neq \{0\}$ for $a, b \in \dot{R} := R \setminus \{0\}$. Each right Ore ring can be embedded in a "field of fractions" as follows. Define an equivalence relation on $R \times \dot{R}$ as follows: $(a,b) \sim (c,d)$ if and only if $bd' = db'$ & $(b',d' \in \dot{R})$ entails $ad' = cb'$. The equivalence classes can be added and multiplied as follows: $(a,b) + (c,d) := (ad'+cb', bd')$ with $bd' = db'$; $(a,b)(c,d) := (ac'', db'')$ with $bc'' = cb''$. The inverse of $(a,b)$ is $(b,a)$ (for $a \neq 0$). One checks that everything is well defined and the set of classes is a division ring. It is denoted by $R_{\dot{R}}$.

An integral domain $R$ which is not right Ore contains elements $a, b$ with $aR \cap bR = \{0\}$. It follows that the elements $a^n b$ $(n = 0,1,2,\ldots)$ are right linearly independent, hence $R$ contains right ideals of arbitrary rank. In particular, every right Noetherian integral domain is a right Ore ring.

If a right Ore ring is also a left Ore ring $(Ra \cap Rb \neq \{0\}$ for $a, b \in \dot{R})$ then there are two fields of fractions, the right field of fractions $R_{\dot{R}}$ as described and - with the obvious changes - a left field of fractions ${}_{\dot{R}}R$. Straight forward verification shows that

(2) $\quad {}_{\dot R}R$ and $R_{\dot R}$ are isomorphic under the map

$\dot R \times R \ni (a,b) \mapsto (d,c) \in R \times \dot R$ where $ad = bc$ .

5.2 Involutions. If $\sigma: R \to R$ is an involution of the right Ore ring then $\sigma$ can be extended to an involution of the division ring $R_{\dot R}$ ([4]).

Proof. Since $aR \cap bR \neq \{0\}$ if and only if $R\sigma(a) \cap R\sigma(b) \neq \{0\}$ we see that a right Ore ring with involution is always a left Ore ring as well. We consider the two fields of fractions $R_{\dot R}$ and ${}_{\dot R}R$ . If $(a,b) = (c,d) \in R_{\dot R}$ then $bd' = db'$ and $ad' = cb'$ for suitable $b', d' \in \dot R$ and therefore $\sigma(d')\sigma(b) = \sigma(b')\sigma(d)$ and $\sigma(d')\sigma(a) = \sigma(b')\sigma(c)$ with $\sigma(d'),\sigma(b') \in \dot R$ . In other words, $(\sigma(b),\sigma(a)) = (\sigma(d),\sigma(c))$ in ${}_{\dot R}R$ . Thus the assignment $(a,b) \in R \times \dot R \mapsto (\sigma(b),\sigma(a)) \in \dot R \times R$ is a bijection between the equivalence classes of $R \times \dot R$ and $\dot R \times R$ and thus a map of $R_{\dot R}$ onto ${}_{\dot R}R$ ; we call it $\varphi$ . Substitution of the definitions yields $\varphi((a,b)+(c,d)) = \varphi((a,b)) + \varphi((c,d))$ and $\varphi((a,b)\cdot(c,d)) = \varphi((c,d))\varphi((a,b))$ . Thus $\varphi$ is an antiisomorphism of $R_{\dot R}$ onto ${}_{\dot R}R$ . Combination with (2) yields the desired extension of $\sigma$ to $R_{\dot R}$ :

(3) $$\sigma\left(\frac{a}{b}\right) = \frac{c}{d} \quad \text{with} \quad \sigma(a)d = \sigma(b)c \ .$$

5.3 The example. We see that all we need in order to have an ordered division ring with involution is an ordered Ore ring $R$ with an involution (" $\frac{a}{b} > 0 \iff ab > 0$ " extends the ordering of $R$ to $R_{\dot R}$ ). We construct one as follows.

In the polynomial ring $\mathbb{Q}[x]$ we define an injective homomorphism by setting $\omega: \Sigma r_i x^i \mapsto \Sigma 2^i r_i x^i$ . Let then $t$ be a new indeterminate; we consider the set $R$ of all polynomials $f = \Sigma a_i t^i$ with coefficients

$a_i \in \mathbb{Q}[x]$ . A noncommutative multiplication between $f$ , $g \in R$ is defined by $fg = \Sigma c_k t^k$ where

(4) $$c_k = \sum_{i+j=k} a_i \omega^i (b_j) \ .$$

In particular $tx = 2xt$ in $R$ . Addition is defined as usual. $R$ is an integral domain. $\mathbb{Q}[x]$ is noetherian and so is $R$ by the Hilbert basis theorem just as in the commutative case. So $R$ is right Ore by 5.1. Since $\mathbb{Q}$ is ordered $\mathbb{Q}[x]$ is ordered by defining $f > 0$ iff the coefficient of the lowest term in $f$ is positive. An ordering of $R$ is obtained from that of $\mathbb{Q}[x]$ by the same procedure. Finally we obtain an involution $\sigma$ of $R$ by setting

(5) $$\sigma\colon \Sigma a_{ik} x^i t^k \mapsto \Sigma a_{ik} x^k t^i \qquad (a_{ik} \in \mathbb{Q}) \ .$$

We find $\sigma(f+g) = \sigma(f) + \sigma(g)$ , $\sigma(fg) = \sigma(g)\sigma(f)$ , $\sigma^2 = \mathbb{1}$ . Thus we have shown that for this particular polynomial ring $R$ the involutorial division ring $(R_R^{\cdot}, \sigma)$ admits an ordering in the sense of Hilbert ([4]).

5.4 <u>Formal reality</u>. By an induction on $n$ one proves, by using the left Ore condition, the following "common denominator theorem" for $R_R^{\cdot}$ in 5.3 :

(6) For $\lambda_1, \ldots, \lambda_n \in R_R^{\cdot}$ there exist $r, t_1, \ldots, t_n \in R$ such that $\lambda_i = r^{-1} t_i$ $(1 \le i \le n)$ .

We are now able to establish

(7) If $\Sigma \lambda_i \lambda_i^\sigma = 0$ then $\lambda_i = 0$ (all i) .

Assuming that not all $\lambda_i$ in (7) are zero we study the typical $\lambda_i \lambda_i^\sigma$ in that sum. By (6) we assume that all $\lambda_i$ are polynomials $\mu = \Sigma a_{jk} x^j t^k$ ($a_{jk} \in \mathbb{Q}$ and depending on $i$ ) : write the $\lambda_i$ with common denominator, $\lambda_i = r^{-1} \mu_i$ $(\mu_i, r \in R)$ ; $\Sigma \lambda_i \lambda_i^\sigma = r^{-1} (\Sigma \mu_i \mu_i^\sigma)(r^{-1})^\sigma = 0$

implies $\Sigma \mu_i \mu_i^\sigma = 0$ . Let $n_i$ be the smallest exponent $j + k$ with $a_{jk} \neq 0$ in the polynomial $\mu_i$ and $n$ the smallest among the finitely many $n_i$ . Thus by the definition of $n$ :

(8) there is a triple $(i,j,k)$ with $j+k = n$ and $a_{jk} \neq 0$ in $\mu_i$ .

The sum $s_{(2n)}$ of terms of "weight" $2n$ in $\mu_i \mu_i^\sigma$ is $s_{(2n)} = \Sigma\, 2^{(n-j)(n-j')} a_{j,n-j}\, a_{j',n-j'}\, x^{n+j-j'} t^{n-j+j'}$ where the sum extends over all $j,j'$ with $0 \leq j \leq n$ and $0 \leq j' \leq n$ . Now $0 = \Sigma \mu_i \mu_i^\sigma = \Sigma s_{(2n)} + s$ where all nonzero monomials in $s$ have an exponent $> 2n$ . Therefore, $\Sigma s_{(2n)} = 0$ and, if we collect the coefficient of $x^n t^n$ , we see that $\sum_{j=0}^{n} 2^{(n-j)(n-j)} a_{j,n-j}^2 = 0$ . Thus $a_{j,n-j} = 0$ for all $j = 0,1,\ldots,n$ and all $\mu$ contradicting (8).

5.5 Summary. We have shown that there exist noncommutative involutorial division rings $(k,*)$ which are formally real and which admit an ordering $\leq$ in the sense of Hilbert. Owing to the formal reality we can define anisotropic hermitean forms $\Phi = \Sigma \xi_i \xi_i^*$ over such $(k,*)$ . Such forms will, however, be indefinite with respect to $\leq$ . For, as $k$ is skew, there exist $\alpha$ with $\alpha \neq \alpha^*$ ; say $\alpha < \alpha^*$ . Thus if $\beta := \alpha - \alpha^*$ then $\beta\beta^* < 0$ . On the other hand, if $\gamma$ is symmetric, then $\gamma\gamma^* = \gamma^2 > 0$ . Thus the anisotropic $\Phi$ changes sign on every 1-dimensional subspace.

For an account on Ore rings, based on the work in [13] and [14], one may consult [2].

*

## References to Appendix I

[1] A.A. Albert, Structure of Algebras. (3rd print. of rev. ed.) AMS Coll. Publ. XXIV, NY 1968.

[2] P.M. Cohn, Free Rings and their Relations. Academic Press London 1971.

[3] A. Fröhlich, Quadratic forms à la local theory. Proc. Camb. Phil. Soc. 63 (1967) 579-586.

[4] H. Glauser, Schiefkörper mit Anordnung und Involution. Master's Thesis, University of Zurich 1972.

[5] H. Gross and H.R. Fischer, Nonreal fields $k$ and infinite dimensional k-vector spaces. Math. Ann. 159 (1965) 285-308.

[6] I. Kaplansky, Forms in infinite-dimensional spaces. Anais Acad. Bras. Ci 22 (1950) 1-17.

[7] M. Krasner, Review on a paper by I. Kaplansky. Math. Rev. AMS vol. 15 (1954) p. 500.

[8] T.Y. Lam, The algebraic theory of quadratic forms. W.A. Benjamin Inc., Reading, Massachusetts 1973.

[9] S. Lang, On quasi algebraic closure. Ann. of Math. 55 (1952) 373-390.

[10] G. Maxwell, A note on Artin's Diophantine Conjecture. Canad. Math. Bull. 13 (1970) 119-120.

[11] M. Nagata, Note on a paper of Lang concerning quasi algebraic closure. Mem. Univ. Kyoto Ser. A 30 (1957) 237-241.

[12] O.T. O'Meara, Introduction to quadratic forms. Grundlehren Bd. 117 Springer Berlin 1963.

[13] O. Ore, Linear equations in non-commutative fields. Ann. of Math. 32 (1931) 463-477.

[14] - , Theory of noncommutative polynomials. Ann. of Math. 34 (1933) 481-508.

[15] A. Pfister, Letter to the author, dated 9.7.1968. (It contains a detailed proof for the fact that $\dot{k}/\dot{k}^2$ is infinite when $k = \bar{\mathbb{Q}}(\sqrt[3]{2})$ , $\bar{\mathbb{Q}}$ the quadratic closure of $\mathbb{Q}$.)

[16] A. Pfister, Systems of quadratic forms. Bull. Soc. Math. de France Mémoire 59 (1979).

[17] G. Terjanian, Dimension arithmetique d'un corps. J. Algebra 22 (1972) 517-545.

## CHAPTER THREE

## WITT DECOMPOSITIONS FOR HERMITEAN $\aleph_o$-FORMS

### 1. Introduction

Forms are $\varepsilon$-hermitean ; if $\alpha \longmapsto \alpha^*$ is the antiautomorphism of the underlying division ring we let $T := \{\alpha + \varepsilon\alpha^* | \alpha \in k\}$ be the additive subgroup of "traces" in $k$. Traces are symmetric elements of $k$ but the converse does not hold when the characteristic is two.

*

Isotropic vectors play a distinguished role in the theory of forms and so do totally isotropic subspaces in a sesquilinear space. The theorem on Witt decompositions is an eminent tool in the classical theory of finite dimensional quadratic spaces. Results on Witt decompositions of infinite dimensional spaces $E$ turn out to be of equal prominence for the handling of totally isotropic subspaces. Let us first give a definition.

For $(E,\Phi)$ a non-degenerate sesquilinear space and $R \subset R^{\perp} \subset E$ a totally isotropic subspace we say that $E$ admits a metabolic decomposition for $R$ if $E$ contains an orthogonal family of planes $P_i = k(r_i, r_i')$ such that their sum possesses an orthogonal supplement in $E$ and $R$ is the span of the $r_i$ . If we let $R'$ be the span of the $r_i'$ we have in particular a decomposition

(0) $$E = (R \oplus R') \overset{\perp}{\oplus} E_o \ .$$

Since $E$ is non-degenerate all planes $P_i$ must be non-degenerate and thus $R^{\perp} = R \oplus E_o$ and $R^{\perp\perp} = R$ in (0). We see that a metabolic decomposition of $E$ for $R$ forces $R$ to be $\perp$-closed (Remark: if the totally isotropic $R$ is not $\perp$-closed one will use metabolic decompositions for the totally isotropic $R^{\perp\perp}$ and then study the location of $R$ within $R^{\perp\perp}$). We say that $E$ admits a Witt decomposition for the totally isotropic $R$ if there is a metabolic decomposition with all $P_i$ hyperbolic planes. In particular $R'$ may then be chosen totally isotropic in (0). In fact, we can equally define a Witt decomposition of $E$ (for $R$) by requiring the existence of a decomposition (0) with $R'$ some totally isotropic space: If $E$ is non-degenerate and $R'$ totally isotropic then it follows from (0) that $R = R^{\perp\perp}$ and that therefore $R \oplus R'$ is an orthogonal sum of planes $P_i$ as specified. If the form on $E$ is trace-valued (always the case when char $k \neq 2$) then the planes $P_i$ are invariably hyperbolic and one need not distinguish between metabolic decompositions and Witt decompositions.

In this chapter we are interested in the existence of Witt decompositions. Kaplansky has shown ([3], p.13,Thm.7) that $\perp$-closed totally isotropic $R \subset E$ always admit metabolic decompositions when dim $E \leq \aleph_0$. The short proof of this fact will be reproduced below (Theorem 3 ). By what we have said before this answers the question of Witt decompositions for <u>trace-valued</u> forms. When char $k = 2$ then forms need no longer be trace-valued and the problem becomes considerably more difficult. We shall solve here the general problem by giving appropriate conditions for Witt decompositions to exist. This settles the issue in countable dimensions.

## 2. The lattice that belongs to the problem

For $X$ any subspace of the non-degenerate $\varepsilon$-hermitean space $(E,\Phi)$ we let $X^*$ be the linear subspace $\{x \in X \mid \Phi(x,x) \in T\}$. Obviously $X^* = X \cap E^*$ . To $E^*$ belong in particular all isotropic vectors of $E$ , e.g. $E^{*\perp} \cap E^{*\perp\perp} \subset E^*$ , so that the radical $E^{*\perp} \cap E^*$ is $\perp$-closed being equal to the closed radical of $E^{*\perp}$ .

If $E$ possesses a Witt decomposition (0) for the totally isotropic subspace $R$ then we read off that

$$R^{\perp} + E^* = E \tag{1}$$

$$R + E^{*\perp} = (R^{\perp} \cap E^*)^{\perp} \tag{2}$$

Observe that $R = R^{\perp\perp}$ by (1) and (2): $R^{\perp\perp} \cap E^{*\perp} = (o)$ by (1) and $R + E^{*\perp}$ is closed by (2) and thus contains $R^{\perp\perp}$.

<u>Theorem 1</u>. Assume that $R \subset R^{\perp} \subset E$ satisfies (1) and (2). Then the lattice $V(R,E^*)$ orthostably generated in the lattice $L(E)$ of all subspaces in $E$ by $R$ and $E^*$ looks as follows

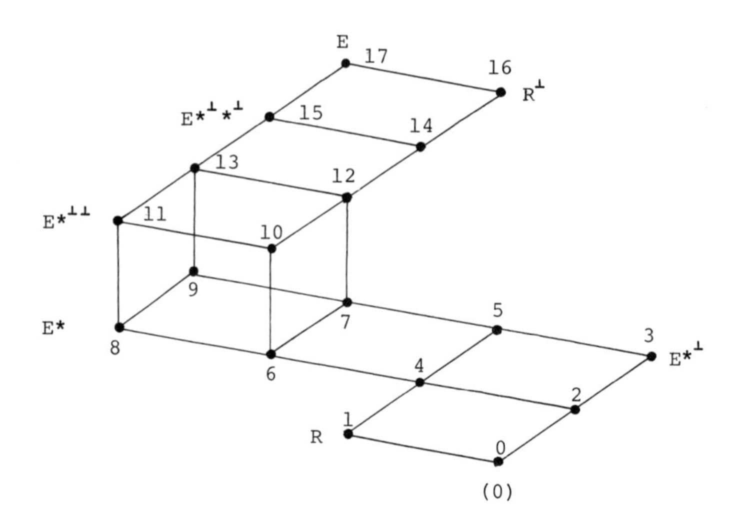


| $X$ | 0 | 1 | 2 | 3 | 4 | 5 | 6 | 7 | 8 | 9 | 10 | 11 | 12 | 13 | 14 | 15 | 16 | 17 |
|---|---|---|---|---|---|---|---|---|---|---|---|---|---|---|---|---|---|---|
| $X^{\perp}$ | 17 | 16 | 15 | 11 | 14 | 10 | 5 | 4 | 3 | 2 | 5 | 3 | 4 | 2 | 4 | 2 | 1 | 0 |

The proof of the theorem is left to the reader. In contrast to dis-discovering the diagram it is routine to verify its correctness by making use of the modular law etc.

The interesting point about this lattice is that it looks just as it ought to look if $E$ had a Witt decomposition for the subspace $R$. Indeed, if (0) is assumed with a totally isotropic $R'$ then it is very easy to jot down $V(R,E^*)$. This fact motivated our search for a proof of sufficiency of (1) & (2). Our result is the following

Theorem 2. (Witt decomposition). Let $R$ be a totally isotropic subspace in the non-degenrate $\varepsilon$-hermitean space $E$ over the division ring $k$. Assume that $\dim E \leqslant \aleph_0$. Then (1) and (2) are necessary and sufficient in order that $E$ possesses a Witt decomposition with respect to $R$.

Remarks 1. Whenever $\Phi$ is trace-valued, $E^* = E$, then condition (1) is empty and (2) reduces to $R = R^{\perp\perp}$ so that we obtain Kaplansky's result of unconditional existence of Witt decompositions for closed totally isotropic subspaces in trace-valued spaces.

2. Let $\Phi$ be symmetric, $E$ spanned by an orthonormal basis, $\dim E = \aleph_0$, $k$ algebraically closed and $\operatorname{char} k = 2$. Then $E^*$ is a hyperplane with $E^{*\perp} = (0)$. No maximal (with respect to $\subset$) totally isotropic $R \subset E$ satisfies (1) & (2) whereas all finite dimensional totally isotropic $R \subset E$ do satisfy (1) & (2). Hence the spaces $R$ admitting a Witt decomposition are not inductively ordered.

3. There are examples where either of (1) and (2) fails but not the other; there are also examples where (1) and (2) hold and all eighteen elements of the lattice are different spaces.

## 3. Metabolic decompositions

We reproduce the anounced result of Kaplansky (Thm.7 in [3], p.13; cf. [1], p.78, exercise 13).

Theorem 3. Let $E$ be a non-degenerate sesquilinear space of countable dimension and $R \subset R^{\perp} \subset E$ a $\perp$-closed subspace. Then $E$ admits a metabolic decomposition

$$E = \left(\overset{\perp}{\bigoplus_{I}} k(r_i, r_i')\right) \overset{\perp}{\oplus} E_0 \quad , \quad k(r_i)_{i \in I} = R \tag{3}$$

Proof. We construct such a decomposition recursively.

Suppose we have constructed finite dimensional <u>non-degenerate</u> subspaces $S, T$ with the following (induction) properties

$$S \perp T \quad , \quad T \subset R^{\perp} \quad , \quad S = \overset{m}{\underset{i=1}{\oplus^{\perp}}} k(r_i, r_i') \quad , \quad (S + T) \cap R = k(r_1, \ldots, r_m) \subset R .$$

Let $(e_i)_I$ be some fixed basis of $E$ and $e_n$ its first member not contained in $S + T$. We shall determine spaces $K$, $L \subset (S + T)^{\perp}$ such that $S' := S \oplus K$ and $T' := T \oplus L$ will again satisfy the induction assumptions ( with $S'$, $T'$ in lieu of $S$, $T$ ) and such that that $e_n \in S' + T'$. In this fashion we achieve a decomposition (3) with $E_0 := \cup T$.

Since $E = (S + T) + (S + T)^{\perp}$ we may decompose $e_n$, $e_n = x + y$ and "adjoin" only the component $y \perp S \oplus T$ to the space $S + T$ in the following construction step.

<u>Case 1</u>: $y \in R$. $(S + T)^{\perp}$ is non-degenerate and thus contains $y'$ with $\Phi(y', y) = 1$. Set $S' = S \oplus k(y, y')$, $T' = T$. How large can $(S' + T') \cap R$ get? Let $d = s+\lambda y+\lambda' y'+t \in S'+T'$ be an element of the intersection. Since $t \in T' = T \subset R^{\perp}$ and $d \in R \subset T^{\perp}$ we find $0 = \Phi(d, T) = 0 + 0 + 0 + \Phi(t, T)$ so $t = 0$ as $T$ is non-degenerate. Hence $0 = \Phi(d, y) = 0 + \lambda \cdot 0 + \lambda' \cdot 1$. Therefore $d - \lambda y = s \in S \cap R \subset k(r_1, \ldots, r_m)$ by induction assumption. We see that $(S' + T') \cap R = k(r_1, \ldots, r_m, y)$. The remaining induction assumptions are obvious in this case.

<u>Case 2</u>: $y \in R^{\perp} \smallsetminus R$. Notice that $y \notin R + S + T$ (indeed, by modularity, $((R + S + T) \cap T^{\perp}) \cap R^{\perp} = (R + S) \cap R^{\perp} = R + S \cap R \subset R)$. Therefore, and by $\perp$-closedness of $R + S + T$, $(R + S + T)^{\perp} \not\subset y^{\perp}$. Pick a vector $t \in (R + S + T)^{\perp}$ with $\Phi(t, y) \neq 0$. If $y$ is isotropic we set $T' = T \oplus k(y, t)$ otherwise simply $T' = T \oplus k(y)$ ; in both cases $S' = S$. It is not difficult to again verify the induction assumptions.

Case 3: $y \in E \smallsetminus R^{\perp}$. Notice that $y \notin R^{\perp} + S + T = R^{\perp} + S$ (since $(R^{\perp} + S) \cap k(r_1, \ldots, r_m)^{\perp} = R^{\perp})$. Therefore, and by the closedness of $R^{\perp} + S + T$, we can pick a vector $r \in (R^{\perp} + S + T)^{\perp} \smallsetminus y^{\perp}$. We set $S' = S \oplus k(r,y)$, $T' = T$, and easily verify all induction assumptions.

Corollary 1: If the space $E$ in Thm.3 is trace-valued then there is a Witt decomposition for every $\perp$-closed totally isotropic $R$. The orbit of $R$ in $L(E)$ under the orthogonal group of $E$ is therefore characterised by $\dim R$ and the isometry class of $R^{\perp}$.

For the proof of Theorem 2 we need a lemma to which we now turn.

## 4. A lemma on orthogonal separation of totally isotropic subspaces

The proof of the following lemma is valid for arbitrary characteristic. We shall need it when $\operatorname{char} k = 2$. For characteristic not 2 there are much more general results in this direction (Chapter VI).

Lemma 1. Let $(E,\Phi)$ be non-degenerate and $R,S$ totally isotropic subspaces with $R \perp S$; $\dim E \leqslant \aleph_0$. There exists an orthogonal decomposition $E = E_1 \overset{\perp}{\oplus} E_2$ with $R \subset E_1$ and $S \subset E_2$ if and only if $(R + S)^{\perp\perp} = R^{\perp\perp} + S^{\perp\perp}$ and $R^{\perp} + S^{\perp} = E$.

Proof. The necessity of the two conditions is obvious. For the proof of the converse we may assume without loss of generality that $R$ and $S$ be closed. We choose a metabolic decomposition of $E$ with respect to $R$, $E = (R \oplus R') \overset{\perp}{\oplus} E_0$ and let $S_1$ be the projection of $S$ onto $E_0$. As $R + S = R + S_1$ is closed we obtain that $S_1$ is closed; so there is a metabolic decomposition of $E_0$ for $S_1$. Thus there are mutually orthogonal planes $k(r_i, r_i')$ $(i \in I)$ and $k(s_j, s_j')$ $(j \in J)$ with $R$ the span of the $r_i$ and $S_1$ the span of the $s_j$ and (for $R'$ the span of the $r_i'$ and $S_1'$ the span of the $s_j'$) we have a decomposition $E = (R \oplus R') \overset{\perp}{\oplus} (S_1 \oplus S_1') \overset{\perp}{\oplus} E_1$. Without loss of generality

$\Phi(r_i,r_i') = 1 = \Phi(s_j,s_j')$ for all $i \in I$ and $j \in J$.

For each $j \in J$ there is $x_j \in R$ such that the family $s_j + x_j$ $(j \in J)$ spans $S$. Set $x_j = \sum \alpha_{jn} r_n$. Since by assumption $R^\perp + S^\perp$ is all of $E$ we have in particular $R' \subset (R + S_1 + S_1' + E_1) + S^\perp = S_1' + S^\perp$. In other words, for each $i \in I$, there exists some $y_i \in S_1'$ such that $r_i' + y_i \perp S$. Set $y_i = \sum \beta_{im} s_m'$. It follows that $\beta_{ij} + \alpha_{ji} = 0$ which shows that matrices $(\alpha_{ij})$ and $(\beta_{ij})$ are both row- and column-finite. If we let $R''$ be the span of the $r_i' + y_i$ $(i \in I)$ then we have that $R + R'' \perp S + E_1$. Does there exist an orthogonal projection of $S_1'$ onto $R + R''$? We see that $s_j' + \sum \xi_{jn} r_n \perp R + R''$ if and only if $\xi_{in} + \beta_{ni} \Phi(s_i',s_i') = 0$. Since $(\beta_{ni})$ is column-finite we can solve for $\xi_{in}$ and get <u>finite</u> sums $\sum \xi_{in} r_n$. Hence $R + R''$ admits an orthogonal supplement that contains $S$. QED.

## 5. Reducing the proof of Thm.2 to the case of a non-degenerate $E^*$

We call $S$ the radical $E^{*\perp} \cap E^*$ of $E^*$.

Let $R \subset E$ be the subspace of Theorem 2 and assume (1) and (2). We have $R^{\perp\perp} = R$ and $S^{\perp\perp} = S$. Furthermore $R \perp S$ as $R \subset E^* \subset S^\perp$ and by (1) we get $R^\perp + S^\perp \supset R^\perp + E^* = E$. Thus in order to show that the pair $R,S$ qualifies for the lemma there remains to prove closedness of $R + S$. We choose some fixed metabolic decomposition for $R$,

$$(4) \qquad E = (R \oplus R') \overset{\perp}{\oplus} E_0 \qquad \text{(notation: } X^{\perp\circ} := X^\perp \cap E_0 \text{)}$$

Intersection with $E^*$ of both sides in (2) gives us the equality $R + S = (R^\perp \cap E^*)^\perp \cap E^*$, hence by (4) $R + S = [(R + E_0) \cap E^*]^\perp \cap E^* = [R + (E_0^*)]^\perp \cap E^* = R + \mathrm{rad}\,(E_0^*)$. Since $E_0$ is an orthogonal summand in $E$ the closure of the space $\mathrm{rad}\,(E_0^*)$ can be computed in $E_0$ with respect to $\Phi | E_0 \times E_0$. By the remark made at the beginning of Sec.2 $\mathrm{rad}\,(E_0^*)$ is therefore closed and hence the sum $R + \mathrm{rad}\,(E_0^*)$ is closed in view of the splitting (4). We have thus shown that

(5) $\qquad R + S = R + \text{rad}\,(E_0^*) = (R + S)^{\perp\perp}$

Incidentally, we have also seen that (2) implies

(6) $\qquad R + S = (R^{\perp}*)^{\perp}*$

which has a natural interpretation: The right hand side in (6) is the closure of $R \subset E^*$ with respect to the restriction $\Phi^* := \Phi|E^* \times E^*$. A subspace $X \subset E^*$ closed with respect to $\Phi^*$ will, of course, contain the radical of the space $E^*$, $S \subset X$. Hence the conclusion (6) of (2) forces the closure of $R$ in $(E^*, \Phi^*)$ to be as small as it can possibly be. In particular,

(7) $\qquad$ if $\text{rad}\, E^* = (0)$ then $R$ is closed with respect to $\Phi^*$.

Since $R$ and $S$ qualify for the lemma there is a splitting $E = E_2 \overset{\perp}{\oplus} E_3$ with $R \subset E_2$ and $S \subset E_3$. $E_3$ has a metabolic decomposition for $S$. Relabeling summands we thus have a decomposition

(8) $\qquad E = (S \oplus S') \overset{\perp}{\oplus} E_1$, $R \subset E_1$ (notation: $X^{\perp_1} := X^{\perp} \cap E_1$)

By using (8) and the fact that $E^* \subset S^{\perp} = S \oplus E_1$, and hence $E^* = S \oplus E_1^*$, it is easy to justify the following

<u>Conclusion</u>. Let $\Phi_1$ be the restriction of $\Phi$ to $E_1 \times E_1$. If $E_1$ admits a Witt decomposition with respect to $R$ then so will $E$. Since $E_1^*$ is non-degenerate and since it follows from (1) and (2) that we have $(1_1)$ $R^{\perp_1} + E_1^* = E_1$ and $(2_1)$ $R + E_1^{*\perp_1} = (R^{\perp_1} \cap E_1^*)^{\perp_1}$ it therefore suffices to prove Theorem 2 under the assumption that $E^*$ is non-degenerate.

## 6. <u>Discussion of properties (1) and (2) when E* is non-degenerate</u>

The following result is the fundament of the proof in the next section.

<u>Lemma 2</u>. Let $\dim E \leqslant \aleph_0$ and assume (1),(2) and $E^{*\perp} \cap E^* = (0)$. Let $A, B \subset E$ be finite dimensional non-degenerate subspaces with the

following properties:

$A \perp B$ , $A \subset E^*$ , $A$ is hyperbolic with $A \cap R$ a maximal totally isotropic subspace of $A$ ,

$B \subset R^{\perp}$ , $B \cap (R + E^{*\perp}) = B \cap E^{*\perp}$ .

Then we have

(I) for every $d \in A^{\perp} \cap E^*$ there exists $e \in E^* \cap R^{\perp}$ such that $d + e \in (A + B)^{\perp}$

(II) for each $z \in (A + B)^{\perp}$ there exists $x \in R^{\perp} \cap (A + B)^{\perp}$ , $y \in E^* \cap (A + B)^{\perp}$ with $z = x + y$

<u>Proof</u>. Let $G$ be a supplement of $E^* \oplus E^{*\perp}$ in $E$ . (1) says that we may pick $G \subset R^{\perp}$. By (7) there is a metabolic decomposition of $E^*$ with respect to $R$ ; since $A$ is the sum of finitely many hyperbolic planes $k(r_i,r_i')$ $(i = 1,\ldots,m)$ with $r_i \in R$ we can use $A$ as an initial stump of a metabolic decomposition of $E$

$$E^* = (R \oplus R') \overset{\perp}{\oplus} H = \overset{\perp}{\underset{i}{\oplus}} k(r_i,r_i') \overset{\perp}{\oplus} H = A \overset{\perp}{\oplus} \overset{\perp}{\underset{i>m}{\oplus}} k(r_i,r_i') \oplus H \tag{9}$$
$$E = (E^* \oplus E^{*\perp}) \oplus G , \quad G \subset R^{\perp} , \quad R = \underset{i}{\oplus} k(r_i)$$

What does now condition (2) mean for (9)? We have $R^{\perp} = R \oplus H \oplus E^{*\perp} \oplus G$ so by (2) $R + E^{*\perp} = (R^{\perp} \cap E^*)^{\perp} = (R \oplus H)^{\perp} = R \oplus E^{*\perp} \oplus (H \oplus G) \cap H^{\perp}$ . Therefore

$$(H \oplus G) \cap H^{\perp} = (0) \tag{10}$$

Let $(b_j)$ be a basis of a supplement of $B \cap (R \oplus E^{*\perp})$ in $B$. $B \subset R^{\perp} = R \oplus H \oplus E^{*\perp} \oplus G$ ; so we break the $b_j$ accordingly into components, $b_j = r_j + h_j + e_j + g_j$ and let $C$ be the span of all projections $h_j + g_j$ (they are linearly independent). Thus $C \subset H \oplus G$ . We try to pick the required vector $e \in H$ , hence it will have to satisfy $\Phi(e,b_j) = -\Phi(d,b_j)$ , i.e. $\Phi(e,h_j + g_j)$ ought to have prescribed values. We can certainly find a vector $e'$ in $E$ which has the required angles with $C$ . Since $C$ is finite dimensional $C^{\perp} + H$

is closed and so by (10) $C^{\perp} + H = (C^{\perp} + H)^{\perp\perp} = (C \cap H^{\perp})^{\perp} = (0)^{\perp} = E$. Hence we decompose $e'$, $e' = e'' + e \in C^{\perp} + H$. Since $B \cap (R + E^{*\perp}) = B \cap E^{*\perp}$ by the assumption of the lemma and $d \in E^*$ we see that $d + e$ is orthogonal to all of $B$. This proves (I). To prove (II) we may start out with vectors $x_1 \in R^{\perp}$, $y_1 \in E^*$ with $z = x_1 + y_1$ by virtue of (1). There is $r \in R \cap A$ such that $x_1 + r \perp A$. Set $c = x_1 + r$, $d = y_1 - r$. We have $d, c \in A^{\perp}$ since $d + c \in A^{\perp}$ by assumption. By part (I) there exists $e \in E^* \cap R^{\perp}$ with $d + e \in (A + B)^{\perp}$. With $x := d + e$ and $y := c - e$ we satisfy (II).

By the lemma we have one half of the following

Corollary. Let $R \subset E$ be totally isotropic, $\dim E \leq \aleph_0$ and $A$, $B$ as in Lemma 2. Then (1) & (2) is equivalent with the validity of (I) & (II) for all $A$, $B$.

To show the remaining (easy) half of the corollary one chooses $A = B = (0)$ to obtain (1) from (II). Hence we may use a decomposition (9) and set $A = (0)$ in order to obtain (10) from (II). But (10) and (2) are equivalent if (1) holds.

## 7. Proof of Theorem 2 when $E^*$ is non-degenerate

Assume that $\dim E \leq \aleph_0$ and $E^* \cap E^{*\perp} = (0)$ and, furthermore, that (1) and (2) holds.

We build up a Witt decomposition for $R$ recursively. Assume that we already have determined finite dimensional non-degenerate subspaces $A$, $B \subset E$ with

(11) $A \perp B$, $A \subset E^*$, $A$ hyperbolic and $A \cap R$ a maximal totally isotropic subspace in $A$, $B \subset R^{\perp}$, $B \cap (R + E^{*\perp}) = B \cap E^{*\perp}$

Let $(e_i)$ be a fixed auxiliary basis of $E$ and $e_r$ the first member not in $A + B$ . We shall determine finite dimensional subspaces $X, Y \subset (A + B)^{\perp}$ such that $A_1 := A \oplus X$ , $B_1 := B \oplus Y$ will again satisfy (11) and have $e_r \in A_1 + B_1$ . In this fashion we obtain a decomposition $E = (R + R') \overset{\perp}{\oplus} E_0$ , $R \oplus R' \subset E^*$ . Although the plan of the proof is quite obvious there will be a number of details to pay attention to.

As $E = (A + B) + (A + B)^{\perp}$ we may decompose $e_r$ and adjoin its component $z$ in $(A + B)^{\perp}$ to the space $(A + B)$ . We shall distinguish between three major cases just as in the proof of Thm. 3.

Case 1. $z \in R \setminus \{0\}$ . Since $E^* = A + (A^{\perp} \cap E^*)$ is non-degenerate we find $d \in E^* \cap A^{\perp}$ with $\Phi(z,d) = 1$ . By (I) of Lemma 2 there is $e \in E^* \cap R^{\perp}$ such that $e + d \in (A + B)^{\perp}$ . Hence $k(z,d + e)$ is a hyperbolic plane in $(A + B)^{\perp}$ . If we set $A' := A + k(z,d + e)$ and $B' := B$ then we have $z \in A' + B'$ and all induction assumptions are easily seen to hold again.

Case 2. $z \in R^{\perp} \setminus R$ . It follows that

$$z \notin R + A + B \tag{12}$$

It may happen that $z \in B + R + E^{*\perp}$ , $z = b + r + e'$ ($b \in B$, $r \in R$, $e' \in E^{*\perp}$) . It then follows that $r \perp A + B$ . If $r \neq 0$ then by Case 1 there is a hyperbolic plane $k(r,r') \perp A + B$ . What do we then know about $z-r$ ? Since $r \perp B$ we have $z-r \perp B$ ; as $z-r \in B+E^{*\perp}$ also $z - r \perp A$ . Since $k(r,r') \subset E^* \cap B^{\perp}$ we have $z - r \perp k(r,r')$ . In total we have $z - r \in (R^{\perp} \setminus R) \cap (A + k(r,r') + B)^{\perp}$ which shows that with $z - r$ we are still in Case 2. What we have shown is that (in the present case)

(13) if $z$ happens to be in $B + R + E^{*\perp}$ then without loss of generality $z \in B + E^{*\perp}$ .

Now if $z$ is non-isotropic we set $A' := A$ and $B' := B + (z)$. If $z$ is isotropic then by (12) we may pick $z' \in (R + A + B)^{\perp} \setminus z^{\perp}$.

We set $A' := A$ and $B' := B \oplus k(z,z')$ .

Let us verify the induction assumption on $B' \cap (R + E^{*\perp})$ in this last case. For a typical vector $q$ of this intersection write $q := b + \lambda z + \mu z' = r + e'$ $(b \in B , r \in R , e' \in E^{*\perp})$ . $z$ is isotropic (hence in $E^*$) and multiplying it with $r + e'$ therefore yields $0 = \mu$ . Hence $\lambda z \in B + R + E^{*\perp}$ ; so $\lambda z = b_1 + e''$ $(b_1 \in B, e'' \in E^{*\perp})$ by (13). Since $b + \lambda z \in R + E^{*\perp}$ we obtain $b + b_1 \in (R + E^{*\perp}) \cap B = E^{*\perp} \cap B$ by induction assumption. Therefore $q := b + \lambda z = (b + b_1) + e'' \in E^{*\perp}$ Ergo $q \in B' \cap E^{*\perp}$ . The remaining induction assumptions are readily seen to hold. This terminates Case 2.

<u>Case 3</u>. $z \notin R^{\perp}$ . By (II) of Lemma 2 $z = x + y$ for some $x \in R^{\perp} \cap (A + B)^{\perp}$ and $y \in E^* \cap (A + B)^{\perp}$ . Since $z \notin R^{\perp}$ we have $y \notin R^{\perp}$ and thus we find $r \in R \cap A^{\perp}$ with $\Phi(r,y) = 1$ so that $k(r,y)$ is a hyperbolic plane in $(A + B)^{\perp}$ . Set $A' := A \oplus k(r,y)$ , $B' := B$ . If $x$ happens to be zero then we are done: $z \in A' + B'$ and all induction assumptions are easily seen to hold. If $x \neq 0$ then the vector $z_1 := x - \Phi(x,y)r$ is in $R^{\perp} \cap (A' \oplus B')^{\perp}$ and thus qualifies either for Case 1 or Case 2 so that by some additional steps we may adjoin $z_1$ to $A' + B'$ . This terminates Case 3.

The proof of Thm 2. in the case of non-degenerate $E^*$ is thus complete. By Sec.5 this proves Theorem 2 in full generality.

## 8. Some general remarks on the proof of Theorem 2

We see fit to include a comment on the rationale behind the proof of Theorem 2 as against some proofs given in later chapters. Note first the following obvious consequence of Theorem 2:

(14) Let $R$ be a totally isotropic subspace of the non-degenerate $\varepsilon$-hermitean space $E (\dim E \leq \aleph_0)$ with (1) and (2). Then the orbit of $R$ (under the orthogonal group of $E$) is characterized by the cardinal $\dim R$ and the isometry class of $R^{\perp}$ .

Assuming that we had a <u>direct</u> proof of (14) could we then obtain Theorem 2 as a corollary ? Given a $\perp$-closed $R \subset R^{\perp} \subset (E,\Phi)$ we certainly have a metabolic decomposition $E = (R \oplus R') \overset{\perp}{\oplus} E_0$ . Define a new form $\Psi$ on $E \times E$ by setting $\Psi$ identically zero on $R' \times R'$ and have it coincide with $\Phi$ on $R \times R$ , $R \times R'$ , $R \times E_0$ , $R' \times E_0$ and $E_0 \times E_0$ The space $(E,\Psi)$ is non-degenerate and Witt decomposed with respect to $R$ . Obviously, if $(E,\Phi)$ does admit a Witt decomposition for $R$ then $(E,\Phi)$ and $(E,\Psi)$ are isometric. Conversely: <u>provided that</u> $(E,\Psi)$ , as defined, <u>turns out to be isometric to</u> $(E,\Phi)$ then it follows without further ado from (14) that $(E,\Phi)$ admits a Witt decomposition with respect to $R$ . We would have liked to present a proof of Theorem 2 along this line instead of directly setting up the required decomposition as we have done in the present chapter. In fact, all the many normal forms effected in subsequent chapters are obtained by applying the strategy just outlined, to wit,

1. characterize orbits by certain invariants ("uniqueness"),
2. construct a specific normal form with prescribed invariants

in order to conclude that in <u>all</u> instances we can exhibit a normal form. We have not succeeded in finding such a proof of Theorem 2 in the general case. However, when $\dim S/T$ is finite then there is such a proof [2]; it will be given in Chapter VIII as a special case of a more general result. (The finiteness assumption on $\dim S/T$ implies that $(E,\Phi)$ is quasistable in the sense of Chapter VII if it contains a totally isotropic $R$ . If, in addition, $R$ satisfies (1) and (2) then we are also able to establish the isometry $(E,\Psi) \cong (E,\Phi)$ .)

For further remarks on decomposition theorems in general see [4], 5-6.

## References to Chapter III

[1] N. Bourbaki, Formes sesquilinéaires et formes quadratiques. ASI 1272 Hermann Paris (1959).

[2] H. Gross, Formes quadratiques et formes non traciques sur les espaces de dimension dénombrable. Bull. Soc. Math. de France, Mémoire 59 (1979).

[3] I. Kaplansky, Forms in infinite dimensional spaces. An. Acad. Brasil. Ci. 22 (1950) 1-17.

[4] -, Infinite abelian groups. Univ. of Mich. Press, Ann Arbor, fourth printing 1962.

# CHAPTER FOUR

# ISOMORPHISMS BETWEEN LATTICES OF LINEAR SUBSPACES WHICH ARE INDUCED BY ISOMETRIES

## 1. Introduction

Let $E$ be a vector space over the division ring $k$ and $L(E)$ the lattice of all linear subspaces of $E$. If $\bar{E}$ is a vector space over the division ring $\bar{k}$ and $\tau: L(E) \to L(\bar{E})$ a lattice isomorphism then by the Fundamental Theorem of Projective Geometry ([1] p. 44) $\tau$ is induced by a semilinear map $T: E \to \bar{E}$ if we assume that $\dim E \geq 3$.

Here we shall investigate questions of the following kind. Assume that $\tau: V \to \bar{V}$ is a lattice isomorphism merely between sublattices of $L(E)$ and $L(\bar{E})$ respectively but assume $E$ and $\bar{E}$ equipped with non-degenerate (orthosymmetric) sesquilinear forms $\Phi$ and $\bar{\Phi}$ respectively. Then ask: is $\tau$ induced by a similitude $T: E \to \bar{E}$ ?

It is not to be expected that these problems have an easy and uniform answer. We shall start out with the special case of alternate forms and look for $\tau: V \to \bar{V}$ which <u>are</u> induced by isometries. The result is Theorem 1 below. If there are "enough" isotropic vectors available in $E$ and $\bar{E}$ then our construction holds also for nonalternate forms. Our main result is Theorem 2 in Section 8. Chapters V, VI and VIII bring applications to this result. Of course we restrict ourselves here to countable dimensions ([4], p. 3 contains a nice example in arbitrary dimensions).

## 2. The kind of lattices admitted

If $\tau: V \to \bar{V}$ is induced by an isometry $T: E \to \bar{E}$, i.e., $X^{\tau} = \{Tx \mid x \in X\}$ for all $X \in V$, then $\tau$ extends to all of $L(E)$ and obviously commutes with orthogonal complementation. Therefore, it is natural to restrict our attention to orthostable lattices:

$$(0) \in V \quad \text{and} \quad E \in V \quad \text{and if} \quad X \in V \quad \text{then} \quad X^{\perp} \in V .$$

Furthermore, if it is induced, then $\tau$ must preserve rank, in particular it must preserve indices, i.e. dimensions of quotient spaces $Y/X$ for neighbouring elements $X \subsetneq Y$ of $V$. Since $\tau$ - when it is induced - also extends to a lattice isomorphism between the complete orthostable lattices generated by $V$, $\bar{V}$ in $L(E)$ and $L(\bar{E})$ respectively it is appropriate to assume at the outset that $V$ is complete as a sublattice of $L(E)$. Indeed, easy examples show that, in general, the indices of $V$ will not determine the indices in the complete sublattice generated by $V$. As we assume that $\dim E \leq \aleph_0$ all chains in $V$ will be denumerable so that it is sufficient to check $V$ for being a $\sigma$-lattice in the sense that every denumerable subset has an infimum and a supremum (see also Section 7 below).

Definition. An element $D$ of a lattice $V$ is called compact if $D \subsetneq \Sigma A_\iota$ for an arbitrary family of $A_\iota \in V$ implies that $D \subsetneq A_1 + \cdots + A_n$ for finitely many among the $A_\iota$. A complete lattice is called algebraic (or compactly generated) if every element is the join of compact elements. An element $D \in V$ is called join-irreducible if $D = A+B$, $A \in V$, $B \in V$ implies that $D = A$ or $D = B$; it is called join-prime if $D \subsetneq A+B$, $A \in V$, $B \in V$ implies $D \subsetneq A$ or $D \subsetneq B$.

We now put down the conditions on the lattice $V$ which make possible the geometric constructions in this chapter. They are:

(1) join-irreducible compact elements are join-prime,

(2) compact elements are joins of join-irreducibles.

We shall often have to deal with finite lattices $V$ satisfying (1) and (2). The adjective "compact" is then redundant and by (1) & (2) every element of $V$ is a join of join-prime elements. Hence $V$ must be distributive by the following

Lemma 1 ([5]). A lattice in which every element is a (finite or infinite) join of join-prime elements is distributive.

Proof. We first show that $a \leq b+c$ implies $a \leq (a \wedge b) + (a \wedge c)$ in such a lattice. Indeed, let $X(a)$ be the set of all join-prime elements $\leq a$. $a = \Sigma\{x | x \in X(a)\}$ by assumption. If $a \leq b+c$ then $x \leq a \leq b+c$ for all $x \in X(a)$ hence $x \leq b$ or $x \leq c$ and thus $x \leq (a \wedge b) + (a \wedge c)$. Ergo $a = \Sigma x \leq (a \wedge b) + (a \wedge c)$ as asserted. Now if our lattice were not distributive it would contain a five element non-distributive sublattice 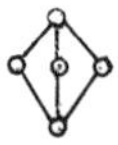 or a five element nonmodular lattice 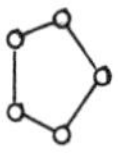 both not enjoying the property just established.

However, the same conclusion holds for the nonfinite case as well. We shall list the pertinent results here.

Lemma 2 ([5]). Let $L$ be an algebraic modular lattice and $c$ a compact element of $L$ which is a join of join-irreducibles. Then $c$ is a join of join-irreducible compact elements.

For a proof we refer to [5]. Clearly, by the two lemmas every algebraic modular lattice with (1) and (2) is distributive.

A lattice is called completely distributive if (for $I, J \neq \emptyset$) the following identity holds,

$$\wedge\{\Sigma\{a_{ij} | j \in J\} | i \in I\} = \Sigma\{\wedge\{a_{i\varphi i} | i \in I\} | \varphi: I \to J\} .$$

In [6] (Statement 3.2 page 320) the following result is proved:

Lemma 3. An algebraic lattice is completely distributive if and only if each compact element is a join of join-prime compact elements.

Hence we see, by Lemma 2 and 3, that modular algebraic lattices with (1) and (2) are even completely distributive. We said before that we would stick to <u>complete</u> sublattices of $L(E)$ . Now $L(E)$ is clearly algebraic and in order to round off our digression on lattices we shall prove

<u>Lemma 4</u>. A complete sublattice $S$ of an algebraic lattice $L$ is algebraic.

<u>Proof</u>. For $c \in L$ define $\hat{c} = \wedge\{x \in S \mid x \geq c\}$ . If $c$ is compact in $L$ then $\hat{c}$ is compact in $S$ . Furthermore if $a \in S \subset L$ is a join of compact elements $c_\iota$ in $L$ then $c_\iota \leq a$ and so $\hat{c}_\iota \leq a$ . Since thus $a \leq \sum c_\iota \leq \sum \hat{c}_\iota \leq a$ we obtain $a = \Sigma \hat{c}_\iota$ .

<u>We summarize</u>: The lattices $V$ which we are going to deal with in this chapter, to wit, complete sublattices of $L(E)$ that satisfy (1) and (2), are completely distributive. Conversely, each completely distributive complete sublattice of $L(E)$ satisfies (1) and (2).

## 3. Statement of Theorem 1 and an outlay of its proof

No assumption is made on the characteristic of the field.

<u>Theorem 1</u>. Let $E$ be an alternate space with $\dim E \leq \aleph_0$ and $V$ , $\bar{V}$ orthostable complete sublattices of the lattice $L(E)$ of all subspaces of $E$ . Assume that $\tau\colon V \to \bar{V}$ is a lattice isomorphism which commutes with orthogonal complementation and which preserves indices. In order that $\tau$ be induced by an isometry $T\colon E \to E$ it is sufficient that $V$ satisfies (1) and (2).

In order to prove this theorem we are going to construct by induction two ascending chains $W_0 \subseteq W_1 \subseteq W_2 \subseteq \ldots$ , $\bar{W}_0 \subseteq \bar{W}_1 \subseteq \bar{W}_2 \subseteq \ldots$ of finite dimensional subspaces of $E$ and a sequence of isometries $T_i\colon W_i \to \bar{W}_i$ such that $T_{i+1}$ is an extension of $T_i$ . We shall arrange for $\bigcup W_i$ and $\bigcup \bar{W}_i$ to be all of $E$ so that the $T_i$ define an isometry $T: E \to E$ . Furthermore the $T_i$ are going to be "compatible" with the given lattice isomorphism $\tau: A \mapsto \bar{A}$ $(A \in V , \bar{A} \in \bar{V})$ :

$$(3) \qquad T_i(W_i \cap A) = \bar{W}_i \cap \bar{A} \quad \text{for all } A \in V .$$

If such sequences $W_i$ , $\bar{W}_i$ exist at all then there will be many possibilities of selecting them. We try to construct them in such a manner that the following distributivity holds for arbitrary families of elements $A_\iota \in V$ :

$$(4) \qquad \bigcap_\iota (W_i + A_\iota) = W_i + \bigcap_\iota A_\iota$$

and, of course, the corresponding property for each $\bar{W}_i$ of the second sequence.

In order to describe this construction it is sufficient to discuss the ith step. If we drop subscripts we are left with the following task

## 4. The construction problem

We are given finite-dimensional subspaces $W$ , $\bar{W} \subset E$ and an isometry $T: W \to \bar{W}$ with

$$(3') \qquad T(W \cap A) = \bar{W} \cap A^\tau \quad (\text{for all } A \in V)$$

$$(4') \qquad \bigcap_\iota (W + A_\iota) = W + \bigcap_\iota A_\iota$$

$$(\bar{4}') \qquad \bigcap_\sigma (\bar{W} + \bar{A}_\sigma) = \bar{W} + \bigcap_\sigma A_\sigma$$

where $(A_\iota)$ , $(\bar{A}_\sigma)$ are arbitrary families in $V$ and $\bar{V}$ respectively. There is furthermore given a vector $x \in E \setminus W$ . One then has to (I) specify in $E$ a finite dimensional subspace $W_1 \supset W \oplus (x)$ , (II) construct a subspace $\bar{W}_1$ such that $T$ extends to an isometry $T_1: W_1 \to \bar{W}$ and, (III) verify that $W_1$ , $\bar{W}_1$ , $T_1$ satisfy again (3'), (4'), ($\bar{4}'$).

We shall now adduce some equalities for dimensions of spaces that will show up in the course of subsequent constructions. These equalitie are consequences of (3') and the properties of $\tau$ . For $A \in \mathcal{V}$ let again $\bar{A} \in \bar{\mathcal{V}}$ be the image of $A$ under the given lattice isomorphism $\tau: \mathcal{V} \to \bar{\mathcal{V}}$ . We have

(i) For all $A \in \mathcal{V}$ : $\dim A/A \cap W^{\perp} = \dim \bar{A}/\bar{A} \cap \bar{W}^{\perp}$

Proof. $A/(A \cap W^{\perp}) \cong (A + W^{\perp})/W^{\perp}$ . Since for $\perp$-closed $V$ we have $\dim U/V = \dim V^{\perp}/U^{\perp}$ we obtain $\dim A/(A \cap W^{\perp}) = \dim W/W \cap A^{\perp}$ and similarly $\dim \bar{A}/(\bar{A} \cap \bar{W}^{\perp}) = \dim \bar{W}/(\bar{W} \cap \bar{A}^{\perp})$ . The lattices are orthostable so $A^{\perp} \in \mathcal{V}$ , $\bar{A}^{\perp} \in \bar{\mathcal{V}}$ and, since $T: W \to \bar{W}$ satisfies (3'), we have $W/W \cap A^{\perp} \cong \bar{W}/\bar{W} \cap \bar{A}^{\perp}$ . Therefore (i).

(ii) For all $A , A_0 \in \mathcal{V}$ with $A_0 \subseteq A$ : $\dim((A_0+W) \cap A)/A_0 = \dim((\bar{A}_0+\bar{W}) \cap \bar{A})/\bar{A}_0$ .

Proof. By (3') $(W \cap A)/(W \cap A_0) \cong (\bar{W} \cap \bar{A})/(\bar{W} \cap \bar{A}_0)$ so $(A_0 + (W\cap A))/A_0 \cong (\bar{A}_0 + (\bar{W}\cap\bar{A}))/\bar{A}_0$ hence (ii) by modularity.

(iii) For all $A , A_0 \in \mathcal{V}$ with $A_0 \subseteq A$ :
$\dim((A_0+W)\cap A)/((A_0+W)\cap A) \cap W^{\perp} = \dim((\bar{A}_0+\bar{W})\cap\bar{A})/((\bar{A}_0+\bar{W})\cap\bar{A}) \cap \bar{W}^{\perp}$.

Proof. By (3') $T$ maps $(A_0^{\perp} \cap W) \cap (W \cap (W \cap A)^{\perp})$ onto the analogously built space so the quotient space $W/A_0^{\perp} \cap W \cap (W\cap A)^{\perp}$ is isomorphic to $\bar{W}/\bar{A}_0^{\perp} \cap \bar{W} \cap (\bar{W} \cap \bar{A})^{\perp}$ . Denominators are $\perp$-closed so $\dim(W^{\perp}+A_0+(W\cap A))/W^{\perp} = \dim(\bar{W}^{\perp}+\bar{A}_0+(\bar{W}\cap\bar{A}))/\bar{W}^{\perp}$ .

(iv) For all $A , A_0 \in \mathcal{V}$ with $A_0 \subseteq A$ :
$\dim A/(A_0 + W) \cap A = \dim \bar{A}/(\bar{A}_0 + \bar{W}) \cap \bar{A}$ .

Proof. $\tau$ preserves indices by the assumption of Thm. 1 so $\dim A/A_0 = \dim \bar{A}/\bar{A}_0$ . From this and (ii) we get (iv) by a simple subtraction.

(v) For all $A , A_0 \in \mathcal{V}$ with $A_0 \subseteq A$ :
$\dim A/((A_0+W)\cap A)\cap W^{\perp} = \dim \bar{A}/((\bar{A}_0+\bar{W})\cap\bar{A})\cap\bar{W}^{\perp}$ .

Proof. Combine (iii) and (iv).

From (i) and (v) we obtain the important equality

(5) for all $A$, $A_0 \in V$ with $A_0 \subseteq A$ we have
$\dim(A\cap W^{\perp})/((A_0+W)\cap A)\cap W^{\perp} = \dim(\bar{A}\cap\bar{W}^{\perp})/((\bar{A}_0+\bar{W})\cap\bar{A})\cap\bar{W}^{\perp}$ .

Remark. In Section 7 below we shall somewhat relax the assumption on completeness of $V$. For $A \in V$ we let then $A_0$ be the sum in $E$ of all proper antecedents of $A$ in $V$, $A_0 = \Sigma\{Z \in V | Z \subsetneqq A\}$. Define $\bar{A}_0 := \Sigma\{Z^{\tau} \in \bar{V} | Z \subsetneqq A\}$. $A_0$ and $\bar{A}_0$ need not be lattice elements. Nevertheless, if we assume that $\tau$ preserves the dimensions of quotient $A/A_0$, $\dim A/A_0 = \dim \bar{A}/\bar{A}_0$, for $A \in V$ and $A_0$, $\bar{A}_0$ as defined ther (5) still holds on the basis of (3') alone: $W \cap A_0$ is still mapped ont $\bar{W} \cap \bar{A}_0$ under $T$ as can be seen by considering a vector of the intersection, etc.

## 5. Solution of the construction problem in the irreducible case

Let $x$, $W$ be the objects mentioned in the above construction problem. Define $M(x,W) := \{Z \in V | x \in W+Z\}$, $M$ is a closed sublattice of $V$ by (4') (it contains $\inf M$ whenever $M \subseteq M$) and, as it clearly is a filter, it is a principal filter. We let $D = D(x,W)$ be its generator. $D$ is compact, obviously.

In this section we shall assume that $D$ is join-irreducible, hence $D$ will be join-prime by (1) of Section 2. This property will, however, only be used at the end when discussing the validity of the induction assumptions (4'), ($\bar{4}$') after the construction step.

We have $x \in W+D$. Without loss of generality $x \in D$. We set $W_1 = W \oplus (x)$ which takes care of (I) in the construction problem formulated in the previous section. We turn to (II). We have to pick

$\bar{x} \in E$ and set $\bar{W}_1 = W \oplus (\bar{x})$ . Where should we pick $\bar{x}$ so that (3') continues to hold for $W_1$ , $\bar{W}_1$ ?

Let $D_0 := \Sigma\{Z \in \mathcal{V} | Z \subsetneqq D\}$ . We have $D_0 \neq D$ for otherwise $D = Z_1 + \cdots + Z_n$ by the compactness of $D$ , thus $n = 1$ , and so $D \subsetneqq D$ , contradiction. Thus $x \in D$ and $x \notin W + D_0$ . It is therefore certainly necessary to select $\bar{x}$ such that

(6) $\qquad \bar{x} \in \bar{D} \quad , \quad \bar{x} \notin \bar{W} + \bar{D}_0$

holds. Suppose we pick $\bar{x}$ such that (6) holds. If we let $\mathcal{M}(\bar{x},\bar{W})$ be the filter associated with the pair $\bar{x}$ , $\bar{W}$ (defined in analogy to $\mathcal{M}(x,W)$ at the beginning of this section) then the first half of (6) tells that $\bar{D}$ is an element of $\mathcal{M}(\bar{x},\bar{W})$ and the second half tells that $\bar{D}_0$ is not. Since $\bar{D}$ covers $\bar{D}_0$ (i.e. $\bar{D}_0 \subsetneqq X \subsetneqq \bar{D}$ is not satisfied by any $X$ ) we see that (6) implies

(7) $\qquad \bar{D}$ is the generator of the filter $\mathcal{M}(\bar{x},\bar{W})$ .

In other words, $\tau$ maps $\mathcal{M}(x,W)$ onto $\mathcal{M}(\bar{x},\bar{W})$ ; this makes it very plausible that $\bar{x}$ has the correct order theoretic ubiety relative to the lattice $\bar{\mathcal{V}}$ . We now have to turn to metric requirements imposed on the choice of $\bar{x}$ .

Let $F$ be a supplement of $W \cap D^{\perp}$ in $W$ and spanned by a basis $f_1, f_2, \ldots, f_n$ . Then $Tf_1, Tf_2, \ldots, Tf_n$ span a supplement $\bar{F}$ of $\bar{W} \cap \bar{D}^{\perp}$ in $\bar{W}$ by (3'). Since $\bar{D}^{\perp} \cap \bar{F} = (0)$ there exists $\bar{y} \in \bar{D}$ that has the requisite products $\Phi(\bar{y},Tf_i) = \Phi(x,f_i)$ $(1 \leq i \leq n)$ . We next convince ourselves that we can move $\bar{y}$ into $\bar{D} \setminus (\bar{D}_0 + \bar{W})$ , in case we should have $\bar{y} \in \bar{D}_0 + \bar{W}$ , by adding a suitable "correction" $t$ to $\bar{y}$ . We shall specify such a $t$ now.

Assume then that $\bar{y} \in \bar{D} \cap (\bar{D}_0 + \bar{W}) = \bar{D}_0 + \bar{W} \cap \bar{D}$ , say $y = \bar{d}_0 + \bar{w}$ . Let $\bar{G}$ be a supplement of $\bar{W} \cap \bar{D}_0^{\perp}$ in $\bar{W}$ spanned by a basis $\bar{g}_1, \bar{g}_2, \ldots, \bar{g}_r$ . We can find $d_0 \in D_0$ with $\Phi(d_0, T^{-1}g_i) = \Phi(\bar{d}_0, \bar{g}_i)$ $(i = 1, 2, \ldots, r)$ . The vector $x_0 := -x + d_0 + T^{-1}\bar{w}$ is in $W^{\perp}$ by (3'). Also by (3') $x_0 \notin W + D_0$ . Therefore, $\dim(D \cap W^{\perp})/((D_0+W) \cap D) \cap W^{\perp} \neq 0$. Hence we conclude by (5) that there exists a vector $t \in \bar{D} \cap \bar{W}^{\perp}$ with $t \notin \bar{D}_0 + \bar{W}$ . This is what we were looking for. Set $\bar{x} = \bar{y} + t$ and define $\bar{W}_1 := \bar{W} \oplus (\bar{x})$ . Extend $T$ to $W_1$ by sending $x$ into $\bar{x}$ .

We are left with the task (III) in the construction problem of Section 4. Let us consider (3') and $A$ an element of $V$ . If $A \in M$ then $D \subset A$ and thus $\bar{D} \subset \bar{A}$ by (7). Therefore, $\bar{x} \in \bar{A}$ and we find $W_1 \cap A = (W \cap A) \oplus (x)$ , $\bar{W}_1 \cap \bar{A} = (\bar{W} \cap \bar{A}) \oplus (\bar{x})$ . If, however, $D \not\subset A$ then $\bar{D} \not\subset \bar{A}$ by (7) and $W_1 \cap A = W \cap A$ , $\bar{W}_1 \cap \bar{A} = \bar{W} \cap \bar{A}$ . In both cases $T_1(W_1 \cap A) = \bar{W}_1 \cap \bar{A}$ .

In order to verify (4') let $(A_\iota)_I$ be an arbitrary family of elements in $V$ . If we should have that $A_\iota \in M$ for all $\iota$ then $\bigcap_I (W_1 + A_\iota) = \bigcap_I (W + A_\iota)$ and similarly for the $\bar{A}_\iota$ by (7) so we may quote the induction assumption. Assume therefore that $A_0 \notin M$ for at least one subscript $0 \in I$ . Let then $d = w_\iota + \lambda_\iota x + a_\iota$ $(\iota \in I)$ be a vector of the intersection $\bigcap_I (W_1 + A_\iota)$ . If $A_\mu \notin M$ for some $\mu \in I$ then $\lambda_\mu = \lambda_0$ for otherwise we could solve for $x$ and obtain $x \in W + (A_0 + A_\mu)$ so $A_0 + A_\mu \in M$ ; this is a contradiction since we assume in this section that $D$ is join-prime (i.e. $M$ is a prime filter). Thus $\lambda_\mu = \lambda_0$ for all $\mu \in I$ with $A_\mu \notin M$ and therefore $d - \lambda_0 x \in \bigcap_I (W + A_\iota) = W + \bigcap_I A_\iota$ . As the other inclusion $\bigcap_I (W_1 + A_\iota) \supseteq W_1 + \bigcap_I A_\iota$ is trivial we have established (4'). To verify $(\bar{4}')$ is mutatis mutandis the same.

We have shown: if $D$ is join-irreducible then the construction problem can be solved. Let us turn to

## 6. Solution of the construction problem in the reducible case (end of the proof of Theorem 1)

We keep the notations of the previous section and assume that $D(x,W)$ is join-reducible. By the fundamental assumption (2) therefore $D(x,W) = D_1 + \cdots + D_n$ $(n>1)$ for finitely many join-irreducibles $D_i \in \mathcal{V}$. Choose a representation where $n$ is minimal and write $x$ as a sum of components $x_i \in D_i$. If we let $D_{11} := D(x_1,W)$ then $D_{11} \subset D$ and $D_{11} + D_2 + \cdots + D_n \in \mathcal{M}(x,W)$. Therefore, $D_{11} + D_2 + \cdots + D_n = D_1 + \cdots + D_n$ and by modularity $D_{11} + D_1 \cap (D_2 + \cdots + D_n) = D_1$. We conclude that $D_{11} = D_1$ by the join-irreducibility of $D_1$ and by the minimality of $n$. This tells us that $D(x_1,W)$ is irreducible and that we may therefore by the previous section "adjoin" $x_1$ to $W$, $W_1 := W \oplus (x_1)$ and $\bar{W}_1 := \bar{W} \oplus (\bar{x}_1)$. Our argument can be repeated to show that $D(x_r,W_{r-1}) = D_r$ $(1 \leq r \leq n)$ so that by a n-fold application of the solution in the previous section the construction problem is solved in the present case.

## 7. Remarks on the case of not complete sublattices

At the end of Section 4 we have shown that (5) holds even if $\mathcal{V}$, $\bar{\mathcal{V}}$ are not complete sublattices but satisfy

$$(*) \qquad \text{for all } A \in \mathcal{V} : \dim A/\Sigma\{Z \in \mathcal{V} \mid Z \subsetneqq A\} = \dim A^{\tau}/\Sigma\{Z^{\tau} \mid Z \subsetneqq A\}$$

If, in addition, we assume that $\mathcal{V}$ satisfies the <u>descending chain condition</u> (DCC) then it is possible to maintain the proof of Sections 5 and 6. Conditions (4) and (4') have to be replaced by the corresponding conditions on <u>finite</u> families. $\mathcal{M}(x,W)$ of Section 5 will still be a sublattice of $\mathcal{V}$. By DDC we have an induction principle: if a subset $\mathcal{N}$

of $\mathcal{V}$ is such that it must contain X provided it contains all proper antecedents of X then $\mathcal{N}$ is all of $\mathcal{V}$ . Hence in solving the construction problem in Section 5 we may assume it to be solvable for all instances with generators D smaller than D(x,W) . This makes the transfer of the vector $\bar{y}$ in Section 5 practically trivial. We summarize:

Theorem 1'. Let E be an alternate space with $\dim E \leq \aleph_0$ and $\mathcal{V}$ , $\bar{\mathcal{V}}$ orthostable sublattices of $\mathcal{L}(E)$ . Assume that $\tau\colon \mathcal{V} \to \bar{\mathcal{V}}$ is a lattice isomorphism which commutes with orthogonal complementation and which preserves indices in the strong sense of (*) above. In order that $\tau$ be induced by an isometry $T\colon E \to E$ it is sufficient that $\mathcal{V}$ satisfy (1) and DCC.

It is not necessary to require (2) since in a lattice with DCC each element is the join of finitely many join-irreducibles.

## 8. Nonalternate forms : Theorem 2

The problem formulated in the introduction to the chapter becomes considerably more difficult if the form is not alternate. Results in Chapters XII and XIII show the complexity of the question for orthostable lattices as simple as [three-element chain diagram] .

Nevertheless a very useful theorem can be extracted from the proof of Theorem 1 for certain nonalternate forms. We continue to assume that the lattices satisfy conditions (1) and (2). (Nondistributive lattices are investigated in Chapter VIII for sesquilinear and in Chapter XVI for quadratic forms.) We now turn to the description of the additional difficulties that arise.

In what follows $\Phi\colon E \times E \to k$ is a nondegenerate $\varepsilon$-hermitean form with respect to some antiautomorphism $\kappa : k \longrightarrow k$ of the division ring $k$ ; $\dim E = \aleph_0$ . We consider the additive subgroup $T := \{\xi+\varepsilon\xi^* | \xi \in k\}$ of $k$ and for each subspace $X \subset E$ we set

$$X^* := \{x \in X | \Phi(x,x) \in T\} .$$

$X^*$ is the trace-valued part of $X$ and, obviously, $X^* = X \cap E^*$ . When $\Phi$ is not trace-valued, possible only when char $k = 2$ , then each isometry of $E$ respects the operation $X \mapsto X^*$ . It is therefore natural to require that the lattices $V$ and $\bar{V}$ contain the element $E^*$ and, as a matter of course, that the lattice isomorphism $\tau: V \to \bar{V}$ satisfy

(8) $$\tau(E^*) = E^* .$$

$V$ and $\bar{V}$ will then be stable under the operation $X \mapsto X^*$ and the latter commutes with $\tau$ . We assert

Theorem 2. Let $(E,\Phi)$ be a nondegenerate $\varepsilon$-hermitean space of dimension $\aleph_0$ and $\tau: V \to \bar{V}$ a lattice isomophism between complete sublattices of $L(E)$ (the lattice of all subspaces of $E$ ). Assume that $V$ and $\bar{V}$ are orthostable and that $\tau$ commutes with orthogonal complementation and satisfies (8) and preserves indices (dimensions of quotients of neighbouring elements in the lattices). In order that $\tau$ be induced by an isometry of $E$ it is sufficient that $V$ , $\bar{V}$ satisfy (1) and (2) and the following condition (C) on their join-irreducible compact elements $Y$ (we set $\operatorname{rad} Y := Y \cap Y^{\perp}$ ) :

(C)

(i) $\dim Y/\operatorname{rad} Y < \infty \Rightarrow \|Y\| = \{0\} = \|Y^{\tau}\|$ .

(ii) $\dim Y/\operatorname{rad} Y = \infty$ & $Y = Y^* \Rightarrow Y$ contains an infinite dimensional totally isotropic subspace $V$ with $V \cap \operatorname{rad} Y = (0)$ , and similarly for $Y^{\tau}$ .

(iii) $\dim Y/\operatorname{rad} Y = \infty$ & $Y \neq Y^* \Rightarrow \|Y\| = \|Y^{\tau}\|$ and for all finite dimensional $F \subset E$ we have $\|Y \cap F^{\perp}\| = \|Y\|$ , $\|Y^{\tau} \cap F^{\perp}\| = \|Y^{\tau}\|$ .

Remarks. 1. If char $k \neq 2$ then (iii) is vacuous; furthermore there is an impressive list of fields $k$ with antiautomorphism $\kappa$ such that there is precisely one isometry class of nondegenerate $\varepsilon$-hermitean $\aleph_0$-forms over $k$. Since each such form admits a symplectic basis we see that (ii) is automatically satisfied. Hence (i) is the only restriction in these cases.

2. For special lattices one can give better results. E.g. if $\mathcal{V}$ is the lattice $\{(0)\subset D^{\perp}\subset D = D^{\perp\perp}\subset E\}$ then every $\tau\colon \mathcal{V} \to \bar{\mathcal{V}}$ which is orthostable and preserves indices and has $\|D\| = \|D^{\tau}\|$ is induced by an isometry $T$ of $E$ (see [3], Sec. 2.5 for a proof). From this follow anew the result that every totally isotropic $\perp$-closed subspace in a non degenerate space $(E,\Phi)$ admits a metabolic decomposition no matter what the form or the field (as long as $\dim E \leq \aleph_0$).

## 9. Proof of Theorem 2

If $Y$ satisfies (ii) of (C) then for each finite dimensional $F \subset E$ the space $Y \cap F^{\perp}$ contains hyperbolic planes so that $\|Y\cap F^{\perp}\| = \mathcal{T}$ By Section 2 we know that every element $X \in \mathcal{V}$ is a join of join-irreducible compact elements $Y$, $X = \Sigma Y$, hence we have by (ii) and (iii) of (C) that

$$\|X\| = \|X^{\tau}\| \pmod{\mathcal{T}} \quad \text{for all } X \in \mathcal{V} . \tag{9}$$

In order to use the scheme of the proof of Thm. 1 we only have to arrange (in Section 5) for $\bar{x}$ to satisfy the additional requirement

$$\Phi(\bar{x},\bar{x}) = \Phi(x,x) \tag{10}$$

besides condition (6) and the condition

$$\Phi(\bar{x},Tw) = \Phi(x,w) \quad \text{for all } w \in W . \tag{11}$$

All this concerns (II) of the construction problem in Section 4.

We keep the notations of Sec. 5. In particular $D_0$ is the join of all proper antecedents of $D \in \mathcal{V}$. We have to show how to satisfy (10) when $D$ is join-irreducible. In this case $D_0$ is the immediate antecedent of $D$.

Case i. $\dim D/D \cap D^{\perp} < \infty$. By (Ci) of the Theorem both $x$ and $\bar{x}$ are isotropic and (10) holds.

Case ii. $\dim D/D \cap D^{\perp} = \infty$ and $D = D^*$. First pick some $\bar{x} \in D^{\tau}$ with (6) and (11) (such $\bar{x}$ exist by the arguments of Sec. 5). Now, if $D/D_0$ should be finite dimensional, then $\| D_0^{\tau} \cap F^{\perp} \| = T$ by (Cii). Hence there is $\bar{z} \in D_0^{\tau} \cap (\bar{W} + (\bar{x}))^{\perp}$ with $\Phi(\bar{z},\bar{z}) = \Phi(x,x) - \Phi(\bar{x},\bar{x})$ ; $\bar{x} + \bar{z}$ then satisfies (6), (10) and (11). If, on the other hand, $\dim D/D_0$ is infinite then we pick such a $\bar{z}$ in $D^{\tau}$ ; then $\bar{x} + \bar{z}$ satisfies (10) and (11) but may violate (6). Suppose we had $\bar{x} + \bar{z} \in \bar{W} + D_0^{\tau}$. It will be easy to manoeuvre the vector out of $\bar{W} + D_0^{\tau}$ since this latter is of infinite codimension in $D^{\tau}$. Let $S$ be a supplement of $D \cap D^{\perp}$ in $D$ that contains a subspace $V$ as specified in (Cii) of Thm. 2. The space $S_1 := S \cap (W \oplus (\bar{x}+\bar{z}))^{\perp}$ must contain hyperbolic planes as its radical is finite dimensional. Therefore, the isotropic vectors $s$ of $S_1$ generate the trace-valued $S_1$. For infinitely many $s$ we have $s \notin \bar{W} + D_0^{\tau}$ since otherwise $\dim D/D_0$ would be finite. For each such $s$, $\bar{x}+\bar{z}+s$ satisfies (6), (10) and (11).

Case iii. $\dim D/D \cap D^{\perp} = \infty$ and $D \neq D^*$. First pick some $\bar{x}_0 \in D^{\tau}$ with (6) and (11). By (Ciii) there is $\bar{z} \in D^{\tau} \cap (\bar{W} + (\bar{x}_0))^{\perp}$ with $\Phi(\bar{z},\bar{z}) = \Phi(x,x) - \Phi(\bar{x}_0,\bar{x}_0)$ (by (Ciii) the set $\|D\|$ is closed under addition and as $D \neq D^*$ we must have char $k = 2$, hence $\|D\|$ is an additive subgroup of $k$ ) so $\bar{x} := \bar{x}_0 + \bar{z}$ satisfies (10) and (11). We claim that it has to satisfy (6). For, assume that $\bar{x} = \bar{w}+\bar{d}_0 \in \bar{W}+D_0^{\tau}$. By (9) we may pick a vector $d_0 \in D_0$ with $\Phi(d_0,d_0) - \Phi(\bar{d}_0,\bar{d}_0) \in T$. Set $y := T^{-1}w+d_0$ ; $y \in W+D_0$ and $\Phi(y,y) \equiv \Phi(\bar{x},\bar{x}) \bmod T$ so $x-y \in D^*$. Since $D^* \subset D_0$ in the present case we obtain $x = (x-y)+y \in W+D_0$, contradiction.

The proof of Theorem 2 is therefore complete.

## 10. Remarks on the method

The method used to prove theorems 1 and 2 can often be applied to situations where not all join-irreducible compact elements of the lattice $V$ satisfy condition (C) of Section 8. This possibility stems from the fact that we need not start the recursive construction of the sequences $(W_i)_i$, $(\bar{W}_i)_i$, $T_i\colon W_i \to \bar{W}_i$ with $W_0 = (0) = \bar{W}_0$ in order to make the scheme of Sec. 3 work. In order to get off ground we may start with any finite dimensional $W_0$, $\bar{W}_0$ such that there is an isometry $T_0\colon W_0 \to \bar{W}_0$ with

(12) $$T_0\colon W_0 \cap A \to \bar{W}_0 \cap A \quad \text{for all } \bar{A} \in V$$

(13) $$\bigcap_{\iota} (W_0 + A_{\iota}) = W_0 + \bigcap_{\iota} A_{\iota}, \quad (A_{\iota} \in V)$$

(14) $$\bigcap_{\iota} (\bar{W}_0 + \bar{A}_{\iota}) = \bar{W}_0 + \bigcap_{\iota} \bar{A}_{\iota}, \quad (\bar{A}_{\iota} \in \bar{V})$$

The idea behind many applications of our method is to arrange the choice of the initial triple $(W_0, \bar{W}_0, T_0)$ in such a way that the compact join-irreducible elements $Y$ which do not satisfy condition (C) are excluded from the role as generators of the filters $M(x,W_n)$ associated with the construction problem in Sec. 4. We formulate this stratagem as

<u>Principle I</u>. Let $(E,\Phi)$ and $\tau: \mathcal{V} \to \bar{\mathcal{V}}$ be as in Theorem 2. Assume that there are finite dimensional spaces $W_0$, $\bar{W}_0 \subset E$ and an isometry $T_0: W_0 \to \bar{W}_0$ which satisfy (12), (13), (14) and the following condition

(15) if $Y$ is a join-irreducible compact element in $\mathcal{V}$ which does not satisfy condition (C) of Sec. 8 then there exists a subspace $H \subset W_0$ which is a linear supplement in $Y$ of the immediate antecedent $Y_0$ of $Y$.

Then $T_0$ can be extended to an isometry of $E$ which induces the lattice isomorphism $\tau$.

A typical application of this principle will be contained in the study of totally isotropic subspaces in Chapter VIII. In that chapter we shall also formulate a second principle which allows us to deal with situations where the lattices $\mathcal{V}$, $\bar{\mathcal{V}}$ are <u>not</u> distributive. It is based on the same idea as Principle I. The range of applications is thereby greatly increased.

We conclude with a remark. The formulation of principles such as the ones mentioned is certainly useful for the conveyance of certain general ideas that concern proofs which, by necessity, are loaded with details. However, according to needs, the strategy as exhibited may be varied in countless ways. In specific situations subtler observations can be made than is possible in a model situation as described in Thm. 2.

*

## References to Chapter IV

[1] R. Baer, Linear Algebra and Projective Geometry. Academic Press, New York 1952.

[2] H. Gross, On Witt's Theorem in the Denumerably Infinite Case. Math. Ann. 170 (1967) 145-165.

[3] H. Gross, Isomorphisms between lattices of linear subspaces which are induced by Isometries. J. Algebra 49 (1977) 537-546.

[4] H. Gross and H.A. Keller, On the definition of Hilbert Space. manuscripta math. 23 (1977) 67-90.

[5] C. Herrmann, On a condition sufficient for the distributivity of lattices of linear subspaces. To appear.

[6] P. Pudlak and J. Tůma, Yeast graphs and fermentation of algebraic lattices. Coll. Math. Soc. J. Bolyai, 14 (1976) Lattice Theory 301-342 ed. by A.P. Huhn and E.T. Schmidt, North Holland Publ. Company, Amsterdam.

## CHAPTER FIVE

## SUBSPACES IN TRACE-VALUED SPACES WITH MANY ISOTROPIC VECTORS

### 1. Introduction

The classical Theorem of Witt says that any isometry $T_0: F \to \bar{F}$ between finite dimensional subspaces $F, \bar{F}$ of a non degenerate trace-valued space $(E,\Phi)$ can be extended to an isometry $T: E \to E$ ([4], Satz 4 and Anmerkung p. 31).

If $\dim F$ is not restricted then this assertion becomes substantially false. Here is an illustration. If $\dim E$ is infinite and non degenerate then it contains subspaces $F \neq E$ with $F^{\perp} = (0)$. If we assume that $\dim E$ is countable then a simple recursive argument will show that $F$ splits, $F = U \overset{\perp}{\oplus} V$, in such a fashion that $U^{\perp} = V$, $V^{\perp} = U$ in $E$. Define $T_0: F \to F$ by setting $T_0 u = \lambda u$, $T_0 v = -\lambda^* v$ $(u \in U,\ v \in V)$ for suitably chosen $\lambda$ in the field $k$. If $\operatorname{char} k \neq 2$ let $\lambda = 1$. If $\operatorname{char} k = 2$ we have to assume that the involution $*$ of $k$ is of the "second kind", i.e. there is an element $\mu$ in the center with $\mu^* \neq \mu$; we then set $\lambda = \mu^{-1}\mu^*$. $T_0$ is then an isometry and it cannot be extended to any proper overspace of $F$ in $E$.

Thus it may well happen that a specific isometry $T_0$ between $F$ and $\bar{F}$ cannot be extended to $E$ whereas some other isometry $T_1: F \to \bar{F}$ does have an extension. In other words we have here two different problems.

Problem 1. For given $F, \bar{F} \subset E$ list necessary and sufficient conditions for an isometry $T: E \to E$ to exist with $TF = \bar{F}$. In other words, give complete sets of invariants for the orbits of subspaces under the orthogonal group of $E$.

<u>Problem 2.</u> Describe conditions which are sufficient for a given isometry $T_0: F \to \bar{F}$ to admit an extension to all of $E$.

<u>Remark.</u> In finite dimensions Witt's Theorem is <u>equivalent</u> with the apparently weaker statement "If $F, \bar{F}$ are isometric then they belong to the same orbit". For the last statement obviously implies the Cancellation Theorem. Hence in order to extend a given $T_0: F \to \bar{F}$ one simply decomposes $F = F_0 \oplus \text{rad } F$ and has $F_0^\perp \cong (T_0F_0)^\perp$. Since there are Witt decompositions of $F_0^\perp$ for $\text{rad } F$ and of $(T_0F_0)^\perp$ for $\text{rad } \bar{F}$ it is evident that $T_0$ can be extended to $E$.

We consider two extreme situations of Problem 1. In this chapter we solve it for spaces which contain "sufficiently" many isotropic vectors. In Chapters XII and XIII we investigate it for definite spaces over ordered fields. Of Problem 2 we treat in Chapter X.

## 2. Classification of a single subspace

The non degenerate sequilinear form $\Phi: E \times E \to k$ is assumed to be trace-valued and $\dim E = \aleph_0$. No assumptions on the field are made.

*

A nice application of Theorem 2 in Chap. IV is to the lattice $\mathcal{V}$ orthostably generated by a single subspace $V \subset E$

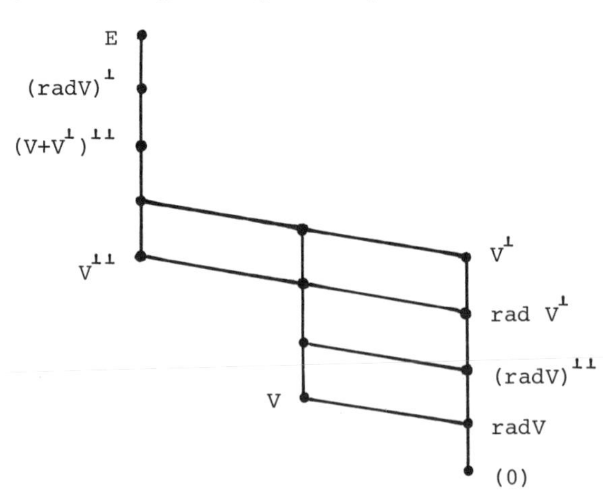

$\mathcal{V}(V)$ is distributive and has 14 elements ([3]). All indices that can be read off from the above lattice occur among the following seven:

$$\begin{array}{ll} & \dim[\text{rad } V] \text{ , } \dim[V/\text{rad } V] \text{ , } \dim[(\text{rad } V)^{\perp\perp}/\text{rad } V] \text{ ,} \\ (1) & \dim[V^{\perp}/\text{rad}(V^{\perp})] \text{ , } \dim[\text{rad}(V^{\perp})/(\text{rad}V)^{\perp\perp}] \text{ , } \dim[(V^{\perp}+V^{\perp\perp})/(V^{\perp}+V)], \\ & \dim[(\text{rad } V^{\perp})^{\perp}/(V^{\perp}+V^{\perp\perp})] \text{ .} \end{array}$$

<u>Remark.</u> In the case of an $\aleph_0$-dimensional $E$ we have $\dim[E/(\text{rad } V)^{\perp}] = \dim[\text{rad } V]$ and $\dim[(\text{rad } V)^{\perp}/(V+V^{\perp})^{\perp\perp}] = \dim[\text{rad}(V^{\perp})/(\text{rad}V)^{\perp\perp}]$ . For $E$ of larger dimension $\dim[E/(\text{rad } V)^{\perp}]$ and $\dim[(\text{rad } V)^{\perp}/(V+V^{\perp})^{\perp\perp}]$ have to be added to the above seven in order to obtain a complete list of indices. It is not difficult to construct examples where all these indices are nonzero.

<u>Theorem 1</u> ([1]). Let $V$ , $\bar{V}$ be isometric subspaces of $E$ which satisfy the following conditions

(i) : $V^{\perp}$ and $\bar{V}^{\perp}$ are isometric,

(ii) : $\dim[(\text{rad } V)^{\perp\perp}/\text{rad } V] = \dim[(\text{rad } \bar{V})^{\perp\perp}/\text{rad } \bar{V}]$ ,

(iii) : $\dim[\text{rad}(V^{\perp})/(\text{rad } V)^{\perp\perp}] = \dim[\text{rad}(\bar{V}^{\perp})/(\text{rad } \bar{V})^{\perp\perp}]$ ,

(iv) : $\dim[V^{\perp\perp}/(V+\text{rad}(V^{\perp}))] = \dim[\bar{V}^{\perp\perp}/(\bar{V}+\text{rad}(\bar{V}^{\perp}))]$ ,

(v) : $\dim[(\text{rad}(V^{\perp}))^{\perp}/(V^{\perp}+V^{\perp\perp})] = \dim[(\text{rad}(\bar{V}^{\perp}))^{\perp}/(\bar{V}^{\perp}+\bar{V}^{\perp\perp})]$ .

There exists an isometry of $E$ which maps $V$ onto $\bar{V}$ whenever the following (sufficient) conditions are satisfied:

(vi) : if $\dim V/\text{rad } V$ is infinite then $V$ contains an infinite dimensional totally isotropic $W$ with $W \cap \text{rad } V = (0)$ ,

(vii) : if $\dim V^{\perp}/\text{rad}(V^{\perp})$ is infinite then $V^{\perp}$ contains an infinite dimensional totally isotropic $W$ with $W \cap \text{rad}(V^{\perp}) = (0)$ .

<u>Proof.</u> We first consider the possibility that $\dim V/\text{rad } V$ is finite. Let $V_0$ be a supplement of $\text{rad } V$ in $V$ . Since $V \cong \bar{V}$ by

assumption there is a supplement $\bar{V}_0$ of rad $\bar{V}$ in $\bar{V}$ which is isometric to $V_0$ . $E = V_0 \oplus V_0^{\perp} = \bar{V}_0 \oplus \bar{V}_0^{\perp}$ and by Witt's Theorem of the finite dimensional case (here we use trace-valuedness of E) we get $V_0^{\perp} \cong \bar{V}_0^{\perp}$ . One easily verifies that the assumptions of the theorem on V and $\bar{V}$ in E imply the corresponding assumptions on rad V , rad $\bar{V}$ in $V_0^{\perp}$ and $\bar{V}_0^{\perp}$ respectively. Hence the argument does in fact reduce the problem to the case where $\dim V/\text{rad } V = 0$ . A similar argument applies to $V^{\perp}$ so that we may assume without loss of generality that $\dim V/\text{rad } V$ , $\dim V^{\perp}/\text{rad}(V^{\perp}) \in \{0, \aleph_0\}$ . We may, of course, also assume that dim V , dim $\bar{V}$ are infinite by Witt's Theorem. Hence each Y among the join-irreducibles in $V$ (the elements labelled in the diagram) has $\dim Y/\text{rad } Y \in \{0, \aleph_0\}$ . Thus we see that condition (C) of Theorem 2 in Chap. IV Sect. 8 is satisfied. Since $V$ is finite and distributive we may quote Thm. 2 to finish our proof.

Corollary 1. Let E be a trace-valued sesquilinear space with contains an infinite dimensional totally isotropic subspace. Then for each $n \in \mathbb{N}$ the set of $\perp$-dense subspaces V with $\dim E/V = n$ forms a single orbit under the orthogonal group of E .

Proof. V and $\bar{V}$ are non degenerate if $\perp$-dense and contain infinite dimensional totally isotropic subspaces if $\dim E/V$ , $\dim E/\bar{V}$ are finite. Hence any such V , $\bar{V}$ are isometric because they are hyperbolic. We may now quote the above theorem.

A particularly nice application is obtained if we consider alternate forms. Isometries $V \cong \bar{V}$ and $V^{\perp} \cong \bar{V}^{\perp}$ can be expressed in terms of dimensions, to wit indices,

$$\dim(\text{rad } V) = \dim(\text{rad } \bar{V}) \ , \ \dim V/\text{rad } V = \dim \bar{V}/\text{rad } \bar{V} \ ,$$
$$\dim V^{\perp}/\text{rad } V^{\perp} = \dim \bar{V}^{\perp}/\text{rad } \bar{V}^{\perp} \ , \ \dim(\text{rad } V^{\perp}) = \dim(\text{rad } \bar{V}^{\perp}) \ .$$

Because (vi) and (vii) of the theorem are empty in the alternate case we see that the indices of $\mathcal{V}(V)$ fix $V$ in $E$ up to metric automorphisms of $E$ :

<u>Corollary 2</u>. If $E$ is alternate then the set of indices of $\mathcal{V}(V)$ is a complete set of orthogonal invariants for the orbit (under the orthogonal group of $E$ ) of the (arbitrary) subspace $V$ .

Corollary 2 becomes false if $\dim E > \aleph_0$ ; a counterexample is given in [2] (Sec. VI, pp. 131-134). See also [0] for the uncountable situation. The special case of a $\perp$- dense $V$ in Corollary 2 was treated by Kaplansky ([3] , Theorem 6. See also question 3 on p. 16 ).

## 3. An application to Witt decompositions

Let $R$ be the radical of a subspace $V \subset E$ . Does there exist an orthogonal decomposition

(2) $$E = E_1 \overset{\perp}{\oplus} E_2 \quad \text{with} \quad R \subset E_1 \ , \ V = R \oplus (V \cap E_2) \quad ?$$

If $R$ is $\perp$- closed the answer is positive since then $E$ has a metabolic decomposition for the subspace $R$ , $E = (R \oplus R') \overset{\perp}{\oplus} E_2$ ; therefore $V \subset R^{\perp} = R \overset{\perp}{\oplus} E_2$ and thus $V = R \oplus (V \cap E_2)$ .

We shall now prove that there is a decomposition (2) even when $R \neq R^{\perp\perp}$ provided that conditions (vi) and (vii) in Theorem 1 are fulfilled. (Bäni pointed out that this proviso is unnecessary; see the postscript p. 135 .)

<u>Theorem 2</u>. Let $(E,\Phi)$ be of dimension $\aleph_0$ , trace-valued and non degenerate. Let $V = R_0 \oplus V_0$ be a subspace that satisfies (vi) and (vii) in Theorem 1 and has $R_0 \subset \operatorname{rad} V$ . There is another decomposition $V = F \oplus G$ with $F \subset \operatorname{rad} V$ , $\dim F = \dim R_0$ and $G$ isometric to $V_0$ such that $E$ splits as follows: $E = E_1 \overset{\perp}{\oplus} E_2$ , $F \subset E_1$ , $G \subset E_2$ .

Corollary. Let $(E,\Phi)$ be as in the Theorem and $R_0$ the radical of a subspace $V \subset E$ that satisfies (vi) and (vii) in Theorem 1. There always exists a Witt decomposition of $E$ for the totally isotropic space $R_0^{\perp\perp}$, $E = (R_0^{\perp\perp} \oplus R') \overset{\perp}{\oplus} E_2$ such that $V = R_0 \oplus (V \cap E_2)$.

Proof of the Corollary. By Thm. 2 we have a Witt decomposition $E = (F^{\perp\perp} \oplus R') \overset{\perp}{\oplus} E_2$ and $V = F \oplus G$, $F \subset R_0$, $G \subset E_2$, $G \cong V_0$ for $V_0$ any supplement of $R_0$ in $V$. Therefore $F = R_0$.

Proof of the Theorem. There is no problem when $\dim R_0$ is finite so let $\dim R_0 = \aleph_0$. There is a Witt decomposition $E = (R^{\perp\perp} \oplus R') \overset{\perp}{\oplus} E_0$ $R = \text{rad } V$. Let $V_1$ be a supplement of $R$ in $V$. The projection $V_2$ of the space $V_1 (\subset R^{\perp\perp} \oplus E_0)$ onto $E_0$ is injective and, of course, isometric to $V_1$. Set $\bar{V} := R \oplus V_2$. We have $\text{rad } V = \text{rad } \bar{V}$, furthermore $V^{\perp} = R^{\perp} \cap V_1^{\perp} = (R^{\perp\perp} \oplus E_0) \cap V_1^{\perp} = R^{\perp\perp} + (E_0 \cap V_1^{\perp}) = R^{\perp\perp} + (E_0 \cap V_2^{\perp}) =$ $= \bar{V}^{\perp}$. Hence the lattices $\mathcal{V}(V)$ and $\mathcal{V}(\bar{V})$ (Sec. 2) have all irreducible elements in common except possibly for $V$, $\bar{V}$. We may quote Theorem 1: There is an isometry $T$ of $E$ which maps $\bar{V}$ onto $V$. Thus we may assume from the outset that $V_1 \subset E_0$ for a suitable Witt decomposition.

It remains to locate $R_0$ in the space $E' := R^{\perp\perp} \oplus R'$. The orthostable lattice generated by $R \subset E'$ is the chain $(0) \subset R \subset R^{\perp} =$ $= R^{\perp\perp} \subset E'$. $\dim R^{\perp\perp}/R$ fixes $R$ in $E'$ modulo metric automorphisms (Theorem 1). One can easily give canonical bases: If $E'$ is spanned by a symplectic basis $\{e_i, e_i'\}_{i \in \mathbb{N}}$ and $R$ by all $e_i - e_1$ then we have an example with $\dim R^{\perp\perp}/R = 1$. The external orthogonal sum of $C$ copies of this kind yields an example with $\dim R^{\perp\perp}/R = C$. By the uniqueness just mentioned (Theorem 1) all instances are of this kind, i.e. there always are symplectic bases in $E'$ which exhibit $R$ in this way (cf. the following section). In the same manner one can give an example where $R = F \oplus R_1$, $\dim F = \dim R_0$ $(= \aleph_0)$, $\dim R_1 = \dim R/R_0$

and where $F$ and $R_1$ can be "separated" in $E'$, $E' = E_1 \overset{\perp}{\oplus} E_3$, $F \subset E_1$, $R_1 \subset E_3$. If we set $E_2 := E_0 \oplus E_3$ and $G := R_1 \oplus V_1$ we have proved Theorem 2.

To round off the picture we contrast the Corollary with

(3) If for <u>each</u> supplement $V_0$ of $\mathrm{rad}\, V$ in $V$ there is a splitting $E = E_1 \overset{\perp}{\oplus} E_2$ with $\mathrm{rad}\, V \subset E_1$ and $V_0 \subset E_2$ then $\dim V/\mathrm{rad}\, V$ is finite.

<u>Proof.</u> Assume that $\dim V_0$ is infinite. We can pick a sequence $(g_i)_{i\in\mathbb{N}}$ in $V_0$ of linearly independent vectors that converges to zero with respect to the weak linear topology $\sigma(\Phi)$. We complete it to a basis $(e_i)_{i\in\mathbb{N}}$ of $V_0$. For $r$ a fixed nonzero element of $\mathrm{rad}\, V$ the family $(e_i+r)_{r\in\mathbb{N}}$ spans another supplement $V_1$ of $\mathrm{rad}\, V$ in $V$. The $g_i + r$ converge to $r$, i.e. $r \in V_1^{\perp\perp} \setminus V_1$. Since $\mathrm{rad}\, V \cap V_1^{\perp\perp} \neq (0)$ the space $E$ cannot split in the required fashion, obviously.

## 4. <u>Remarks on canonical bases</u>

In explicit calculations it is often advantageous to introduce canonical bases. We shall give two examples of what we mean.

<u>Example 1</u>. Let $E$ be spanned by an orthogonal basis $(e_i)_{\mathbb{N}}$. The hyperplane $V = k(e_i - e_0)_{\mathbb{N}}$ has $V^{\perp} = (0)$. Assume that $V$ contains an infinite dimensional totally isotropic subspace and $\mathrm{char}\, k \neq 2$. Then if $\bar{V}$ is any other hyperplane in $E$ with $\bar{V}^{\perp} = (0)$ there is by Theorem 1 an isometry $T: E \to E$ with $TV = \bar{V}$. If we set $\bar{e}_i = Te_i$ we see that $(\bar{e}_i)_{\mathbb{N}}$ is an orthogonal basis of $E$ such that $(\bar{e}_i - \bar{e}_0)_{\mathbb{N}}$ spans $\bar{V}$: All $\perp$-dense hyperplanes in $E$ can be exhibited by introducing a basis of this type.

The same argumentation shows that if $V$ has $V^{\perp} = (0)$ and $\dim E/V = \mathfrak{a} \neq 0$ then $E$ is an orthogonal sum of $\mathfrak{a}$ copies $E_1$ of $E$

and $V$ is an orthogonal sum of $\alpha$ hyperplanes $V_\iota \subset E_\iota$, $V_\iota^\perp = (0)$.

Whenever the "arithmetical" situation is such that we can apply the "uniqueness" theorem of the previous section then we can also introduce bases of the space which "exhibit" the subspace in some standard fashion.

Example 2. The same idea applies to situations where we can apply Theorem 2 of Chap. IV. Let $J \subset \mathbb{N}$ and $\varphi_i : I_i \to J$ ($i \in \mathbb{N}$) be a family of epimorphic maps where $\{I_i | i \in \mathbb{N}\}$ is a partitioning of $\mathbb{N}$ with card $I_i = \mathbb{N}$. Let $E = (R \oplus R') \overset{\perp}{\oplus} E_0$ be a Witt decomposition of $E$ for the totally isotropic subspace $R$ and $(r_i)_{i\in\mathbb{N}}$, $(r_i')_{i\in\mathbb{N}}$, $(e_i)_{i\in J}$ bases of $R$, $R'$, $E_0$ respectively with $\Phi(r_i, r_j') = 1$, $\Phi(e_i, e_i) \neq 0$ and all other products zero.

Let $W$ be the span of all families of vectors $(r_1 + r_i' + e_{\varphi_1(i)})_{i\in I_1}$, $(r_2 + r_i' + e_{\varphi_2(i)})_{i\in I_2}$, ... .

It is routine to verify that

$$(4) \qquad W^\perp = (0), \quad W \cap R^\perp = (0), \quad W + R^\perp = E,$$

in other words, the $\perp$-stable lattice $V = V(R,W)$ generated by $R$ and $W$ is

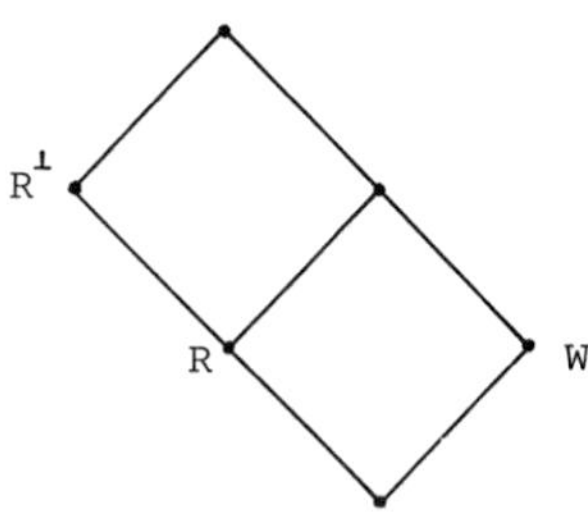


Assume e.g. that the field is algebraically closed and $\Phi$ is symmetric, then Theorem 2 of Chap. IV can be applied to $V$ and any $\bar{V} = V(R,\bar{W})$ where $\bar{W}$ satisfies (4). In other words, each $\perp$-dense

supplement $\bar{W}$ of $R^{\perp}$ in $E$ is spanned in the above fashion.

*

References to Chapter V

[0] P. Amport, Teilraumverbände in überabzählbar dimensionalen Sesquilinearräumen. Ph.D.Thesis Univ. of Zurich 1978.

[1] H. Gross, On Witt's Theorem in the Denumerably Infinite Case. Math. Ann. 170 (1967) 145-165.

[2] -, Der euklidische Defekt bei quadratischen Räumen. Math. Ann. 180 (1969) 95-137.

[3] I. Kaplansky, Forms in infinite dimensional spaces. An. Acad. Bras. Ci. 22 (1950) 1-17.

[4] E. Witt, Theorie der quadratischen Formen in beliebigen Körpern. J. reine angew. Math. 176 (1937) 31-44.

*

Postscript. This concerns the corollary to Theorem 2 on page 132. By refining our method in Chapter IV W. Bäni has been able to eliminate (vi) and (vii) from the corollary so that there always exists a decomposition (2) (p. 131). See his forthcoming paper "Inner product spaces of infinite dimension; on the lattice method".

## CHAPTER SIX

## ORTHOGONAL AND SYMPLECTIC SEPARATION

### 1. Introduction

When one has to handle two mutually perpendicular subspaces $F$ and $G$ in a sesquilinear space $(E,\Phi)$ it is often a great advantage if $E$ splits orthogonally such that $F$ and $G$ are contained in summands,

$$(0) \qquad E = E_1 \overset{\perp}{\oplus} E_2 \ , \quad F \subset E_1 \ , \quad G \subset E_2$$

If this happens then we say that $F$ and $G$ can be orthogonally separated in $E$. From (0) we read off that

$$(1) \qquad (F+G)^{\perp\perp} = F^{\perp\perp} + G^{\perp\perp} \ ,$$

$$(2) \qquad F^{\perp} + G^{\perp} = E \ .$$

Orthogonal separation presents no problem if $\dim E$ is finite. Here we shall treat the case of trace-valued spaces in dimension $\aleph_0$. We shall prove that (1) & (2) is sufficient for a splitting (0) to exist provided that there are enough isotropic vectors available in $E$. This proviso is not due to the fact that our proof will be via Theorem 2 of Chap. IV; if isotropic vectors do not abound then (1) & (2) actually ceases to be sufficient for the existence of a splitting. ([2] contains an example; for further details see the postscript on p. 150 )

The problem of orthogonal separation is equivalent to the following problem if the characteristic is not two: Let $T_0 : H \to H$ be a metric involution $(T_0^2 = \mathbb{1})$ on the subspace $H$ of $E$; does $T_0$ admit an involutory extension $T$ to all of $E$? The answer is positive if and only if $F := \mathrm{Ker}(T_0 - \mathbb{1})$ and $G := \mathrm{Ker}(T_0 + \mathbb{1})$ can be orthogonally separated in $E$.

As remarked our proof will be by Theorem 2 of Chap. IV. In the next section we compute the lattice $\mathcal{V}(F,G)$ generated by an orthogonal pair $F$, $G$ that satisfies (1) and (2). In order to separate $F$ and $G$ in $E$ we shall construct an isometric space $\bar{E}$ with subspaces $\bar{F}$, $\bar{G}$ that are separated in $\bar{E}$ and such that there is a lattice isomorphism $\tau\colon \mathcal{V}(F,G) \to \mathcal{V}(\bar{F},\bar{G})$ which sends $F$ in $\bar{F}$ and $G$ in $\bar{G}$ and which qualifies for Theorem 2: if $\tau$ is induced by $T\colon E \to \bar{E}$ then $T^{-1}$ will transport the splitting of $\bar{E}$ into a splitting that separates $F$ and $G$ in $E$. The proof will make it evident that a fully worked out diagram of the lattice $\mathcal{V}(F,G)$ is indispensable both for finding and for grasping the arguments.

In the last section we treat an analogous problem termed symplectic separation. It has a geometric reformulation quite similar to that of orthogonal separation and will be needed in Chapter X.

## 2. On the lattice $\mathcal{V}(F,G)$ of an orthogonal pair

Let $\mathcal{V}(F)$ and $\mathcal{V}(G)$ be Kaplansky's lattices orthostably generated by $F$ and $G$ respectively in the lattice $\mathcal{L}(E)$ of all linear subspaces of $E$ (Chap. V Sec. 2). Decompose $\mathcal{V}(F) = \mathcal{J}_1 \cup \mathcal{F}_1$ where

$$(3)\qquad \begin{aligned} \mathcal{J}_1 &:= \{F^{\perp\perp},\ F+\operatorname{rad}(F^{\perp}),\ \operatorname{rad}(F^{\perp}),\ F+(\operatorname{rad} F)^{\perp\perp},\ (\operatorname{rad} F)^{\perp\perp},\ F,\\ &\qquad \operatorname{rad} F,\ (0)\}\\ \mathcal{F}_1 &:= \{F^{\perp},\ F+F^{\perp},\ F^{\perp\perp}+F^{\perp},\ (F+F^{\perp})^{\perp\perp},\ (\operatorname{rad} F)^{\perp},\ E\} \end{aligned}$$

$\mathcal{J}_1$ is the principal ideal generated by $F^{\perp\perp}$ in $\mathcal{V}(F)$ ; $\mathcal{F}_1$ is the principal filter generated by $F^{\perp}$ . Let $\mathcal{V}(G) = \mathcal{J}_2 \cup \mathcal{F}_2$ be the analogous decomposition of $\mathcal{V}(G)$ . Then the following lemma is very easy to prove.

<u>Lemma 1.</u> Let $\mathcal{V}(F,G)$ be the orthostable lattice generated in $\mathcal{L}(E)$ by a pair $F$ , $G$ of subspaces of $E$ . If $F$ and $G$ are orthogonally separated in $E$ then $\mathcal{V}(F,G)$ is the set $\mathcal{J} \cup \mathcal{F}$ where

(4) $$\mathcal{J} := \{X_1 + X_2 | X_i \in \mathcal{J}_i \ (i=1,2)\} \ ;$$
$$\mathcal{F} := \{Y_1 \cap Y_2 | Y_i \in \mathcal{F}_i \ (i=1,2)\} \ .$$

In particular card $V(F,G) \leq 8^2 + 6^2 = 10^2$. Furthermore $V(F,G)$ is distributive. There are examples where all 100 elements are different.

We now prove

Theorem 1. Let $(E,\Phi)$ be a nondegenerate sesquilinear space of arbitrary dimension. If two subspaces $F,G \subset E$, $F \perp G$, satisfy (1) and (2) then the lattice $V(F,G)$ is distributive and $V(F,G) = \mathcal{J} \cup \mathcal{F}$.

Proof. 1(Distributivity). We first show that the lattice generated in $L(E)$ by $V(F) \cup V(G)$ is distributive whenever $F \perp G$. By Theorem 6 of [5] and symmetry it is sufficient to verify that $(B+B') \cap C = (B \cap C) + (B' \cap C)$ for all $B,B' \in V(F)$ and all $C \in V(G)$. Since $F \perp G$ we have $Y \supset F^{\perp} \supset G^{\perp\perp} \supset X$ for all $X \in \mathcal{J}_2$, $Y \in \mathcal{F}_1$. This and the symmetric fact we express by

(5) $$\mathcal{J}_1 \leq \mathcal{F}_2 \ , \ \mathcal{J}_2 \leq \mathcal{F}_1 \ .$$

The only elements in $V(F)$ which are not join-irreducible are $Z_1 = F + (F \cap F^{\perp})^{\perp\perp}$, $Z_2 = F + (F^{\perp} \cap F^{\perp\perp})$, $Z_3 = F + F^{\perp}$, $Z_4 = F^{\perp\perp} + F^{\perp}$. When $i = 3, 4$ we obtain the distributivity of $Z_i \cap Y$ using (5) and modularity. The same works for $i = 1, 2$ and $Y \in \mathcal{F}_2$. Finally (5) implies that $Y = F^{\perp} \cap Y$ for $Y \in \mathcal{J}_2$ therefore

$$Z_1 \cap Y = [F+(F \cap F^{\perp})^{\perp\perp}] \cap F^{\perp} \cap Y = (F \cap F^{\perp})^{\perp\perp} \cap Y \subset (F \cap Y) + [(F \cap F^{\perp})^{\perp\perp} \cap Y]$$
$$Z_2 \cap Y = [F+(F^{\perp} \cap F^{\perp\perp})] \cap F^{\perp} \cap Y = F^{\perp} \cap F^{\perp\perp} \cap Y \subset (F \cap Y) + [(F^{\perp} \cap F^{\perp\perp}) \cap Y]$$

This proves our contention bearing in mind the distributive inequality. We see in particular that the set $\mathcal{J} \cup \mathcal{F}$ is a sublattice of $L(E)$.

2(Orthostability of $\mathcal{J} \cup \mathcal{F}$). We prove that

(6) $$X_1^{\perp\perp} + X_2^{\perp\perp} = (X_1 + X_2)^{\perp\perp} \quad \text{for all } X_i \in \mathcal{J}_i \ .$$

First we remark that by (2), (5) and modularity $X_1^\perp = X_1^\perp \cap (F^\perp + G^\perp) = F^\perp + (X_1^\perp \cap G^\perp) = F^\perp + (X_1^{\perp\perp} + G^{\perp\perp})^\perp$ so that (taking orthogonals) we obtain

$$(7) \qquad X_1^{\perp\perp} = (X_1^{\perp\perp} + G^{\perp\perp})^{\perp\perp} \cap F^{\perp\perp} .$$

Now we turn to (6). Since $F^{\perp\perp} + G^{\perp\perp}$ is closed by (1) we may write $(X_1^{\perp\perp} + X_2^{\perp\perp})^{\perp\perp} = (X_1^{\perp\perp} + X_2^{\perp\perp})^{\perp\perp} \cap (F^{\perp\perp} + G^{\perp\perp}) \leq (X_1^{\perp\perp} + G^{\perp\perp})^{\perp\perp} \cap (F^{\perp\perp} + G^{\perp\perp}) = [(X_1^{\perp\perp} + G^{\perp\perp})^{\perp\perp} \cap F^{\perp\perp}] + G^{\perp\perp}$ (the last equality by modularity). Hence by (7) we see that $(X_1^{\perp\perp} + X_2^{\perp\perp})^{\perp\perp} \leq X_1^{\perp\perp} + G^{\perp\perp}$. By a symmetric argumentation also $(X_1^{\perp\perp} + X_2^{\perp\perp})^{\perp\perp} \leq X_2^{\perp\perp} + F^{\perp\perp}$. Thus $(X_1^{\perp\perp} + X_2^{\perp\perp})^{\perp\perp} \leq (X_1^{\perp\perp} + G^{\perp\perp}) \cap (X_2^{\perp\perp} + F^{\perp\perp}) = X_1^{\perp\perp} + X_2^{\perp\perp}$ which establishes (6).

Let us now prove orthostability of $\mathcal{J} \cup \mathcal{F}$. Since $\mathcal{J}^\perp \subset \mathcal{F}$ is trivial it remains to show that

$$(8) \qquad (Y_1 \cap Y_2)^\perp \in \mathcal{J} \quad \text{for all} \quad Y_i \in \mathcal{F}_i$$

As a consequence of (6) we see that (8) holds for all cases $Y_i = X_i^\perp$ with $X_i \in \mathcal{J}_i$. Direct inspection shows that all $Y_1 \in \mathcal{F}_1$ which are not of this shape are of the kind

$$(9) \qquad Y_1 = X_1' + X_1^\perp \qquad (X_1 , X_1' \in \mathcal{J}_1) .$$

By (5) and modularity we obtain for such a $Y_1$ and for $Y_2 = X_2^\perp$ $(X_2 \in \mathcal{J}_2)$ that $(Y_1 \cap Y_2)^\perp = (X_1' + (X_1^\perp \cap X_2^\perp))^\perp = (X_1'^\perp \cap X_1^{\perp\perp}) + X_2^{\perp\perp} \in \mathcal{J}$. If $Y_2$ is of shape (9) as well then the calculations immediately reduce to the case just treated.

As $\mathcal{J} \cup \mathcal{F}$ is a lattice and orthostable it is the lattice $\mathcal{V}(F,G)$. The proof of Theorem 1 is thus complete.

Remarks. 1) In [4] we have computed the orthostable lattice $\mathcal{V}(F,G)$ of an orthogonal pair under conditions considerably more general than (1) and (2). First of all the lattice $\mathcal{V}$ generated by $\mathcal{V}(F) \cup \mathcal{V}(G)$

(no $\perp$-stability assumed) is distributive and has at most 258 elements under the sole assumption that $F \perp G$. In order that $V$ or a small extension of it be orthostable it is necessary to introduce some restrictions such as we have here by dint of (1) and (2). Proofs can be formulated in any modular lattice with an antitone mapping $\perp$.

2) The crucial point in the above proof is (6). With a knack for topological arguments one might reason as follows. The assertion certainly holds for $X_1 = F^{\perp\perp}$ and $X_2 = G^{\perp\perp}$ by (1). $(X_1+X_2)^{\perp\perp}$ is the closure of $X_1 + X_2$ with respect to the weak linear topology $\sigma(\Phi)$ on $E$. Since $F^{\perp\perp} + G^{\perp\perp}$ is closed we have that $(X_1+X_2)^{\perp\perp}$ is the closure of $X_1 + X_2$ with respect to the restriction $\sigma_1 = \sigma|_{F^{\perp\perp}+G^{\perp\perp}}$. But (2) tells that $\sigma_1$ is the product topology of $\sigma|_{F^{\perp\perp}}$ and $\sigma|_{G^{\perp\perp}}$. Since $X_1 \subset F$, $X_2 \subset G$ we conclude (6).

3) Theorem 1 was first proved in [1] by setting up recursively an orthogonal splitting (0). The fruitfulness of the method applied here is based on a reversal of steps. First lattices are computed and then general theorems on induced lattice isomorphisms are applied.

For the discussion of orthogonal separation we need only consider the case of $\perp$-closed subspaces $F$ and $G$. The lattice $V(F,G)$ then looks as follows

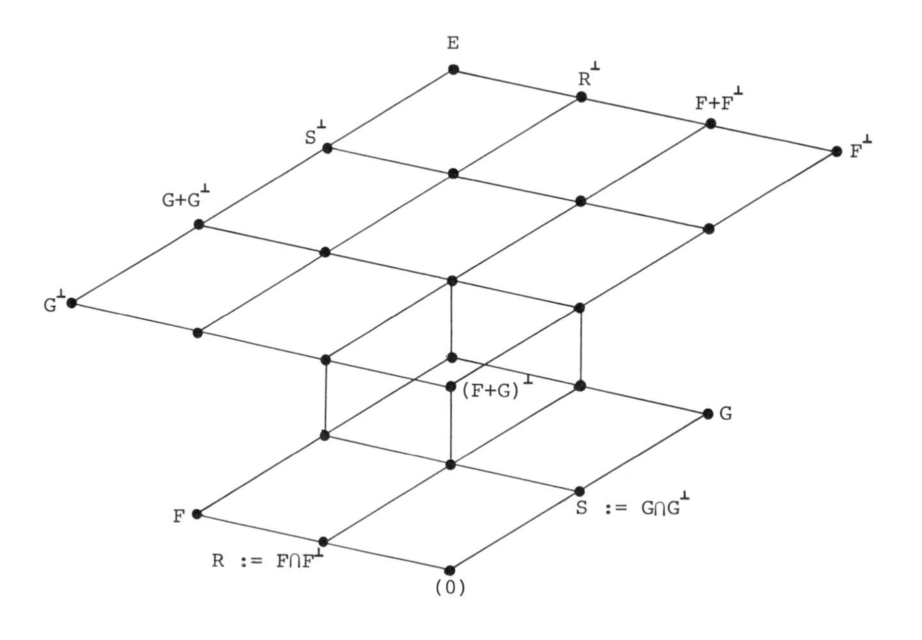


The orthostable lattice of a pair $F \perp G$ with $F^{\perp\perp} = F$, $G^{\perp\perp} = G$, $(F+G)^{\perp\perp} = F + G$, $F^{\perp} + G^{\perp} = E$.

## 3. Orthogonal separation in trace-valued spaces

In this section $(E,\Phi)$ will be nondegenerate, trace-valued and of dimension $\aleph_0$.

*

We now prove the following

Theorem 2 [(1)]. Assume that the subspaces $F, G \subset E$ satisfy (1), (2) and $F \perp G$. In order that $F$ and $G$ are orthogonally separated in $E$, i.e. that there is a decomposition (0), it is sufficient that the following condition holds

(10) If $\dim X/\mathrm{rad}\, X$ is infinite for $X$ one of $F$, $G$, $(F+G)^{\perp}$ then $X$ contains a totally isotropic subspace $V$ with $V \cap \mathrm{rad}\, X = (0)$.

<u>Proof</u>. We examine first the possibility of simplifications. In the statements (1), (2), (10), "$F \perp G$" we may replace $F$ and $G$ by $F^{\perp\perp}$ and $G^{\perp\perp}$ without falsifying them. Hence if we prove the theorem for $\perp$-closed $F$ and $G$ we can, in the general case, separate $F^{\perp\perp}$ and $G^{\perp\perp}$. But thereby we separate $F$ and $G$. Thus we may and shall assume that

(11) $F^{\perp\perp} = F$ , $G^{\perp\perp} = G$ .

Furthermore we want to have

(12) $\dim F = \dim G = \aleph_0$ .

If it should happen that $\dim F$ is finite we write $F = \mathrm{rad}\, F \oplus F_0$. $F_0$ can be chopped off from $E$ being nondegenerate and of finite dimension. Assume thus $F_0 = (0)$. If $\mathrm{rad}\, F = (f)$ is 1-dimensional pick $f' \in G^{\perp} \setminus (G \oplus (f))^{\perp}$ (the set is not empty as $G = G^{\perp\perp}$) so that the decomposition $E = k(f,f') \oplus k(f,f')^{\perp}$ separates $F$ and $G$ in this case. An inductive argument takes care of the case where $1 < \dim F < \infty$. $G$ can be treated similarly; hence (12) is in force.

<u>Case I</u>: $F$ and $G$ totally isotropic. The lattice $V(F,G)$ reduces to

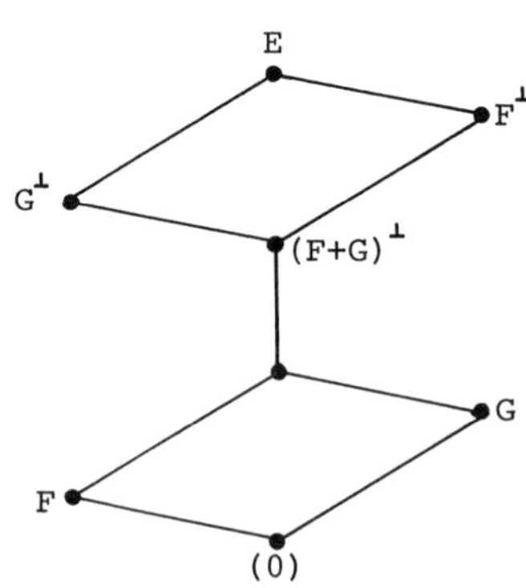

<u>Case Ia</u>: $\dim(F+G)^{\perp}/F+G = 0$ . Consider two hyperbolic spaces $\bar{E}_1 = \bar{F} \oplus L_1$ , $\bar{E}_2 = \bar{G} \oplus L_2$ with $\aleph_0$ the dimension of all four totally isotropic spaces $\bar{F}$ , $\bar{G}$ , $L_1$ , $L_2$ . Let $\bar{E}$ be the orthogonal sum $\bar{E}_1 \overset{\perp}{\oplus} \bar{E}_2$ . $E$ and $\bar{E}$ are both hyperbolic and hence isometric. There is an obvious lattice isomorphism $V(F,G) \to V(\bar{F},\bar{G})$ which, by Theorem 2 of Chap. IV, is induced by an isometry $T$ that sends $F$ on $\bar{F}$ and $G$ on $\bar{G}$ . The decomposition $E = T^{-1}(\bar{E}_1) \overset{\perp}{\oplus} T^{-1}(\bar{E}_2)$ separates $F$ and $G$ .

<u>Case Ib</u>: $\dim(F+G)^{\perp}/F+G = n < \infty$ . This can be reduced to the previous case by chopping off a linear supplement of $F+G$ $(=\mathrm{rad}((F+G)^{\perp}))$ in $(F+G)^{\perp}$ . Observe that conditions (1), (2) can always be "transferred" to the reduced situation.

<u>Case Ic</u>: $\dim(F+G)^{\perp}/F+G = \aleph_0$ . Here we pick spaces as follows. $\bar{E}_1 = (\bar{F} \oplus L_1) \overset{\perp}{\oplus} H_1$ , $\bar{E}_2 = (\bar{G} \oplus L_2) \overset{\perp}{\oplus} H_2$ where $\bar{F}$ , $\bar{G}$ , $L_1$ , $L_2$ are as in Case Ia and where $H_1$ , $H_2$ are hyperbolic and of dimension $\aleph_0$ . Again we form the orthogonal sum $\bar{E} = \bar{E}_1 \overset{\perp}{\oplus} \bar{E}_2$ . Since (10) holds we see again that $V(F,G)$ and $V(\bar{F},\bar{G})$ qualify for Theorem 2 of Chapter IV. We conclude as in Case Ia.

<u>Case II</u>: $F$ is totally isotropic. The lattice $V(F,G)$ reduces to

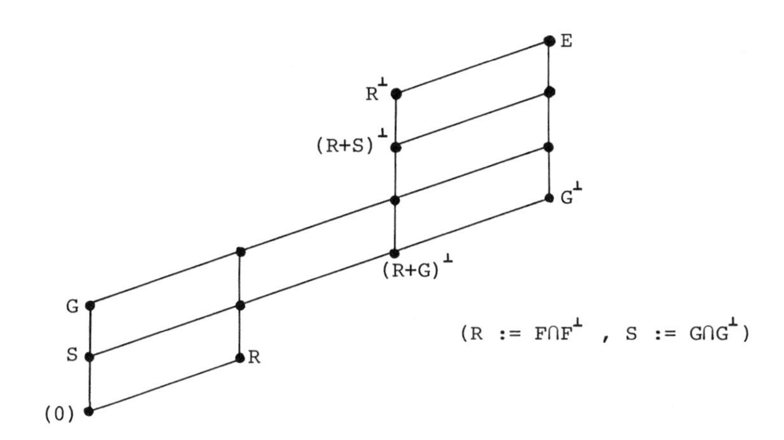

If $\dim G/S$ is finite we fall back into Case I by chopping off a finite dimensional orthogonal summand $U$ of $E$ with $R, S \subset U^{\perp}$. Assume therefore that $\dim G/S = \aleph_0$. If $\dim(R+G)^{\perp}/R+S$ should happen to be finite we can reduce the problem to the situation where $(R+G)^{\perp} = R + S$ by chopping off a finite dimensional subspace of $E$ and proceed as follows. Let $\bar{E}_1 := \bar{F} \oplus L_1$, $\bar{E}_2 := (\bar{S} \oplus L_2) \overset{\perp}{\oplus} H_2$ where $\bar{F}$, $\bar{S}$, $L_1$, $L_2$ are totally isotropic, $\dim \bar{F} = \dim F$, $\dim \bar{S} = \dim(\text{rad } G)$, $\bar{F} \oplus L_1$ and $\bar{S} \oplus L_2$ nondegenerate, $H_2$ hyperbolic and of dimension $\aleph_0$. We set $\bar{G} := \bar{S} \oplus H_2$ and obtain an obvious lattice isomorphism $\tau: V(F,G) \to V(\bar{F},\bar{G})$ that preserves indices and qualifies for an application of Theorem 2 of Chapter IV. We are left with the case of an infinite $\dim(R+G)^{\perp}/R+S$. Here we may choose $\bar{E}_1$ as in Case Ia and for $\bar{E}_2$ we simply take a copy of $E$ with $G$ playing the part of $\bar{G}$. Setting $\bar{E} = \bar{E}_1 \overset{\perp}{\oplus} \bar{E}_2$ gives us again a suitable isomorphism $\tau: V(F,G) \to V(\bar{F},\bar{G})$.

<u>Case III</u>: $F/\text{rad } F$ and $G/\text{rad } G$ are infinite dimensional. Hence we have $\dim X/\text{rad } X = \aleph_0$ when $X = G^{\perp}$, $(R+G)^{\perp}$, $F^{\perp}$, $(F+S)^{\perp}$. Hence (C) of Theorem 2 of Chapter IV is satisfied by dint of (10) for these particular join-irreducible $X$.

We now specify $\bar{F}, \bar{G} \subset \bar{E}$. In case $\dim(F+G)^{\perp}/R+S$ is zero we let $\bar{E}_1 = (\bar{R} \oplus L_1) \overset{\perp}{\oplus} \bar{F}_0$, $\bar{E}_2 = (\bar{S} \oplus L_2) \overset{\perp}{\oplus} \bar{G}_0$ where $\bar{R}$, $\bar{S}$, $L_1$, $L_2$ are totally isotropic, $\bar{R} \oplus L_1$, $\bar{S} \oplus L_2$ nondegenerate, $\bar{F}_0$, $\bar{G}_0$ hyperbolic spaces. We let the dimension of all spaces be $\aleph_0$. Setting $\bar{E} = \bar{E}_1 \overset{\perp}{\oplus} \bar{E}_2$ we can again define a lattice isomorphism $\tau: V(F,G) \to V(\bar{F},\bar{G})$ that preserves indices and sends $F$ into $\bar{F} = \bar{R} \oplus \bar{F}_0$ etc. We quote Theorem 2 of Chap. IV to finish the proof in this situation. The case of finite $\dim(F+G)^{\perp}/R+S$ is reduced to the case just treated by chopping off a suitable finite dimensional subspace of $E$. If, on the other hand, $\dim(F+G)^{\perp}/R+S$ in infinite matters are even simpler. Let $\bar{E}_1$, $\bar{E}_2$ be each a different copy of $E$ with $F$ playing the part of

$\bar{F}$ in $\bar{E}_1$ and G playing the part of $\bar{G}$ in $\bar{E}_2$. Setting $\bar{E} = \bar{E}_1 \overset{\perp}{\oplus} \bar{E}_2$ we can draw the same conclusion as before. The proof of the theorem is thus complete.

Remark. Let $F, G \subset E$ be a "disjoint" pair, $F \cap G = (0)$. Then by the corollary in Chap.X, Sec.3 (1) and (2) are equivalent with F, G being a dual modular and a modular pair in the lattice $L_{\perp\perp}(E)$ of $\perp$-closed subspaces of E (By a Theorem of H.A. Keller $L_{\perp\perp}$ is modular if and only if dim E is finite [6], [3].). Still, as the above proof makes evident, arguments for proving Theorem 2 cannot be kept "inside $L_{\perp\perp}$".

## 4. Symplectic separation in trace-valued spaces

In this section $(E,\Phi)$ will be nondegenerate, trace-valued and of dimension $\aleph_0$ unless specified otherwise. "$A \overset{\circ}{\oplus} B$" denotes a direct sum $A \oplus B$ of totally isotropic spaces A, B that is nondegenerate.

*

If F and G are totally isotropic subspaces of E then we say that they are symplectically separated in E if there exists a Witt decomposition $E = (W \overset{\circ}{\oplus} W') \overset{\perp}{\oplus} E_0$ with $F \subset W$ and $G \subset W'$. If such is the case then, obviously, (1) and (2) hold. Moreover, we can prove the following companion to Theorem 1:

Theorem 3. Let $(E,\Phi)$ be a nondegenerate sesquilinear space of arbitrary dimension. If two totally isotropic subspaces $F, G \subset E$ satisfy (1) and (2) then the lattice $V(F,G)$ orthostably generated by F and G is distributive and $V(F,G)$ is the set $J \cup F$ where

$$J := \{X_1+X_2 \mid X_i \in J_i \ (i=1,2)\}, \quad F := \{Y_1 \cap Y_2 \mid Y_i \in F_i \ (i=1,2)\},$$
$$J_1 := \{F^{\perp\perp}, F+(F^{\perp\perp}\cap G^{\perp}), F^{\perp\perp}\cap G^{\perp}, F+(F\cap G^{\perp})^{\perp\perp}, (F\cap G^{\perp})^{\perp\perp}, F, F\cap G^{\perp}, (0)\},$$
$$F_1 := \{G^{\perp}, F+G^{\perp}, F^{\perp\perp}+G^{\perp}, (F+G^{\perp})^{\perp\perp}, (G\cap F^{\perp})^{\perp}, E\},$$

and $J_2$ , $F_2$ are obtained from $J_1$ , $F_1$ by interchanging F and G . In particular card $V(F,G) \leq 8^2 + 6^2 = 10^2$ .

Thus we see - just as in the orthogonal case - that $V(F,G)$ of a pair with (1) and (2) looks as it ought to look if the pair is to be separated (if F and G <u>are</u> symplectically separated then it is quite easy to see that $V(F,G)$ is the set $J \cup F$ ). The proof of Theorem 3 is mutatis mutandis the proof of Theorem 1 (cf. Remark 1 in [4]) and will not be written out here. We shall need only the case where (11) holds. Here is the diagram:

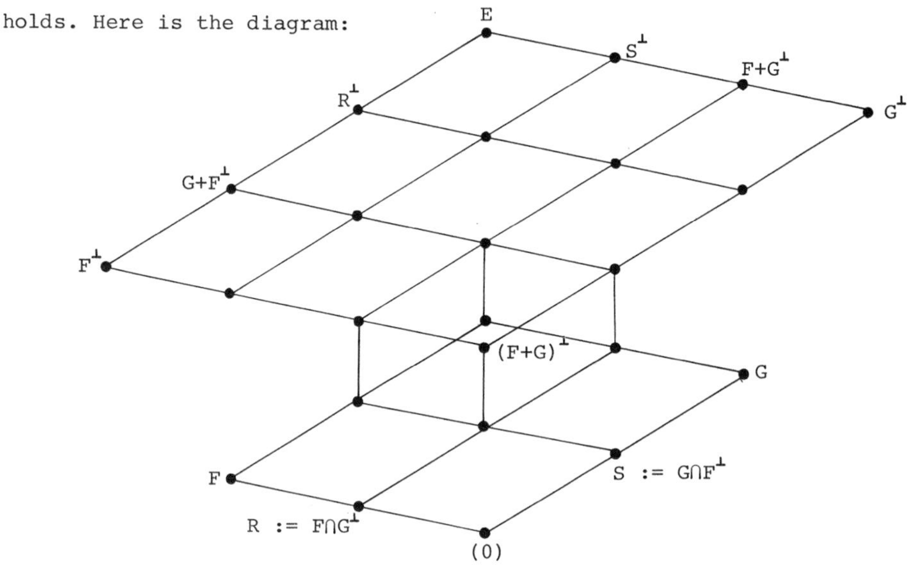


The orthostable lattice of a pair F , G with

$F \perp F$ , $G \perp G$ , $F^{\perp\perp} = F$ , $G^{\perp\perp} = G$ , $(F+G)^{\perp\perp} = F+G$ , $F^{\perp} + G^{\perp} = E$

As is to be expected we can prove that (1) and (2) are <u>sufficient</u> for a pair F , G of totally isotropic subspaces to be symplectically separated whenever the "arithmetical" situation is such that Theorem 2 of Chap. IV can be applied.

In the orthogonal case of the previous section we have used the given spaces $F , G \subset E$ to construct a separated situation $\bar{F} , \bar{G} \subset \bar{E}$ with naturally isomorphic lattices $V(F,G)$ , $V(\bar{F},\bar{G})$ . We could do the same over again or leave it as an exercise. However, we intend to illustrate the possibility of a different attitude towards the lattice. Namely, we shall try to build it up from simpler lattices by taking orthogonal sums. To this end we start with

<u>Example 1</u>. Let $A = \overset{\perp}{\underset{\mathbb{N}}{\oplus}} k(r_i, r_i')$ be an orthogonal sum of hyperbolic planes and $E = A \oplus (a)$ an overspace with the line (a) isotropic; let the product between $a$ and all $r_i$ be equal to $1$ and all products of $a$ with $r_i'$ be zero. Set $F := k(r_{2i})_{i \in \mathbb{N}}$, $G := k(r_{2i}')_{i \in \mathbb{N}}$. The orthostable lattice $V(F,G)$ looks as follows

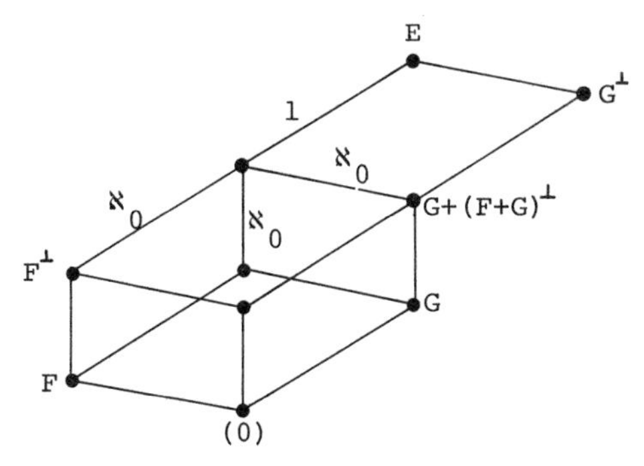


By taking an orthogonal sum of $\alpha \leq \aleph_0$ copies of such spaces we obtain exactly the same diagram but with $\dim E/(F^\perp + G) = \alpha$. Still, by taking the orthogonal sum of such a space ( $\alpha$ copies of the original one) and a similar object which is obtained by taking $\beta$ copies of the original one but with roles of $F$ and $G$ interchanged we obtain a lattice $V(F,G)$ which looks as follows

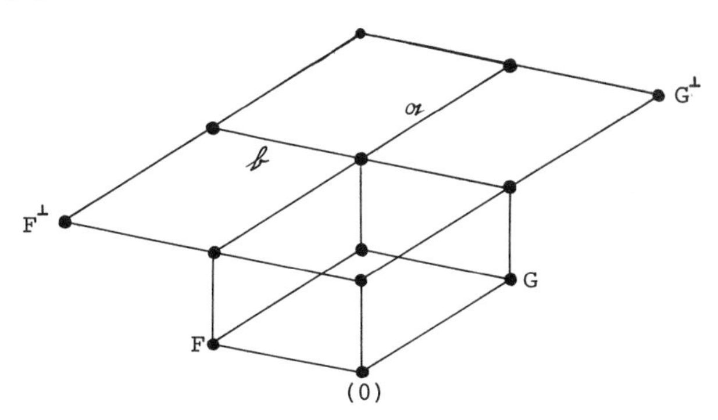

Example 2. Let $E = (L \overset{\circ}{\oplus} L') \overset{\perp}{\oplus} (M \overset{\circ}{\oplus} M') \overset{\perp}{\oplus} (N \overset{\circ}{\oplus} N')$ and put $F := L \oplus M$, $G := M' \oplus N'$. $V(F,G)$ looks as follows:

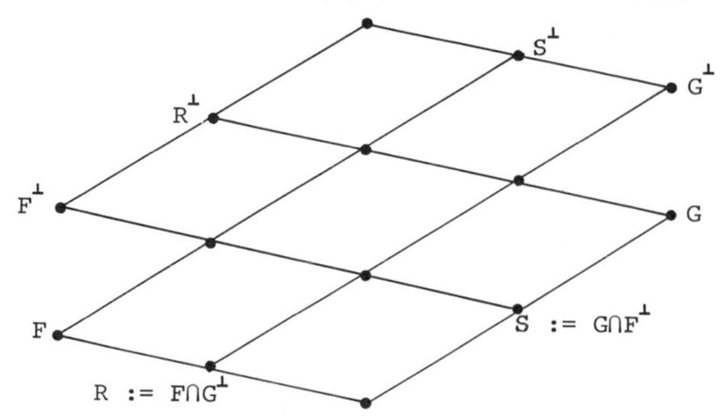


Example 3. By adding an orthogonal summand $P$ to the space in Example 2 and by leaving $F$ and $G$ unchanged, we obtain instead of the above lattice the following one

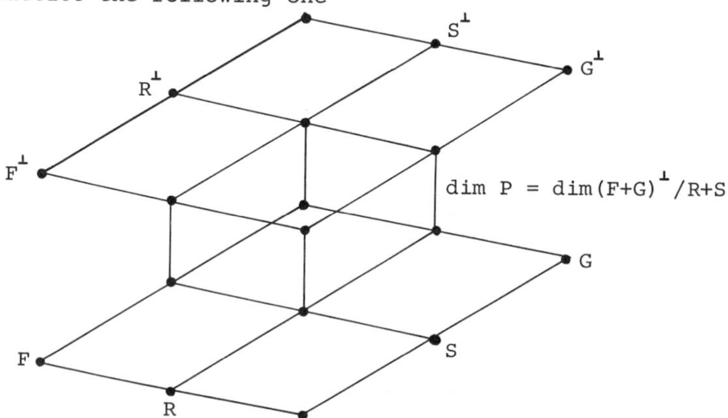


It is furthermore easily seen, that by taking an orthogonal sum of a space $E_1$ as given by Example 1 and a space $E_2$ as given in Example 2 provides us with a pair $F, G \subset E$ whose lattice $V(F,G)$ looks just as the general case. We allow also one of $E_1$, $E_2$ to be $(0)$.

Let us verify that there are enough "parameters" to adjust all indices to the indices prescribed by the given $V(F,G)$. There is only one critical situation: Superposition of two examples $E_1$ and $E_2$ yields a space which has $\dim(F+G)^{\perp}/R+S = \aleph_0$ due to the "infinite cube" in Example 1 (unless we let $E_1$ be $(0)$). Now, if in the lattice put before us $n = \dim(F+G)^{\perp}/R+S$ is finite then (since $n = \dim(F+G+(F+G)^{\perp})/(F+G)$ the space $F+G+(F+G)^{\perp}$ is $\perp$-closed and therefore equal to $(R+S)^{\perp}$. Hence we are in the situation of Example 3 which <u>does</u> allow for prescribed finite "height" of the cube. Thus, by means of the given examples, we can produce a separated pair which gives a lattice that is naturally isomorphic to the lattice of the pair in Theorem 3. It is now not too difficult a matter to apply Theorem 2 of Chap. IV in order to obtain the following

<u>Theorem 4</u> ([1]). Assume that the totally isotropic subspaces $F, G \subset E$ satisfy (1) and (2). In order that $F$ and $G$ are symplectically separated in $E$, i.e. that there is a Witt decomposition $E = (W \mathring{\oplus} W') \overset{\perp}{\oplus} E_0$ with $F \subset W$, $G \subset W'$, it is sufficient that the following condition holds

(12) If $\dim(F+G)^{\perp}/\mathrm{rad}((F+G)^{\perp})$ is infinite then $(F+G)^{\perp}$ contains a totally isotropic subspace $V$ with $V \cap \mathrm{rad}((F+G)^{\perp}) = (0)$.

*

## References to Chapter VI

[1] L. Brand, Erweiterung von algebraischen Isometrien in sesquilinearen Räumen. PhD Thesis, Univ. of Zurich 1974.

[2] F.K. Fischer, Orthogonale und symplektische Zerlegung sesquilinearer Räume. Master's Thesis, Univ. of Zurich 1977.

[3] A. Frapolli, Generalizzazione di un teorema di H.A. Keller sulla modularità del reticolo dei sottospazi ortogonalmente chiusi di uno spazio sesquilineare. Master's Thesis, Univ. of Zurich 1975. (This concerns some technicalities when $\mathrm{char}\, k = 2$; in [6] it is assumed that $\mathrm{char}\, k \neq 2$.)

[4] H. Gross and P. Hafner, The sublattice of an orthogonal pair in a modular lattice. Ann. Acad. Sci. Fenn. vol. 4 1978/1979.

[5] B. Jónsson, Distributive sublattices of a modular lattice. Proc. Amer. Math. Soc. 6 (1955) 682-688.

[6] H.A. Keller, Ueber den Verband der orthogonal abgeschlossenen Teilräume eines hermiteschen Raumes. Letter to the author of Nov. 7, 1973, pp. 1-6.

*

Postscript (concerning orthogonal separation). The paper mentioned in the postscript on page 135 contains a criterion for orthogonal separability in positive definite spaces over $\mathbb{R}$ . Furthermore, it is shown that condition (10) in Theorem 2 (p. 141) is superfluous provided that one of the subspaces to be separated is totally isotropic.

## CHAPTER SEVEN

## CLASSIFICATION OF HERMITEAN FORMS IN CHARACTERISTIC 2

### 1. Introduction

All forms considered in this chapter are $\varepsilon$-hermitean forms over a field $k$ of characteristic 2 equipped with antiautomorphism $\xi \mapsto \xi^*$. In $k$ we consider the additive subgroups $S := \{\alpha \in k | \alpha = \varepsilon\alpha^*\}$ and $T := \{\alpha + \varepsilon\alpha^* | \alpha \in k\}$ of "symmetric" elements and of "traces" respectively. The factor group $S/T$ is a k-left vectorspace under the composition $\lambda(\sigma+T) = \lambda\sigma\lambda^* + T$ $(\sigma \in S, \lambda \in k)$ . $\hat{}: S \to S/T$ is the canonical map.

In this chapter we classify in dimension $\aleph_0$ a kind of sesquilinear forms termed weakly stable. Let us say a word about the philosophy behind this classification. As we have pointed out earlier (in Chap. II)it is _not_ necessary for $\aleph_0$-forms that they represent all (or nearly all) elements of the base field $k$ in order to have a trivial classification. In contrast to the finite dimensional situation it suffices for this that the elements $\alpha \in k$ which _are_ represented are being represented "often". In view of recursive constructions in the realm of the countable "often" means that there should be vectors with inner product $\alpha$ in the orthogonals of all _finite_ dimensional subspaces. Spaces (or forms) with this kind of homogeneity were called stable in Chap. II. Now a sesquilinear space $(E,\Phi)$ qualifies as weakly stable if $E$ contains some $\aleph_0$-dimensional subspace $F$ with a stable restriction of $\Phi$ to $F$ . How can such a stable subspace help to make classify all of $(E,\Phi)$ more easy? It will help because in most cases we shall have $T \subset \|F\|$ which means that vectors with prescribed traces as inner products are "freely" available in $E$ . Because for all vectors $x,y$ we have that $\Phi(x+y,x+y)$ equals $\Phi(x,x) + \Phi(y,y)$ modulo $T$ weak stability amounts to a certain mobility when trying to break a space into simpler pieces.

This program had been carried out in [2] for symmetric bilinear forms over fields with finite degree $[k:k^2]$ . The assumption on the field forced all forms over $k$ to be quasistable in the sense of [3]. Quasistable forms are special weakly stable forms (cf. the definition in section 4 below); they are treated in [3].

For the discussion of the classical situation in finite dimensions the reader should consult [4].

## 2. Multiples of rigid spaces

Let $(E,\Phi)$ be a non degenerate sesquilinear space. We shall abbreviate $\Phi(x,x)$ as $\|x\|$ and call $\varphi$ the linear map

$$x \mapsto \|x\|\hat{} = \|x\| + T$$

from $E$ into $S/T$ ("value map"). For $F$ a subspace of $E$ we set $\|F\| = \{\|f\| \mid f \in F \setminus \{0\}\}$ and $F_* := \{f \in F \mid \|f\| \in T\}$ . $F_*$ is a linear subspace of $F$ and $\dim F/F_* \leq \dim S/T$ .

Definition. $F$ is called rigid if $F_* = (0)$ .

The orthogonal group of a rigid space reduces to the identity (hence the terminology) since any non trivial isometry $x \mapsto x'$ would produce some non zero $x + x'$ with inner product a trace. Rigid spaces are easily checked for isometry: We have $E \cong \bar{E}$ if and only if the value spaces in $S/T$ are equal, $\varphi E = \bar{\varphi}\bar{E}$ , and the bijection $\bar{\varphi}^{-1} \circ \varphi : E \to \bar{E}$ happens to be an isometry (in fact the unique isometry). We shall consider rigid spaces as building blocks and not analyze them any further.

In our first theorem we discuss spaces of the shape $2A := A \overset{\perp}{\oplus} A$ where $A$ is a rigid space of dimension $\leq \aleph_0$ . An equation $A = \langle \alpha_1, \alpha_2, \ldots \rangle$ will mean that $A$ is spanned by an orthogonal basis $(a_i)_J$ with $\|a_i\| = \alpha_i$ for all $i$ .

Theorem 1. Let $A = \langle\alpha_1,\dots\rangle$ and $B = \langle\beta_1,\dots\rangle$ be rigid spaces over $k$ of dimensions $\leq \aleph_0$. In order that there exists an isometry $A \overset{\perp}{\oplus} A \cong B \overset{\perp}{\oplus} B$ it is necessary and sufficient that the families $(\hat{\alpha}_i)$, $(\hat{\beta}_i)$ are "dual" in the vector space $S/T$, i.e. that there exists a matrix $(A_{ij})$ over $k$ such that

$$(1) \qquad \hat{\beta}_i = \sum_j A_{ij}\hat{\alpha}_j \ , \quad (A_{ij}) \text{ invertible and both row- and column-finite}$$

Clearly, if $A$ and $B$ are of finite dimension then the condition on the families $(\hat{\alpha}_i)$, $(\hat{\beta}_i)$ is satisfied if and only if they span the same value spaces in $S/T$; if in addition the forms are symmetric the condition reduces to the equality $\|A\| = \|B\|$.

Proof of the necessity of (1). Let I be an initial segment of $\mathbb{N}$ and $a_i$, $b_i$ $(i \in I)$ orthogonal bases in $A$ and $B$ respectively with $\|a_i\| = \alpha_i$, $\|b_i\| = \beta_i$. Choose congruent bases $a_i'$, $b_i'$ in copies of $A$ and $B$ respectively. Since we assume $A \overset{\perp}{\oplus} A \cong B \overset{\perp}{\oplus} B$ we have equations

$$b_i = \Sigma B_{ij}a_j + \Sigma C_{ij}a_j' \ , \quad b_i' = \Sigma D_{ij}a_j + \Sigma E_{ij}a_j'$$

for some invertible row-finite matrix $\begin{pmatrix} B & C \\ D & E \end{pmatrix}$. Since orthogonal matrices with respect to orthogonal bases are both row- and column-finite we get the asserted relation by applying the linear map $\varphi$ to both sides of one of the above equations, $\hat{\beta}_i = \Sigma(B_{ij}+C_{ij})\hat{\alpha}_j$.

We now turn to the converse assertion of the theorem. The 1-dimensional case $A = \langle\alpha\rangle$, $B = \langle\beta\rangle$ is very easy. By assumption we have $\alpha + \beta = \mu + \varepsilon\mu^* \in T$. We choose a new orthogonal basis $a_1, a_1'$ in $A \overset{\perp}{\oplus} A$

$$(2) \qquad \begin{aligned} a_1 &= a + \mu\alpha^{-1}(a+a') \\ a_1' &= a + (\mu\alpha^{-1}+\beta\alpha^{-1})(a+a') \end{aligned}$$

Since $\|a_1\| = \|a_1'\| = \|a\| + \mu + \varepsilon\mu^*$ we have $\langle\alpha,\alpha\rangle \cong \langle\beta,\beta\rangle$.

The next lemma still deals with a rather special case; it is, however, the key to the general situation.

<u>Lemma 1</u>. Let $\gamma,\gamma_o,\gamma_1,\ldots,\gamma_n \in S$ and assume that in $S/T$ the element $\hat{\gamma}$ is in the span of $\hat{\gamma}_o,\ldots,\hat{\gamma}_n$ but not in the span of $\hat{\gamma}_1,\ldots,\hat{\gamma}_n$ . Then $\langle\gamma,\gamma\rangle \overset{\perp}{\oplus} \langle\gamma_1,\ldots,\gamma_n\rangle \cong \langle\gamma_o,\gamma_o\rangle \overset{\perp}{\oplus} \langle\gamma_1,\ldots,\gamma_n\rangle$ .

<u>Proof of the lemma</u>. By assumption $\gamma = \lambda_o\gamma_o\lambda_o^* + \sum_1^n \lambda_i\gamma_i\lambda_i^* + \tau$ for suitable $\lambda_i \in k$ with $\lambda_o \neq 0$ and $\tau \in T$ . Set $\bar{\gamma} = \gamma + \tau$ so that $\langle\bar{\gamma},\bar{\gamma}\rangle = \langle\gamma,\gamma\rangle$ by the one dimensional case just treated. Let $e_o,e_o',e_1,\ldots,e_n$ be a diagonal basis in $\langle\gamma_o,\gamma_o\rangle \oplus \langle\gamma_1,\ldots,\gamma_n\rangle$ with inner products as indicated. Introduce a new orthogonal basis by

$$
\begin{aligned}
f_o &= \xi_o e_o + (\xi_o+\lambda_o)e_o' + \sum_1^n \lambda_i e_i \\
f_o' &= \lambda_o e_o' + \sum_1^n \lambda_i e_i \qquad\qquad (3)\\
f_i &= \xi_i(e_o + e_o') + e_i \qquad (1\leq i\leq n)
\end{aligned}
$$

where $\quad \xi_i^* := \gamma_o^{-1}\lambda_o^{-1}\lambda_i\gamma_i \quad$ and $\quad \xi_o := \bar{\gamma}\lambda_o^{*-1}\gamma_o^{-1}$ .

We have $\|f_o\| = \|f_o'\| = \bar{\gamma}$ , $\|f_i\| = \gamma_i$ so $\langle\gamma_o,\gamma_o\rangle \overset{\perp}{\oplus} \langle\gamma_1,\ldots,\gamma_n\rangle \cong$ $\cong \langle\bar{\gamma},\bar{\gamma}\rangle \overset{\perp}{\oplus} \langle\gamma_1,\ldots,\gamma_n\rangle$ . Since $\langle\bar{\gamma},\bar{\gamma}\rangle \cong \langle\gamma,\gamma\rangle$ the proof of the lemma is complete.

<u>Proof of the theorem</u> (sufficiency of (1)). Assume first that $\dim A$ is finite and that we have constructed an orthogonal basis $(e_i)$ in $A$ such that for a congruent basis in a copy of $A$ we have

$$
(e_i) \overset{\perp}{\oplus} (e_i') \cong \langle\beta_i,\beta_i\rangle \quad \text{for all } i \leq r . \qquad (4)
$$

We shall show that we can find another orthogonal basis $(f_i)$ in $A$ such that $(f_i) \overset{\perp}{\oplus} (f_i') \cong \langle\beta_i,\beta_i\rangle$ for all $i \leq r+1$ . We have $\hat{\alpha}_i = \Sigma B_{ij}\|e_j\|^\wedge$ for an invertible matrix over $k$ . Hence by (1) the element $\hat{\beta}_{r+1}$ is in the span of the $\|e_i\|^\wedge$, $\hat{\beta}_{r+1} = \Sigma\lambda_i\|e_i\|^\wedge$. By rigidity and by (4) we may assume that $\lambda_{r+1} \neq 0$ . We now apply the lemma with $\beta_{r+1}$ and $\|e_{r+1}\|$ in the roles of $\gamma$ and $\gamma_o$ respectively and with the $\|e_i\|$ $(i \in I \setminus \{r+1\})$ in the role of the $\gamma_1,\ldots,\gamma_n$ . More precisely, we introduce an orthogonal basis $(f_i)$ in $A$ such that $(f_{r+1}) \overset{\perp}{\oplus} (f'_{r+1}) \overset{\perp}{\oplus} (f_1) \overset{\perp}{\oplus} \cdots \overset{\perp}{\oplus} (f_r) \overset{\perp}{\oplus} (f_{r+2}) \oplus \cdots$ appears as

$$\langle\beta_{r+1},\beta_{r+1}\rangle \overset{\perp}{\oplus} \underset{i\neq r+1}{\oplus^{\perp}} \langle\|f_i\|\rangle \cong \langle\beta_{r+1},\beta_{r+1}\rangle \overset{\perp}{\oplus} \langle\beta_1,\beta_2,\dots,\beta_r\rangle \overset{\perp}{\oplus} \underset{i>r+1}{\oplus^{\perp}} \langle\|f_i\|\rangle.$$

Therefore the basis $(f_i)$ of $A$ has the required properties. If we repeat this step at most $\dim A$ times, at each step introducing a new orthogonal basis in $A$, then we can arrive at a basis of $A$ which shows that $A \overset{\perp}{\oplus} A \cong \underset{I}{\oplus^{\perp}}\langle\beta_i,\beta_i\rangle \cong B \oplus B$.

Assume then that $\dim A$ is infinite but that we are in the rather special situation where there exists a partitioning $I = \bigcup_r I_r$ with all $I_r$ finite and such that

$$\text{(5)} \qquad \text{for all } i \in I_r \text{ we have } \hat{\beta}_i = \sum_{j\in I_r} A_{ij}^{(r)}\hat{\alpha}_j \qquad (r \in \mathbb{N})$$

with all $(A_{ij}^{(r)})$ invertible matrices over $k$. We then have $\underset{I_r}{\oplus^{\perp}}\langle\alpha_i\alpha_i\rangle \cong \underset{I_r}{\oplus^{\perp}}\langle\beta_i,\beta_i\rangle$ by the finite dimensional case and thus again $A \overset{\perp}{\oplus} A = \underset{r}{\oplus^{\perp}}(\underset{I_r}{\oplus}\langle\beta_i,\beta_i\rangle) \cong B \overset{\perp}{\oplus} B$. To be in the situation of (5) means that the matrix $(A_{ij})$ in (1) appears as diagonal if partitioned into suitable blocks of finite sizes. Such a special $\mathbb{N}\times\mathbb{N}$-matrix is called a <u>diagonal string</u>. Now every invertible both row- and column-finite matrix is a finite product of invertible diagonal strings, in fact, a product of two invertible diagonal strings ([1], [5]) so that a chain of isometries leads again to the desired isometry $A \overset{\perp}{\oplus} A \cong B \overset{\perp}{\oplus} B$. This remark finishes the proof of Theorem 1.

As a consequence of Thm. 1 it is possible to give simple conditions on rigid spaces $A,B$ which are necessary and sufficient for multiples $nA$ and $nB$ to be isometric $(1\leq n\leq\aleph_0)$. We shall defer the formulation of these conditions to the end of the next section. For the issue of our chapter we need only the case $n = 2$ treated in Thm. 1.

## 3. The relation $\sim$ on the forms of countable dimension

Every non degenerate space $(E,\Phi)$ of dimension $\leq \aleph_0$ admits

<u>almost orthogonal</u> bases, i.e. bases $(e_i)_{i\in I}$ such that the associated inner product matrix is both row- and column-finite. We claim that if $(f_i)_{i\in I}$ is another such basis in $E$ then (for all $i\in I$) $e_i = \sum_j A_{ij} f_j$ with $(A_{ij})$ an invertible $I \times I$ matrix over $k$ that is both row- and column-finite. Indeed, let $M$ be the inner product matrix of $(e_i)$ and $N$ the inner product matrix of $(f_i)$. We certainly have $e_i = \Sigma A_{ij} f_j$ for a row-finite matrix $(A_{ij})$ and $M = ANA^{tr}$. $A$ has a (unique) row-finite (two-sided) inverse, $M$ and $N$ are row- and column-finite and have (unique) row- and column-finite(two-sided) inverses (as can be seen by choosing one of the two bases a union of an orthogonal and a symplectic system). Therefore the product $N^{-1}A^{-1}ANA^{tr}$ exists and is associative and hence $A^{tr} = N^{-1}A^{-1}M$. Thus $A^{tr}$ is row-finite and so $A$ is column-finite.

<u>Definition</u>. Let $E$ and $\bar{E}$ be non degenerate spaces over $k$ of dimension $\leq \aleph_0$. We set $E \sim \bar{E}$ if and only if for some (and hence for all) almost orthogonal bases $(e_i)_{i\in I}$, $(f_i)_{i\in I}$ of $E$ and $\bar{E}$ respectively there exists an invertible row- and column-finite $I \times I$ matrix $(A_{ij})$ over $k$ such that

(6) $\qquad \varphi e_i = \sum_j A_{ij}\, \varphi f_j \qquad$ (for all $i \in I$) ;

here $\varphi$ is the value map $x \longmapsto \|x\|^\wedge = \|x\| + T$ introduced in section 2.

$\sim$ is an equivalence relation on the class of all non degenerate k-spaces of at most countable dimension. Theorem 1 may be formulated as

(7) $\qquad$ if $E$ and $\bar{E}$ are rigid then $E \overset{\perp}{\oplus} E \cong \bar{E} \overset{\perp}{\oplus} \bar{E}$ if and only if $E \sim \bar{E}$.

Thus if e.g. $E = \langle\alpha_1,\alpha_2,\ldots\rangle$ is rigid then the space $\bar{E} := \langle\alpha_1,\alpha_2+\alpha_1,\alpha_3+\alpha_1,\ldots\rangle$ is rigid as well and has the same value space as $E$, $\varphi\bar{E} = \varphi E$; yet $E$ and $\bar{E}$ are not isometric since obviously not $E \sim \bar{E}$ . A generalization of Theorem 1 is given by the following corollary announced at the end of the last section.

Corollary. Let $A$ and $B$ be rigid spaces and $1 \leq n \leq \aleph_0$ . In order that $nA \cong nB$ the following condition is necessary and sufficient: $A \cong B$ if $n$ is finite and odd, $A \sim B$ if $n$ is finite and even, $\varphi A = \varphi B$ if $n = \aleph_0$ .

Proof. If $n$ is finite and odd then we use the fact that

$$(8) \quad \langle\beta\rangle \overset{\perp}{\oplus} \langle\beta,\beta\rangle \cong \langle\beta\rangle \overset{\perp}{\oplus} P \ , \ P \text{ hyperbolic} .$$

Indeed, if $b_1, b_2, b_3$ is an orthogonal basis with $\|b_i\| = \beta$ we introduce a new basis $b_1' = b_1 + b_2 + b_3$ , $b_2' = \beta^{-1}(b_1 + b_2)$, $b_3' = b_2 + b_3$ and see that (8) holds. Thus if $A = \langle\alpha_1, \alpha_2, \ldots\rangle$ then since $(2r+1)\langle\alpha_i\rangle \cong rP \oplus \langle\alpha_i\rangle$ we obtain $(2r+1)A \cong \aleph_0 P \overset{\perp}{\oplus} A \cong [(2r+1)A]_* \oplus [(2r+1)A]_*^{\perp}$ and an analogous isometry for $(2r+1)B$ . Since any isometry $U \cong V$ maps $U_*^{\perp}$ onto $V_*^{\perp}$ we obtain the asserted isometry $A \cong B$ from $(2r+1)A \cong (2r+1)B$ and, of course, the converse implication as well.

If $n = 2r$ then by (8) $n\langle\alpha_i\rangle \cong (r-1)P \overset{\perp}{\oplus} \langle\alpha_i, \alpha_i\rangle$ . We may assume $r > 1$ since Thm. 1 takes care of the case $r = 1$ . Hence $nA \cong \aleph_0 P \overset{\perp}{\oplus} A \overset{\perp}{\oplus} A$ . The assertion that $A \sim B$ is proved by (13) below.

If $n = \aleph_0$ then from $\aleph_0 A \cong \oplus_i^{\perp} \langle\alpha_i, \alpha_i, \ldots\rangle \cong \aleph_0 B \cong \oplus_i \langle\beta_i, \beta_i, \ldots\rangle$ follows $\oplus_i \varphi\langle\alpha_i\rangle = \oplus_i \varphi\langle\beta_i\rangle$ , i.e. $\varphi A = \varphi B$ . Conversely, from $\varphi A = \varphi B$ it follows that $\|\aleph_0 A\| = \|\aleph_0 B\|$ ; since the two multiples are stable we conclude from this that $\aleph_0 A \cong \aleph_0 B$ .

## 4. Weakly stable spaces

Assume that $(E, \Phi)$ is non degenerate and of dimension $\aleph_0$ . For $F \subset E$ we set

$$\|F\|_\infty := \cap \{\|F \cap X^{\perp}\| \mid X \subset E \text{ and } \dim X < \dim E\} .$$

The sets $\|F\|$ and $\|F\|_\infty$ are obviously orthogonal invariants attached to the orbits of $F$ under the orthogonal group of $E$. If $\|F\|_\infty \neq \emptyset$ then $\|F\|_\infty$ is an additive subgroup of $k$; if $\|F\|_\infty \not\subseteq \{0\}$ then $E$ contains for every non zero $\alpha \in \|F\|_\infty$ an $\aleph_0$-dimensional subspace $G \cong \langle\alpha,\alpha,\ldots\rangle$ with $E = G \oplus G^\perp$ and thus we have $T \subset \|G\|_\infty \subset \|F\|_\infty$. We define

$$F_\infty := \{f \in F \mid \|f\| \in \|F\|_\infty\}$$

If $T \subset \|F\|_\infty$ then $F_\infty$ is a subspace of $F$.

<u>Lemma 2</u>. If $T \subset \|F\|_\infty$ then $F \cap F_*^{\perp\perp} = F_\infty$

<u>Proof</u>. Let $f \in F_\infty$ and a finite dimensional $X \subset E$ be given. Since $\|f\|$ is in $\|F\|_\infty$ there exists $f_1 \in F \cap X^\perp \cap f^\perp$ with $\|f_1\| = \|f\|$. Thus $\|f+f_1\| = 0 \in T$ and we see that the translate $f + X^\perp$ (weak linear neighbourhood of f) meets $F_*$. Hence $f \in F_*^{\perp\perp}$. Conversely, let $f \in F \cap F_*^{\perp\perp}$ and $X$ be as before. There is a vector $y \in F_*$ (depending on X) such that $f - y \in X^\perp$. Since $T \subset \|F\|_\infty$ there is $z \in F_\infty \cap X^\perp \cap (f-y)^\perp$ with $\|z\| = \|y\| + \Phi(f,y) + \Phi(y,f) \in T$. The vector $f - y + z$ is in $X^\perp$ and has inner product $\|f\|$. Since $X$ was arbitrary this shows that $\|f\| \in \|F\|_\infty$ i.e., $f \in F_\infty$.

If $T \not\subseteq \|F\|_\infty = \{0\}$ then the set $F_\infty$ is not, in general, a linear subspace of $F$. Here we have

<u>Lemma 3</u>. Let $\|F\|_\infty \neq \emptyset$. We have $T \not\subseteq \|F\|_\infty$ if and only if $\Phi$ is not symmetric and $F$ has a decomposition of the following kind: $F = (F\cap F^\perp) \overset{\perp}{\oplus} H \overset{\perp}{\oplus} (R\oplus R') \overset{\perp}{\oplus} G$ where $H$ is hyperbolic and finite dimensional, $(F\cap F^\perp) \oplus R$ is totally isotropic and infinite dimensional, $(R'+G)_* = G_*$, $R' + G$ is anisotropic (and hence $\dim H = 2\dim(F_*/F_*\cap V^\perp)$, with $V$ any maximal totally isotropic subspace of $F$, is an invariant of $F$). In particular, if $T \not\subseteq \|E\|_\infty = \{0\}$ then $E$ is of the shape $E \cong 2\cdot(\text{inf. dim. rigid}) \overset{\perp}{\oplus} (\text{fin.dim.hyperb.}) \overset{\perp}{\oplus} (\text{anisotropic})$.

<u>Proof</u>. Assume that $F$ admits a decomposition as indicated and let $\gamma \in \|F\|_\infty$. There exists $x \in F \cap H^\perp$ with $\|x\| = \gamma$ and $y \in F \cap H^\perp \cap x^\perp$

with $\|y\| = \gamma$ ; so $\|x+y\| = 0$ . Write $x + y = f + r + r' + g$ ($f \in F \cap F^{\perp}$, $r \in R$, $r' \in R'$, $g \in G$) ; $\|x+y\| = 0 = \Phi(r,r') + \Phi(r',r) + \|r'\| + \|g\|$ so $\|r'+g\| \in T$ . By assumption therefore $r' = 0$ hence $g = 0$ as $G$ is anisotropic. We have shown that $x \equiv y \bmod (F \cap F^{\perp}) \oplus R$ . Express analogously $x = f_o + r_o + r_o' + g_o$ and pick $z$ with $\|z\| = \gamma$ in $F \cap H^{\perp} \cap r_o'^{\perp} \cap g_o^{\perp} \cap x^{\perp}$ . Since $x \equiv z \bmod (F \cap F^{\perp}) \oplus R$ we must have $g_o = 0$ , furthermore, for $r_1$ the component in $R$ of the vector $z$, $0 = \Phi(z,r_o') = \Phi(r_1,r_o') + \|r_o'\|$ . So $\Phi(r_1,r_o')$ is constant for all $z \in F \cap H^{\perp} \cap r_o'^{\perp} \cap x^{\perp}$ with $\|z\| = \gamma$ . By choosing $z$ suitably we see that $\|r_o'\| = 0$ : Ergo $r_o' = 0$ and we have $\|F\|_{\infty} \subset \{0\}$ .

Assume conversely that $T \not\subset \|F\|_{\infty} \neq \emptyset$ (Since $0 \in \|F\|_{\infty}$ the form cannot be symmetric and $\|F\|_{\infty} = \{0\}$ ). Let $F_o$ be a supplement of $F \cap F^{\perp}$ in $F$ and $L$ a maximal totally isotropic subspace of $F_o$ . There is (see Chapter III) a decomposition $F_o = (L \oplus L') \overset{\perp}{\oplus} F_1$ . Let $B$ be the projection onto $L'$ of $(L'+ F_1)_*$ . If $B$ were of infinite dimension, then one could construct an infinite sum $S$ of pairwise orthogonal hyperbolic planes in $F_o$ , in fact, one can arrange for $S$ to be an orthogonal summand of $E$ . But this is a contradiction as $T \not\subset \|F\|_{\infty}$ . Hence the number $n := \dim B$ is finite and precisely $n$ hyperbolic planes can be chopped off, $F_o = H \overset{\perp}{\oplus} H_1$, $\dim H = 2n$ , $H$ hyperbolic. $H_1$ can now be decomposed as $H_1 = (R \oplus R') \overset{\perp}{\oplus} G$ , $R \subset L$ , $(R'+ G)_* = G_*$ . This is the required decomposition of $F$ . The proof is thus complete; it also proves

<u>Lemma 4</u>. If $\|F\|_{\infty} = \emptyset$ then $F \cong 2\cdot$(fin.dim.rigid) $\overset{\perp}{\oplus}$ (fin.dim. hyperb.) $\overset{\perp}{\oplus}$ (fin.dim.totally isotropic) $\overset{\perp}{\oplus}$ (anisotropic).

However, we shall not pursue the spaces $E$ with empty $\|E\|_{\infty}$ any further.

<u>Definition</u>. Assume that $(E,\Phi)$ is non-degenerate and $\dim E = \aleph_o$. A subspace $F$ is called <u>weakly stable in</u> $E$ if $\|F\|_{\infty} \neq \emptyset$ . $F$ is

called <u>stable in</u> E if $\|F\|_\infty = \|F\|$ and it is called quasistable ([3]) <u>in</u> E if $F = F_o \overset{\perp}{\oplus} F_1$ with $\dim F_o$ finite and $F_1$ stable in E .

If $(F_\iota)_{\iota \in J}$ is a family of subspaces stable in E and if for at least one $\iota \in J$ we have $T \subset \|F\|_\iota$ then the sum $\sum_J F_\iota$ is stable in E . $E_\infty$ invariably is the largest subspace stable in E . An easy recursive argument shows that

(9) If $T \subset \|E\|_\infty$ then the isometry class of $E_\infty$ is characterized by $\|E\|_\infty$ and $\dim(E_\infty \cap E_\infty^\perp)$ .

## 5. Fitting together stable and rigid spaces

Let A be a non-degenerate stable space and B, C rigid spaces with disjoint value-spaces in $S/T$ , $\varphi B \cap \varphi C = (0)$ . We wish to extend the given forms (on A,B,C) to all of the vector space $E = A \oplus B \oplus C$ in such a way that we shall have

$$A^\perp = B \quad \text{and} \quad B^\perp = A \ .$$

As all spaces considered here are of dimensions $\leq \aleph_o$ this is possible if and only if dim B and dim C satisfy

$$\dim B < \infty \Rightarrow \dim C = 0 \ .$$

If dim B is finite then we have no choice but to form the external orthogonal sum $E = A \overset{\perp}{\oplus} B$ . If $\dim B = \aleph_o$ then we may partition an orthogonal basis $\mathcal{B}$ of B into dim C many infinite subfamilies $\mathcal{B}_\gamma$ and likewise an orthogonal basis $\mathcal{A}$ of A into dim C many infinite subfamilies $\mathcal{A}_\gamma$ . If $\mathcal{C} = (c_\gamma)$ is an orthogonal basis of C then we let the inner products of $c_\gamma$ with all members of $\mathcal{A}_\gamma \cup \mathcal{B}_\gamma$ be 1 and set zero all other products between members of different bases $\mathcal{A},\mathcal{B},\mathcal{C}$. This will give us $A^\perp = B$ and $B^\perp = A$ in $E = A \oplus B \oplus C$ . Obviously $E_\infty = A$ . Although there are, of course, other possibilites of extending the given forms to $A \oplus B \oplus C$ when dim B is infinite we shall invariably end up with the same isometry class by the following

Theorem 2. Assume that $(E,\Phi)$ is weakly stable and that span $E_\infty$ is non degenerate. Then we have $T \subset \|E\|_\infty$ (in particular span $E_\infty = E_\infty$ ). The isometry class of $E$ is determined by the collection of the following invariants: $\|E\|_\infty$ , the isometry class of the rigid space $E_\infty^\perp$ , the isometry class of an arbitrary supplement of $E_\infty$ in $E$ that contains $E_\infty^\perp$ (every supplement of $E_\infty$ is rigid).

Proof. Lemma 3 (Section 4) shows that we cannot have $\|E\|_\infty = \{0\} \not\supset T$ when span $E_\infty$ is non degenerate. Let then $\bar{E}$ be another space with the same invariants. Let $\psi : S \to \bar{S}$ be an isometry between supplements of $E_\infty$ and $\bar{E}_\infty$ in $E$ and $\bar{E}$ respectively, and $E_\infty^\perp \subset S$ , $\bar{E}_\infty^\perp \subset \bar{S}$ . As $E_\infty^\perp \cong \bar{E}_\infty^\perp$ we have, by rigidity, that $\psi$ maps $E_\infty^\perp$ onto $\bar{E}_\infty^\perp$ . We now show that the restriction $\psi_o : E_\infty^\perp \to \bar{E}_\infty^\perp$ of $\psi$ can be extended to an isometry $E \cong \bar{E}$ .

Assume that we have constructed finite dimensional spaces $W$ , $\bar{W}$ and an isometry $\psi_1 : E_\infty^\perp \oplus W \to \bar{E}_\infty^\perp \oplus \bar{W}$ which extends $\psi_o$ and such that

(10) $\quad (E_\infty^\perp \oplus W) \cap E_\infty = W \cap E_\infty, \; (\bar{E}_\infty^\perp \oplus \bar{W}) \cap \bar{E}_\infty = \bar{W} \cap \bar{E}_\infty \; , \; \psi_1(W \cap E_\infty) = \bar{W} \cap \bar{E}_\infty .$

We now extend $\psi_1$ on $E_\infty^\perp \oplus W \oplus (x)$ for arbitrarily prescribed $x \in E$.

CASE I: $x \in E_\infty \setminus (E_\infty^\perp \oplus W)$ . Since $\bar{W} \cap \bar{E}_\infty^\perp = (0)$ there is $\bar{x} \in \bar{E}_\infty$ with $\bar{\Phi}(\bar{x},\psi_1 w_i) = \Phi(x,w_i)$ for all members $w_i$ of a basis of $W$ . Since $\bar{E}_\infty$ is stable in $\bar{E}$ , we find $\bar{e} \in \bar{W}^\perp \cap \bar{x}^\perp \cap \bar{E}_\infty$ with $\|\bar{e}\| = \|x\| + \|\bar{x}\| \in \|E\|_\infty$. $\psi_1$ is extended by sending $x$ into $\bar{x} + \bar{e}$ . The induction assumptions (10) are easily seen to hold again for $W \oplus (x)$ in lieu of $W$ .

CASE II: $x \in E_\infty + (E_\infty^\perp \oplus W)$ . This brings us back to case I.

CASE III: $x \notin E_\infty + (E_\infty^\perp \oplus W)$ . Let $S_o$ be a supplement of $E_\infty^\perp$ in $S$ and set $\bar{S}_o = \psi S_o$ . We decompose $x = e + e' + s$ according to $E = E_\infty \oplus E_\infty^\perp \oplus S_o$. Without loss of generality $e = e' = o$ , i.e. $x = s \in S_o$ . Set $\bar{s} := \psi s$ . In $\bar{E}_\infty$ we choose an element $\bar{f}$ with

$\bar{\Phi}(\bar{f},\psi_1 w_i) = \bar{\Phi}(\bar{s},\psi_1 w_i) + \Phi(s,w_i)$ and in $\bar{E}_\infty$ we pick furthermore a vector $\bar{g} \in \bar{E}_\infty \cap \bar{W}^\perp \cap \bar{s}^\perp \cap \bar{f}^\perp$ with $\|\bar{g}\| = \|s\| + \|\bar{s}+\bar{f}\| \in \|E\|_\infty$. We extend $\psi_1$ on $E_\infty^\perp \oplus W \oplus (s)$ by sending $x = s$ into $\bar{x} := \bar{s} + \bar{f} + \bar{g}$. To finish the step let us compute $(\bar{E}_\infty^\perp \oplus \bar{W} + (\bar{x})) \cap \bar{E}_\infty$. Assume that $\bar{e} + \bar{w} + \lambda(\bar{s}+\bar{f}+\bar{g})$ is in the intersection $(\bar{e} \in \bar{E}_\infty^\perp, \bar{w} \in \bar{W})$. Hence $\lambda\bar{s} \in \bar{E}_\infty + \bar{E}_\infty^\perp \oplus \bar{W}$. If $\lambda \neq 0$ we can find, by CASE II, a vector $s_1 \in E_\infty + E_\infty^\perp \oplus W$ with $\|s_1\| = \|\bar{s}\|$. Therefore $\|s_1 + s\| \in T \subset \|E\|_\infty$ which means that $s \equiv s_1 \mod E_\infty$. But this is impossible in the present case. Therefore $\lambda = 0$ and thus $(\bar{E}_\infty^\perp \oplus \bar{W} \oplus (\bar{x})) \cap \bar{E}_\infty = (\bar{E}_\infty^\perp \oplus \bar{W}) \cap \bar{E}_\infty$. In the present case it is trivial that $(E_\infty^\perp \oplus W \oplus (x)) = (E_\infty^\perp \oplus W) \cap E_\infty$. This proves that the induction assumptions (10) are again valid after extending $\psi_1$.

We can therefore extend $\psi_o$ to all of $E$ and the proof of the theorem is complete.

## 6. The classifiaction of weakly stable spaces

We have already investigated two special cases of non degenerate weakly stable spaces $E$. Theorem 1 in Section 2 treats the case where $E_\infty$ is totally isotropic and satisfies $E_\infty = E_\infty^\perp$. Theorem 2 in the previous section discusses in detail the case where $E_\infty$ is non degenerate. The next theorem shows that it is possible to break the general case into these two special cases. Thus we shall then be able to fuse theorems 1 and 2 (Scholion).

Theorem 3. Let $E$ be weakly stable, non degenerate and of dimension $\aleph_o$. If $T \subset \|E\|_\infty$ then the isometry class of $E$ is characterized by the collection of the following invariants of $E$, (i) the $\sim$ class of $E$, (ii) the isometry class of $R^\perp$ where $R := E_\infty \cap E_\infty^\perp$. If $T \not\subset \|E\|_\infty$ then the same statement holds provided $R$ is defined to be the radical of the span of the set $E_\infty$. In either case, if $\dim R$ is finite then the invariant (i) may be replaced by the invariant (i'): $\|E\|$.

Proof. We shall first discuss the case where $T \subset \|E\|_\infty$. Clearly an isometry $E \to \bar{E}$ maps $R$ onto $\bar{R}$, $R^\perp$ onto $\bar{R}^\perp$ and it induces a relation $E \sim \bar{E}$. Assume conversely that $\bar{E}$ has the same invariants (i) and (ii) as $E$. The spaces $R$ and $\bar{R}$ are $\perp$-closed by lemma 2 so we may quote lemma 1 and obtain decompositions

$$(11) \qquad E = (R \oplus R') \overset{\perp}{\oplus} L \,, \quad \bar{E} = (\bar{R} \oplus \bar{R}') \overset{\perp}{\oplus} \bar{L} \,.$$

Since $R \oplus L = R^\perp \cong \bar{R}^\perp = \bar{R} \oplus \bar{L}$ the spaces $L$ and $\bar{L}$ (or, for that matter, any supplements of $R,\bar{R}$ in $R^\perp$, $\bar{R}^\perp$ respectively) are isometric. Furthermore $R'$ and $\bar{R}'$ are rigid for, if we had a non zero element in $\|R'\| \cap T$ then we could specify a hyperbolic plane $P$ not in $R^\perp$ which is a contradiction as $P$ would also belong to $E_\infty$ (recall that we are in the case where $T \subset \|E\|_\infty$).

Let $(r_i')_{i\in I}$ be an almost orthogonal basis in $R'$ and $(r_i)_{i\in I}$ a congruent basis in a copy of $R'$ and $\bar{r}_i, \bar{r}_i'$ $(i \in I)$ analogous objects in $\bar{E}$. Let $(\ell_j)_{j\in J}$ be an almost orthogonal basis in $L$ and $(\bar{\ell}_j)_{j\in J}$ a congruent basis in $\bar{L}$. From the assumption that $E \overset{\sim}{=} R' \overset{\perp}{\oplus} R' \overset{\perp}{\oplus} L \sim \bar{E} \overset{\sim}{=} \bar{R}' \overset{\perp}{\oplus} \bar{R}' \overset{\perp}{\oplus} \bar{L}$ we obtain equations

$$(12) \qquad \varphi r_i' = \sum_I B_{ij}\, \varphi \bar{r}_j' + \sum_J C_{ij}\, \varphi \ell_j \,, \quad (i \in I)$$

where $(B_{ij})$ and $(C_{ij})$ are both row- and column-finite. Set $r_i'' := r_i' + \Sigma C_{ij}\ell_j$. The vectors $r_i''$ $(i \in I)$ span a supplement $R''$ of $R^\perp$ in $E$. Further, since $(C_{ij})$ is column-finite we see that $\ell_j \in (R \oplus R'') + (R \oplus R'')^\perp$, in other words, $R \oplus R''$ admits an orthogonal supplement $L_1$ in $E$, $E = (R \oplus R'') \overset{\perp}{\oplus} L_1$. We now have $R'' \sim \bar{R}'$ as the $r_i''$ form an almost orthogonal basis with $\varphi r_i'' = \Sigma B_{ij} \varphi \bar{r}_j'$ for we claim that $(B_{ij})$ is invertible. Clearly $\varphi R'' \subset \varphi R'$ and $\varphi E = \varphi R'' + \varphi L \subset \varphi R' + \varphi L$. To obtain equality, $\varphi R'' = \varphi R'$, and thence invertibility it suffices to show that $\varphi E = \varphi R' \oplus \varphi L$. Assume by way of contradiction that the sum is not direct, say $u \in R'$, $v \in L$ with $\|u\| = \|v\| \neq 0$. Then $\|u+v\| \in T \subset \|E\|_\infty$ so $u + v \in E_\infty \subset R^\perp$; but $v \in R^\perp$ and $u \notin R^\perp$, contradiction. We have, in effect, proved the following when $T \subset \|E\|_\infty$

(13) If $E \sim \bar{E}$ then $R' \sim \bar{R}'$ for suitably chosen supplements $R'$, $\bar{R}'$ of $R^\perp := (E_\infty \cap E_\infty^\perp)^\perp$ and $\bar{R}^\perp$ respectively.

Since $R'' \sim \bar{R}'$ is seen to hold we may quote Thm. 1 in order to obtain an isometry $(R \oplus R'') \cong (\bar{R} \oplus \bar{R}')$. Since $L_1 \cong \bar{L}$ we have an isometry $E \cong \bar{E}$ as asserted by the theorem.

If $T \not\subset \|E\|_\infty$ and $R$, $\bar{R}$ are the radicals of the <u>spans</u> of the sets $E_\infty$, $\bar{E}_\infty$ respectively we reach the same conclusions by using Lemma 3 of Section 4 and Theorem 1.

Obviously, if $\dim R$ is finite then without assuming (i) we obtain (12) with $B_{ij}$ of finite size and $(C_{ij})$ row- and column-finite provided we have $\|E\| = \|\bar{E}\|$. This completes the proof of Theorem 3.

We shall now summarize our results by combining Theorem 2 (Lemma 3, Section 4 respectively) with theorem 3 in the following scholion. Since all supplements of the radical $F \cap F^\perp$ of a subspace $F$ are isometric we may without any risk of confusion speak of the isometry class of <u>the non degenerate part</u> of $F$.

<u>Scholion</u>. Let $(E,\Phi)$ be weakly stable, non degenerate and of dimension $\aleph_0$ and set $R := E_\infty \cap E_\infty^\perp$. We distinguish between the two cases $T \subset \|E\|_\infty$ and $T \not\subset \|E\|_\infty$.

If $T \subset \|E\|_\infty$ then the isometry class of $E$ is characterized by the set of the following invariants of $E$: (i) the $\sim$ class of $E$ (if $\dim R$ is finite $\|E\|$ suffices), (ii) $\|E\|_\infty$, (iii) the isometry class of the non degenerate part of $E_\infty^\perp$, (iv) the isometry class of anyone supplement $X$ of $E_\infty$ in $R^\perp$ such that $R + X \supseteq E_\infty^\perp$. The classes in (iii) and (iv) are classes of rigid spaces.

If $T \not\subset \|E\|_\infty$ then the isometry class of $E$ is characterized by the collection of the following three invariants: (j) the $\sim$ class of $E$, (jj) the isometry class of the non degenerate part of $(\text{span } E_\infty)^\perp$ which is anisotropic, (jjj) the natural number $n = \dim (E_*/(E_* \cap V^\perp))$ where $V$ is any maximal totally isotropic subspace of $E$. $E$ splits off an orthogonal sum $H$ of $n$ hyperbolic planes, $E = H \overset{\perp}{\oplus} E_1$ and $E_1$ is uniquely determined by the invariants (j) and (jj) of $E$; furthermore $E_1 \cong 2 \cdot (\text{inf.dim.rigid}) \overset{\perp}{\oplus} (\text{anisotropic})$ so that the set $(E_1)_\infty$ is a linear subspace, $(E_1)_\infty = (E_1)_* \cap (E_1)_*^\perp$.

If $E$ is quasistable we always have $T \subset \|E\|_\infty$ and finite dimension of $R = E_\infty \cap E_\infty^\perp$. Thus $E_\infty + E_\infty^\perp = R^\perp$ and the invariant (iv) in the scholion is trivial. Thus

Corollary ([3]). If $E$ is quasistable then the isometry class of $E$ is characterized by $\|E\|$, $\|E\|_\infty$, the isometry class of the (finite dimensional rigid) non degenerate part of $E_\infty^\perp$.

Remark. In the case where $T \subset \|E\|_\infty$ the following diagram helps to locate the roles of the invariants (i), (ii), (iii), (iv) in the scholion; it gives the sublattice (in the lattice of all subspaces of E) generated by the space $E_*$ under the operations $+$, $\cap$, orthogonal complementation.

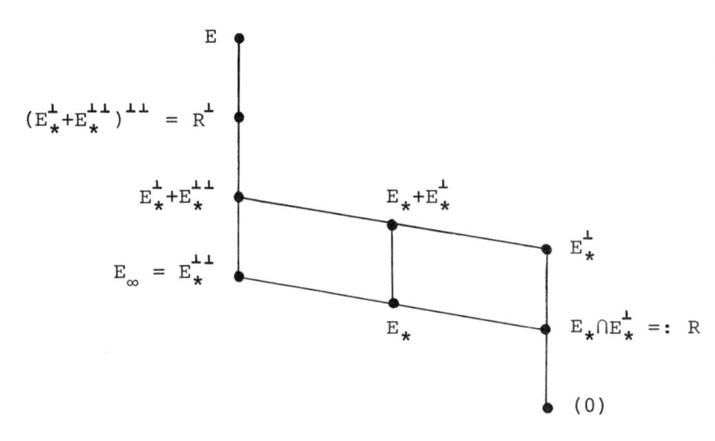

## 7. Representatives

If a weakly stable space $E$ has $T \subset \|E\|_\infty$ then it can be decomposed as $E = (R \oplus R') \overset{\perp}{\oplus} (A \oplus B \oplus C)$ where $E_\infty = R \oplus A$ and $E_\infty^\perp = R \oplus B$. Because $E_\infty = E_\infty^{\perp\perp}$ we have $C = (0)$ in case $\dim B$ is finite (e.g. whenever $E$ is quasistable). In other words

(14) If $\dim E_\infty^\perp / E_\infty \cap E_\infty^\perp$ is finite then the two isometry classes of (iii) and (iv) in the scholion coincide.

Are there any other relations among the invariants (i), (ii), (iii), (iv) of the scholion? There is the trivial condition that the set $\|E\|$ given by (i) contain the set $\|E\|_\infty$ as given by (ii) as well as the analogous $\|E\|$-sets belonging to the objects in (iii) and (iv). As to (ii) it is obvious that the inverse image in $S$ under $\hat{}: S \to S/T$ of any subspace $Y \subset S/T$ of dimension $\leq \aleph_0$ may serve as $\|E\|_\infty$. Assume then that this set and the other invariants of the scholion are prescribed and are such that the relations we have listed are satisfied. We set $A := \overset{\perp}{\bigoplus}_{\hat{\alpha} \in Y} \langle \alpha, \alpha, \ldots \rangle$ if $Y \neq (0)$ and $A$ an orthogonal sum of $\aleph_0$ hyperbolic planes if $Y = (0)$. We pick $B$ and $B \oplus C$ of the types prescribed by (iii) and (iv) of the scholion and extend the forms onto all of $A \oplus B \oplus C$ in the manner explained in Section 5. From the $\sim$ class given in (i) we get in particular the prescribed value space $\varphi E$. Choose a rigid space $R'$ with $\varphi E = \varphi R' \oplus \varphi(A \oplus B \oplus C)$ and <u>defin</u>

$$E := R' \overset{\perp}{\oplus} R' \overset{\perp}{\oplus} (A \oplus B \oplus C)$$

$R'$ can be chosen such that $E$ drops into the prescribed $\sim$ class. $E$ has now all the prescribed invariants.

The corresponding discussion when $T \not\subset \|E\|_\infty$ is yet easier and is left to the reader.

## 8. Suitable fields for weak stability

If we demand that <u>all</u> non degenerate $\aleph_0$-forms over a certain field k be weakly stable then k is such that all such forms are even quasistable by the following

Lemma 5. If in the following statements on k the word "space" means a non-degenerate k-vector space $(E,\Phi)$ of dimension $\aleph_0$ then all five statements are equivalent.

(j) All k-spaces are weakly stable;

(jj) All k-spaces E are weakly stable and have $T \subset \|E\|_\infty$ ;

(jjj) All k-spaces are quasistable;

(jv) All k-spaces contain non zero isotropic vectors;

(v) dim S/T is finite and there is only one trace-valued k-space up to isometry.

Proof. (jjj) $\Rightarrow$ (j) is trivial. (j) $\Leftrightarrow$ (jv) because for anisotropic k-spaces E we have $\|E\|_\infty = \emptyset$ . (jv) $\Rightarrow$ (jj) because of Lemma 3 (Section 4). (jj) $\Rightarrow$ (v): Assume (jj). As (jj) $\Rightarrow$ (j) is trivial we also have (jv). Afortiori there are no rigid k-spaces, in other words $\dim S/T < \infty$. Furthermore, if $\|E\| \subset T$ then (by (jj)) $E_\infty = E$ and E is an orthogonal sum of hyperbolic planes.

(v) = (jjj): If (v) holds then $\dim E/E_* < \infty$ for all k-spaces E . Hence we can find decompositions $E = E_1 \overset{\perp}{\oplus} E_2$ with $\dim E_1 = \aleph_0$ and $\|E_1\| \subset T$ . By (5) $E_1$ is an orthogonal sum of $\aleph_0$ hyperbolic planes which shows that $T \subset \|E\|_\infty$ and $E_\infty$ is a linear subspace. As the non degenerate part of $E_\infty^\perp$ is rigid it is of finite dimension so that $E_\infty + E_\infty^\perp = R^\perp$ where $R = E_\infty^\perp \cap E_\infty$ . We have decompositions (cf. (11)) $E = (R \oplus R') \overset{\perp}{\oplus} E_1 \overset{\perp}{\oplus} E_2$ where $E_\infty = R \oplus E_1$ , $E_\infty^\perp = R \oplus E_2$ and where R' is rigid. Therefore $\dim E/E_1 < \infty$ and $E_1$ is stable in E . Q.E.D.

## References to Chapter VII

[1] F. Ayres, The expression of non-singular row-finite matrices in terms of strings. Ann.Univ.Sci. Budapest. Sectio Math. 7 (1964) 91-96.

[2] H. Gross and R.D. Engle, Bilinear forms on k-vector spaces of denumerable dimension in the case of char(k) = 2 , Comment. Math.Helv. 40 (1965) 247-266.

[3] G. Maxwell, Classification of countably infinite hermitean forms over skewfields. Amer. J. Math. 96 (1974) 145-155.

[4] J. Milnor, Symmetric inner products in characteristic 2. In "prospects in Mathematics" Ann. of Math. Studies, No. 70 Princeton Univ. Press 59-75.

[5] P. Vermes, Multiplicative groups of row- and column-finite infinite matrices. Ann. Univ. Sci. Budapest, Sectio Math. 5 (1962) 15-23.

CHAPTER EIGHT

# SUBSPACES IN NON-TRACE-VALUED SPACES

## 1. Introduction

All spaces considered in this chapter are of denumerable dimension and $\varepsilon$-hermitean over a field $k$ of characteristic 2 equipped with an antiautomorphism $\xi \mapsto \xi^*$ . With $k$ is associated the k-vector space $S/T$ ($S := \{\alpha \in k \mid \alpha = \varepsilon\alpha^*\}$ and $T := \{\alpha + \varepsilon\alpha^* \mid \alpha \in k\}$ the additive subgroups in $k$ of symmetric elements and traces respectively); $\varphi: E \to S/T$ is the k-vector space homomorphism which sends $x \in E$ into the coset $\Phi(x,x) + T$ . It is invariably assumed in this chapter that

(0) $\dim_k S/T < \infty$ .

This will enable us to make the most of stability and quasi-stability (Chapter VII). For example, the isometry class of a nondegenerate quasi-stable space $(E,\Phi)$ is determined by the isometry class of $E^{*\perp}$ and the subspaces $\varphi E$ and $\varphi E^{*\perp\perp}$ in $S/T$ (This follows from the corollary to the scholion in VII.10.). Here we set again $X^* := \{x \in X \mid \Phi(x,x) \in T\}$ and have, by (0), that $\dim X/X^* \leq \dim S/T < \infty$ for all subspaces $X \subset E$ . In particular $E^{*\perp}$ and $\varphi X$ will always be finite dimensional. We also recall that $E^{*\perp}$ is left pointwise fixed under any metric automorphism of $(E,\Phi)$ .

*

In this chapter we show how to put to use the results of Chapters IV and VII for the classification of subspaces in the unwieldy situation of non-trace-valued forms. We do this by discussing in detail the case of totally isotropic subspaces. Other cases may then be attacked in a similar fashion provided that the task of readying the indispensable lattice has been accomplished (see [1][2]).

Although the aim of this chapter is to a large extent the conveyance of a method the main result is of independent interest. Theorem 2 and its Corollary 1 in Section 7 give a complete characterization of the orbits of totally isotropic subspaces by means of invariants that can actually be handled in practical applications.

## 2. The lattice of a totally isotropic subspace ($\dim S/T < \infty$)

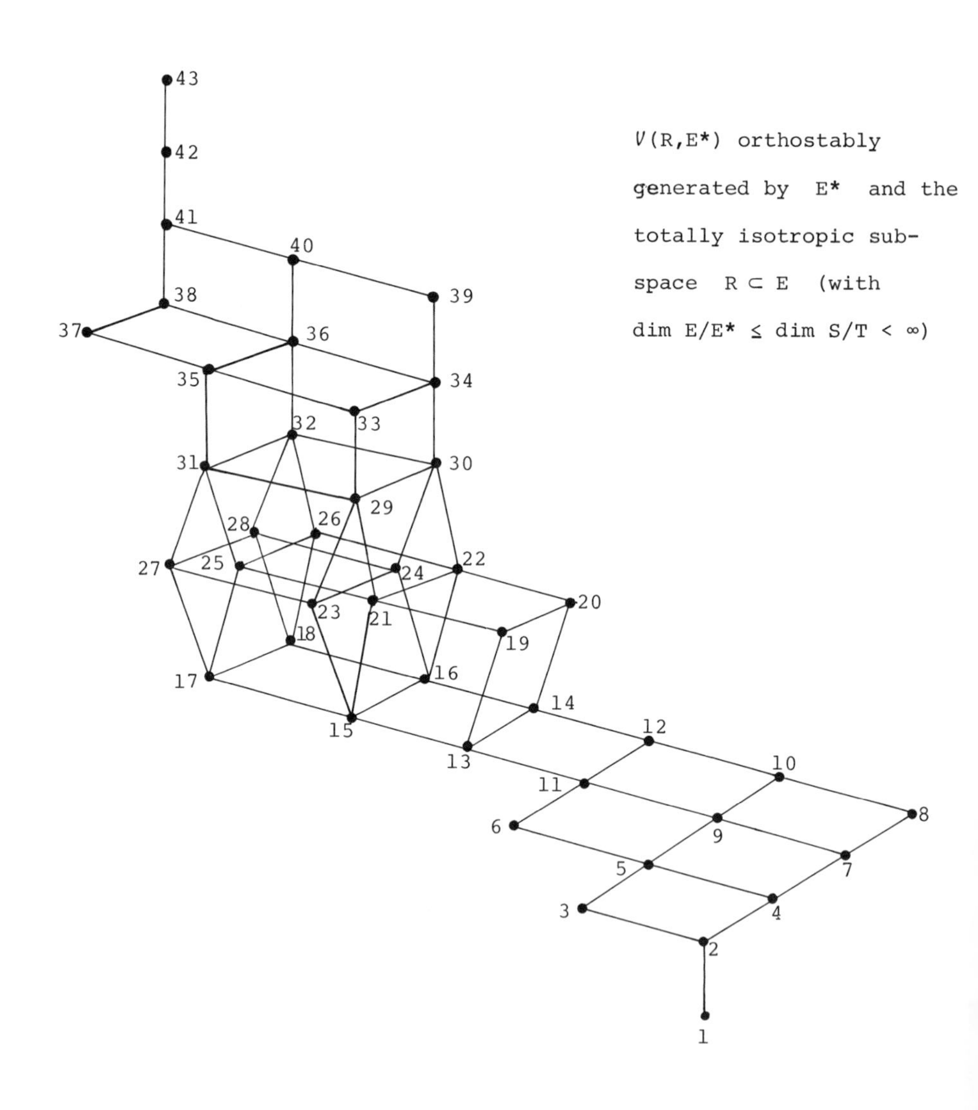


$V(R,E^*)$ orthostably generated by $E^*$ and the totally isotropic subspace $R \subset E$ (with $\dim E/E^* \leq \dim S/T < \infty$)

$1 = (0)$

$2 = R\cap E^{*\perp *}$

$3 = R$

$4 = R^{\perp\perp}\cap E^{*\perp *}$

$5 = R+R^{\perp\perp}\cap E^{*\perp *}$

$6 = R^{\perp\perp}$

$7 = E^{*\perp *}$

$8 = E^{*\perp}$

$9 = R+E^{*\perp *}$

$10 = R+E^{*\perp}$

$11 = R^{\perp\perp}+E^{*\perp *}$

$12 = R^{\perp\perp}+E^{*\perp}$

$13 = R^{\perp *\perp *}$

$14 = R^{\perp *\perp *}+E^{*\perp}$

$15 = R^{\perp *}$

$16 = R^{\perp *}+E^{*\perp}$

$17 = E^{*}$

$18 = E^{*}+E^{*\perp}$

$19 = E^{*\perp\perp}\cap R^{\perp *\perp}$

$20 = R^{\perp *\perp}$

$21 = R^{\perp *}+E^{*\perp\perp}\cap R^{\perp *\perp}$

$22 = R^{\perp *}+R^{\perp *\perp}$

$23 = R^{\perp *\perp\perp}$

$24 = R^{\perp *\perp\perp}+E^{*\perp}$

$25 = E^{*}+E^{*\perp\perp}\cap R^{\perp *\perp}$

$26 = E^{*}+E^{*\perp\perp}\cap R^{\perp *\perp}+E^{*\perp}$

$27 = E^{*}+R^{\perp *\perp\perp}$

$28 = E^{*}+R^{\perp *\perp\perp}+E^{*\perp}$

$29 = E^{*\perp\perp}\cap R^{\perp *\perp *\perp}$

$30 = R^{\perp *\perp *\perp}$

$31 = E^{*}+E^{*\perp\perp}\cap R^{\perp *\perp *\perp}$

$32 = E^{*}+R^{\perp *\perp *\perp}$

$33 = E^{*\perp\perp}\cap R^{\perp}$

$34 = E^{*\perp *\perp}\cap R^{\perp}$

$35 = E^{*}+E^{*\perp\perp}\cap R^{\perp}$

$36 = E^{*}+E^{*\perp *\perp}\cap R^{\perp}$

$37 = E^{*\perp\perp}$

$38 = E^{*\perp *\perp}$

$39 = R^{\perp}$

$40 = E^{*}+R^{\perp}$

$41 = E^{*\perp\perp}+R^{\perp}$

$42 = (R\cap E^{*\perp *})^{\perp}$

$43 = E$

The nice diagram is reproduced from [3]. We shall construct examples where all 43 elements are different spaces (a division ring with $\dim S/T \geqq 8$ is needed for this to be possible).

The utility of the diagram is greatly increased by a table for the operation $\perp$ in $\mathcal{V}(R,E^{*})$ :

| X | 1 | 2 | 4 | 3 5 6 | 7 | 8 | 9 11 | 10 12 | 13 | 19 | 14 | 20 | 15 23 | 16 24 |
|---|---|---|---|---|---|---|---|---|---|---|---|---|---|---|
| $X^{\perp}$ | 43 | 42 | 41 | 39 | 38 | 37 | 34 | 33 | 30 | 24 | 29 | 23 | 20 | 19 |

| 17 25 27 31 35 37 | 18 26 28 32 36 38 | 21 29 | 22 30 | 33 | 34 | 39 | 40 41 | 42 | 43 |
|---|---|---|---|---|---|---|---|---|---|
| 8 | 7 | 14 | 13 | 12 | 11 | 6 | 4 | 2 | 1 |

## 3. Remarks on the verification of diagrams

Although the diagram for $\mathcal{V}(R,E^*)$ in the previous section is still modest in size it is complex enough to require a systematic verification. And verified it must be! For, it has not been devised for purposes of illustration or for interrupting the monotony of the text; the diagram is a tool without the help of which we could not have delved into our problem.

It is for the sake of the novitiate that we adduce some hints for this verification (to find the diagram is another matter).

First, we convince ourselves that the partial order depicted by the unlabelled diagram - call it $\mathcal{D}$ - is a lattice. The legend then sponsors a map $\nu: \mathcal{D} \to L(E)$ into the lattice of all subspaces of the vector space $E$ (it sends the element 7 of $\mathcal{D}$ into $E^{*\perp *} \in L(E)$, etc.) This map $\nu$ must be shown to be a lattice homomorphism and its image in $L(E)$ stable under $\perp: L(E) \to L(E)$. This is accomplished as follows.

Second, verify that $\nu$ respects the ordering by checking all pairs of neighbouring elements in $\mathcal{D}$. For example, we have $22 \leq 26$ by $\mathcal{D}$; do we have $R^{\perp *} + R^{\perp * \perp} \subset E^* + E^{*\perp\perp} \cap R^{\perp * \perp} + E^{*\perp}$ for the images under $\nu$? As an illustration let us verify this in detail.

Since $R^{\perp *} \subset E^*$ (trivially) it suffices to show that $R^{\perp * \perp}$ is contained in the space $E^{*\perp\perp} \cap R^{\perp * \perp} + E^{*\perp} = (E^{*\perp}+R^{\perp *})^{\perp} + E^{*\perp}$ (which is $\perp$-closed because $\dim E^{*\perp} \leq \dim S/T$ is finite). We try to establish the converse inclusion for the respective orthogonals; as both spaces are $\perp$-closed this will be equivalent to the former inclusion. We thus wish to show that $(E^{*\perp}+R^{\perp *})^{\perp\perp} \cap E^{*\perp\perp} \subset R^{\perp *\perp\perp}$. Since $\dim E^{*\perp}$ is finite we have $(E^{*\perp}+R^{\perp *})^{\perp\perp} = E^{*\perp} + R^{\perp *\perp\perp}$ and, by using modularity, our assertion translates as $E^{*\perp} \cap E^{*\perp\perp} + R^{\perp *\perp\perp} \subset R^{\perp *\perp\perp}$. We have reduced the problem to proving that

(1) $$E^{*\perp} \cap E^{*\perp\perp} \subset R^{\perp *\perp\perp} .$$

This is easily done. $R$ belongs to $E^*$ and afortiori $R \subset (\text{rad } E^*)^{\perp}$. Thus $\text{rad } E^* \subset R^{\perp}$. Since $\text{rad } E^* = \text{rad } (E^{*\perp})$ we are done. Most of the remaining seventyfive checks for inclusion will be seen to be quite trivial. Having in this fashion settled the inclusions one is in a position to turn to the next step.

_Third_, verify that the map $\nu$ respects joins. For example, the element 20 of $\mathcal{D}$ is the join of 19 and 8; do we have $(E^{*\perp\perp} \cap R^{\perp *\perp}) + E^{*\perp} = R^{\perp *\perp}$ for the images under $\nu$? One inclusion is trivial and, by $\perp$-closedness, we may turn to the corresponding orthogonals and establish $(E^{*\perp} + R^{\perp}*)^{\perp\perp} \cap E^{*\perp\perp} \subset R^{\perp *\perp\perp}$. But this was verified in the example under step 2 above. In this fashion one checks all joins of pairs of elements in $\mathcal{D}$. There are, in principle, $\binom{43}{2}$ checks to be made. However, the work is cut down drastically by observing that it suffices to establish that each join-irreducible element gives the correct join with _each_ other element.

_Fourth_, verify that $\nu$ respects meets. This step is very similar to the preceeding one (the steps may be interchanged). It suffices to show that each meet-irreducible element gives the right meets with _all_ other elements (in the image of $\nu$). Having already done the joins, the checking of meets can sometimes be simplyfied by writing elements as suitable sums of other elements.

When the last two steps have been accomplished then we know that $\nu$ is a lattice homomorphism, i.e. the _labelled_ diagram gives the correct joins and meets. There remains the last step.

<u>Fifth</u>, check orthostability. Since $(X+Y)^\perp = X^\perp \cap Y^\perp$ only join-irreducible elements have to be looked at. For example, where is the orthogonal of $R^{\perp*\perp*\perp}$ (number 30)? Looking at a metabolic decomposition for $R^{\perp\perp}$, $E = (R^{\perp\perp}\oplus R') \overset{\perp}{\oplus} E_o$, we see that $R^{\perp*\perp*} = R^{\perp\perp} \oplus E_o^{*\perp o*}$. Hence $R^{\perp*\perp*}$ is $\perp$-closed because $\dim(R^{\perp*\perp*}/R^{\perp\perp}) = \dim E_o^{*\perp o*} \leq \dim S/T < \infty$ (as $E_o$ is nondegenerate). We answer our question simply by saying that $30^\perp = 13$.

## 4. Totally isotropic subspaces: the indices

Let $(R_\iota, E_\iota)_{\iota\in I}$ be a family where, for all $\iota\in I$, $E_\iota$ is a non-degenerate $\varepsilon$-hermitean space over $k$ and $R_\iota$ a totally isotropic subspace of $E_\iota$. If $E$ is the (external) orthogonal sum of the $E_\iota$ and $R$ the subspace $\Sigma R_\iota$ then we call $(R,E)$ <u>the sum of the pairs</u> $(R_\iota, E_\iota)$. A pair $(R,E)$ is called <u>reducible</u> if it is a sum of two pairs $(R_1,E_1)$, $(R_2,E_2)$ such that $E^* = E_1^* + E_2^*$ and $E_1 \neq E_1^*$, $E_2 \neq E_2^*$.

In the next section we shall list certain irreducible pairs. In a later section we shall prove that arbitrary pairs are sums of such. Here we shall introduce the indices of a pair needed further on.

Let $R \subset R^\perp \subset E$ and, as usual, $E$ be nondegenerate. Let furthermore $M$ be a supplement of $E^{*\perp*}$ in $E^{*\perp}$ and $N := R \cap E^{*\perp} = R \cap E^{*\perp}$. Since $M$ is nondegenerate and of finite dimension we have $E = M + M^\perp$. $M^\perp$ admits a metabolic decomposition with respect to $N$. As $R \subset E^* \subset M$ and $R \subset N^\perp$ we have therefore a decomposition as follows:

$$(2) \qquad \begin{aligned} E &= M \overset{\perp}{\oplus} (N\oplus N') \overset{\perp}{\oplus} E_o \\ E^* &= \qquad\quad N \qquad\quad \oplus E_o^* \\ R &= \qquad\quad N \qquad\quad \oplus E_o \cap R \end{aligned}$$

Because every metric automorphism of $E$ leaves $E^{*\perp}$ pointwise fixed we shall have to assume that

(3) $$R \cap E^{*\perp *} = \bar{R} \cap E^{*\perp *}$$

i.e. $N = \bar{N}$ if the totally isotropic $\bar{R}$ is to belong to the orbit of $R$ . Therefore, it may be assumed that $M = \bar{M}$ and $N' = \bar{N}'$ in a decomposition for $\bar{R}$ analogous to that in (2). We see that the discussion about $R$ in $E$ is shifted entirely to the discussion about $R_o :=$ $E_o \cap R$ in $E_o$ . In other words, it suffices to consider the case of a pair $(R,E)$ that satisfies

(4) $$E^{*\perp *} = E^{*\perp} \text{ and } R \cap E^{*\perp} = (0) .$$

Assuming (4) the lattice $V(R,E^*)$ is as follows:

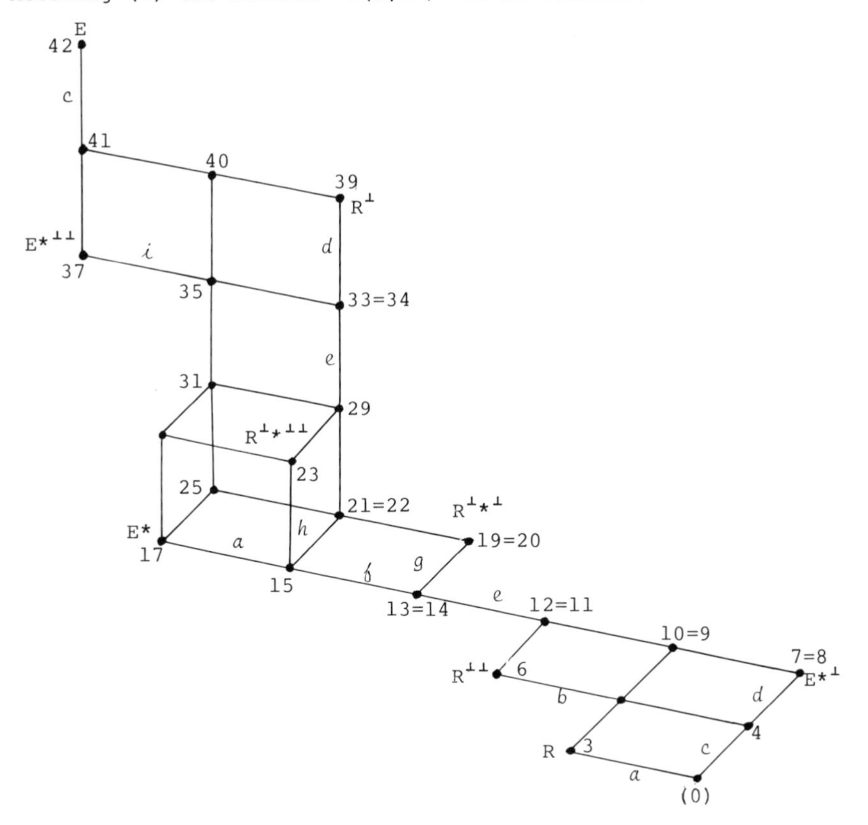

Let $(x/y)$ be the dimension of the quotient space $X/Y$ where $X,Y$ carry the numbers $x$ and $y$ respectively in the legend to the diagram in Section 2. We then define <u>indices</u> by

$$(5)\qquad \begin{aligned} &a := (3/2)\ ,\ b := (6/5)\ ,\ c := (4/2)\ ,\ d := (7/4)\ ,\ e := (13/11) \\ &f := (15/13)\ ,\ g := (19/13)\ ,\ h := (23/15)\ ,\ i := (37/35)\ . \end{aligned}$$

If (4) is not assumed we have to add the following

$$(6)\qquad m := (8/7) = \dim M\ ,\quad n := (2/1) = \dim N\ .$$

One then proves the following equalities for <u>arbitrary</u> $(R,E)$, $\dim E \le \aleph_0$,

$$(17/15) = a\ ,\quad (33/29) = e\ ,\ (39/34) = d\ ,\ (43/42) = n\ .$$

The eleven cardinals introduced in (5) and (6) are obvious invariants of the orbit of $R$ under the action of the orthogonal group of $E$. By modularity all dimensions of quotients of elements in $V(R,E^*)$ can be expressed by these eleven cardinals $a, b, \ldots, i, m, n$. (If the dependence on the pair $(R,E)$ is to be exhibited we write $a(R)$, $b(R)$, etc.)

We have the following relations among the indices:

<u>Lemma 1</u>. (i) $c + d + e + g + h + i + m + n \le \dim S/T < \infty$

(ii) $a < \infty \Rightarrow b = c = e = i = g = \text{zero}$

(iii) $f < \infty \Rightarrow h = \text{zero}$

(iv) $f < \infty \ \&\ T = \{0\} \Rightarrow f$ is even

In order to prove the lemma we need the following fact:

<u>Lemma 2</u>. Let $R \subset R^\perp \subset E$, $E$ nondegenerate, $\dim E \le \aleph_0$. If $\dim R$ is finite then $E$ admits a Witt-decomposition for $R$ if and

only if $R \cap E^{*\perp} = (0)$ .

The easy proof is by induction on dim R and is left to the reader.

<u>Proof</u> of Lemma 1. (i) $\dim E/E^* \leq \dim S/T$ .

(ii) If $a < \infty$ then $\dim R = a + n < \infty$ and $R = R^{\perp\perp}$ . Hence $b + c$ is zero. Furthermore, as space 29 is $\perp$-closed and of codimension $a$ in space 31, we must have $31^{\perp\perp} = 31$ . The $\perp$-table in Section 2 gives $31^{\perp\perp} = 37$ so $e + i$ is zero. In order to compute $g$ one may, by decomposition (2), assume that (4) holds. Since then $R \cap E^{*\perp} = (0)$ we may quote Lemma 2 and assume a decomposition $E = (R \oplus R') \overset{\perp}{\oplus} E_o$ with $R \oplus R' \subset E^*$ . In this setting it is easy to check that the spaces 14 and 20 coincide, i.e. $g = 0$ .

(iii) Since 20 is $\perp$-closed and of codimension $f$ in space 22 we obtain here $22 = 22^{\perp\perp} = 30$ . Thus $h$ is zero.

(iv) Let us use a metabolic decomposition $E = (R^{\perp\perp} \oplus R') \overset{\perp}{\oplus} E_o$ for $R^{\perp\perp}$ . We find $f = \dim E_o^*/\text{rad } E_o^*$ so that $f$ is the dimension of a non-degenerate trace-valued $\varepsilon$-hermitean space. If $T = \{0\}$ , i.e. k commutative and the form symmetric, then such spaces are sums of hyperbolic planes. Hence the assertion.

The proof of Lemma 1 is thus complete. By Theorem 1 in Section 6 below we see that the lemma lists all relations to which the indices are subjected.

## 5. <u>Totally isotropic subspaces: the irreducible objects</u>

<u>5.1</u>. Let E be a hyperbolic plane and R an isotropic line in E . The pair (R,E) has $a = 1$ and all other indices zero.

5.2. Let $E$ be an orthogonal sum of hyperbolic planes $k(r_i, r_i')$ $i \in \mathbb{N}$. Let $R$ be the span of all $r_i + r_o$. The pair $(R,E)$ has $b = 1$ and all other indices zero except for $a$ which has to be $\aleph_0$ by Lemma 1.

5.3. Let $E$ be an orthogonal sum of hyperbolic planes $k(r_i, r_i')$ $i \in \mathbb{N}$, and a space $\langle \alpha, \alpha \rangle$ where $\alpha \in S \setminus T$. Let $R$ be the span of all $r_i + z$ where $0 \neq z$ is an isotropic vector in $\langle \alpha, \alpha \rangle$. The pair $(R,E)$ has $c = 1$ and all other indices zero except for $a$ which is $\aleph_0$ by Lemma 1.

5.4. $R := (0) \subset E := \langle \alpha, \alpha \rangle$ with $\alpha \in S \setminus T$ gives a pair with $d = 1$ and all other indices zero.

5.5. Let $E$ be an orthogonal sum of metabolic planes $k(r_i, r_i') \cong \begin{pmatrix} 0 & 1 \\ 1 & \alpha \end{pmatrix}$, $i \in \mathbb{N}$ and $\alpha \in S \setminus T$, and a plane $\langle \alpha, \alpha \rangle$. Let $R$ be the span of the $r_i$. The pair $(R,E)$ has $e = 1$ and all other indices zero except for $a$ which has to be $\aleph_0$ by Lemma 1.

5.6. If $T = \{0\}$ we let $E$ be a hyperbolic plane, if there is $\gamma \in T \setminus \{0\}$ we let $E = \langle \gamma \rangle$; set $R = (0)$ in either case. The pair $(R,E)$ has $\mathfrak{f} = 2$ or $1$ and all other indices equal to zero.

5.7. Pick $\alpha \in S \setminus T$ and let $E$ be an orthogonal sum of metabolic planes $k(r_i, r_i')$ as in 5.5 and a line $\langle \alpha \rangle$. Let $R$ be as in 5.5. The pair $(R,E)$ has $g = 1$ and all other indices zero except for $a$ which has to be $\aleph_0$ by Lemma 1.

5.8. Here is an example with $h = 1$ and all other indices zero except for $\mathfrak{f}$ which has to be $\aleph_0$ by Lemma 1: $E = \langle \alpha, \alpha, \ldots \rangle$ and $R = (0)$ with $\alpha \in S \setminus T$. Here the nondegenerate part of 15 is hyperbolic. In view of 5.6 we are therefore left with the possibilities of an anisotropic space 15. Let $(e_i)_{i\in\mathbb{N}}$ be an orthogonal basis of an arbitrary anisotropic trace-valued space $E_o$ . Let $E = E_o \oplus (a)$ where all products $\Phi(e_i, a) = 1$ and $\alpha := \Phi(a,a) \in S\setminus T$ . For $R := (0) \subset E$ we find $h = 1$ and $15 = E_o$ . Our list is yet too small! Later on we shall need enough irreducible pairs to enforce isometries $E \cong \bar{E}$ when the pair $(R,E)$ is given and the pair $(\bar{R},\bar{E})$ is to be an orthogonal sum of irreducible pairs from the list such that $(\bar{R},\bar{E})$ coincides with $(R,E)$ indexwise. This presents no problem if $\dim R$ is infinite for then we have stability. If $\dim R$ is finite we are at a loss with our examples since e.g. a space with $h = 2$ need not be an orthogonal sum of two spaces with $h = 1$ . Because it would be very awkward to formally exclude the case with finite $\dim R$ we have no choice but to include here, in one lump, all irreducible pairs $((0),E)$ when $E$ is anisotropic, $E^{*\perp\perp} = E$ and $h = \dim E/E^{*\perp\perp} < \infty$ .

5.9. Let $E$ be a sum of metabolic planes as in 5.5 and let $R$ be as in 5.5. The pair $(R,E)$ has $i = 1$ and all other indices zero except for $a$ which has to be $\aleph_0$ by Lemma 1.

5.10. For $\alpha \in S\setminus T$ set $R := (0) \subset E := \langle\alpha\rangle$ . Here $m = 1$ and all other indices zero.

5.11. For $\alpha \in S\setminus T$ set $E := \langle\alpha,\alpha\rangle$ and let $R$ be the isotropic line in $E$ . Then $n = 1$ and all other indices are zero.

We summarize by giving the following table of irreducible pairs $(R,E)$ constructed above:

| | $a$ | $b$ | $c$ | $d$ | $e$ | $f$ | $g$ | $h$ | $i$ | $m$ | $n$ | $E^*$ | dim R $(a+n)$ | dim E |
|---|---|---|---|---|---|---|---|---|---|---|---|---|---|---|
| 1 | 1 | | | | | | | | | | | $E=E^{*\perp\perp}=E^*$ | 1 | 2 |
| 2 | $\aleph_0$ | 1 | | | | | | | | | | $E=E^{*\perp\perp}=E^*$ | $\aleph_0$ | $\aleph_0$ |
| 3 | $\aleph_0$ | | 1 | | | | | | | | | $E\neq E^{*\perp\perp}=E^*$ | $\aleph_0$ | $\aleph_0$ |
| 4 | | | | 1 | | | | | | | | $E\neq E^{*\perp\perp}=E^*$ | 0 | 2 |
| 5 | $\aleph_0$ | | | | 1 | | | | | | | $E=E^{*\perp\perp}\neq E^*$ | $\aleph_0$ | $\aleph_0$ |
| 6 | | | | | | 1,2 | | | | | | $E=E^{*\perp\perp}=E^*$ | 0 | 1,2 |
| 7 | $\aleph_0$ | | | | | | 1 | | | | | $E=E^{*\perp\perp}\neq E^*$ | $\aleph_0$ | $\aleph_0$ |
| 8 | | | | | | $\aleph_0$ | | 1 | | | | $E=E^{*\perp\perp}\neq E^*$ | 0 | $\aleph_0$ |
| 9 | $\aleph_0$ | | | | | | | | 1 | | | $E=E^{*\perp\perp}\neq E^*$ | $\aleph_0$ | $\aleph_0$ |
| 10 | | | | | | | | | | 1 | | $E\neq E^{*\perp\perp}=(0)$ | 0 | 1 |
| 11 | | | | | | | | | | | 1 | $E\neq E^{*\perp\perp}=E^*$ | 0 | 2 |

Notice that in all cases $E^*$ is either $\perp$-closed or $\perp$-dense (or both). Examples 1, 2, 6 are trace-valued. If we are not in the symmetric case $(T = \{0\})$ an element $\gamma \in T\setminus\{0\}$ enters the description of Example 6. In the remaining examples 3, 4, 5, 7, 8, 9, 10, 11 invariably $\dim E/E^* = 1$ ; in their description enters the "parameter" $\alpha \in S\setminus T$ . If $\alpha$ is varied in $S\setminus T$ we obtain different irreducible pairs. Their isometry types correspond uniquely with the different lines in the k-vector space $S/T$ . If we vary $\gamma$ in $T\setminus\{0\}$ and $\alpha$ in $S\setminus T$ we get <u>all</u> the building blocks needed to build an arbitrarily given pair (R,E). Of this we shall treat in the next two sections.

## 6. The invariants of a totally isotropic subspace

The k-vector space homomorphism $\varphi: E \to S/T$ with

$$x \mapsto \Phi(x,x) + T$$

induces a map $L(E) \to L(S/T)$ and a lattice homomorphism which we call $\varphi$ too,

$$V(R,E^*) \to L(S/T) \ .$$

We are particularly interested in the image of the lattice $V(R,E^*)$ in Section 1 because the image lattice $\varphi V(R,E^*)$ is an obvious invariant of the orbit of $R$ under the action of the orthogonal group. In order to obtain a diagram of the most general homomorphic image of $V(R,E^*)$ we simply contract the diagram in Section 2 "along the direction of E*" (as if we were to push together the bellows of a camera). The result is

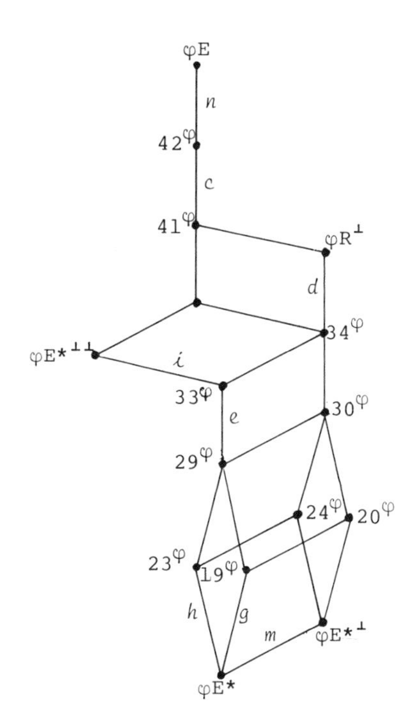

Referring to the above diagram we introduce subsets

(7) $\mathcal{M}$ , $\mathcal{G}$ , $\mathcal{H}$ , $\mathcal{E}$ , $\mathcal{J}$ , $\mathcal{D}$ , $\mathcal{C}$ , $\mathcal{N}$

of S in the following way. Let $\mathcal{M}$ be any set of $m$ elements $\Phi(e,e) \in S$ where e runs through an orthogonal basis of a supplement M of $E^{*\perp *}$ in $E^{*\perp}$ . (We consider <u>rigid</u> spaces as "building blocks" here and do not analyze them any further.) Similarly, let $\mathcal{G}$ be a set of elements $\Phi(e,e)$ , where e runs through an orthogonal basis of a supplement G of $R^{\perp *\perp *} + E^{*\perp}$ in $R^{\perp *\perp}$ ; card $\mathcal{G} = g$ . In other words, we are going to elect the rigid spaces M and G as members of a certain collection of invariants; the spaces M and M+G are the isometry classes of the "nondegenerate part" of the spaces $E^{*\perp}$ and $R^{\perp *\perp}$ . The remaining sets $\mathcal{H}$ , $\mathcal{E}$ , $\mathcal{J}$ , $\mathcal{D}$ , $\mathcal{C}$ , $\mathcal{N}$ are <u>arbitrary</u> subsets of S such that the images in S/T span - in turn - a supplement of $15^{\varphi}$ in $23^{\varphi}$ , of $29^{\varphi}$ in $33^{\varphi}$ , of $33^{\varphi}$ in $\varphi E^{*\perp\perp} = 37^{\varphi}$ , of $34^{\varphi}$ in $\varphi R^{\perp} = 39^{\varphi}$ , of $41^{\varphi}$ in $42^{\varphi}$ , of $42^{\varphi}$ in $\varphi E$ .

Let then (R,E) be an arbitrarily given pair and (7) the subsets associated with it. We wish to define a sum $(\bar{R},\bar{E})$ of irreducible pairs from the list in Section 5 which has the same indices and the same invariants (7) as (R,E) and such that an isometry $E \cong \bar{E}$ maps $R \cap E^{*\perp}$ onto $\bar{R} \cap \bar{E}^{*\perp}$ . We start by taking a sum of $m$ copies of $(R_{10},E_{10})$ as described in Section 5.10 where the parameter $\alpha$ runs through $\mathcal{M}$ . Let $(R(\mathcal{M}),E(\mathcal{M}))$ be the resulting pair. Similarly we build pairs $(R(\mathcal{G}),E(\mathcal{G}))$ , ... , $(R(\mathcal{N}),E(\mathcal{N}))$ by taking orthogonal sums of pairs of the kind described - in turn - in Section 5.7, 5.8, 5.5, 5.9, 5.4, 5.3, 5.11. In each sum we let the parameter $\alpha$ used in the description of the irreducible pairs run through the corresponding sets (7). In 5.8 we have, in the situation of finite dim R , to select an irreducible example with the correct space $\overline{15}$ , i.e. with $\overline{15}$ isometric to the

corresponding subspace of E . There is no problem with $\dim(\mathrm{rad}\ 15) = \dim 13 = n + e + d + c + b + a$ . If we add up everything, $(\tilde{R},\tilde{E}) := (R(M),E(M)) \overset{\perp}{\oplus} \cdots \overset{\perp}{\oplus} (R(N),E(N))$ we have a pair $(\tilde{R},\tilde{E})$ whose lattice $\varphi V(\tilde{R},\tilde{E}^*)$ coincides with the lattice $\varphi V(R,E^*)$ in S/T of the given pair. (To this one only has to remark that the union $M \cup G \cup \ldots \cup N$ has an image family under $\varphi$ that is linearly independent in S/T so that $\tilde{E}^*$ is the sum $\Sigma E_l^*$ of all the $E_l$ used in the construction of the sum; all indices then behave additively.) By adding to the pair $(\tilde{R},\tilde{E})$ a sum of $a(R)$ copies of the irreducible pair in Section 5.1 we can furthermore adjust the invariant $a$ to the given $a(R)$ . Similarly we adjust $b$ by adding $b(R)$ copies of the pair described in Section 5.2 Call $(\bar{R},\bar{E})$ this final sum and observe that it has $\bar{E}$ isometric to E . This follows at once from the scholion in VII.6 when $\dim \bar{R} = \aleph_0$ : since $T \subset \|E\|_\infty = \|E^{*\perp\perp}\|$ the invariants $\|E\|$ and $\|E^{*\perp\perp}\|$ are made up of full equivalence classes of S/T so that they, together with $M$ , fix the isometry class of E according to the scholion .
We summarize:

<u>Theorem 1</u>. Let E be a nondegenerate $\varepsilon$-hermitean space of dimension $\leq \aleph_0$ and $R \subset R^{\perp} \subset E$ . Then there exists an orthogonal sum $(\bar{R},\bar{E})$ of irreducible pairs as listed in Section 5 such that the following hold.

(8)

(i) E and $\bar{E}$ are isometric under an isometry which maps $R \cap E^{*\perp}$ onto $\bar{R} \cap \bar{E}^{*\perp}$ .

(ii) $a(R) = a(\bar{R})$ , $b(R) = b(\bar{R})$ .

(iii) $R^{\perp *}$ and $\bar{R}^{\perp *}$ are isometric.

(iv) $R^{\perp * \perp}$ and $\bar{R}^{\perp * \perp}$ are isometric.

(v) There is a lattice isomorphism $\tau: V(R,E^*) \to V(\bar{R},\bar{E}^*)$ which sends R in $\bar{R}$ and renders commutative the diagram:

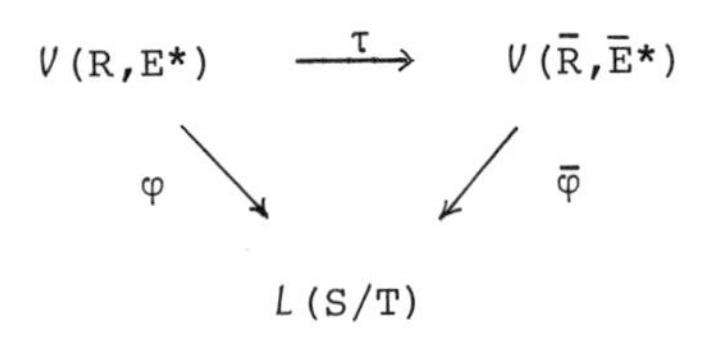


Remark 1. Statements (i) to (v) contain some redundancy. If, instead of (v), we only know that the lattice $\varphi V(R,E^*)$ is identical with the lattice $\bar{\varphi}V(\bar{R},\bar{E}^*)$ then we have equality of all corresponding indices, except for $a$ , $b$ and $\mathfrak{f}$ . But $a$ , $b$ and $\mathfrak{f}$ are taken care of by (ii) and (iii). We see that there exists an index preserving lattice isomorphism $\tau$ with $\varphi = \bar{\varphi}\circ\tau$ .

Remark 2. Let us look at the irreducible pairs in Section 5 once more. The trace-valued types in 5.1, 5.2, 5.6 are irreducible in the strict sense that they cannot be further decomposed into summands $\neq ((0),(0))$ . On the other hand, arbitrary sums of such pairs are irreducible by our definition (at the beginning of Section 4) because $E = E^*$ . Since the indices not $a$ , $b$ , $\mathfrak{f}$ are finite we have

(9) For given $(R,E)$ there is a sum $(\bar{R},\bar{E})$ of finitely many irreducible pairs that satisfies (8) in Theorem 1.

Obviously, the summands in (9) are not - and cannot be - uniquely determined.

Remark 3. Let in $L(S/T)$ be given any sublattice $W$ of the shape depicted by the diagram at the beginning of this section. It is now clear that by introducing supplements between suitable pairs of neighbouring elements in $W$ we can arrive at sets $M$ , $G$ , ... , $N$ (as in (7)) which in turn can be used for the construction of a pair $(\bar{R},\bar{E})$ as in the proof of Theorem 1. We list this result as a

Corollary. For any sublattice $\mathcal{W}$ in $L(S/T)$ of the shape

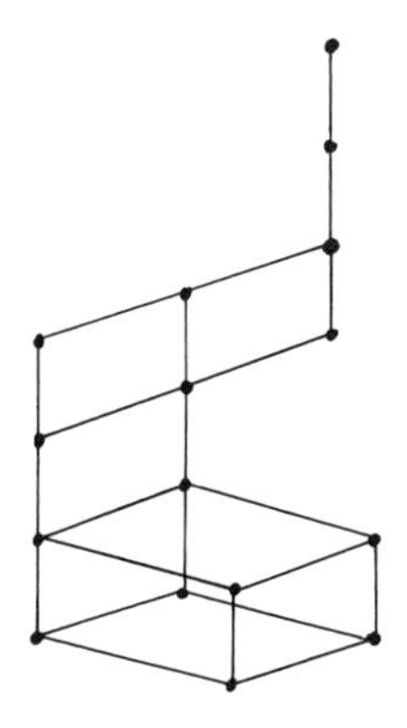

we can define a nondegenerate $\varepsilon$-hermitean space $(E,\Phi)$ with the following property. E contains a totally isotropic subspace R such that $\varphi\mathcal{V}(R,E^*) = \mathcal{W}$ . By adding to $(R,E)$ pairs of the kind described in Section 5.1, 5.2, 5.6 we can furthermore alter the isometry type of $R^{\perp *}$ and enlarge the cardinals $a(R)$ , $b(R)$ . In particular, we see that there are no other relations between the invariants $a(R)$ , $b(R)$ , $\mathfrak{f}(R)$ , $\varphi\mathcal{V}(R,E^*)$ than are listed in Lemma 1.

Remark 4. Notice that if (i) in (8) holds then each isometry $E \cong \bar{E}$ will map $R \cap E^{*\perp}$ onto $\bar{R} \cap \bar{E}^{*\perp}$ .

We are now ready to insert the keystone.

## 7. The decomposition theorem

In order to prove that the pair $(\bar{R},\bar{E})$ in Theorem 1 - or in (9) - is equivalent to the pair $(R,E)$ , i.e. in order to construct an isometry $E \cong \bar{E}$ which maps R onto $\bar{R}$ we have to make the following incisive assumption:

(10) If $\dim R^{\perp *}/\mathrm{rad}(R^{\perp *})$ is infinite, then $R^{\perp *}$ contains an infinite dimensional totally isotropic subspace disjoint from the radical.

Remark. Condition (10) says that in the case of an infinite index $\mathfrak{s}$ the space 13 possesses a quasistable supplement $F$ in 15 . As $15 \subset E^*$ the space $F$ is then a sum of hyperbolic planes. There are fields $(k,*,\varepsilon)$ such that there is only one isometry class of trace-valued $\varepsilon$-hermitean spaces in dimension $\aleph_0$ ; hence in these cases (10) is automatically satisfied. Condition (10) is not only natural in the light of methods introduced in Chapter IV. Assuming (10) is a necessity if we wish to prove a general result to the effect that the indices and the lattice $\varphi\mathcal{V}(R,E^*)$, in $S/T$, of value-spaces constitute, essentially a complete set of invariants for the pair $(R,E)$. In the absence of isotropic vectors the arithmetical properties of the field play a sensitive part in the characterization of spaces (see Chapters XI, XII, XIII). On the other hand, it may be possible to omit (10) in special cases such as treated in Theorem 3 below.

Theorem 2. Let $E$ and $\bar{E}$ be $\varepsilon$-hermitean $k$-spaces, nondegenerate and of dimension $\aleph_0$ . Assume that $k$ has finite $\dim S/T$ . Let $R \subset E$ and $\bar{R} \subset \bar{E}$ be totally isotropic subspaces such that conditions (8) are satisfied. In order that there is an isometry $E \cong \bar{E}$ which maps $R$ onto $\bar{R}$ it is sufficient that (10) is satisfied.

Proof. Since we can write down decompositions of the kind (2) for both $R$ and $\bar{R}$ and since $E^{*\perp} \cong \bar{E}^{*\perp}$ , $R \cap E^{*\perp} \cong \bar{R} \cap \bar{E}^{*\perp}$ under isometries of the whole spaces $E$ and $\bar{E}$ we can conclude that (4), and the corresponding equality for $(\bar{R},\bar{E})$ , may be assumed without loss of generality. (Chopping off isometric finite dimensional orthogonal summands is awkward since the "Cancellation Theorem" does not hold when $E \neq E^*$ .) Condition (8) is inherited in the reduced situation. This verification and similar ones below are left to the reader. Although such calculations may be not difficult they are by no means to be neglected.

We are thus in the situation of the diagram in Section 4. This diagram and the $\perp$-table of Section 2 will be used freely in what follows. We start out by considering

<u>Case $\alpha = \delta = \aleph_0$</u> . Corresponding to (7) let $G$ , $H$ , $\mathcal{E}$ , $I$ , $D$ , $C$ be supplements - in turn - of 13 in 19 , of 15 in 23 , of 29 in 33 , of 35 in 37 , of 33 in 39 , of 41 in 42 . We show that we can arrange it such that the six spaces are pairwise perpendicular. Let us start with $C$ at the top of the diagram. Pick a preliminary $D$ . We shall change $D$ mod 15 . Suppose the first member $x_1$ of an orthogonal basis of $D$ is not orthogonal to a basis $(y_i)$ of $C$ . Since $C \cap 15^{\perp} = (0)$ we can find $t_1 \in 15$ with $\Phi(t_1, y_i) = \Phi(x_1, y_i)$ for all $y_i$ . We replace $x_1$ by $x_1 + t_1$ . We then treat the next element $x_2$ of the basis of $D$ : we find a vector $t_2 \in 15$ such that $x_2 + t_2$ is orthogonal to all of $C$ and $x_1 + t_1$ . In this fashion we find a space $D_1 \equiv D$ (mod 15) with $D_1 \perp C$ . We may have altered the isometry class of $D$ by swiching to $D_1$ , but $\Phi(x_i, x_i) \equiv \Phi(x_i + t_i, x_i + t_i)$ mod $T$ . In view of (10) we can find in $(C + D_1)^{\perp} \cap 15$ an orthogonal family $s_1$ , $s_2$ , ... such that the system $x_1 + t_1 + s_1$ , $x_2 + t_2 + s_2$ , ... is congruent to the original system $x_1$ , $x_2$ , ... . Call $D_2$ the span of the new system. Now we adjust $I$ to the space $C + D_2$ by modulating it modulo 15 . In this fashion we can work our way down to $H$ . Dropping subscripts which may have accumulated we assume that $C$ , $D$ , ... , $H$ are pairwise orthogonal and of prescribed isometry class. What can be done about $G$ ? Since $13^{\perp} \cap (C + D + I + \mathcal{E}) = (0)$ we can, by the same procedure, achieve $G \perp C + D + I + \mathcal{E}$ by changing it modulo 13 . As $G + 13 \subset 13^{\perp}$ we do not thereby change the isometry type of $G$ . Since invariably $G \perp H$ we have what we wanted. By such changes modulo 15 we can furthermore achieve that any of the sets $\mathcal{H}$ , $\mathcal{E}$ , $\mathcal{J}$ , $\mathcal{D}$ , $\mathcal{C}$ as introduced in (7) are "realized" by an orthogonal basis of $H$ , $\mathcal{E}$ , $I$ , $D$ , $C$ respectively.

This remark allows to arrange for isometries $H \cong \bar{H}$ , $\mathcal{E} \cong \bar{\mathcal{E}}$ , etc.

We now set $W_o := C \oplus D \oplus I \oplus \mathcal{E} \oplus H \oplus G$ and define $\bar{W}_o$ analogously. Let $T_o \colon W_o \to \bar{W}_o$ be the obvious isometry which sends $C$ into $\bar{C}$ etc.(Notice that $G \cong \bar{G}$ by virtue of (iv) in (8).) Now we may quote Principle I: the only join-irreducibles in the lattice which will enter the recursive construction of the required isometry are 17, 15, 13, 8, 6, 4, 3 . The last five are totally isotropic and thus qualify for condition C mentioned in (15) of Principle I in IV.10. Furthermore space 17 qualifies for C(ii) because $E^*$ contains the infinite dimensional $R$ and $R \cap E^{*\perp} = (0)$ ; finally 15 qualifies for C(ii) by virtue of assumption (10) made in Theorem 2. The proof of Theorem 2 is thus complete in this case. The next case will be the

<u>Case $\alpha = \aleph_0$ and $\delta = 0$</u> . As in the previous case we want to define $W_o$ as an orthogonal sum of supplements $C , D , I , \ldots , G$ of prescribed isometry type. We start out with a fixed $C$ and change $I$ modulo 17 ; $17 = E^*$ contains a totally isotropic subspace of infinite dimension and is therefore an orthogonal sum of hyperbolic planes and a finite dimensional radical; therefore, we can arrive at a supplement $I_1$ isometric to $I$ and orthogonal to $C$ . We then change $D$ and $\mathcal{E}$ modulo 15 ; as $\delta = 0$ we have $15 \subset 15^{\perp}$ and we can again arrange it such that we do not thereby change isometry classes. $H = (0)$ by Lemma 1 in Section 4. By a similar procedure we can achieve that $C , D , I , \mathcal{E}$ are isometric to $\bar{C} , \bar{D} , \bar{I} , \bar{\mathcal{E}}$ . We are left with $G$ which we change modulo 14. Then we can apply Principle I of IV.10.

<u>Case $\alpha = \aleph_0$ and $\delta < \infty$</u> . We can reduce this case to the previous one. Let $F$ be a supplement of 13 in 15 and $E_o := F^{\perp}$ . Since $F$ is trace-valued and $E$ is quasistable ($\dim S/T < \infty$ and $T \subset \|E\|_{\infty}$) we obtain from the Corollary to the Scholion in VII.6 that $E_o \cong E$ .

Hence the isometry $E \cong \bar{E}$ implies that $E_o \cong \bar{E}_o$ if $\bar{E}_o := \bar{F}^{\perp}$ for $\bar{F}$ any supplement of $\overline{13}$ in $\overline{15}$. Since $F$ and $\bar{F}$ may furthermore be assumed isometric by assumption of the theorem we only have to study the pairs $(R,E_o)$, $(\bar{R},\bar{E}_o)$ which brings us to the first case.

Case $a < \infty$. By Lemma 1 in Section 4 the diagram looks now as follows

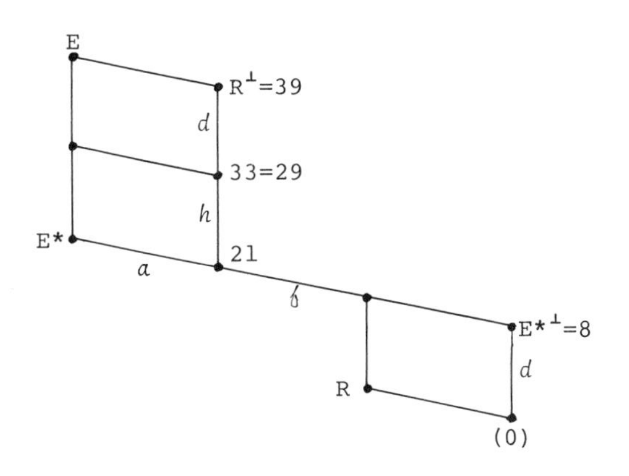


Since $\dim E = \aleph_0$ we have $\delta = \aleph_0$ so (10) is in force. We can now arrive at $W_o := D \overset{\perp}{\oplus} H \cong \bar{W}_o := \bar{D} \overset{\perp}{\oplus} \bar{H}$ as in the first case and quote Principle I of IV.10. This terminates the proof of Theorem 2.

Corollary 1. Let $E$ be a $\aleph_0$-dimensional nondegenerate $\varepsilon$-hermitean space over a division ring with $\dim S/T$ finite. Let $R$ be a totally isotropic subspace. If (10) holds then $(R,E)$ is a sum of finitely many irreducible pairs from the list in Sec. 5. The following objects constitute a <u>complete</u> set of orthogonal invariants for the orbit of $R$ under the action of the orthogonal group of $E$ :

(j) The cardinals $a(R) := \dim R/(R \cap E^{*\perp})$

$b(R) := \dim R^{\perp\perp}/R+R^{\perp\perp} \cap E^{*\perp *}$ ,

(jj) the (totally isotropic, finite dimensional) subspace $R \cap E^{*\perp} \subset E$ ,

(11) (jjj) the isometry class of the (finite dimensional) nondegenerate part of $R^{\perp *\perp}$ (a rigid space),

(jv) the isometry class of the nondegenerate part of $R^{\perp *}$ in case its dimension $\delta$ is finite,

(v) the lattice $\varphi V(R,E^*)$ in the k-vector space $S/T$ .

Remark. One can now discuss various special cases. For example, if the form is symmetric then $k$ is commutative and $S/T$ naturally isomorphic to the $k^2$-vector space $k$ . If $E$ is stable (precisely the case when $E$ is a countable orthogonal sum of spaces $\langle \alpha,\alpha,...\rangle$ and hence $E^{*\perp} = 0$) then the orbits of totally isotropic subspaces $R$ are completely characterized by their indices ((5) and (6) in Section 4) and the isometry class of the finite dimensional (rigid) nondegenerate part of $R^{\perp *\perp}$ . This isometry class can nicely be described by Milnor's Clifford determinant so that $R$ is characterized up to metric automorphisms of $E$ by eleven cardinals and an element of the Clifford algebra associated with the quadratic map $\lambda \mapsto \lambda^2$ (of the $k^2$ vector space $k$ into $k^2$). Hence we have the

Corollary 2. Let $E = \langle 1,1,...\rangle$ be symmetric over a commutative field of characteristic 2. There are $\aleph_0$ different orbits of totally isotropic subspaces under the action of the orthogonal group of $E$ . Each orbit is completely characterized by the indices defined by the lattice in Section 2. The set of all maximal totally isotropic subspaces decomposes into two orbits.

## 8. On closed totally isotropic subspaces

If we assume that in the diagram of Section 2 we have that (cf. (5), (6))

(12) $\qquad b = c = e = g = 0$

then the lattice $V(R,E^*)$ reduces to

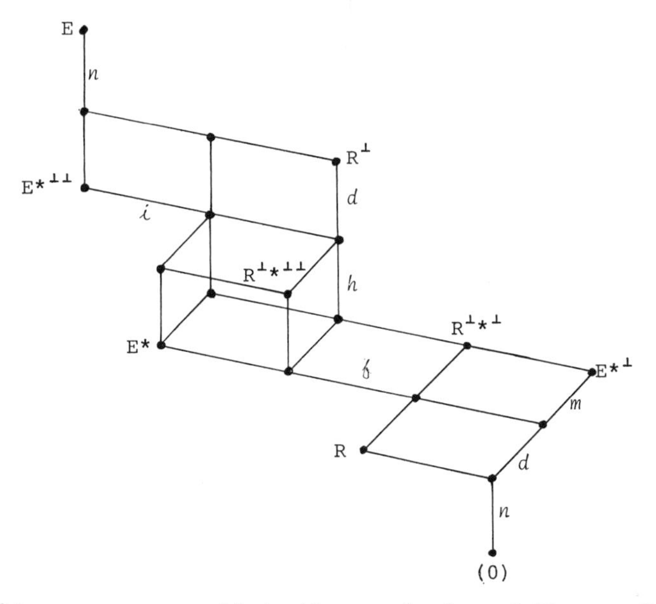


In this case we can obtain the conclusion of Theorem 2 without having to assume (10):

Theorem 3. Let $E$ be a $\aleph_0$-dimensional nondegenerate $\varepsilon$-hermitean space over a division ring with $\dim S/T$ finite. Let $R$, $\bar{R}$ be infinite dimensional totally isotropic subspaces which satisfy (12), i.e. $R = R^{\perp\perp}$, $R^{\perp *\perp} = R + E^{*\perp}$ (always the case when $E^*$ is $\perp$-closed) and analogously for $\bar{R}$. In order that there exists an isometry $E \cong E$ which maps $R$ onto $\bar{R}$ the following are necessary and sufficient.

$$\text{(j)} \quad E^{*\perp} \cap R = E^{*\perp} \cap \bar{R} \quad , \quad \text{(jj)} \quad R^{\perp} \cong \bar{R}^{\perp} .$$

Remark. The necessity of (j) & (jj) is obvious. In order to make easy the comparison between (j) & (jj) and the invariants (11) of Cor.1 we first remark that $\varphi V(R,E^*)$ in $L(S/T)$ is identical with $\varphi V(\bar{R},E^*)$ if and only if the chain in $L(S/T)$

$$(0) \subset \varphi R^{\perp *\perp\perp} \subset \varphi R^{\perp *\perp\perp} + \varphi E^{*\perp} \subset \varphi R^{\perp} \subset \varphi E^{*\perp\perp} + \varphi R^{\perp} \subset \varphi E$$

is identical with the corresponding chain defined by $\bar{R}$ . This follows from the above diagram. Secondly, from (jj) we obviously get an isometry $R^{\perp *} \cong \bar{R}^{\perp *}$ . However, by making use of metabolic decomposition of $E$ for $R$ and $\bar{R}$ respectively we see that from (jj) we get an isometry

$$(13) \qquad R^{\perp *\perp\perp} \cong \bar{R}^{\perp *\perp\perp} .$$

Therefore, the two chains are seen to coincide and thus

$$(14) \qquad \varphi V(R,E^*) = \varphi V(\bar{R},E^*) .$$

Our proof will make it clear that it is (13) which replaces assumption (10) and the invariant (jv) in Theorem 2. (In the setting of Thm. 2 (13) is an immediate consequence via stability.)

Proof (of Theorem 3) 1. If we write down decompositions of the kind (2) both for $R$ and $\bar{R}$ we see (by the assumptions of the theorem) that we may assume without loss of generality that (4) holds and the corresponding equalities for $(\bar{R},\bar{E})$ as well. We are thus once more in the situation of the diagram in Section 4 (with $b = c = e = g = 0$ by assumption of the theorem).

2. Let $D := E^{*\perp}$ and $D'$ be a supplement of space 33 in 39 . $D \oplus D'$ is nondegenerate by looking at the diagram (reason as follows if necessary: $D'$ is rigid so any vector $d$ of the radical of $D \oplus D'$ belongs to $D$ . As $d \in 7 \subset 11 = 33^{\perp}$ the element $d$ would be orthogonal to all of $39 = 33 \oplus D'$ ; hence $d \in 39^{\perp} \cap D = 6 \cap D = (0)$).

Let $\bar{D} := \bar{E}^{*\perp}$ and $\bar{D}'$ be a supplement of $\overline{33}$ in $\overline{39}$. By (14) we may choose $\varphi\bar{D}' = \varphi D'$. Furthermore, by changing $D'$ modulo 13 we can enforce that $D' \cong \bar{D}'$. Thus $D \oplus D' \cong \bar{D} \oplus \bar{D}'$; let $E_1$ and $\bar{E}_1$ be the orthogonals of the two sums in $E$ and $\bar{E}$ respectively.

We have $D \overset{\perp}{\oplus} E_1 = D^{\perp} = E^{*\perp\perp} \cong \bar{E}^{*\perp\perp} = \bar{D}^{\perp} = \bar{D} \overset{\perp}{\oplus} \bar{E}_1$ therefore $E_1 \cong \bar{E}_1$ because $D$ and $\bar{D}$ are the radicals of the two isometric spaces.

<u>3</u>. Let $E_1$, $\bar{E}_1$ be as at the end of the previous step. There are metabolic decompositions

$$(15) \qquad E_1 = (R \oplus R') \overset{\perp}{\oplus} E_2 , \quad \bar{E}_1 = (\bar{R} \oplus \bar{R}') \overset{\perp}{\oplus} \bar{E}_2 , \quad E_1 \cong \bar{E}_1 .$$

Since $e = 0$ we find that $E_2^*$ is $\perp$-dense. Furthermore $\varphi E_2 = \varphi(R+E_2) = 30^{\varphi} = \overline{30}^{\varphi} = \varphi\bar{E}_2$. Yet we cannot conclude that $E_2 \cong \bar{E}_2$ by a stability argument since we won't have $T \subset \|E_2\|$, $\|\bar{E}_2\|$ (both spaces can be anisotropic because we do not assume (10)). However, by taking orthogonals in the spaces at the end of step 1 we find that

$$(16) \qquad \begin{aligned} R^{\perp *} &= D \oplus R \oplus E_2^* \\ R^{\perp * \perp\perp} &= D \oplus R \oplus E_2 \end{aligned}$$

and, of course, an analogous expression for $\bar{R}^{\perp * \perp\perp}$. Since $D \oplus R = \mathrm{rad}(R^{\perp * \perp\perp})$ by (16) we can conclude from (13) that

$$(17) \qquad E_2 \cong \bar{E}_2 .$$

Since thus (jj) is inherited by $(R,E_1)$, $(\bar{R},\bar{E}_1)$ we have achieved a reduction to the case where $b = c = d = e = g = m = n = 0$. From now on we drop the subscript 1 in "$E_1$", "$\bar{E}_1$".

<u>4</u>. (17) and the fact that $E_2 = E_2^{*\perp\perp}$ can be used to elaborate on the metabolic decompositions (15). We shall make use of the following simple fact

Lemma 3. Let $E$ be a $\aleph_0$-dimensional nondegenerate $\varepsilon$-hermitean space over $k$ with $E = E^{*\perp\perp}$. Let $(\lambda_j)_{j\in\mathbb{N}}$ be any family of elements $\lambda_j \in \|E\| \smallsetminus T$. Then $E$ has an orthogonal basis $(e_i)_{i\in\mathbb{N}}$ such that for each $\lambda_j$ there are infinitely many $e_{i'}$ among the $e_i$ with $\Phi(e_{i'}, e_{i'}) \equiv \lambda_j \pmod T$.

Proof of Lemma 3. Assume that $e_1, \ldots, e_m$ have been determined. Let $\lambda \in \|E\|\smallsetminus T$ be prescribed. There is $x_o \in E$ with $\lambda = \Phi(x_o,x_o)$. Since $E^*$ is $\perp$-dense it is met by the translate $x_o + k(e_1,\ldots,e_m,x_o)^\perp$: there exists $e_{m+1} \in k(e_1,\ldots,e_m,x_o)^\perp$ such that $x_o + e_{m+1} \in E^*$. Hence $\Phi(e_{m+1},e_{m+1}) \equiv \lambda \pmod T$. It is clear that a systematic scheme can be devised such that each $\lambda_i$ will infinitely often show up when carrying through a Gram-Schmidt orthogonalization process for $E$. Q.E.D.

Let then $R \oplus R'$ in (15) be the orthogonal sum of the metabolic planes $k(r_j,r_j')$ $(j\in J)$. We have $\varphi R' + \varphi R^\perp = \varphi E$. We show how to arrange it such that $\varphi R'$ equals a previously fixed supplement $S$ of $\varphi R^\perp$ in $\varphi E$. Since $\varphi\Phi(r_j',r_j') \in S + \varphi E_2 = \varphi E_1$ we have $\varphi\Phi(r_j',r_j') = \sigma_j + \lambda_j$, $\sigma_j \in S$, $\lambda_j \in \varphi E_2$ $(j\in J)$. We now quote Lemma 3 and may assume that $E_2$ has an orthogonal basis $\mathfrak{B}$ where for each plane $P_j = k(r_j,r_j')$ with $\lambda_j \notin T$ there is a different member $e_{j'}$ of $\mathfrak{B}$ with $\Phi(e_{j'}, e_{j'}) \equiv \lambda_j \pmod T$. For each plane $P_j$ with $\lambda_j \notin T$ we carry through a change of basis in the 3-dimensional space $P_j \oplus (e_{j'})$ according to the schema

$$k(r_j,r_j') \overset{\perp}{\oplus} (e_{j'}) = k(r_j,r_j'+e_{j'}) \overset{\perp}{\oplus} (e_{j'}+\Phi(e_{j'},e_{j'})r_j) .$$

We obtain thus a new metabolic decomposition $E = (R\oplus R'') \overset{\perp}{\oplus} E_2'$ with $\varphi E_2' = \varphi E_2$ and $\varphi R'' \subset S$ (and hence equality). We may therefore assume that in (15) the spaces $\varphi R'$ and $\varphi\bar{R}' \subset S/T$ are prescribed supplements of $\varphi E_2$ and $\varphi\bar{E}_2$ in $\varphi E = \varphi\bar{E}$ respectively.

5. Let $I$ and $\bar{I}$ be supplements of 35 in 37 = E and $\overline{35}$ in $\bar{E}$ respectively with $\varphi I = \varphi \bar{I}$. By what we said at the end of the previous step there exist metabolic decompositions (15) with

$$\varphi R' = \varphi I = \varphi \bar{I} = \varphi \bar{R}' . \tag{18}$$

Because $\dim R$ is infinite the spaces $R \oplus R'$ and $\bar{R} \oplus \bar{R}'$ are stable, hence by (18) they are isomorphic. In view of (17) we have reduced the problem to the case $E = R \oplus R'$, $\bar{E} = \bar{R} \oplus \bar{R}'$, i.e.,

$$b = c = d = e = f = g = h = m = n = 0$$

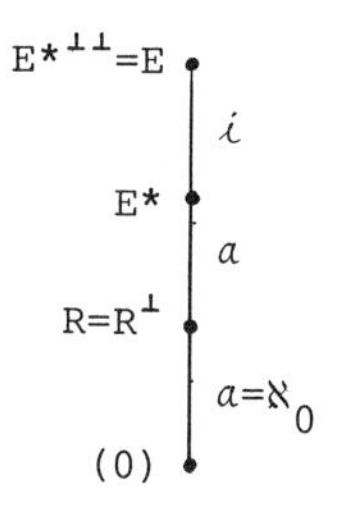


We choose supplements $I$, $\bar{I}$ of $E^*$ in $E$ and $\bar{E}^*$ in $\bar{E}$ respectively; they can be changed mod $E^*$, if necessary, to achieve $I \cong \bar{I}$. If we let $T_o$ be any isometry between $I$, $\bar{I}$ we may quote Principle I of IV.10 and obtain an isometry $E \cong \bar{E}$ which maps $R$ onto $\bar{R}$. This finishes the proof of Theorem 3.

## 9. The case of Witt decompositions reviewed

Let again $R$ be a totally isotropic subspace in a $\varepsilon$-hermitean nondegenerate space $(E,\Phi)$. Assume $\dim R$ to be infinite. We fix some metabolic decomposition of $E$,

$$E = (R^{\perp\perp} \oplus R') \overset{\perp}{\oplus} E_o \tag{19}$$

for the totally isotropic $R^{\perp\perp}$ and define a new form $\Psi$ on $E \times E$ as follows: Set $\Psi$ identically zero on $R' \times R'$ and let $\Psi$ coincide with $\Phi$ on $R^{\perp\perp} \times (R^{\perp\perp} \oplus R' \oplus E_o)$ and $(R' \oplus E_o) \times E_o$ . $(E,\Psi)$ is nondegenerate and Witt decomposed for $R^{\perp\perp}$ .

We continue to assume that the division ring has finite $\dim S/T$ . Hence both $(E,\Phi)$ and $(E,\Psi)$ are quasistable. Their isometry classes are characterized by the isometry classes of $E^{*\perp}$ and by $\varphi E$ , $\varphi E^{*\perp\perp}$ in $S/T$ . We can prove equality of these invariants if $(E,\Phi)$ satisfies

$$(20) \qquad R^{\perp} + E^{*} = E$$

$$(21) \qquad R^{\perp\perp} + E^{*\perp} = R^{\perp *\perp}$$

<u>Lemma 4</u>. Let $\dim S/T < \infty$ and assume that $R$ in $(E,\Phi)$ satisfies (20) and (21). Then the space $(E,\Psi)$ defined above by means of the decomposition (19) is isometric to $(E,\Phi)$ .

<u>Proof</u>. We show that all the pertinent invariants can be expressed in $(E_o,\Phi)$ of (19). As $\Psi$ coincides with $\Phi$ on $E_o$ the assertion will follow. Since $R^{\perp} = R^{\perp\perp} + E_o$ by (19) we get from (20) that

$$(22) \qquad \varphi E = \varphi E_o .$$

We turn to the isometry class of $E^{*\perp}$ . Notation: if $X \subset E_o$ we set $X^{\perp_o} = X^{\perp} \cap E_o$ ; for example $R^{\perp *\perp} = (R^{\perp\perp}+E_o)^{*\perp} = R^{\perp\perp} \oplus E_o^{*\perp_o}$ by (19). Therefore, by (21)

$$(23) \qquad R^{\perp\perp} + E^{*\perp} = R^{\perp\perp} \oplus E_o^{*\perp_o} .$$

Since $E$ is nondegenerate it follows from (20) that the sum $R^{\perp\perp} + E^{*\perp}$ is direct and orthogonal; we obtain from (23) that

$$(24) \qquad E^{*\perp} \cong E_o^{*\perp_o} .$$

From (20) we get $E^{*\perp\perp} = E^{*} + E^{*\perp\perp} \cap R^{\perp} = E^{*} + (R^{\perp\perp}+E^{*\perp})^{\perp}$ . Hence by (23) we obtain $E^{*\perp\perp} = E^{*} + (R^{\perp\perp}+E_o^{*\perp_o})^{\perp} = E^{*} + R^{\perp\perp} + E_o^{*\perp_o\perp_o}$ . Therefore,

(25) $$\varphi E^{*\perp\perp} = \varphi E_o^{*\perp_o\perp_o} .$$

From (22), (24), (25) we obtain $(E,\Phi) \cong (E,\Psi)$ as asserted. Since $E^{*\perp} \cap R^{\perp\perp} = (0)$ by (20) and $(R^{\perp},\Phi) = (R^{\perp\perp}+E_o,\Phi) \cong (R^{\perp},\Psi)$ we obtain from Theorem 3 that there is an isometry $(E,\Phi) \cong (E,\Psi)$ which maps $R^{\perp\perp}$ onto $R^{\perp\perp}$. Hence $(E,\Phi)$ admits a Witt decomposition for the totally isotropic $R^{\perp\perp}$.

Of course, when (20) holds then the index $i = \dim E^{*\perp\perp}/E^{*\perp\perp}\cap(R^{\perp}+E^{*})$ vanishes and the proof of Theorem 3, in its last step, sets up a Witt decomposition for $R$. A direct proof of Theorem 3 which does not manufacture a Witt decomposition in its course seems possible only when (10) is assumed (such a proof is the proof of Theorem 2).

## 10. Remarks on related results (Principle II)

The sample which makes up the present chapter is atypical in one aspect: The relevant lattice is distributive. A nice example of a non-distributive lattice is given by Moresi in [3] :

Theorem 4. Let the division ring have finite $\dim S/T$. Assume that $F$ is a $\perp$-dense subspace in the nondegenerate $\varepsilon$-hermitean space $E$ with $\perp$-closed $E^*$. The orthostable lattice $V(F,E^*)$ generated by $F$ and $E^*$ in $L(E)$ is finite; it has 37 elements in general and is given by the following diagram

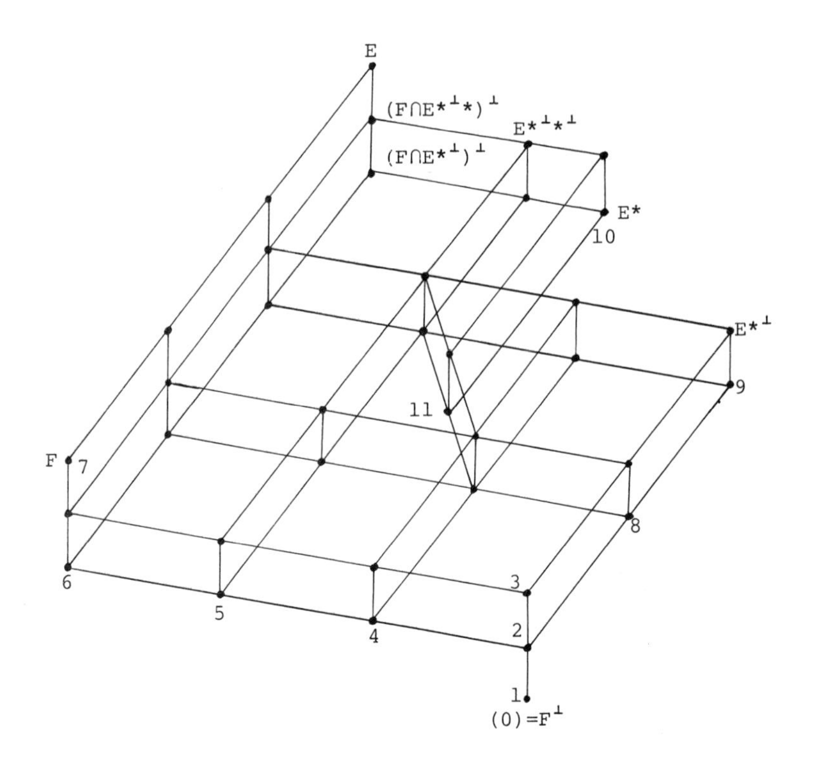


The join-irreducibles are numbered from 1 to 11. Their orthogonals determine $\perp$ on $V(F,E^*)$ . $1^{\perp}$, $2^{\perp}$, $3^{\perp}$, $7^{\perp}$, $8^{\perp}$, $10^{\perp}$ are listed in the diagram. By the $\perp$-density of $F$ we have $(F\cap X^{\perp})^{\perp} = X$ for all finite dimensional spaces $X$ ; this gives $4^{\perp} = E^{*\perp}$ , $5^{\perp} = 3+8$ , $6^{\perp} = 3$ . We have $9^{\perp} = 3+10$ since $10^{\perp\perp} = 10$ by assumption of the theorem and because $\dim 3$ is finite. To find $11^{\perp}$ observe that $4+8 \subset 11 \subset 10$ so that $E^{*\perp} = 10^{\perp} \subset 11^{\perp} \subset 4^{\perp} \cap 8^{\perp} = E^{*\perp} \cap E^{*\perp *\perp} = E^{*\perp}$ . The verification of the fact that the labelled diagram gives the correct joins and meets is left as an exercise. (Whenever $\dim S/T \geq 4$ there are examples with all 37 elements of the lattice different spaces.)

One can prove the following

Theorem 5 ([2]). Let $F$ and $E$ be as in Thm. 4 and $\dim E = \aleph_0$. Assume that $F$ is quasistable (the case if and only if $F$ contains an infinite dimensional totally isotropic subspace). Then the orbit of $F$ under the orthogonal group is completely characterized by the space $F \cap E^{*\perp}$ (left pointwise fixed) and the indices $\dim E/(F+E^{*\perp})$, $\dim(F+E^{*\perp})/(F+E^{*\perp *})$, $\dim(F+E^{*\perp *})/F$, $\dim(F\cap E^{*\perp})/(F\cap E^{*\perp *})$.

The proof is given with all the necessary details in [2]. It has the remarkable feature that the lattice $\varphi V(F,E^*)$ in the value space $S/T$ does not enter the scene. It has the new feature of having to deal with a nondistributive lattice. For the benefit of the reader who wants to investigate similar cases we shall formulate here a strategy which has proved helpful to us in many cases ([2]). A model application of this strategy is given in the course of proving the "Arf theorem" in Chapter XVI.

Let $V$, $\bar{V} \subset L(E)$ be not distributive and assume that a lattice isomorphism $\tau: V \to \bar{V}$ has to be squeezed through to a linear map $\tilde{\tau}$ in the underlying vector space. Since $V$ contains (A, B, C diamond) as a sublattice we do have a real obstacle. For, if we define $\tilde{\tau}$ on $A$ and $B$ - say by recursion - such that $\tilde{\tau}(A) = A^\tau$, $\tilde{\tau}(B) = B^\tau$ then $\tilde{\tau}(C)$ is automatically decided upon as $C \subset A + B$. Will we have $\tilde{\tau}(C) \subset C^\tau$? As $\tilde{\tau}$ has to be an isometry on top of it we see that the problem may be overdetermined in various ways and prohibit a solution.

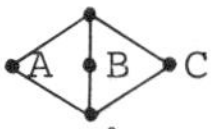

If the indices in (A, B, C diamond) are finite and if there are only finitely many such nondistributive spots in $V$ it may happen that one is able to start the recursive construction of $\tilde{\tau}$ by a $\tilde{\tau}_o: W_o \to \bar{W}_o$ where the finite dimensional $W_o$ covers all the nondistributive places in $V$; then one is done in view of the result in the distributive case.

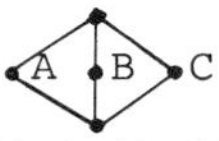

We shall now give a precise version of this natural idea. The verification of its validity is left to the reader.

Principle II. Let $(E,\Phi)$ be a nondegenerate $\varepsilon$-hermitean space of dimension $\leq \aleph_0$. Let $V \subset L(E)$ be a complete sublattice which is stable under the operations $X \mapsto X^{\perp}$ and $X \mapsto X^* := \{x \in X \mid \Phi(x,x) \in T\}$ and in which compact elements are joins of join-irreducibles. Let $\bar{V} \subset L(E)$ be a second lattice of this kind and $\tau: V \to \bar{V}$ a lattice isomorphism which preserves indices (= dimensions of quotients of neighbouring elements in the lattices) and which commutes with $\perp$ and $*$ (defined on the lattices).

Assume that there exist finite dimensional subspaces $W_o$, $\bar{W}_o \subset E$ and an isometry $\tilde{\tau}_o: W_o \to \bar{W}_o$ with

$$(26) \qquad \tilde{\tau}_o(W_o \cap A) = (\bar{W}_o \cap A^{\tau}) \qquad \text{for all } A \in V$$

$$(27) \qquad \bigcap_{\iota} (W_o + A_{\iota}) = W_o + \bigcap_{\iota} A_{\iota} \qquad (A_{\iota} \in V)$$

$$(28) \qquad \bigcap_{\iota} (\bar{W}_o + \bar{A}_{\iota}) = \bar{W}_o + \bigcap_{\iota} \bar{A}_{\iota} \qquad (\bar{A}_{\iota} \in \bar{V})$$

and such that condition (15) of Principle I in IV.10 is satisfied.

In order that $\tilde{\tau}_o$ admits an isometric extension $\tilde{\tau}: E \to E$ that induces the lattice isomorphism $\tau$ it is sufficient that the following condition holds.

(29) If $X \in V$ is join-irreducible and compact but not join-prime then there exists a subspace $H \subset W_o$ which is a linear supplement in $X$ of the immediate antecedent of $X$.

## References to Chapter VIII

[1] H. Gross, Formes quadratiques et formes non traciques sur les espaces de dimension dénombrable. Bull. Soc. Math. de France Mémoire 59 (1979).

[2] - and H.A. Keller, On the non trace - valued forms. To appear in Adv. in Math.

[3] R. Moresi, Studio su uno speciale reticolo consistente in sotto-spazi di uno spazio sesquilineare nel caso caratteristica due. Master's Thesis, Univ. of Zurich 1977.

CHAPTER NINE

# INVOLUTIONS IN HERMITEAN SPACES IN CHARACTERISTIC TWO

## 1. Introduction

Fields and forms are as specified under the caption of Chapter VIII. In addition we shall often assume that the field is such that

(0) there is only one isometry class in dimension $\aleph_0$ of nondegenerate trace-valued $\varepsilon$-hermitean forms.

Hence all nondegenerate $\aleph_0$-forms will be quasistable here if (0) is in force.

*

We shall classify the involutions $I$ in the orthogonal group of an $\varepsilon$-hermitean space $(E,\Phi)$ in dimension $\aleph_0$. The problem is trivial in characteristic not two because then $(E,\Phi)$ splits orthogonally into the spaces $\ker(I-\mathbb{1})$ and $\ker(I+\mathbb{1})$. In characteristic 2 involutions are puzzling. We shall treat in full generality the case where $(E,\Phi)$ is trace-valued and the field satisfies (0) (Theorem 2 in Section 7). In finite dimensions or, more generally, when $\text{im}(I-\mathbb{1})$ is $\perp$-closed, then the problem is tantamount to classifying <u>quasistable</u> spaces over the given field; in general, the question is more complex but still of the same degree of difficulty. The solution which we develop here makes it quite evident that without investing some serious work into lattice computation there can be no hope of mastering the problem.

As a side result we obtain the missing half to a result by Kaplansky. In Theorem 3 of [1] he showed that the self-adjoint linear transformations $U$ with $U^2 = 0$ in a space of countable dimension, equipped with a symmetric form over a quadratically closed field of chacateristic not 2, are determined up to orthogonal similarity by three cardinal number invariants. An entirely analogous result holds in characteristic 2 but then seven cardinal number invariants are needed.

## 2. The form derived from an involution

Let $\operatorname{char} k = 2$ and $(E,\Phi)$ be nondegenerate. If $I: E \to E$ is an isometry with $I^2 = \mathbb{1}$ (involution) we define $U: E \to E$ by

(1) $$I = \mathbb{1} + U .$$

Since $I^2 = \mathbb{1}$ we find

(2) $$U^2 = 0 .$$

Since $I$ preserves $\Phi$ we find by (2) that $\Phi(Ux,Uy) = 0$ for all $x, y \in E$ and

(3) $$\Phi(Ux,y) + \Phi(x,Uy) = 0 .$$

($U$ is self-adjoint with respect to $\Phi$.) Conversely, if $U: E \to E$ is a linear map with (2) and (3) then (1) defines an involution.

We now define a new sesquilinear form $\Psi$ on $E \times E$ by

(4) $$\Psi(x,y) := \Phi(Ux,y) .$$

Since $U$ is self-adjoint the form $\Psi$ will be $\varepsilon$-hermitean provided $\Phi$ is $\varepsilon$-hermitean. The form $\Psi$ is highly degenerate; its radical is the kernel of $U$, hence $\operatorname{rad} \Psi$ has the same dimension as $E$ if $\dim E$ is infinite and is at least of dimension $\frac{1}{2} \dim E$ when $E$ is finite dimensional. We call $\Psi$ <u>the form derived from the involution</u> $I$ of $(E,\Phi)$.

## 3. Orthogonal similarity

If $I$ is an involution of $(E,\Phi)$ and $D \in \mathcal{O}(E,\Phi)$ an arbitrary element of the orthogonal group then

(5) $$\bar{I} = D \cdot I \cdot D^{-1}$$

is likewise an involution. In this chapter we are interested in the "types" of involutions which the space $(E,\Phi)$ admits. Therefore, we shall not consider $I$ and $\bar{I}$ as different since they have like definitions with respect to congruent bases $\mathfrak{B}$ and $\mathfrak{B}^D$ of $E$. Thus the question arises: when are two given involutions orthogonally similar? Now if (5) holds then $D$ is an isometry $E \cong E$ with respect to either form $\Phi$ and $\Psi$, $(E,\Phi,\Psi) \cong (E,\Phi,\bar{\Psi})$. The converse is equally obvious; if $D\colon (E,\Phi,\Psi) \to (E,\Phi,\bar{\Psi})$ is an isometry then $\Phi(DUD^{-1}x,y) = \Phi(UD^{-1}x,D^{-1}y) = \Psi(D^{-1}x,D^{-1}y) = \bar{\Psi}(x,y) = \Phi(\bar{U}x,y)$ for all $x$, $y \in E$. Hence we can conclude (5) if $(E,\Phi)$ is nondegenerate. Let us state this simple but important fact as

Lemma 1. Two involutions $I$, $\bar{I}$ of the nondegenerate space $(E,\Phi)$ are orthogonally similar if and only if $(E,\Phi,\Psi)$ and $(E,\Phi,\bar{\Psi})$ are isometric.

## 4. A special case

Results of Chapter VII will allow us to decide the question of orthogonal similarity in important special cases. Let $I = \mathbb{1} + U$ be an involution of $(E,\Phi)$ and set $J := \operatorname{im} U$, $K := \ker U$. By (2) and (3) we have

(6) $$J \subset J^{\perp\perp} \subset J^{\perp} = K .$$

Since from now on we shall officially assume $\dim E$ to be countable we can use metabolic decompositions for the totally isotropic $J^{\perp\perp}$,

$E = (J^{\perp\perp} \oplus L) \overset{\perp}{\oplus} E_o$ . Since $E_o \subset J^\perp = K$ we see that $I$ is the identity on the orthogonal summand $E_o$ . For certain investigations it will be no loss if we neglect $E_o$ , i.e. if we assume that

(7) $\qquad J^{\perp\perp} = K$ .

Lemma 2. Let $(E,\Phi)$ be a nondegenerate $\varepsilon$-hermitean trace-valued space of dimension $\leq \aleph_0$ . If the involution $I = \mathbb{1} + U$ has (7) and $\perp$-closed $J = \text{im } U$ then, for $\bar{I} = \mathbb{1} + \bar{U}$ a second involution of this kind, an isometry $(E,\Psi) \cong (E,\bar{\Psi})$ implies an isometry $(E,\Phi,\Psi) \cong (E,\Phi,\bar{\Psi})$.

Proof. As $\Phi$ is trace-valued we have Witt decompositions $E = K \oplus L$ , $E = \bar{K} \oplus \bar{L}$ . Since $K = \text{rad } \Psi$ , $\bar{K} = \text{rad } \bar{\Psi}$ we obtain from $(E,\Psi) \cong (E,\bar{\Psi})$ an isometry $D_1 \colon (L,\Psi) \to (\bar{L},\bar{\Psi})$ . We show that $D_1$ has a transpose $D_2 \colon \bar{K} \to K$ with respect to the pairings $\langle L,K\rangle$ , $\langle \bar{L},\bar{K}\rangle$ sponsored by $\Phi$ . Indeed, for given $\bar{x} \in \bar{K}$ there is a unique $\bar{y} \in \bar{L}$ with $\bar{x} = \bar{U}\bar{y}$ (since $\bar{J} = \bar{J}^{\perp\perp} = \bar{K}$ ). Therefore, $\Phi(\bar{x},D_1 z) = \Phi(\bar{U}\bar{y},D_1 z) = \bar{\Psi}(\bar{y},D_1 z) = \Psi(D_1^{-1}\bar{y},z) = \Phi(UD_1^{-1}\bar{y},z)$ for all $z \in L$ . Thus $\bar{x} \mapsto UD_1^{-1}\bar{y}$ is the transpose $D_2$ of $D_1$ . Set $\check{D}_1 := D_2^{-1}$ . It is now clear that if we define a linear map $D \colon E \to E$ by $D|_L = D_1$ , $D|_K = \check{D}_1$ then $D$ is an isometry for both $\Phi$ and $\Psi$ . The mapping $D$ just constructed is not, of course, the only solution.

We can now apply the results in Chapter VII on the classification of quasistable spaces $(E,\Psi)$ . Since such spaces are infinite dimensional we should have that $\dim J = \aleph_0$ in the situation of Lemma 2. If $\dim J$ is <u>finite</u> then $I$ is the identity on a subspace of finite codimension in $E$ . The discussion of such isometries is actually a problem in finite dimensions (by virtue of Lemma 2). We shall dismiss it for the moment and assume in this section that

(8) $\qquad \dim E/\ker(I-\mathbb{1}) = \aleph_0$ .

We shall need the k-vector space $S/T$ associated with the division ring $(k,*,\varepsilon)$. With the involution $I$ of $(E,\Phi)$ we can associate the k-vector space homomorphism $\psi: E \to S/T$ defined by

$$(9) \qquad \psi: x \mapsto \Psi(x,x) + T = \Phi((\mathbb{1}+I)x,x) + T \in S/T$$

$\psi$ has to be evaluated on the closure (with respect to $\Psi$) of the trace-valued part of $(E,\Psi)$ which we formally introduce by

$$(10) \qquad A := \{x \in E \mid \Phi(Ux,x) \in T\} .$$

The closure of $A$, with respect to $\Psi$, is $(U(UA)^{\perp})^{\perp}$ which equals (cf. Lemma 3 in Section 6 below) $(J \cap A^{\perp})^{\perp}$ and hence $A^{\perp\perp}$ as $J$ is assumed closed in $(E,\Phi)$. Thus, if the division ring enjoys property (0) then $(E,\Psi)$ and $(E,\bar{\Psi})$ are isometric if and only if the following hold

$$(11) \qquad ((UA)^{\perp},\Psi) \cong ((\bar{U}\bar{A})^{\perp},\Psi) \qquad (\perp \text{ with respect to } \Phi)$$

$$(12) \qquad \psi E = \bar{\psi} E \qquad (\psi \text{ as defined in (9)})$$

$$(13) \qquad \psi A^{\perp\perp} = \bar{\psi}\bar{A}^{\perp\perp} \qquad (\perp \text{ with respect to } \Phi)$$

Finally we remark that it is possible to get rid of the restriction (7) if we assume

$$(14) \qquad (J^{\perp},\Phi) \cong (\bar{J}^{\perp},\Phi) .$$

We summarize our considerations in the following

Theorem 1. Let $k$ be a division ring with property (0) and $E$ a nondegenerate $\varepsilon$-hermitean trace-valued $\aleph_0$-dimensional k-space. Let $I$ and $\bar{I}$ be involutions of $(E,\Phi)$ with (8) and assume that

$$J := \operatorname{im}(I-\mathbb{1}) \quad , \quad \bar{J} := \operatorname{im}(\bar{I}-\mathbb{1}) \quad \text{are } \perp\text{-closed.}$$

In order that $I$ and $\bar{I}$ are orthogonally similar it is necessary and sufficient that (11), (12), (13) and (14) hold.

It is not difficult to supplement Theorem 1 by listing canonical representatives for the involutions with closed $J$ . We shall give a complete list after the general case has been discussed.

## 5. A lattice material to the solution of the general problem

The method of the previous section to establish orthogonal similarity of two involutions $I$ , $\bar{I}$ in $(E,\Phi)$ could be pushed one step further - instead of assuming $J = J^{\perp\perp}$ it would suffice to assume that $J + A^{\perp} = J^{\perp\perp}$ . However, if no assumptions are made on $J$ then the isometry $D: (E,\Phi) \cong (E,\Phi)$ which gives $\bar{I} = DID^{-1}$ has to be constructed anew from scratch. The problem arises which obvious invariants should be observed in order that the recursive construction of $D$ is not, at the outset, doomed to fail. We observe that the involution $I: (E,\Phi) \to (E,\Phi)$ is also an isometry for its derived form $\Psi$ , $\Psi(Ix,Iy) = \Phi(U^2x+Ux,y+Uy) = \Phi(Ux,y) = \Psi(x,y)$ . Hence we have the following commutative diagram of isometries

$$
(15) \qquad \begin{array}{ccc}
(E,\Phi,\Psi) & \xrightarrow{D} & (E,\Phi,\bar{\Psi}) \\
I \downarrow & & \downarrow \bar{I} \\
(E,\Phi,\Psi) & \xrightarrow[D]{} & (E,\Phi,\bar{\Psi})
\end{array}
$$

This leads us to the following consideration:

(16) Definition. Let $I$ be an involution of $(E,\Phi)$ . $\mathcal{V}(I)$ is the smallest sublattice in the lattice $\mathcal{L}(E)$ of all subspaces of $E$ which contains $(0)$, $E$ and $A$ (defined in (10)) and which is stable under $\perp$ and $U := \mathbb{1} + I$ .

$\mathcal{V}(I)$ is an invariant attached to the (orthogonal) similarity class of the involution $I$ . Notice that the derived form of the involution does not intrude itself upon the definition of $V(I)$ .

The lattice $V(I)$ admits two operations, $\perp$ and $U$ . If $\bar{I}$ is orthogonally similar to $I$ then the map $D$ in (15) induces a lattice isomorphism $\tau: V(I) \to V(\bar{I})$ which preserves indices (dimensions of quotients of neighbouring elements in the lattices) and which commutes with $\perp$ and $U$ . Furthermore $\tau$ commutes with the maps $\psi$ and $\bar{\psi}$ , or rather, with the maps from $L(E)$ into $L(S/T)$ induced by $\psi$ and $\bar{\psi}$ . We shall not make a difference between $\psi$ , $\bar{\psi}$ and these induced maps, thus

$$(17) \qquad \begin{array}{ccc} V(I) & \xrightarrow{\ \tau\ } & V(\bar{I}) \\ \psi \searrow & & \swarrow \bar{\psi} \\ & L(S/T) & \end{array}$$

is commutative.

In the situation of closed $J$ , as assumed by Theorem 1, the lattice $V(I)$ looks as follows

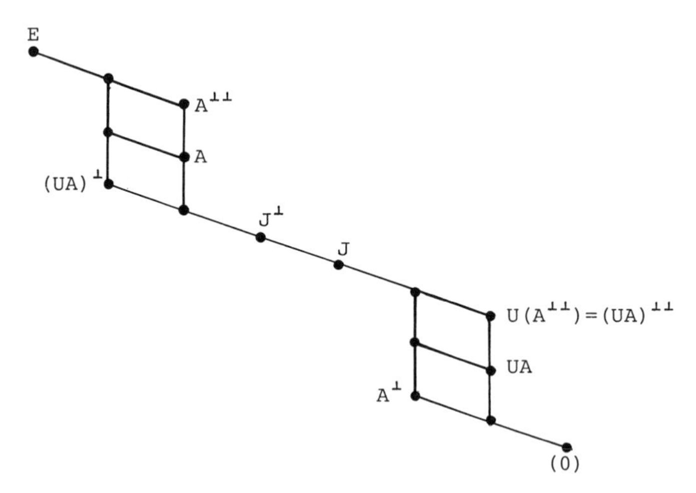


With this lattice in view let us look at Theorem 1 again (we may omit (14))when trying to grasp what is at stake). Condition (11) & (12) & (13)

is equivalent with (11) and the existence of a $\tau$ which makes the diagram (17) commutative. Clearly, the existence of a commutative (17) appears to be the vital part in Theorem 1. In this setting the assumption that $J$ and $\bar{J}$ be closed seems rather ad hoc and superfluous. And superfluous it is! We handed over the problem to Studer who succeeded in [2] to prove our conjecture that (11) and a commutative (17) suffice for the general case as well. To accomplish this it was necessary to compute the lattice $V(I)$ in order to be able to construct an isometry $D: (E,\Phi,\Psi) \to (E,\Phi,\bar{\Psi})$ that induces the given $\tau$ (cf. Section 8 below).

We terminate this section by presenting Studer's lattice $V(I)$ as given in [2]. $V(I)$ turned out to be finite and distributive; it consists of the following spaces (cf. the diagram on the next page).

$1=(0)$

$2=UA\cap U(UA)^{\perp}=A^{\perp}\cap UA$

$3=A^{\perp}\cap(UA)^{\perp\perp}$

$4=U(UA)^{\perp}=J\cap A^{\perp}$

$5=A^{\perp}$

$6=UA$

$7=(UA+A^{\perp})\cap(UA)^{\perp\perp}$

$8=UA+U(UA)^{\perp}$

$9=A^{\perp}+UA$

$10=U(A^{\perp\perp})$

$11=(U(A^{\perp\perp})+A^{\perp})\cap(UA)^{\perp\perp}$

$12=U(A^{\perp\perp})+U(UA)^{\perp}$

$13=A^{\perp}+U(A^{\perp\perp})$

$14=J\cap(UA)^{\perp\perp}$

$15=(J+A^{\perp})\cap(UA)^{\perp\perp}$

$16=(UA)^{\perp\perp}$

$17=U(UA\cap U(UA)^{\perp})^{\perp}$

$18=(J+A^{\perp})\cap(A^{\perp}+(UA)^{\perp\perp})$

$19=A^{\perp}+(UA)^{\perp\perp}$

$20=J=UE$

$21=J+A^{\perp}$

$22=J^{\perp\perp}$

$23=J^{\perp}=K$

$24=A\cap(UA)^{\perp}$

$25=(UA)^{\perp}$

$26=A$

$27=A+(UA)^{\perp}$

$28=A^{\perp\perp}$

$29=A^{\perp\perp}+(UA)^{\perp}$

$30=(U(UA)^{\perp})^{\perp}=(J\cap A^{\perp})^{\perp}$

$31=(A^{\perp}\cap UA)^{\perp}$

$32=E$

| X | 32 | 31 | 30 | 29 | 28 | 27 | 26 | 25 | 24 | 23 |
|---|---|---|---|---|---|---|---|---|---|---|
| UX | 20 | 17 | 14 | 12 | 10 | 8 | 6 | 4 | 2 | 1 |

$\alpha := \dim 32/31$ , $\beta := \dim 31/29$ , $\gamma := \dim 29/27$

$\delta := \dim 25/24$ , $\mu := \dim 22/21$ , $\nu := \dim 23/22$

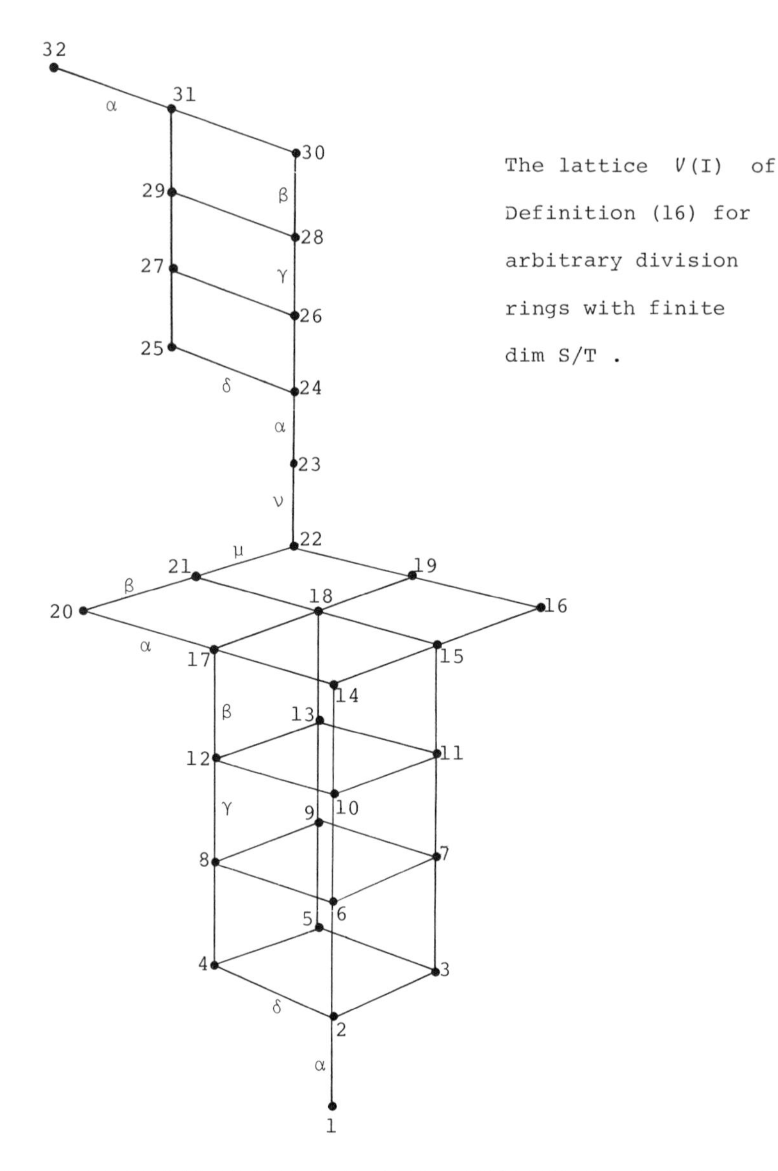


The lattice $V(I)$ of Definition (16) for arbitrary division rings with finite dim $S/T$ .

| $X$ | 1 | 2 | 3 | 4 | 5 | 6 | 7 | 8 | 9 | 10 | 11 | 12 | 13 | 14 | 15 | 16 |
|---|---|---|---|---|---|---|---|---|---|---|---|---|---|---|---|---|
| $X^{\perp}$ | 32 | 31 | 29 | 30 | 28 | 25 | 25 | 24 | 24 | 25 | 25 | 24 | 24 | 25 | 25 | 25 |

| $X$ | 17 | 18 | 19 | 20 | 21 | 22 | 23 | 24 | 25 | 26 | 27 | 28 | 29 | 30 | 31 | 32 |
|---|---|---|---|---|---|---|---|---|---|---|---|---|---|---|---|---|
| $X^{\perp}$ | 24 | 24 | 24 | 23 | 23 | 23 | 22 | 19 | 16 | 5 | 3 | 5 | 3 | 4 | 2 | 1 |

In the next section we shall give clues for the verification of the correctness of the above diagram.

*

## 6. Remarks on the lattice

The first thing to check in the diagram is, of course, the inclusions. As an illustration we shall consider the following two cases:

$$(18) \qquad 10 \subset 14 \quad , \quad 17 \subset 18 \; .$$

(See VIII.3 as far as general remarks on the procedure are concerned.) The first inclusion is easy. $14 = 20 \cap 16$ by definition and $10 \subset 20$ is obvious. The map $U = \mathbb{1} + I$ is continuous in the weak linear topology $\sigma(\Phi)$ and therefore preserves accumulation; hence $10 \subset 16$ . In order to prove the second inclusion (and many other relations in the lattice) we need first that

$$(19) \qquad 28 \cap 25 \subset 26 \quad (\text{hence } 28 \cap 25 = 24 ) \; .$$

Let $t \in A^{\perp\perp} \cap (UA)^{\perp}$ and compute $\Phi(Ut,t)$ . Since, by adjointness, $(UA)^{\perp} = U^{-1}(A^{\perp})$ we see that $Ut \in UU^{-1}(A^{\perp}) \subset A^{\perp}$ so $\Phi(Ut,t) = 0$ . Thus $t \in A$ by the definition of $A$ . This proves (19). We have used a simple property of $U$ and continue to use it and similar properties, such as are expressed in the following

Lemma 3. If $U: E \to E$ is self-adjoint and has $U^2 = 0$ then for all subspaces $F \subset E$ we have (notation $K = \ker U$ , $J = \operatorname{im} U$ ) :

(i) $UUF = (0)$ (ii) $U^{-1}U^{-1}F = E$

(iii) $U^{-1}UF = F+K$ (iv) $UU^{-1}F = F\cap J$

(v) $(UF)^{\perp} = U^{-1}(F^{\perp})$ (vi) $U((F\cap J)^{\perp}) = (U^{-1}F)^{\perp}\cap J$

(vii) $U(F\cap(UF)^{\perp}) = UF\cap U((UF)^{\perp})$

The proof of the lemma is very easy.

Let us now turn to the second inclusion in (18). We have $18 = 19 \cap 21$ by definition of $18$. Since $17 \subset 20 \subset 21$ is entirely obvious we are left with the verification of $17 \subset 19$. We rewrite $17$ by using in turn (vii), (v), (iv) of the lemma and find $17 := U(UA \cap U(UA)^{\perp})^{\perp} = U((U(A \cap (UA)^{\perp}))^{\perp}) = UU^{-1}((A \cap (UA)^{\perp})^{\perp}) = (A \cap (UA)^{\perp})^{\perp} \cap J \subset (A\cap(UA)^{\perp})^{\perp}$ $24^{\perp}$. On the other hand, since $\dim A^{\perp} \leq \dim S/T < \infty$, we have that $19$ is $\perp$-closed, $19 = (A^{\perp\perp} \cap (UA)^{\perp})^{\perp}$, and thus by (19) we read off that $19 = 24^{\perp}$. As we have shown that $17 \subset 24^{\perp}$ we are through with verifying $17 \subset 19$.

The remaining inclusions are easier and are left to the reader. Stability under $U$ is quite obvious: The interval [1,23] is mapped onto 1 and the interval [23,32] is mapped onto the interval [1,20]. E. g., $U(30) = U(U(UA)^{\perp})^{\perp} = UU^{-1}((UA)^{\perp\perp}) = (UA)^{\perp\perp} \cap J =: 14$ by the above lemma.

The legend to the lattice is drawn up in a fashion as to make apparent stability under $U$ and $\perp$. Therefore it is appropriate to check $\perp$-stability before turning to the sums and intersections. We find $X^{\perp}$ without any problem except for $X = 19$. In order to find $19^{\perp}$ let us first verify that $24$ is $\perp$-closed,

$$(20) \qquad 24^{\perp\perp} = 24 .$$

All we need for this is to show that $\dim 24/23 < \infty$ (as $23$ is $\perp$-closed). Now $\dim 24/23 \leq \dim 25/23$ and $\dim 25/23$ can be estimated by using a Witt decomposition of $E$ for the totally isotropic $J^{\perp\perp}$, $E = (J^{\perp\perp} \oplus L) \overset{\perp}{\oplus} E_{o}$. We can read off $K = J^{\perp} = J^{\perp\perp} \oplus E_{o}$, $A = (J^{\perp\perp} \oplus E_{o}) \oplus (A\cap L)$ and $\dim 25/23 = \dim((U(A\cap L))^{\perp} \cap L) \leq \dim J/U(A\cap L) = \dim L/A \cap L = \dim E/A \leq \dim S/T < \infty$. This establishes (20). Now we had already shown that $19 = 24^{\perp}$ when checking inclusions. Therefore we have $19^{\perp} = 24^{\perp\perp} = 24$ as asserted by the table.

We now turn to the intersections. The following assertion is crucial for all that follows

(21) $$3 \cap 4 = 2 .$$

The shortest way to get (21) is as follows: By the definition of $8$ we have $8^{\perp} = (6+4)^{\perp} = 25 \cap 30$ , hence

(22) $$25 \cap 30 = 24 .$$

On the other hand we have

(23) $$\dim 25/24 = \dim 31/30 < \infty$$

because the chain $32 \supset 31 \supset 30$ is mapped in the chain $1 \subset 2 \subset 4$ by $\perp$ - which preserves the two indices as all three spaces are $\perp$-closed - and because the chain $23 \subset 24 \subset 25$ is thrown on the same chain $1 \subset 2 \subset 4$ by $U$ (which preserves the two indices since $\ker U = 23$ ). From (22) and (23) we obtain

(24) $$25 + 30 = 31 .$$

If we pass to the orthogonals in (24) we get $16 \cap 4 = 2$ and therefore $2 \subset 3 \cap 4 \subset 16 \cap 4 = 2$ . This proves (21). The remaining intersections $X \cap Y$ ( $X$ meet irreducible, $Y$ arbitrary) are now quite easy to handle.

There remain the sums. So many relations have accumulated by now that this verification presents no problems. E. g. from $30 \cap 29 = 28$ we obtain, by taking orthogonals, that

(25) $$3 + 4 = 5 .$$

Hence - in order to have yet another example - $4 + 7 = 4 + (9 \cap 16) = 9 \cap (4+16) = 9 \cap (3+4+16) = 9 \cap (5+16) = 9 \cap 19 = 9$ .

Remark. We end this section by observing that $V(I)$ happens to be stable under

$$X \mapsto U^{-1}(X) = U^{-1}(X \cap J) .$$

Each $X \cap J$ is of the shape $UY$ with $K \subset Y \in V(I)$ so $U^{-1}X = Y$ . We have

$$\perp \circ U = U^{-1} \circ \perp \tag{26}$$

since this holds in all of $L(E)$ by (v) of Lemma 3.

## 7. The classification theorem

We shall now state the general result hinted at in Section 5. It is assumed that the division ring $(k,*,\varepsilon)$ satisfies condition (0). We jot down once more the assumptions accumulated thus far:

> The division ring $k$ is of characteristic 2, it has $\dim S/T < \infty$ , and there is only one isometry class of nondegenerate trace-valued $\varepsilon$-hermitean spaces over $k$ in dimension $\aleph_0$ .

The space $(E,\Phi)$ will always be assumed nondegenerate, trace-valued and of dimension $\aleph_0$ . For $I$ an involution of $(E,\Phi)$ the derived form $\Psi$ induces a map $L(E) \to L(S/T)$ via the homomorphism $\psi$ in (9). It throws the lattice $V(I) \subset L(E)$ homomorphically into the lattice

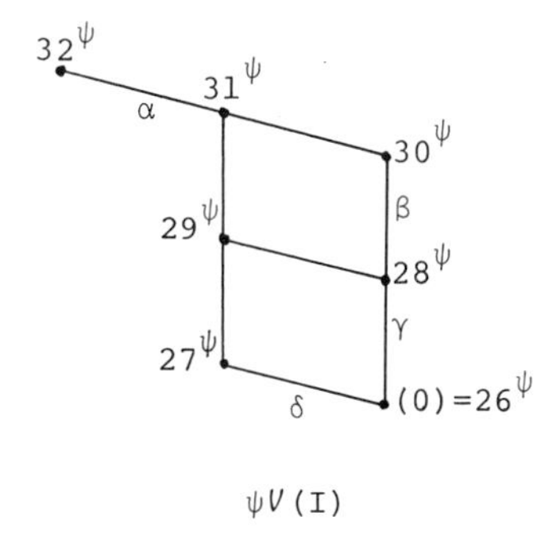


$\psi V(I)$

We pause to recall the case - dismissed earlier by (8) - where $\dim J$ is finite (it would be inconvenient to formally exclude this case). By the Lemmas 1 and 2 the similarity class of $I$ is still characterized by the isometry classes of $(E,\Psi)$ and $(J^{\perp},\Phi)$ respectively (cf. (14)). The first was characterized by (11), (12), (13) for $\dim J$ infinite; when $\dim J < \infty$ we have to replace (13) by the assumption $(A,\Psi) \cong (\bar{A},\bar{\Psi})$. The nondegenerate part of $(J^{\perp},\Phi)$ is of dimension $\nu = \aleph_0$ when $\dim J$ is finite. Hence by virtue of (0) the isometry class of $(J^{\perp},\Phi)$ is fixed by the dimension of its radical (which is $J$).

We are now ready to state the classification theorem

<u>Theorem 2</u>. Let $I$ and $\bar{I}$ be involutions of $(E,\Phi)$. We shall distinguish between the cases "$\dim J < \aleph_0$" (Case I) and "$\dim J = \aleph_0$" (Case II). In order that $I$ and $\bar{I}$ be orthogonally similar the following are necessary and sufficient.

In Case I:

(j) $((UA)^{\perp},\Psi) \cong ((\bar{U}\bar{A})^{\perp},\bar{\Psi})$ , i.e. $(25,\Psi) \cong (\overline{25},\bar{\Psi})$

(jj) $(A,\Psi) \cong (\bar{A},\bar{\Psi})$ , i.e. $(26,\Psi) \cong (\overline{26},\bar{\Psi})$

(jjj) $\psi E = \bar{\psi} E$ ,

(jv) $\dim J = \dim \bar{J}$ .

In Case II: (j) and the following two

(v) $(J^{\perp},\Phi) \cong (\bar{J}^{\perp},\Phi)$ ,

(vj) there exists an index preserving lattice isomorphism $\tau: V(I) \to V(\bar{I})$ which maps $A$ on $\bar{A}$ and is compatible with the operations $\perp$ , $U$ and $\perp$ , $\bar{U}$ on the lattices and which renders commutative the diagram (7).

Condition (j) & (v) & (vj) is equivalent with the conjunction of (j), (v) and

$$(vj_1) \qquad \psi E = \bar{\psi} E$$

$$(vj_2) \qquad \psi(J \cap A^{\perp})^{\perp} = \bar{\psi}(\bar{J} \cap \bar{A}^{\perp})^{\perp} \quad , \text{ i.e. } \quad 30^{\psi} = \overline{30}^{\bar{\psi}}$$

$$(vj_3) \qquad \psi A^{\perp\perp} = \bar{\psi}\bar{A}^{\perp\perp} \quad , \text{ i.e. } \quad 28^{\psi} = \overline{28}^{\bar{\psi}}$$

$$(vj_4) \qquad \mu = \bar{\mu}$$

A proof of Theorem 2 for symmetric forms is given in [2]. We shall not reproduce it here in detail but we shall sketch the principal features of such a proof. This brings us to the next section.

## 8. Remarks on the proof of the classification theorem

Case I of Theorem 2 presents no problems; what is at issue is Case II. A proof for this case has to produce an isometry

$$D : (E,\Phi,\Psi) \to (E,\Phi,\bar{\Psi})$$

which induces a prescribed lattice isomorphism $\tau: V(I) \to V(\bar{I})$ . If $J \neq J^{\perp\perp}$ then the method used to prove Theorem 1 still yields an isometry

$$D_o : (J \oplus L \oplus E_o,\Phi,\Psi) \to (\bar{J} \oplus \bar{L} \oplus \bar{E}_o,\Phi,\bar{\Psi})$$

defined on $J \oplus L \oplus E_o \subset E = (J^{\perp\perp} \oplus L) \overset{\perp}{\oplus} E_o$ ( $E = (\bar{J}^{\perp\perp} \oplus \bar{L}) \overset{\perp}{\oplus} \bar{E}_o$ is an analogous Witt decomposition for the totally isotropic $\bar{J}^{\perp\perp}$ ). Could we not extend $D_o$ to all of $E$ ? The answer is that we can always extend $D_o$ to $(J \oplus L \oplus E_o) + A^{\perp}$ but not any further in general.

In order to prove the first assertion recall that $D_o$ arose from joining an isometry $D_1 : (L,\Psi) \to (\bar{L},\bar{\Psi})$ with its contragradient $\check{D}_1 : J \to \bar{J}$ (with respect to the pairings by $\Phi$ ). As $A^{\perp} \subset J^{\perp\perp}$ we seek to extend $\check{D}_1$ to $J + A^{\perp}$ such that products with elements of $L$ will be preserved. Introduce supplements:

$$A^{\perp} = (A^{\perp} \cap J) \oplus A_1 = 4 \oplus A_1$$

$$4^{\perp} = A^{\perp\perp} \oplus A_2 = 28 \oplus A_2 \quad , \quad A_2 \subset L$$

$$L = L \cap 4^{\perp} \oplus L_1$$

Since $4^{\perp} \cap L_1 = (0)$ and $\dim L_1$ is finite we can arrange for $A_1 \perp L_1$ . $A_1 \oplus A_2$ is nondegenerate and hence hyperbolic for $\Phi$ (look at the diagram: $\dim A_2 = \dim 4^{\perp}/28 = \dim 30/28 = \beta = \dim 5/4 = \dim A_1$ ). Now it is obvious how to proceed. Let $\bar{A}_2 := D_1(A_2)$ and $\bar{L}_1 := D_1(L_1)$ ; then pick $\bar{A}_1$ in analogy to $A_1$ . The restriction $D_{11}$ of $D_1$ to $A_2$ can be extended to $A_1 \oplus A_2 \to \bar{A}_1 \oplus \bar{A}_2$ be joining it with the contragradient $\check{D}_{11}$ of $D_1$ . Since $D_1$ maps $A^{\perp\perp} \cap L$ onto $\bar{A}^{\perp\perp} \cap \bar{L}$ (by virtue of $(vj_3)$) it is clear that we can extend $D_1 \oplus \check{D}_1$ , by joining it with $\check{D}_{11}$ , to an isometry for $\Phi$ (isometry with respect to $\Psi$ is trivial since $J^{\perp\perp}$ is in the radical of $\Psi$ ).

As to the second assertion: if $x \in J^{\perp\perp} \setminus (J+A^{\perp})$ where should $x$ be mapped? We have no control whatever over the image of $(x)^{\perp} \cap L$ under $D_1$ because $D_1$ is immune from any intrinsic property that relates to the embedding $J \subset J^{\perp\perp}$ . That is why the recursive construction for $D$ has to be repeated from scratch.

Thus, we see that we can get a proof for Theorem 2 almost free via earlier results if $\mu = 0$ (i.e. $J + A^{\perp} = J^{\perp\perp}$). If $\mu \neq 0$ then one has to start all over again and - as we shall see - the old and the new construction run entirely <u>alike</u>; one merely has to <u>adapt</u> the course of recursion to $\mu$ being nonzero. This nicely illustrates a sigh by Bäni: Change a trifle and you can't <u>reduce</u> the problem to the former situation but have to start allover again! We think that this true remark pinpoints an intrinsic feature of infinite dimensional linear algebra and not some inadequacy of the method.

We shall now sketch the construction of a $D$ with (15) ab ovo. The first thing to do is to reduce the problem to the case where

(28) $$\alpha = \delta = \nu = 0 .$$

"$\nu = 0$" is achieved by chopping off the boring summand $E_o$ . How to get rid of $\delta$ ? Pick a supplement $P$ with $24 \oplus P = 25$ . Hence $2 \oplus UP = 4$ . A glance at the diagram and the $\perp$-table reveals that $H := P \oplus UP$ is nondegenerate. Hence $E = H \oplus H^{\perp}$ is a decomposition into U-invariant subspaces. By the assumptions (j), $(vj_1)$,$(vj_2)$ in Theorem 2 $(E,\Psi) \cong (E,\bar{\Psi})$ . Pick an isometry and let $\bar{P}$ be the image of $P$ . Then $\overline{24} \oplus \bar{P} = \overline{25}$ and we let $\bar{H} = \bar{P} \oplus \overline{UP}$ . We can change $P$ and $\bar{P}$ mod UP and $\overline{UP}$ respectively such that $P$ and $\bar{P}$ become totally isotropic for $\Phi$ . Now we join a $\Psi$-isometry $P \to \bar{P}$ with its contragradient $UP \to \overline{UP}$ and obtain an isometry $(H,\Phi,\Psi) \cong (\bar{H},\Phi,\bar{\Psi})$ ; further $(H^{\perp},\Phi) \cong (\bar{H}^{\perp},\Phi)$ by Witt's Theorem for finite dimensions and $(H^{\perp},\Psi) \cong (\bar{H}^{\perp},\bar{\Psi})$ by construction. The assumptions of Theorem 2 are inherited by the reduced situations $H^{\perp}$ , $\bar{H}^{\perp}$ and here we find $\delta = 0$ . To then arrive at $\alpha = 0$ one chops off a summand $X \oplus Y \oplus UX \oplus UY$ where $X := 25 \cap L$ $(= 24 \cap L$ as $\delta = 0$ ) and $Y \oplus (30 \cap L) = L$ . In this manner we arrive at a situation with (28) in force; this means that $(L,\Psi)$ has a dense trace-valued part, i.e. that $(L,\Psi)$ is stable. The lattice $V(I)$ then is

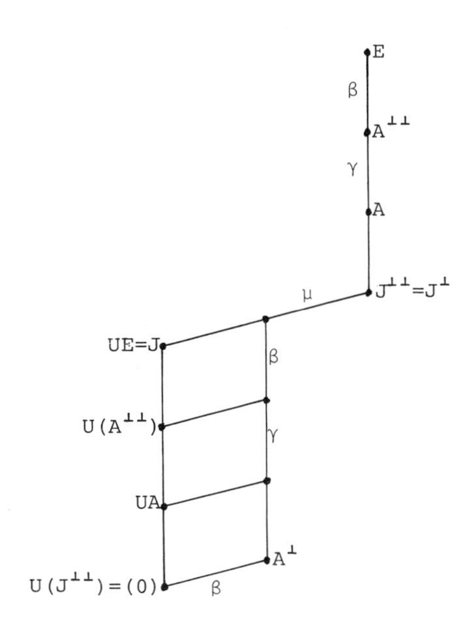


$V(I)$ when $\alpha = \delta = \nu = 0$

The stable spaces $(L,\Psi)$ and $(\bar{L},\bar{\Psi})$ - totally isotropic for $\Phi$ - are isometric by assumption $(vj_1)$ of Theorem 2; $A \cap L$ is mapped onto $\bar{A} \cap \bar{L}$ and, because of assumption $(vj_3)$, it follows that thereby $A^{\perp\perp} \cap L$ is mapped onto $\bar{A}^{\perp\perp} \cap \bar{L}$ . Therefore one starts the recursive construction of a $D$ with (15) on the U-invariant finite dimensional space

$$W_o := B \oplus C \oplus UB \oplus UC \oplus A^{\perp}$$

where $(A \cap L) \oplus C = A^{\perp\perp} \cap L$ and $(A^{\perp\perp} \cap L) \oplus B = E$ ; $\bar{W}_o$ is defined by adding analogous objects isometric under $(L,\Psi) \cong (\bar{L},\bar{\Psi})$ . One then extends the isometry $(W_o,\Phi,\Psi) \cong (\bar{W}_o,\Phi,\bar{\Psi})$ step by step by adjoining vectors $x$ from $A \smallsetminus J^{\perp\perp}$ or from $J^{\perp\perp} \smallsetminus (A^{\perp}+J)$ . In the first case one tears off 2-dimensional U-invariant spaces that contain $x$ and in the second case one simply tears off $(x)$ ($J^{\perp\perp}$ is U-invariant). There are no problems; the reader may, if equipped with the requisite patience, supply the details.

## 9. On the classification of nilpotent self-adjoint transformations

There are nontrivial instances where $(k,*,\varepsilon)$ has $S = T$ (e.g. in the case of so called involutions of the second kind). In such situations the classification theorem is akin to that of characteristic not 2. We shall again distinguish the two cases of Theorem 2.

Theorem 3. Assume that $k$ is as in Section 7 and has $S = T$ . Let $(E,\Phi)$ be nondegenerate and of dimension $\aleph_0$ . In order that two involutions $I$ , $\bar{I}$ be orthogonally similar the following are necessary and sufficient.

In Case I (dim $J < \infty$ ; $V(I)$ reduces to the chain $(0) \subset J \subset J^{\perp} \subset E$ ) :

(i) $\quad (E,\Psi) \cong (E,\bar{\Psi})$ .

In Case II (dim $J = \aleph_0$ ; $V(I)$ reduces to $(0) \subset J \subset J^{\perp\perp} \subset J^{\perp} \subset E$ ) :

(ii) $\quad \dim J^{\perp}/J^{\perp\perp} = \dim \bar{J}^{\perp}/\bar{J}^{\perp\perp}$ ("$\nu = \bar{\nu}$") ,
$(J^{\perp},\Phi) \cong (\bar{J}^{\perp},\Phi)$ in case $\nu$ is finite ,

(iii) $\quad \dim J^{\perp\perp}/J = \dim \bar{J}^{\perp\perp}/\bar{J}$ ("$\mu = \bar{\mu}$") .

Another special case of Theorem 2 worth of mention is described in

Theorem 4. Let $k$ be a perfect commutative field of characteristic 2 and $\Phi$ a nondegenerate alternate bilinear form on the k-vector space $E$ ; $\dim E = \aleph_0$ . In order that two involutions $I$ and $\bar{I}$ of $(E,\Phi)$ be orthogonally similar it is necessary and sufficient that there is an index-preserving lattice isomorphism between the lattices of the involutions that commutes with $\perp$ , $U$ and $\perp$ , $\bar{U}$ . In other words (cf. the diagram in Section 5) the (orthogonal) similarity class of an involution I is characterized by the following seven cardinal number invariants

$$\begin{aligned}&\dim J^{\perp}/J^{\perp\perp}\ ,\quad \dim J^{\perp\perp}/J+A^{\perp}\ ,\quad \dim J+A^{\perp}/J\ ,\\ (29)\qquad &\dim U(A^{\perp\perp})/UA\ ,\quad \dim UA/(A^{\perp}\cap UA)\ ,\ \dim A^{\perp}\cap UA\ ,\\ &\dim J\cap A^{\perp}/(A^{\perp}\cap UA)\ .\end{aligned}$$

Here $A := \{x \in E \mid \Phi(Ux,x) = 0\}$ , $U := I - \mathbb{1}$ , $J := \operatorname{im} U$ .

In [1] Kaplansky had proved that over commutative fields of characteristic not 2 in which every element is a square the self-adjoint linear transformations $U$ with $U^2 = 0$ in a space of countable dimension equipped with a nondegenerate inner product are characterized by the three invariants

$$(30)\qquad \dim J^{\perp}/J^{\perp\perp}\ ,\quad \dim J^{\perp\perp}/J\ ,\quad \dim J\ .$$

As Theorem 4 also classifies all self-adjoint $U$ with $U^2 = 0$ we have here the companion in characteristic 2 to Kaplansky's result. We see that the second invariant in (30) has to be refined in two and the third invariant in (30) into five invariants when $\operatorname{char} k = 2$ .

## 10. Canonical representatives

In finite dimensions we have the following three well known "irreducible" types of involutions:

First, the nondegenerate spaces $E_1$ of dimensions $\leq 2$ which cannot be broken into proper orthogonal summands equipped with the involution $\mathbb{1}$.

Second, the hyperbolic planes $E_2 = k(e,e')$ equipped with the involution $I_2$: $e \mapsto \lambda^{-1}e'$ , $e' \to \lambda e$ where the parameter $\lambda$ can vary in $S \setminus \{0\}$ .

Third, orthogonal sums of two hyperbolic planes $E_3 = k(e,e') \overset{\perp}{\oplus} k(f,f')$ equipped with the involution $I_3$ : $e \mapsto f$ , $e' \mapsto f'$ ; $f \mapsto e$ , $f' \mapsto e'$ .

With these three types one can build a representative in each similarity class when $\dim J < \aleph_0$ (Case I of Theorem 2). By taking infinite sums one gets - trivially - certain cases with $\dim J = \aleph_0$ . We shall see that there are just two more genuinely new possibilities in dimension $\aleph_0$ (Types 1 and 5 below). Let us give a complete list of the "irreducible" types when $\dim J = \aleph_0$ (Case II in Theorem 2).

<u>Type 0</u>: $\alpha = \beta = \gamma = \delta = \nu = \mu = 0$ . Let $E_o$ be an orthogonal sum of $\aleph_0$ spaces of the kind $E_3$ above. Joining all involutions $I_3$ on $E_3$ defines $I_o$ on $E_o$ .

<u>Type 1</u>: $\mu = 1$ and the other indices are zero. Let $E_o$ , $I_o$ be as in the previous example and set $E = E_o \oplus (a)$ where the vector a is isotropic and has product 1 with all basis vectors of the defining basis for $I_o$ . Set $Ia = a$ and let $I$ coincide with $I_o$ on $E_o$ .

Type 2: all indices are zero except for $\nu$. Pick a space $E_1$ as above with $\mathbb{1}$ as the involution. Join it with a space of Type 0 ( $\nu = 2$ iff $(k,*,\varepsilon) = (k,\mathbb{1},1)$).

Type 3: $\delta = 1$ and the remaining indices vanish. Pick a space $E_2$ as above with the parameter $\lambda \in S \setminus T$ and join it with an example of Type 0.

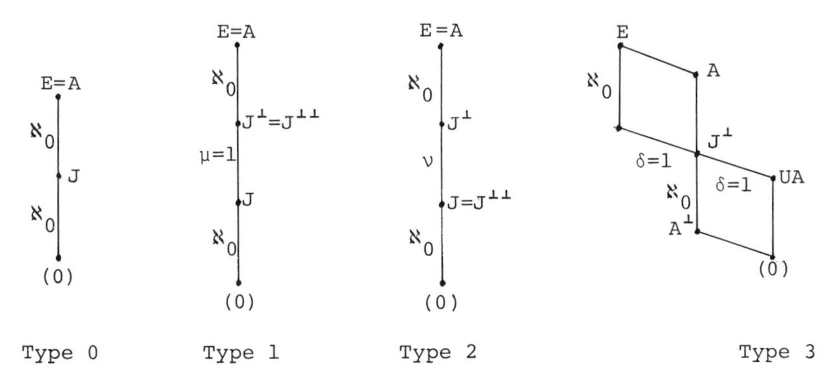


Type 4: all indices but $\gamma$ are zero. For a fixed $\lambda \in S \setminus T$ form an orthogonal sum of $\aleph_0$ copies of $E_2$ as given above.

Type 5: $\beta$ is the only nonzero index. Let $E_4$, $I_4$ be as in the previous case. Set $E = E_4 \oplus (a)$ where the isotropic $a$ has all products 1 with the elements of the first half of the defining symplectic basis for $I_4$ and has products $\lambda$ with the elements of the second half. Set $Ia = a$ and let $I$ coincide with $I_4$ on $E_4$.

Type 6: $\alpha = 1$, $\beta = \gamma = \delta = \nu = \mu = 0$. Let $E_o$, $I_o$ be of Type 0 and add two copies of $E_2$ with equal $\lambda \in S \setminus T$, $E = E_o \overset{\perp}{\oplus} E_2 \overset{\perp}{\oplus} E_2$. Join the involutions on the summands to get one on $E$.

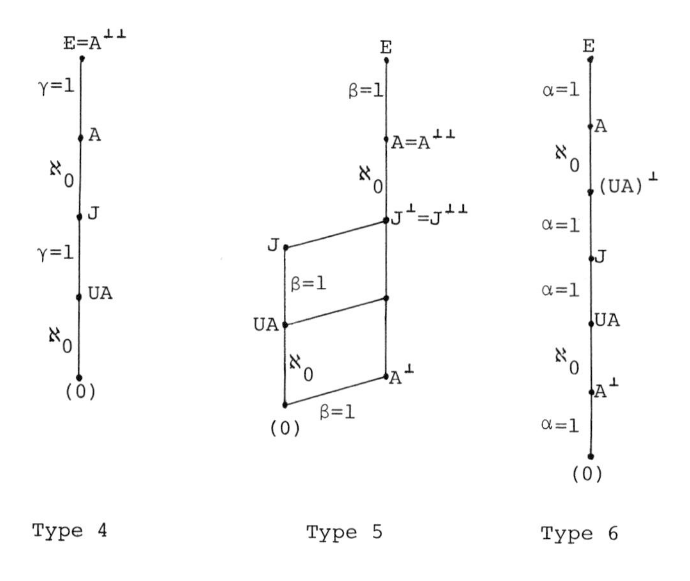


Type 4 Type 5 Type 6

*

## References to Chapter IX

[1] I. Kaplansky, Orthogonal similarity in infinite-dimensional spaces. Proc. Amer. Math. Soc. 3 (1952), 16-25.

[2] M. Studer, Involutionen in abzählbardimensionalen alternierenden Räumen bei Charakteristik zwei. Ph. D. Thesis, University of Zurich 1978.

# CHAPTER TEN

# EXTENSION OF ISOMETRIES

## 0. Introduction

The main result in this chapter is a theorem in [1] on the extension of isometries $\varphi : V \to \bar{V}$ between $\perp$-closed subspaces of a sesquilinear space $E$ (Theorems 5 and 9 below). The crucial assumptions for an extension to exist turn out to be equality of the isometry types of $V^{\perp}$ and $\bar{V}^{\perp}$ and homeomorphy of $V$ and $\bar{V}$ under $\varphi$ with respect to the weak linear topology $\sigma(\Phi)$ attached to the form on $E$.

In discussing extension problems we have made moderate use of the concept of a dual pair. It is possible to eliminate the concept but only at the cost of perspicuity. We think that the student of forms in infinite dimensional vector spaces should be well acquainted with the concept of dual pair (without, of course, espousing any beliefs into orgies of duality). We have therefore included in the first sections some of the classical notions introduced in [7]. Mackey's characterization of modular and dual modular pairs of closed subspaces is of interest to us in the light of the results on orthogonal and symplectic separation. (This topic of Chapter VI is taken up again, though in a different vein, in Section 7 below.)

In Section 8 we have included a short discussion on the log frame Theorem. By applying this theorem to the results of Section 5 we show how these results can be extended to certain uncountable spaces. This application is representative for a host of applications that can be made to results obtained in countable dimensions.

## 1. Recall of dual pairs (algebraic formulation)

Let $k$ be a division ring, $E$ a k-left vector space and $E'$ a k-right vector space, $\langle\ ,\ \rangle : E \times E' \to k$ a nondegenerate bilinear form (i.e. if $\langle x,y\rangle = 0$ for all $x \in E$ then $y = 0$ and, if $\langle x,y\rangle = 0$ for all $y \in E'$ then $x = 0$ ). We say that $E$ and $E'$ are dually paired by $\langle\ ,\ \rangle$ or that $E$ and $E'$ form a dual pair and the like.

Example 1. Let $E'$ be the set $E^* := \mathrm{Hom}_k(E,k)$ of all k-linear maps $f : E \to k$ ; $E^*$ is a k-right vector space under pointwise addition and right multiplication by scalars from $k$ . For $x \in E$ , $f \in E^*$ we define $\langle x,f\rangle := f(x)$ .

Example 2. Let $\Phi : E \times E \to k$ be a nondegenerate $\perp$-symmetric sesquilinear form with respect to the antiautomorphism $\kappa : k \to k$ . We can convert the k-left space $E$ into a k-right space $E' = E$ by defining $x\lambda = \kappa^{-1}(\lambda)x$ for all $x \in E$ and all $\lambda \in k$ . Then $\langle x,y\rangle := \Phi(x,y)$ is a bilinear form on $E \times E' = E \times E$ .

Example 3. Let $(E,\Phi)$ and the pairing $\langle\ ,\ \rangle$ be as in Example 2. Consider a $\perp$-dense subspace $V \subset (E,\Phi)$ , $V^{\perp} = (0)$ . The restriction of $\langle\ ,\ \rangle$ to $V \times E \subset E \times E$ sets in duality the pair of spaces $V$ , $E$ .

If $E$ and $E'$ are in duality and $X$ is a subspace of $E$ then we set

(1) $$X^{\circ} := \{y \in E' \mid \langle x,y\rangle = 0 \text{ for all } x \in X\} .$$

For $Y$ a subspace in $E'$ we set $Y^{\circ} = \{x \in E \mid \langle x,y\rangle = 0 \text{ for all } y \in Y\}$. The operator $^{\circ}$ shares properties familiar with $\perp$ such as: $X \subset (X^{\circ})^{\circ}$, $X_1 \subset X_2 \Rightarrow X_2^{\circ} \subset X_1^{\circ}$ , $(X_1 + X_2)^{\circ} = X_1^{\circ} \cap X_2^{\circ}$ . Thus a number of arguments and calculations on $\perp$ can be carried over to $^{\circ}$ without any changes; e.g., subspaces $X \subset E$ which are "orthogonals", i.e. $X = Y^{\circ}$ for some

$Y \subset E'$ , are "closed" in the sense that $(X^\circ)^\circ = X$ . In particular we have:

(2) If $U \subset V \subset E$ and $(U^\circ)^\circ = U$ and $\dim V/U < \infty$ then $(V^\circ)^\circ = V$.

Example 3 above illustrates how "subpairs" can be formed from given pairs. The general definition runs as follows. If $V \subset E$ is an arbitrary subspace in $E$ then the pairing $\langle\ ,\ \rangle : E \times E' \to k$ induces a nondegenerate bilinear form on $V \times (E'/V^\circ)$ . We mean this induced form when we say that $\langle\ ,\ \rangle$ dually pairs $V$ and $E'/V^\circ$ ; this dual pair is called a <u>subpair</u> of the dual pair $(E,E')$ . Symmetrically, if $W$ is a subspace of $E$ then $\langle\ ,\ \rangle$ induces a bilinear form on $(E/W) \times W^\circ$ ; however, in order to obtain a nondegenerate induced form we have to require that $W$ be closed, $(W^\circ)^\circ = W$ . With this proviso added we call $(E/W, W^\circ)$ a <u>quotient pair</u> of $E, E'$ . Finally, if $(E,E')$ and $(D,D')$ are two dual pairs over $k$ we can put into duality the product spaces $E \times D$ , $E' \times D'$ by the definition $\langle (x,z),(x',z')\rangle := \langle x,x'\rangle + \langle z,z'\rangle$ ; the dual pair thus defined is the <u>direct product</u> or the <u>direct sum</u> of the dual pairs $(E,E')$ and $(D,D')$ . (In the general case of an arbitrary family $(E_\iota, E_\iota')_{\iota \in I}$ of dual pairs we define a direct sum by setting into duality the spaces $\bigoplus_I E_\iota$ and $\prod_I E_\iota'$ via $\langle (x_\iota),(y_\iota)\rangle := \sum_{\iota \in I} \langle x_\iota, y_\iota\rangle$ .)

We finish this short rappel with the definition of homomorphisms between pairs. If $E$ and $E'$ are in duality then the two homomorphisms $E' \to E^*$ , $E \to (E')^*$ defined by $y \mapsto \langle .,y\rangle$ and $x \mapsto \langle x,.\rangle$ are injective. It is therefore possible - and useful - to think of $E'$ and $E$ as subspaces in the full algebraic duals $E^*$ and $(E')^*$ respectively (cf. Example 1). Let then $(E,E')$ and $(D,D')$ be given dual pairs over $k$ and

(3) $\qquad E \xrightarrow{\varphi} D$

a k-linear map. There is the uniquely determined "dual" map $\varphi^* : D^* \to E^*$ defined by $\langle e, \varphi^* d^* \rangle = \langle \varphi e, d^* \rangle$ (where $\langle\ ,\ \rangle$ is as in Example 1), $e \in E$, $d^* \in D^*$. We can restrict $\varphi$ to the subspace $D' \subset D^*$ and get a map

$$E^* \xleftarrow{\varphi^*} D' .$$

If $D'$ happens to be mapped inside $E' \subset E^*$ under $\varphi^*$, i.e., if we actually have a map

(4) $\qquad E' \xleftarrow{\varphi^*} D' \quad (\text{with } \langle e, \varphi^* d' \rangle = \langle \varphi e, d' \rangle)$

then $\varphi$ in (3) is called a <u>homomorphism of the pair</u> $(E,E')$ <u>into the pair</u> $(D,D')$. If $\varphi$ is bijective and $\varphi^* D' = E'$ then $\varphi$ is called an <u>isomorphism of the pairs</u> $(E,E')$ and $(D,D')$. We readily check that if $\varphi$ in (3) is a homomorphism of pairs then $K := \ker \varphi$ is closed, $(K^\circ)^\circ = K$. Notice that by nondegeneracy of the pairings there can be at most one map $\varphi^*$ satisfying (4); if it exists (i.e. if $\varphi$ is a homomorphism of pairs) then $\varphi^*$ is called the <u>transpose</u> of $\varphi$ and often denoted by ${}^t\varphi$.

<u>Example 4</u>. Let $(E,E')$ be a dual pair and $K \subset E$ a closed subspace, $K = (K^\circ)^\circ$. The canonical map $\mu : E \to E/K$ is a homomorphism of the dual pair $(E,E')$ into the quotient pair $(E/K, K^\circ)$ (because $\mu^*$ is the map $K^\circ \hookrightarrow E'$).

<u>Example 5</u>. Let $\varphi : E \to D$ be a homomorphism of the dual pairs $(E.E')$ and $(D,D')$, $K := \ker \varphi$, $\mu : E \to E/K$ the canonical map, $\hat{\varphi} : E/K \to \operatorname{im} \varphi$ the uniquely determined k-linear map with $\hat{\varphi} \circ \mu = \varphi$. Then $\hat{\varphi}$ is a homomorphism of the quotient pair $(E/K, K^\circ)$ into the pair $(\operatorname{im} \varphi, D'/(\operatorname{im}\varphi)^\circ)$ (because the dual map $\varphi^* = \mu^* \circ \hat{\varphi}^*$ maps into $E'$ by assumption).

Definition 1. Let $\varphi : E \to D$ be as in Example 5. Then the map $\varphi$ is an open homomorphism of pairs if and only if the associated map $\hat{\varphi}$ is an isomorphism of pairs.

## 2. Topological setting

Just as we have defined the weak linear topology $\sigma(\Phi)$ in a sesquilinear k-space $(E,\Phi)$ we can define "weak" linear topologies on the spaces $E$ , $E'$ of a dual pair $(E,E')$ . The set $\{ Y^\circ \mid Y \subset E' \ \& \ \dim Y < \infty \}$ serves as a 0-neighbourhood basis for a linear topology - called $\sigma(E,E')$ - on $E$ ; symmetrically, the linear topology $\sigma(E',E)$ on $E'$ , induced by $E$ , has $\{ X^\circ \mid X \subset E \ \& \ \dim X < \infty \}$ as a 0-neighbourhood basis. If $k$ is endowed with the discrete topology then $\sigma(E,E')$ and $\sigma(E',E)$ are the coarsest linear topologies that render the pairing $\langle \ , \ \rangle$ separately continuous. The closures of linear subspaces $Z$ with respect to these topologies are given by $(Z^\circ)^\circ$ . The subspace in $E^*$ (the full algebraic dual of $E$ ) consisting of all continuous linear maps from the topological space $(E,\sigma(E,E'))$ into the discrete $k$ coincides with $E'$ . This follows directly from (2). Now we prove

Theorem 1. Let $(E,E')$ and $(D,D')$ be dual pairs over the division ring $k$ and $\varphi : E \to D$ k-linear. Then the following are equivalent: (i) $\varphi$ is a homomorphism of pairs, (ii) $\varphi$ is weakly continuous, i.e. continuous with respect to the linear topologies $\sigma(E,E')$ and $\sigma(D,D')$ on $E$ and $D$ respectively.

Proof. Assume (i) and let ${}^t\varphi$ be the transpose of $\varphi$ . It suffices to establish continuity of $\varphi$ at the origin. Let $X^\circ$ be a typical neighbourhood of $0 \in D$ . Then $({}^t\varphi X)^\circ$ is a neighbourhood of $0 \in E$ and its image, under $\varphi$ , is contained in $X^\circ$ . Thus $\varphi$ is continuous. Conversely, if $\varphi$ is assumed continuous then we want to show that

${}^t\varphi x \in E^*$ actually is in $E'$ for all $x$ in $D'$ . But ${}^t\varphi x = x \circ \varphi$ is continuous, so ${}^t\varphi x \in E'$ by the remark preceding this theorem. Q.E.D. We have the following

<u>Corollary 1</u>. Let $\varphi$ be as in the theorem. If $\varphi$ is weakly continuous then so is its transpose ${}^t\varphi$ , i.e. ${}^t\varphi : D' \to E'$ is continuous with respect to the linear topologies $\sigma(D',D)$ , $\sigma(E',E)$ induced by $D$ and $E$ on $D'$ and $E'$ respectively. Further, $\varphi$ is an isomorphism of pairs if and only if $\varphi$ is a homeomorphism between the topological spaces $(E,\sigma(E,E'))$ and $(D,\sigma(E,E'))$ .

If $(E,E')$ is a dual pair and $V \subset E$ a subspace then there is the linear topology $\sigma(V, E'/V^\circ)$ on $V$ induced via the pairing that defines the subpair $(V, E'/V^\circ)$ . This topology coincides with the topology induced on $V$ by the topology $\sigma(E,E')$ on the overspace $E$ .

If $(E,E')$ is a dual pair and $W \subset E$ a closed subspace then there is the linear topology $\sigma(E/W, W^\circ)$ on $E/W$ induced via the pairing that defines the quotient pair $(E/W, W^\circ)$ . This topology coincides with the quotient topology on $E/W$ induced by $\sigma(E,E')$ on $E$ .

If $(E,E')$ and $(D,D')$ are dual pairs then the linear topology $\sigma(E \times D, E' \times D')$ on $E \times D$ induced via the pairing that defines the product pair $(E \times D, E' \times D')$ is the product topology of $\sigma(E,E')$ and $\sigma(D,D')$ on $E$ and $D$ respectively.

From these remarks we obtain the following

<u>Corollary 2</u>. Let $\varphi$ be as in the theorem. Then $\varphi$ is an open homomorphism of pairs if and only if $\varphi$ is continuous and carries open sets of the topological space $(E, \sigma(E,E'))$ into open sets of $(\mathrm{im}\varphi, \sigma(D,D'))$ (i.e. $\varphi$ is open in the usual topological sense).

Remark 1. If a nondegenerate sesquilinear space $(E,\Phi)$ is conceived of as a space in duality with itself (as explained in Example 2 of Sec. 1) then the linear topology $\sigma(E,E)$ on $E$ via duality is precisely the weak linear topology $\sigma(\Phi)$ associated with the form.

The reader who is interested to learn more about the linear topologies that can be associated with a pairing is referred to §§ 10 - 13 in [6].

## 3. Mackey's theorem on modular pairs

Consider a dual pair $(E,E')$ and two arbitrary subspaces $M$, $N$ in $E$. We have the three subpairs $(M,M')$, $(N,N')$, $(M+N,(M+N)')$ where $M' := E'/M^\circ$ etc. The map $\varphi : M \times N \to M + N$ defined by

$$(5) \qquad \varphi : (m,n) \longmapsto m + n$$

is weakly continuous, i.e. a homomorphism of the product pair $(M,M') \times (N,N')$ into the pair $(M+N, (M+N)')$.

Lemma 1. ([7], p. 167). The following are equivalent: (i) $\varphi$ is open, (ii) for each pair $y_1', y_2' \in E'$ with $y_1' - y_2' \in (M \cap N)^\circ$ there is $z' \in E'$ such that $z' - y_1' \in M^\circ$ and $z' - y_2' \in N^\circ$.

Proof. By the definition in Sec. 1 $\varphi$ is open if and only if for each "functional" $f \in M' \times N' \subset (M \times N)^*$ that vanishes on the kernel $K = \text{im}\,\varphi = \{(z,-z) \mid z \in M \cap N\}$ there exists $z' \in E'$ such that $\varphi^* z' = f$, i.e. $z' \circ \varphi = f$. Furthermore, by definition of $M'$, $N'$ the functional $f$ is induced by a pair $(y_1',y_2') \in E' \times E'$ with $\langle (m,n), f\rangle = \langle m,y_1'\rangle + \langle n,y_2'\rangle$ for all $(m,n) \in M \times N$. We see that $f$ vanishes on $K$ precisely when $y_1' - y_2' \in (M \cap N)^\circ$. Now we can prove the assertions:

Let $y_1', y_2' \in E'$ be given elements with $y_1' - y_2' \in (M \cap N)^\circ$ and

$f \in K^\circ \subset M' \times N'$ the corresponding element $(y_1', y_2')$. If we assume $\varphi$ open there is $z' \in E'$ with $\varphi^* z' = f$, i.e. $\langle m+n, z'\rangle = \langle m, y_1'\rangle + \langle n, y_2'\rangle$ for all $m \in M$, $n \in N$; thus $z' - y_1' \in M^\circ$, $z' - y_2' \in N^\circ$. This proves the implication (i) $\Rightarrow$ (ii).

Conversely, if $f = (y_1', y_2')$ is in $K^\circ$ and if (ii) is assumed then there is $z'$ with $\langle m, z'\rangle = \langle m, y_1'\rangle$ and $\langle n, z'\rangle = \langle n, y_2'\rangle$ for all $m \in M$, $n \in N$. From this we obtain $\varphi^* z' = f$, $\langle (m, n), \varphi^* z'\rangle = \langle m + n, z'\rangle = \langle m, y_1'\rangle + \langle n, y_2'\rangle = \langle (m, n), f\rangle$, i.e. $\varphi$ is open.

The importance of this lemma to us lies in the following particularization (cf. the remark at the end of the previous section).

Theorem 2. Let $(E, \Phi)$ be a nondegenerate sesquilinear space of arbitrary dimension, $M$ and $N$ linear subspaces with $M \cap N = (0)$, $\sigma(M)$, $\sigma(N)$, $\sigma(M \oplus N)$ the topologies induced on $M$, $N$, $M \oplus N$ respectively by the linear topology $\sigma(\Phi)$ associated to the form. Then the following are equivalent: (i) $\sigma(M \oplus N)$ is the product topology of $\sigma(M)$ and $\sigma(N)$, (ii) $M^\perp + N^\perp = E$.

Indeed, as $M \cap N = (0)$ the pairs $(M, M') \times (N, N')$ and $(M + N; (M + N)')$ are isomorphic pairs so that $\varphi$ in (5) is a homeomorphism for the weak topologies. Furthermore, if $M \cap N = (0)$ then (ii) in the lemma is equivalent with $M^\circ + N^\circ = E'$.

If we restrict ourselves to the contemplation of closed subspaces $M, N \subset E$ then there is a purely lattice theoretic version of the (equivalent) properties (i) and (ii) in the lemma. We first need the

Definition 2. Let $A, B$ be elements of a lattice $L$. The ordered pair $(A, B)$ is called a modular pair if

$$(6) \qquad \text{for all } C \leq B, \quad (C \vee A) \wedge B = C \vee (A \wedge B);$$

the pair $(A, B)$ is called <u>dual modular</u> if

(7) for all $C \geq B$ , $(C \wedge A) \vee B = C \wedge (A \vee B)$ .

Clearly, a lattice is modular if and only if all pairs of elements are modular (dual modular). We shall not pursue the purely lattice theoretic aspect here but turn directly to lattices of subspaces in a vector space. (For the general theory we refer to [8] and [9].)

Let $(E,E')$ be a dual pair of vector spaces and $L(E,E')$ , $L(E',E)$ the lattices of closed subspaces in $E$ and $E'$ respectively ( $X_1 \wedge X_2 := X_1 \cap X_2$ , $X_1 \vee X_2 := (X_1 + X_2)^{\circ\circ}$ for $X_1 = X_1^{\circ\circ}$ , $X_2 = X_2^{\circ\circ}$ arbitrary closed subspaces). Since $X \mapsto X^\circ$ is an antiisomorphism of lattices $L(E,E') \to L(E',E)$ it follows that

(8) The pair $(A, B)$ is a modular pair in $L(E,E')$ if and only if the pair $(A^\circ, B^\circ)$ is dual modular in $L(E',E)$ .

We assert

(9) $(A,B)$ is a dual modular pair if and only if $A \vee B = A + B$

<u>Proof</u>. Assume that $A \vee B = A + B$ . We have to show that $(C \wedge A) \vee B \geq C \wedge (A \vee B)$ for all $C \geq B$ (the converse inequality holds in every lattice). $C \wedge (A \vee B) = C \cap (A + B) = (C \cap A) + B \subset (C \wedge A) \vee B$ as asserted. Assume conversely that $A + B \neq (A \vee B) = (A + B)^{\circ\circ}$ and pick $x \in (A + B)^{\circ\circ} \setminus (A + B)$ . Set $C = B \oplus (x)$ ; $C^{\circ\circ} = C$ by (2), thus $C \in L(E,E')$ . Furthermore $C \wedge (A \vee B) = C$ . On the other hand $(C \wedge A) \vee B = B$ for, if $y \in C \wedge A$ then $y = b + \lambda x \in A$ , so $\lambda x \in A + B$ and therefore $\lambda = 0$ by the choice of $x$ , hence $y \in B$ . Thus (7) is violated. Q.E.D.

We terminate with the announced theorem by Mackey ([7], Thm. III-7, p. 167).

<u>Theorem 3</u>. Let $E, E'$ be a dual pair and $M, N$ closed subspaces in $E$. Then the homomorphism $\varphi$ in (5) is open if and only if $(M,N)$ is a modular pair in $L(E,E')$.

<u>Proof</u>. Assume $(M,N)$ modular. By (8) and (9) therefore $M^\circ + N^\circ = (M^\circ + N^\circ)^{\circ\circ} = (M \cap N)^\circ$. We show that property (ii) of the lemma holds. Indeed, if $y_1' - y_2' \in (M \cap N)^\circ$ then $y_1' - y_2' = m' + n'$ $(m' \in M^\circ, n' \in N^\circ)$ by what we just said and the vector $z' := y_1' - m' = n' + y_2'$ has the requisite properties. Conversely, if $\varphi$ is open then (by the lemma) for $x' = x' - 0 \in (M \cap N)^\circ$ there is $z' \in E'$ with $z' - x' \in M^\circ$, $z' - 0 \in N^\circ$ whence $x' \in M^\circ + N^\circ$. We have proved that $(M^\circ + N^\circ)^{\circ\circ} = (M \cap N)^\circ \subset M^\circ + N^\circ$; because the converse inclusion is trivial we have $(M^\circ + N^\circ)^{\circ\circ} = M^\circ + N^\circ$. By (9) and (8) we obtain modularity of the pair $(M, N)$.

<u>Corollary</u>. Let $(E,\Phi)$ be a nondegenerate sesquilinear space of arbitrary dimension, $M$ and $N$ $\perp$-closed subspaces of $E$ with $M \cap N = (0)$. $(M, N)$ is a modular pair in the lattice $L_{\perp\perp}(E)$ of $\perp$-closed subspaces of $E$ if and only if $M^\perp + N^\perp = E$. $(M, N)$ is a dual modular pair in $L_{\perp\perp}(E)$ if and only if $M + N$ is $\perp$-closed.

## 4. <u>Isometries between dense subspaces</u>

Let $(E,\Phi)$ be a nondegenerate sesquilinear space and

$$(10) \qquad V \xrightarrow{\varphi_\circ} \bar{V}$$

an isometry between $\perp$-dense subspaces $V, \bar{V}$ in $E$. If we make use of the dual pairs $(V, E)$, $(\bar{V}, E)$ with the pairings given by $\Phi$ (cf. Example 3 in Sec. 1) then the discussion of an extension $\varphi : E \to E$ of $\varphi_\circ$ becomes quite trivial. For, if there is such a $\varphi$ then $\varphi^{-1}$ coincides with the transpose

$$(11) \qquad E \xleftarrow{{}^t\varphi_\circ} E$$

of the homeomorphism $\varphi_o$ and is thus uniquely determined. Conversely, if we make the (necessary) assumption that $\varphi_o$ be a homeomorphism with respect to $\sigma(\Phi)$ then there exists the transpose ${}^t\varphi_o$ ; the map $\varphi := {}^t\varphi_o{}^{-1}$ coincides on $V \subset E$ with $\varphi_o$ . Hence the vector space automorphism $\varphi$ extends $\varphi_o$ and satisfies, by definition, the orthogonality relations

$$\Phi(\varphi v, \varphi e) = \Phi(v, e) \qquad (v \in V, e \in E). \tag{12}$$

The map $\varphi$ is continuous with respect to the topologies $\sigma(E, V)$ and $\sigma(E, \bar{V})$ and, obviously, need not be continuous with respect to $\sigma(\Phi) = \sigma(E, E)$ (change $\Phi$ without upsetting (12)); a fortiori, $\varphi$ need not be an isometry. However, if $\varphi$ happens to be $\sigma(\Phi)$-continuous then (12) does imply that $\varphi$ is an isometry, $\Phi(\varphi f, \varphi e) = \Phi(f, e)$ for $f, e \in E$ . Indeed, $f$ is an accumulation point of $V$ , $f = \lim \mathcal{F}$ for $\mathcal{F}$ a filter on $V$ and $\varphi f = \lim \varphi\mathcal{F}$ if $\varphi$ is assumed continuous. Hence $\Phi(\varphi f, \varphi e) = \Phi(\lim \varphi\mathcal{F}, \varphi e) = \lim \Phi(\varphi\mathcal{F}, \varphi e) = \lim \Phi(\mathcal{F}, e) = \Phi(f, e)$. Thus we have proved

<u>Lemma 2</u>. Let $(E,\Phi)$ be a nondegenerate sesquilinear space (of arbitrary dimension). Then the following are equivalent: (i) $\varphi_o$ in (10) admits an isometric extension to $E$ , (ii) $\varphi_o$ is a $\sigma(\Phi)$-homeomorphism and its transpose ${}^t\varphi_o$ in (11) is $\sigma(\Phi)$-continuous.

<u>Lemma 3</u>. ([1] Satz 4, p. 16). Let $(E,\Phi)$ be as in Lemma 2. Assume in addition $\Phi$ to be alternate. Then the following are equivalent: (i) $\varphi_o$ in (10) admits an isometric extension to $E$ , (ii) the orthogonals $U^\perp$ and $(\varphi_o U)^\perp$ are isometric for all subspaces $U \subset V$ with $\dim V/U \leq 2$ .

<u>Proof</u>. We only have to prove the implication (ii) $\Rightarrow$ (i). If we let $U$ run through the hyperplanes $V \cap x^\perp$ , $x \in E$ then we learn from

(ii) that $\varphi_\circ$ is a $\sigma(\Phi)$-homeomorphism. Hence there exists the transpose ${}^t\varphi_\circ$ in (11) and $\varphi := {}^t\varphi_\circ^{-1}$ satisfies (12). Let then $f, e \in E$ be arbitrary. Pick $v \in V$ such that $f + v \perp e$. Thus the plane $P$ spanned by $f + v$ and $e$ is totally isotropic and so is the plane $(\varphi_\circ(P^\perp \cap V))^\perp$ by (ii). This means that $\varphi f + \varphi v \perp \varphi e$, thus $\Phi(f + v, e) = \Phi(\varphi f + \varphi v, \varphi e)$. From (12) we get $\Phi(f,e) = \Phi(\varphi f, \varphi e)$. Hence $\varphi$ is an isometry.

## 5. Isometries between closed subspaces

Let us assume in the first place that $(E,\Phi)$ is a nondegenerate infinite dimensional alternate space. Let us call here partial isomorphism in $(E,\Phi)$ an isometry

(13) $$\varphi : V \longrightarrow \bar{V}$$

where $V, \bar{V}$ are $\perp$-closed subspaces in $E$ such that

(14) $$\dim V^\perp / \operatorname{rad} V = \dim \bar{V}^\perp / \operatorname{rad} \bar{V} .$$

and such that $\varphi$ is a homeomorphism for the linear topology $\sigma(\Phi)$.

Remark 2. Since $\varphi$ is an isometry we have $\dim(\operatorname{rad} V) = \dim(\operatorname{rad} \bar{V}$ hence by (14) the spaces $V^\perp$ and $\bar{V}^\perp$ are isometric as $\Phi$ is alternate.

Lemma 4. The set $J$ of partial isomorphisms in $(E,\Phi)$ has the Ping-Pong property: for any $\varphi \in J$ and $x \in E$ there is some $\varphi_1 \in J$ which extends $\varphi$ and which has $x$ in its domain (range).

Proof. Assume that a homeomorphism $\varphi$ with (13) and (14) is given and $x \in E \setminus V$. We first extend $\varphi$ to an isometry $\varphi_1$ on $V_1 := V \oplus (x)$. Then we shall show that $\varphi_1$ is a homeomorphism for $\sigma(\Phi)$ and that (14) holds for $V_1, \bar{V}_1$ in place of $V, \bar{V}$.

Case 1: $x \in V + V^\perp$. Set $x = y + z$, $y \in V$ and $z \in V^\perp$. Since

$z \notin V$ we conclude by (14) that there is $\bar{z} \in \bar{V}^{\perp} \setminus \bar{V}$. In this case we extend $\varphi$ by sending $x$ into $\bar{x} := \varphi y + \bar{z}$.

Case 2: $x \notin V + V^{\perp}$. Since $\varphi$ is an isomorphism of the pairs $(V, E/V^{\perp})$ and $(\bar{V}, E/\bar{V}^{\perp})$ there is $\bar{x} \in E \setminus \bar{V}^{\perp}$ such that ${}^{t}\varphi\bar{x} = x$ (mod $V^{\perp}$), i.e. $\Phi(\varphi v, \bar{x}) = \Phi(v, {}^{t}\varphi\bar{x}) = \Phi(v,x)$ for all $v \in V$. From this relation we can also conclude that $\bar{x} \notin \bar{V} + \bar{V}^{\perp}$ and that we may extend $\varphi$ by sending $x$ into $\bar{x}$.

The isometry $\varphi_1 : V \oplus (x) \longrightarrow \bar{V} \oplus (\bar{x})$ thus defined has $\perp$-closed domain and range; in particular, $V$ and $\bar{V}$ are closed in the topologies on $V \oplus (x)$ and $\bar{V} \oplus (\bar{x})$ induced by $\sigma(\Phi)$, hence also open in these topologies because the quotients are finite dimensional (the quotient topology is discrete so $V$ and $\bar{V}$ are open as inverse images of (0) under the canonical map). Thus $\varphi_1$ maps the zero-neighbourhood $V$ homeomorphically onto the zero-neighbourhood $\bar{V}$; hence $\varphi_1$ and $\varphi_1^{-1}$ are continuous at the origin, hence continuous.

There remains to compute $d = \dim (V \oplus (x))^{\perp} / \operatorname{rad} (V \oplus (x))$. In Case 1 we find (by modularity) $\operatorname{rad} (V \oplus (x)) = (\operatorname{rad} V) \oplus (z)$; thus $d = \dim V^{\perp} / \operatorname{rad} V - 2$. An analogous computation holds for $\bar{V} \oplus (\bar{x})$ so (14) is inherited. In Case 2 we find $\operatorname{rad} (V \oplus x) = (\operatorname{rad} V) \cap x^{\perp}$ so $d = \dim V^{\perp} / \operatorname{rad} V - 1$ if $\operatorname{rad} V \subset x^{\perp}$ and $d = \dim V^{\perp} / \operatorname{rad} V$ if $\operatorname{rad} V \not\subset x^{\perp}$; since $\varphi_1$ is an isometry with $x \mapsto \bar{x}$ we find ourselves with $\operatorname{rad} \bar{V} \subset \bar{x}^{\perp}$ in the former case and with $\operatorname{rad} \bar{V} \not\subset \bar{x}^{\perp}$ in the latter case. Hence (14) is inherited.

The proof of Lemma 4 is thus complete.

If the dimension of $(E,\Phi)$ is denumerable then we obtain from Lemma 4, by using standard arguments, the following

<u>Theorem 4</u>. ([1] Satz 1, p. 8). Let $(E,\Phi)$ be a nondegenerate alternate space of dimension $\aleph_0$. An isometry $\varphi : V \longrightarrow \bar{V}$ where $V, \bar{V} \subset E$ are $\perp$-closed admits an extension to all of $E$ if and only if $\varphi$ is a homeomorphism for the weak linear topology $\sigma(\Phi)$ and $V$, $\bar{V}$ satisfy (14).

Indeed, each partial isomorphism in $J$ can be extended to a metric automorphism of $E$.

The restriction to alternate forms in Theorem 4 is unduly severe. With some additional stability assumptions we reach a wide class of spaces to which our result applies. We first state the generalization intended.

<u>Theorem 5</u>. ([1] Satz 3, p. 14). Let $(E,\Phi)$ be a nondegenerate trace-valued sesquilinear space of dimension $\aleph_0$, $\varphi : V \longrightarrow \bar{V}$ an isometry, $V$ and $\bar{V}$ $\perp$-closed and $\varphi$ homeomorphic with respect to $\sigma(\Phi)$. In order that $\varphi$ admits an isometric extension to the entire space $E$ the following is sufficient: If $\dim V^{\perp}/\operatorname{rad} V$ is finite then $V^{\perp}$ and $\bar{V}^{\perp}$ are isometric spaces; if $\dim V^{\perp}/\operatorname{rad} V$ is infinite then $V^{\perp}$ and $\bar{V}^{\perp}$ contain totally isotropic subspaces $W$, $\bar{W}$ of infinite dimension such that $W \cap \operatorname{rad} V = (0)$, $\bar{W} \cap \operatorname{rad} \bar{V} = (0)$.

<u>Proof</u>. We first take care of the case where $\dim V^{\perp}/\operatorname{rad} V$ is finite. Let $V_1$ be a supplement of $R := \operatorname{rad} V$ in $V^{\perp}$ and $\bar{V}_1$ a supplement of $\bar{R} := \operatorname{rad} \bar{V}$ in $\bar{V}^{\perp}$. $V_1 \cong \bar{V}_1$ since $V^{\perp} \cong \bar{V}^{\perp}$. $V_1{}^{\perp}$ can be Witt decomposed relative to $R$, $V_1{}^{\perp} = (R \oplus R') \overset{\perp}{\oplus} E_1$ with $R'$ totally isotropic. Hence $E = (R \oplus R') \overset{\perp}{\oplus} V_1 \overset{\perp}{\oplus} E_1$ with $V = V^{\perp\perp} = (R \oplus V_1)^{\perp} = R \oplus E_1$. Let $E = (\bar{R} \oplus \bar{R}') \overset{\perp}{\oplus} \bar{V}_1 \overset{\perp}{\oplus} \bar{E}_1$ with $\bar{V} = \bar{R} \overset{\perp}{\oplus} \bar{E}_1$ be an analogous decomposition with respect to $\bar{V}$. Let $\varphi_\circ$ be the restriction of $\varphi$ to $R$. Since $\varphi_\circ$ is a $\sigma(\Phi)$-homeomorphism it is an isomorphism of the pairs $(R,R')$ and $(\bar{R},\bar{R}')$ and we can use the inverse of the

transpose $^t\varphi_\circ : \bar{R}' \longrightarrow R'$ to extend $\varphi_\circ$ to $R \oplus R'$. Thus we have produced an extension of $\varphi$ to all of $E$ in this case.

Assume then that $\dim V^\perp / \operatorname{rad} V$ is infinite. Let us go through the steps in the proof of Lemma 4. If we are in Case 1 then we can pick $\bar{z} \in \bar{V}^\perp \setminus \bar{V}$ with $\Phi(\bar{z},\bar{z}) = \Phi(z,z)$ because $\Phi$ is trace-valued and $\bar{V}^\perp$ contains (infinitely many) hyperbolic planes $P$ with $P \cap \bar{V} = (0)$. If we are in Case 2 we first pick $\bar{x} \in E \setminus (\bar{V} + \bar{V}^\perp)$, as explained, with $\Phi(x,v) = \Phi(\bar{x},\varphi v)$ for all $v \in V$. Then we select $t \in \bar{V}^\perp \cap \bar{x}^\perp$ with $\Phi(t,t) = \Phi(x,x) - \Phi(\bar{x},\bar{x})$ and switch from $\bar{x}$ to $\bar{x} + t$.

With these amendments we get a proof for Theorem 5 out of the proof of Lemma 4.

## 6. Isometries between arbitrary subspaces

Again, let $(E,\Phi)$ be a nondegenerate trace-valued space of dimension $\aleph_0$ and

$$\varphi : V \longrightarrow \bar{V}$$

an isometry between subspaces $V, \bar{V} \subset E$. We continue to assume that $\varphi$ is a homeomorphism for the linear topology $\sigma(\Phi)$. We consider the pairs $(V, E/V^\perp)$ and $(E/V^\perp, V^{\perp\perp})$ induced by $\Phi$ and the corresponding pairs attached to $\bar{V}$. Since $\varphi$ is homeomorphic there exists the transpose $^t\varphi$,

$$\begin{array}{ccc} (V, E/V^\perp) & , & (E/V^\perp, V^{\perp\perp}) \\ \varphi\downarrow \quad \uparrow {}^t\varphi & & {}^t\varphi\uparrow \quad \vdots\downarrow \\ (\bar{V}, E/\bar{V}^\perp) & , & (E/\bar{V}^\perp, \bar{V}^{\perp\perp}) \end{array} ,$$

and we may ask if $^t\varphi$ is a homeomorphism for the quotient topology induced by $\sigma(\Phi)$. This is tantamount to asking if $^t\varphi$ is an isomorphism of pairs for the second pairings above. If such is the case then the transpose $V^{\perp\perp} \to \bar{V}^{\perp\perp}$ extends $\varphi$ and is an isometry (by continuity and by separate continuity of $\Phi$ ). Hence it can be extended to all of $E$ by Theorem 5, provided we add the requisite arithmetic conditions enunciated there, viz,

(15) If $\dim V^{\perp}/V^{\perp} \cap V^{\perp\perp}$ is finite then $V^{\perp}$ and $\bar{V}^{\perp}$ are isometric, if $\dim V^{\perp}/V^{\perp} \cap V^{\perp\perp}$ is infinite then $V^{\perp}$ and $\bar{V}^{\perp}$ contain totally isotropic subspaces $W, \bar{W}$ of infinite dimension such that $W \cap \mathrm{rad}(V^{\perp\perp}) = (0)$ , $\bar{W} \cap \mathrm{rad}(\bar{V}^{\perp\perp}) = (0)$.

Conversely, if an isometric extension $E \to E$ of $\varphi$ exists, then its restriction to $V^{\perp\perp}$ is the transpose of ${}^t\varphi$ in the above diagram. This settles the issue:

<u>Theorem 6</u>. Let $(E,\Phi)$ be a nondegenerate trace-valued space of dimension $\aleph_0$ and $V, \bar{V}$ subspaces which satisfy (15). Let $\varphi : V \to \bar{V}$ be an isometry which is a homeomorphism for the linear topology $\sigma(\Phi)$ . In order that $\varphi$ admits an extension to an isometry $E \to E$ it is necessary and sufficient that the transpose ${}^t\varphi : E/\bar{V}^{\perp} \longrightarrow E/V^{\perp}$ is a homeomorphism for the quotient topologies of $\sigma(\Phi)$ .

<u>Remark 3</u>. Since $\varphi : V \to \bar{V}$ maps $R := \mathrm{rad}\, V$ onto $\bar{R} := \mathrm{rad}\, \bar{V}$ it is tempting to track the two maps $\varphi_o : R \to \bar{R}$ and $\hat{\varphi} : V/R \to \bar{V}/\bar{R}$ induced by $\varphi$ instead of trailing $\varphi$ . This will lead us to Lemma 5 below. We have seen above that the existence of an extension of $\varphi_o$ to $E$ implies:

(16) The transpose ${}^t\varphi_o : E/\bar{R}^{\perp} \longrightarrow E/R^{\perp}$ is a homeomorphism for the quotient topologies induced by $\sigma(\Phi)$ .

Likewise we can pursue $\hat{\varphi}$ by using the pairs $(V/R, R^{\perp}/V^{\perp})$ , $(R^{\perp}/V^{\perp}, V^{\perp\perp}/R^{\perp\perp})$ induced by $\Phi$ and the corresponding pairs attached to $\bar{V}/\bar{R}$ . The existence of an isometric extension of $\varphi$ to $E$ implies

(17) The transpose ${}^t\hat{\varphi} : \bar{R}^{\perp} / \bar{V}^{\perp} \to R^{\perp} / V^{\perp}$ is a homeomorphism with respect to the topologies $\sigma(\bar{R}^{\perp} / \bar{V}^{\perp}, \bar{V}^{\perp\perp} / \bar{R}^{\perp\perp})$ and $\sigma(R^{\perp} / V^{\perp}, V^{\perp\perp} / R^{\perp\perp})$.

From the existence of an extension $\bar{\varphi}$ to $E$ of $\varphi : V \to \bar{V}$ we get (16) and (17); but $\bar{\varphi}$ cannot be recaptured from (16) and (17) because some information on how the objects $R$ and $V / R$ are pieced together inside $E$ is obviously lost. For example, if $X$ and $\bar{X}$ are linear supplements of $R^{\perp\perp}$ and $\bar{R}^{\perp\perp}$ in $V^{\perp\perp}$ and $\bar{V}^{\perp\perp}$ respectively then we can look at the lifting $\varphi_1 : X \to \bar{X}$ of ${}^t({}^t\hat{\varphi}) : V^{\perp\perp} / R^{\perp\perp} \to \bar{V}^{\perp\perp} / \bar{R}^{\perp\perp}$ (${}^t({}^t\hat{\varphi})$ exists if we assume (17)). $\varphi_1$ is a homeomorphism with respect to $\sigma(X, R^{\perp}/V^{\perp}) = \sigma(X, R^{\perp} + X^{\perp}/X^{\perp})$ and $\sigma(\bar{X}, \bar{R}^{\perp}/\bar{V}^{\perp}) = \sigma(\bar{X}, \bar{R}^{\perp} + \bar{X}^{\perp}/\bar{X}^{\perp})$. On the other hand, from the existence of $\bar{\varphi}$ we have first that for each choice of $X$ there is $\bar{X}$ such that $\varphi_1$ turns out homeomorphic for $\sigma(\Phi)$, i.e. for $\sigma(X, E/X^{\perp})$ and $\sigma(\bar{X}, E/\bar{X}^{\perp})$ and, second, the map $\varphi_1$ does coincide with $\varphi$ on $V \cap X$ (everything is such a snug fit here because the map $\varphi_1$ is uniquely determined if the existence of $\bar{\varphi}$ is assumed). It is possible to recapture $\bar{\varphi}$ (from (16) and (17)) if we require slightly less about the liftings of ${}^t({}^t\hat{\varphi})$, namely

(18) For each $\sigma(\Phi)$-topological supplement $X$ of $R^{\perp\perp}$ in $V^{\perp\perp}$ with $V \subset R \oplus X$ the transpose ${}^t({}^t\hat{\varphi}) : V^{\perp\perp}/R^{\perp\perp} \to \bar{V}^{\perp\perp}/\bar{R}^{\perp\perp}$ of ${}^t\hat{\varphi}$ in (17) admits a $\sigma(\Phi)$-continuous lifting $\varphi_1 : X \to \bar{X}$ which coincides with $\varphi$ on $X \cap V$.

We shall see below that the image $\bar{X} = \operatorname{im} \varphi_1$ turns out to be a topological supplement of $\bar{R}^{\perp\perp}$ in $\bar{V}^{\perp\perp}$. Our assertion is formulated in the following

<u>Lemma 5</u>. Let $(E,\Phi)$ be a nondegenerate $\aleph_0$-dimensional alternate space and $V, \bar{V}$ subspaces with $\dim V^{\perp}/V^{\perp} \cap V^{\perp\perp} = \dim \bar{V}^{\perp}/\bar{V}^{\perp} \cap \bar{V}^{\perp\perp}$ and $\dim R^{\perp}/R^{\perp\perp} = \dim \bar{R}^{\perp}/\bar{R}^{\perp\perp}$ ($R$ and $\bar{R}$ the radicals of $V$ and $\bar{V}$ respectively).

Assume that $\varphi : V \to \bar{V}$ is an isometry and a $\sigma(\Phi)$-homeomorphism. Then $\varphi$ admits an isometric extension to all of $E$ if and only if the homeomorphisms $\varphi_o : R \to \bar{R}$ and $\hat{\varphi} : V/R \to \bar{V}/\bar{R}$ satisfy (16),(17),(18).

Proof. We only have to establish sufficiency of the conditions listed. We first quote the corollary to Theorem 2 in Sec. 3 of Chap. V: there is a Witt decomposition as follows:

$$(19) \qquad E = (R^{\perp\perp} \oplus R') \overset{\perp}{\oplus} E_o \quad \text{with} \quad V = R \oplus V_o \quad \text{and} \quad V_o = V \cap E_o .$$

From this decomposition we see that $\sigma(\Phi)$ induces on $V$ the product topology of $\sigma(\Phi)|_R$ and $\sigma(\Phi)|_{V_o}$. Since $\varphi$ is a weak homeomorphism the topology $\sigma(\Phi)$ induces on $\bar{V}$ the product topology of $\sigma(\Phi)|_{\bar{R}}$ and $\sigma(\Phi)|_{\varphi V_o}$. Hence by the corollary to Lemma 1 in Sec. 3 we obtain

$$(20) \qquad \bar{R}^{\perp} + (\varphi V_o)^{\perp} = E .$$

From (19) we obtain furthermore $V^{\perp\perp} = R^{\perp\perp} \oplus V_o^{\perp\perp}$. Thus we may let $X = V_o^{\perp\perp}$ in (18) so $\bar{V}^{\perp\perp} = \bar{R}^{\perp\perp} \oplus \varphi_1(V_o^{\perp\perp})$. By continuity of $\varphi_1$ we have $\varphi_1(V_o^{\perp\perp}) \subset (\varphi_1 V_o)^{\perp\perp} \subset \bar{V}^{\perp\perp}$; by (20) furthermore $\bar{R}^{\perp\perp} \cap (\varphi V_o)^{\perp\perp} = (0)$. Since $\varphi_1 V_o = \varphi V_o$ by (18) therefore $\varphi_1(V_o^{\perp\perp}) = (\varphi_1 V_o)^{\perp\perp}$. By continuity of $\varphi_1$ and separate continuity of $\Phi$ we now see that $\varphi_1$ is an isometry. Furthermore, we now have $\bar{V}^{\perp\perp} = \bar{R}^{\perp\perp} \oplus (\varphi V_o)^{\perp\perp}$ which may be read as

$$(21) \qquad (\bar{R} \oplus \varphi V_o)^{\perp\perp} = \bar{R}^{\perp\perp} \oplus (\varphi V_o)^{\perp\perp} .$$

Because $\bar{R} \perp \varphi V_o$ and $\bar{R} \cap \varphi V_o = (0)$ we can, by (20) and (21), orthogonally separate $\bar{R}$ and $\varphi V_o$; i.e. there is a decomposition $E = E_1 \overset{\perp}{\oplus} E_2$ with $\bar{R} \subset E_1$, $\varphi V_o \subset E_2$ (Theorem 2 in Chapter VI). $E_1$ can be Witt decomposed for $\bar{R}^{\perp\perp}$. After reshuffling of the spaces we obtain a decomposition entirely analogous to (19),

$$E = (\bar{R}^{\perp\perp} \oplus \bar{R}') \overset{\perp}{\oplus} \bar{E}_o \quad \text{with} \quad \bar{V} = \bar{R} \oplus \varphi V_o , \ \varphi V_o \subset \bar{E}_o .$$

Now it transpires that we can extend $\varphi$ to all of $E$. By (16) ${}^t\varphi_o$ is a $\sigma(\Phi)$-homeomorphism; therefore there exists its transpose $\bar{\varphi}_o := {}^t({}^t\varphi_o) : R^{\perp\perp} \longrightarrow \bar{R}^{\perp\perp}$; it is homeomorphic for $\sigma(\Phi)$. Therefore there exists its contragredient $({}^t\bar{\varphi}_o)^{-1} : R' \longrightarrow \bar{R}'$. We join it with $\bar{\varphi}_o$ to get an isometry $\varphi_2 : R^{\perp\perp} \oplus R' \longrightarrow \bar{R}^{\perp\perp} \oplus \bar{R}'$. Finally, by the assumptions in the lemma, it follows that $\dim E_o = \dim R^{\perp}/R^{\perp\perp} = \dim \bar{E}_o$; therefore the alternate spaces $E_o, \bar{E}_o$ are isometric. We may quote Theorem 5 to obtain an extension $\varphi_3 : E_o \longrightarrow \bar{E}_o$ of the $\sigma(\Phi)$-homeomorphic isometry $\varphi_1 : V_o^{\perp\perp} \longrightarrow \bar{V}_o^{\perp\perp}$. We join $\varphi_2$ and $\varphi_3$ to obtain an isometric extension $\bar{\varphi} : E \longrightarrow E$ of $\varphi$.

## 7. The results of Chapter VI as an inference from Theorem 5 in Section 5

We shall illustrate the contention made in the caption by deriving the results on orthogonal separation (Theorem 2 in Chapter VI) from the results in Section 5.

Theorem 7. Let $(E,\Phi)$ be a nondegenerate trace-valued sesquilinear space of dimension $\aleph_0$. Assume that the subspaces $F, G \subset E$ satisfy the conditions (j) $(F+G)^{\perp\perp} = F^{\perp\perp} + G^{\perp\perp}$, (jj) $F^{\perp} + G^{\perp} = E$ and $F \perp G$. In order that $F$ and $G$ are orthogonally separated in $E$, i.e. that there is a decomposition $E = E_1 \overset{\perp}{\oplus} E_2$, $F \subset E_1$, $G \subset E_2$, it is sufficient that the following holds:

If $\dim (F+G)^{\perp}/\operatorname{rad}((F+G)^{\perp})$ is infinite then there exists a totally isotropic subspace $W \subset (F+G)^{\perp}$ of infinite dimension with $W \cap \operatorname{rad}((F+G)^{\perp}) = (0)$.

Remark. We had in fact derived results of this kind via Theorem 5 in Section 5 before the results in Chapter VI had been found (see [3], Bemerkung p. 20). Bäni pointed out that even a more general version than Theorem 2 in Chapter VI can be obtained in this fashion; indeed, in Theorem 7 above we require condition (10) of Chapter VI only when (cf. loc. cit.) $X = (F+G)^{\perp}$ ; we do not, in the proof below, need it for $X = F, G$ . This is very satisfying because here we have complete symmetry with the corresponding Theorem 4 in Chapter VI on symplectic separation.

Proof. (Bäni) We may and we shall assume that $F, G$ are $\perp$-closed.

Case 1: $\dim (F+G)^{\perp}/\operatorname{rad}(F+G)^{\perp}) = \aleph_0$ . Since $(E,\Phi)$ is trace-valued and because $W \subset E$ we have an isometry $E \cong E \overset{\perp}{\oplus} E$ (external orthogonal sum) because both spaces are hyperbolic. We map the subspace $F \oplus G \subset E$ into $E \overset{\perp}{\oplus} E$ by throwing the summand $F$ onto the subspace $F$ contained in the first summand of $E \overset{\perp}{\oplus} E$ and by throwing the summand $G$ onto the subspace $G$ contained in the second summand in $E \overset{\perp}{\oplus} E$ . In $E$ the subspace $F + G$ is closed and, by Theorem 2, it carries the product topology of $\sigma(\Phi)$ on $F$ and $G$ respectively. The same holds true, trivially, for the intended image $F \oplus G \subset E \overset{\perp}{\oplus} E$ . Thus the intended mapping $\varphi : F \oplus G \longrightarrow F \oplus G$ is a weak homeomorphism. By the existence of $W$ we see that $(F+G)^{\perp}$ contains a $\aleph_0$-dimensional hyperbolic space $H$ ; since $H \subset F^{\perp} \cap G^{\perp}$ we see that, in $E \overset{\perp}{\oplus} E$ , $(\operatorname{im}\varphi)^{\perp}$ contains copies of $H$ so that the condition in Theorem 5 (on both $V$ and $\bar{V}$ ) is satisfied. Thus $\varphi$ can be extended to an isometry $E \cong E \overset{\perp}{\oplus} E$ which proves the assertion in this case.

Case 2: $\dim (F+G)^{\perp}/\operatorname{rad}((F+G)^{\perp}) < \aleph_0$ . By chopping off a finite dimensional orthogonal summand we reduce the problem of separation to the case where this dimension is zero, i.e.,

$$(F+G)^{\perp} = \operatorname{rad} F \oplus \operatorname{rad} G$$

Since $F \oplus G$ carries the product topology by (jj) the space $\operatorname{rad} F \oplus \operatorname{rad} G$ is $\perp$-closed, hence there is a Witt decomposition

$E = ((\mathrm{rad}\, F \oplus \mathrm{rad}\, G) \oplus C) \overset{\perp}{\oplus} E_o$ , $C$ totally isotropic. Sums being topological we can rewrite the hyperbolic part, $E = (\mathrm{rad}\, F \oplus C') \overset{\perp}{\oplus} (\mathrm{rad}\, G \oplus C'') \overset{\perp}{\oplus} E_o$ , $C' := C \cap (\mathrm{rad}\, G)^{\perp}$ , $C'' = C \cap (\mathrm{rad}\, F)^{\perp}$ (cf. [6], (6) p. 94). So far we have separated only rad $F$ and rad $G$ .

We now consider the projection pr from $F \oplus G = (F \oplus G)^{\perp\perp} = \mathrm{rad}\, F \overset{\perp}{\oplus} \mathrm{rad}\, G \overset{\perp}{\oplus} E_o$ onto $E_o$ . Since $E_o = \mathrm{pr}\, F + \mathrm{pr}\, G$ and since $\mathrm{pr}\, F \perp \mathrm{pr}\, G$ we obtain $E_o = \mathrm{pr}\, F \overset{\perp}{\oplus} \mathrm{pr}\, G$ as $E_o$ is nondegenerate. Since $F \oplus G$ and $\mathrm{pr}\, F \oplus \mathrm{pr}\, G$ are topological sums and pr is open and, of course, continuous, it follows that the restrictions $F \to \mathrm{pr}\, F$ , $G \to \mathrm{pr}\, G$ are open and continuous.

If we Witt decompose $E$ with respect to rad $F$ , rad $G$ then we find topological decompositions $F = \mathrm{rad}\, F \oplus F_o$ , $G = \mathrm{rad}\, G \oplus G_o$ . This means that the isometry $\omega$ in

$$\begin{array}{ccc} \mathrm{rad}\, F \oplus F_o = F & \longrightarrow & \mathrm{pr}\, F \\ \cup & \overset{\omega}{\nearrow} & \\ F_o & & \end{array}$$

is a homeomorphism. Set $\bar{F} := \mathrm{rad}\, F \oplus \mathrm{pr}\, F$ , $\bar{G} := \mathrm{rad}\, G \oplus \mathrm{pr}\, G$ . Thus we obtain a weakly homeomorphic isometry $\varphi : F \oplus G \longrightarrow \bar{F} \oplus \bar{G}$ . We find $\dim (\bar{F} + \bar{G})^{\perp} / \mathrm{rad}\, ((\bar{F} + \bar{G})^{\perp}) = 0$ so that we may quote Theorem 5. As $\bar{F}$ and $\bar{G}$ are separated we are done. This finishes the proof of Theorem 7.

## 8. Transgression into the uncountable: an application of the log frame

Let $(H, \Phi)$ be a nondegenerate sesquilinear space. We are interested in orthogonal decompositions of $H$ into orthogonal summands of dimensions $\leq \aleph_0$ ; if $H$ admits such decompositions then we call $(H, \Phi)$ euclidean. A nondegenerate sesquilinear space which is isometric to a subspace of a euclidean space is called preeuclidean (see the examples in the introduction of Chapter II).

On $(H, \Phi)$ we introduce the denumerable linear topology $\sigma_1(\Phi)$ attached to the form; it has $\{ X^\perp \mid X \subset H \ \& \ \dim X < \aleph_1 \}$ as a 0-neighbourhood basis (the division ring $k$ is, as usual, endowed with the discrete topology). We have the following decomposition theorem, the so-called log frame *) ( [10] , Satz 3, p. 240; [5] , Corollary 8, p. 18; [2] , Theorem 2, p. 1570 ) :

Theorem 8. Let $(H, \Phi)$ be euclidean and $H = \oplus^\perp_{\nu \in J} H_\nu$ some fixed decomposition with $\dim H_\nu \leq \aleph_0$ . If $E$ is a $\sigma_1(\Phi)$-closed subspace of $H$ then there exists a partitioning $J = \cup J_\iota$ with $\operatorname{card} J_\iota \leq \aleph_0$ and such that $H = \oplus^\perp_\iota G_\iota$ , $G_\iota = \oplus^\perp \{ H_\nu \mid \nu \in J_\iota \}$ and $E = \oplus_\iota (E \cap G_\iota)$ .

Consider an isometry $\varphi : V \longrightarrow \bar{V}$ between $\perp$-closed subspaces $V, \bar{V}$ in a euclidean space $H$ of dimension $> \aleph_0$ . The topology $\sigma_1(\Phi)$ is finer than $\sigma(\Phi)$ thus $V, \bar{V}$ are $\sigma_1(\Phi)$-closed and, by Theorem 8, there are orthogonal decompositions of $H$ into subspaces $G_\iota$ , $\bar{G}_\mu$ such that $0 < \dim G_\iota$ , $\dim \bar{G}_\mu \leq \aleph_0$ (since the number of summands equals $\dim H$ we may let $\iota, \mu$ run through the same index set) and such that $V$ is the sum of all $V \cap G_\iota$ , $\bar{V}$ the sum of all $\bar{V} \cap \bar{G}_\iota$ . In order to fix ideas we formulate the following

Lemma 6. Let $H = \oplus^\perp_I G_\iota = \oplus^\perp_I \bar{G}_\iota$ with $V = \oplus_I (V \cap H_\iota)$ , $\bar{V} = \oplus_I (\bar{V} \cap \bar{H}_\iota)$ . Then there exists a partitioning of $I$ , $I = \cup \{ M_\mu \mid \mu \in J \}$ with $\operatorname{card} M_\mu \leq \aleph_0$ and such that $\varphi$ maps, for each $\mu \in J$ , the intersection $V \cap \oplus^\perp \{ G_\iota \mid \iota \in M_\mu \}$ onto the corresponding intersection $\bar{V} \cap \oplus^\perp \{ \bar{G}_\iota \mid \iota \in M_\mu \}$ .

---

*) Raami saha (Finnish), scie à cadre (French), Gattersäge (German). These classical sawmills have, of course, a gang of strictly less than $\aleph_0$ saws; their forerunner, the famous troncòn da Trentìn has only one blade.

The proof is obvious: We start with $V \cap G_o$ and decompose the images $\varphi e_\nu$ of some basis $(e_\nu)_\nu$ of $V \cap G_o$ according to the second decomposition $H = \overset{\perp}{\underset{I}{\oplus}} \bar{G}_\iota$ . Nonzero components arise only in a countable number of summands $\bar{G}_\iota$ , say for $\iota \in I_1 \subset I$ . Map now bases of all intersections $\bar{V} \cap \bar{G}_\iota$ with subscripts $\iota$ out of $I_1$ by $\varphi^{-1}$ and decompose the images according to the first decomposition $H = \underset{I}{\oplus} G_\iota$ . Nonzero components can arise only in a countable number of summands, say for $\iota \in I_2 \subset I$ , (because there are at most $\aleph_0$ card $I_1 \leq \aleph_0{}^2 = \aleph_0$ images and each producing finitely many components only). We keep switching back and forth ad infinitum (i.e. a countable number of times). We collect all subscripts of $G_\iota$ which we meet in this procedure into the union $M_o := \{0\} \cup I_2 \cup I_4 \cup \ldots$ and all subscripts of $\bar{G}_\nu$ we run into are contained in $\bar{M}_o := I_1 \cup I_3 \cup \ldots$ . Let then $\lambda \in I \setminus M_o$ . We repeat the step with $V \cap G_\lambda$ to get subsets $M_1 \subset I \setminus M_o$ , $\bar{M}_1 \subset I \setminus M_o$ of cardinality $\leq \aleph_0$ . Transfinite iteration yields partitionings $I = M_o \cup M_1 \cup \ldots$ , $I = \bar{M}_o \cup \bar{M}_1 \cup \ldots$ with the property that $\varphi$ throws $V \cap \oplus\{ G_\iota \mid \iota \in M_\kappa \}$ onto $\bar{V} \cap \oplus\{ \bar{G}_\iota \mid \iota \in \bar{M}_\kappa \}$ . A second partitioning shows that we can achieve $M_\kappa = \bar{M}_\kappa$ . This finishes the proof of the lemma.

We have shown that $(H,\Phi)$ , if it is euclidean, admits orthogonal decompositions $H = \overset{\perp}{\underset{I}{\oplus}} G_\iota = \overset{\perp}{\underset{I}{\oplus}} \bar{G}_\iota$ with $\dim G_\iota$ , $\dim \bar{G}_\iota \leq \aleph_0$ and compatible with $\varphi$ ,

$$\text{(22)} \qquad \begin{aligned} \varphi(V \cap G_\iota) &= \bar{V} \cap \bar{G}_\iota \qquad (\iota \in I) \\ V = \Sigma\, V \cap G_\iota \quad &, \quad \bar{V} = \Sigma\, \bar{V} \cap \bar{G}_\iota \end{aligned}$$

A further partitioning shows that it is harmless to assume besides (22) that the dimensions $n_\iota$ of the orthogonals of $V \cap G_\iota$ , taken in $G_\iota$ , modulo their radicals are, for each $\iota \in I$ , either zero or $\aleph_0$ - unless, of course, $n = \dim V^\perp / \operatorname{rad} V$ should be finite and nonzero.

In that case we arrange for $n_o = n$ and $n_\iota = 0$ for all $\iota \in I \setminus \{0\}$. A final reshuffling allows us to assume that the $n_\iota$ equal the corresponding dimensions $\bar{n}_\iota$ $(\iota \in I)$.

After these preparations we can apply Theorem 5 to each summand $G_\iota$. We obtain the following result ([1], Satz 2, p. 12):

Theorem 9. Let $(H,\Phi)$ be a euclidean trace-valued $k$-space with respect to the antiautomorphism $*$ of $k$. Assume that $(k,*,\varepsilon)$ admits only one isometry type in dimension $\aleph_0$ of nondegenerate trace-valued forms. An isometry $\varphi : V \longrightarrow \bar{V}$ between $\perp$-closed subspaces of $E$ admits an extension to all of $E$ if and only if the following two conditions are satisfied: (i) $\varphi$ is a homeomorphism with respect to the weak linear topology $\sigma(\Phi)$, (ii) if $\dim V^\perp / \operatorname{rad} V$ is finite then $V^\perp$ and $\bar{V}^\perp$ are isometric spaces; if $\dim V^\perp / \operatorname{rad} V$ is infinite then it equals $\dim \bar{V}^\perp / \operatorname{rad} \bar{V}$.

Notice that the topology $\sigma_1(\Phi)$ does not enter Theorem 9.

## 9. On the extension of algebraic isometries

Here we consider the situation where $\varphi : V \longrightarrow V$ is a metric automorphism on the subspace $V$ in the (nondegenerate) sesquilinear $k$-space $E$ and we study extensions of $\varphi$ to $E$. We assume that $\varphi$ is algebraic on $V$. This means that there is a polynomial $f$ over the center $C$ of $k$, such that $f(\varphi) = \Sigma a_i \varphi^i = 0$, i.e.

$$\Sigma a_i \varphi^i x = 0 \quad \text{for all } x \in V .$$

We ask if $\varphi$ admits an (isometric) extension $\bar{\varphi}$ to all of $E$ such that $\bar{\varphi}$ is algebraic on $E$, i.e. $\bar{f}(\bar{\varphi}) = 0$ for a multiple $\bar{f}$ of $f$ over $C$.

_Example_. Consider the isometry $v \longmapsto \lambda v$ on the totally isotropic subspace $V \subset E$ , $\lambda$ a fixed element of the center such that $\lambda\lambda^* \neq 1$ . Since $E$ is assumed nondegenerate and $\lambda\lambda^* \neq 1$ no isometry $\bar{\varphi} : E \longrightarrow E$ can satisfy the polynomial equation $f = X - \lambda = 0$ . However, there is an extension $\bar{\varphi}$ of $\varphi$ which satisfies the equation $\bar{f}(\bar{\varphi}) = 0$ on $E$ where $\bar{f} = (X-\lambda)(X-(\lambda^*)^{-1})(X-1) = f \cdot (X-(\lambda^*)^{-1})(X-1)$ ; this follows from the existence of a Witt decomposition of $E$ with respect to $V^{\perp\perp}$ , $E = (V^{\perp\perp} \oplus V') \overset{\perp}{\oplus} E_o$ ( $V' \subset V'^{\perp}$ ) . $\bar{\varphi}$ is the identity on $E_o$ (thus we may delete the factor $X - 1$ in $\bar{f}$ in case $V^{\perp} = V^{\perp\perp}$ ) , $\bar{\varphi}$ dilates by $(\lambda^*)^{-1}$ on $V'$ and dilates by $\lambda$ on $V^{\perp\perp}$.

Actually, we shall consider here only the rather modest situation where $f \in C[X]$ splits into different linear factors. We then sort out the factors in the following fashion:

$$f = \prod_{j=1}^{n} (X-\nu_j) \prod_{i=1}^{m} (X-\lambda_i)(X-\mu_i)^{m_i} \text{ , with pairwise different factors, } \nu_j\nu_j^* = 1 \text{ , } \lambda_i\lambda_i^* \neq 1 \text{ but } \lambda_i\mu_i^* = 1 \text{ , and } m_i \in \{0,1\} . \tag{23}$$

With such a polynomial $f$ we associate the multiple

$$\bar{f} = \prod_{j=1}^{n} (X-\nu_j) \prod_{i=1}^{m} (X-\lambda_i)(X-\mu_i) \quad \text{if } n > 0$$
$$\text{and } \bar{f} = (X-1) \prod_{i=1}^{m} (X-\lambda_i)(X-\mu_i) \quad \text{if } n = 0 . \tag{24}$$

Let us introduce the eigenspaces of $f$ in (23), $K_j = \ker(\varphi - \nu_j)$, $L_i = \ker(\varphi - \lambda_i)$ , $M_i = \ker(\varphi - \mu_i)$ . The $L_i$ and $M_i$ are totally isotropic. Furthermore, any two different $X, Y$ among the eigenspaces are mutually perpendicular, $X \perp Y$ , with the only exceptions $\{X, Y\} = \{L_i, M_i\}$ , $i = 1,\ldots,m$ . It transpires that we can extend $\varphi$ by trying to apply orthogonal and symplectic separation to the eigenspaces; for there will be a splitting of $E$ into the eigenspaces of $\bar{\varphi}$ if such there is. We have the following result (for arbitrary characteristic) :

Theorem 10. ( [3] , Satz 4, p. 32 ). Let $(E,\Phi)$ be a $\aleph_0$-dimensional nondegenerate trace-valued $\varepsilon$-hermitean space over $(k,*,\varepsilon)$. Assume that $(k,*,\varepsilon)$ admits only one isometry type of such spaces in dimension $\aleph_0$. Let $\varphi : V \longrightarrow V$ be an isometry that satisfies the polynomial equation $f(\varphi) = 0$ on $V$ and where the polynomial splits into different linear factors over the center of $k$. Write $f$ as in (23) and associate with it the polynomial $\bar{f}$ in (24). Then the following statements are equivalent : (i) There exists an isometric extension $\bar{\varphi} : E \longrightarrow E$ of $\varphi$ with $\bar{f}(\bar{\varphi}) = 0$ on $E$, (ii) The $n+2m$ eigenspaces $X_i = K_1, \ldots, K_n, L_1, M_1, L_2, M_2, \ldots, L_m, M_m$ of $\varphi$ - some of the $M_i$ may be $(0)$ - satisfy the conditions $(X_1 \oplus \cdots \oplus X_{n+2m})^{\perp\perp} = X_1^{\perp\perp} + \cdots + X_{n+2m}^{\perp\perp}$ and, for all $p$ between $1$ and $n+2m$, $(X_1 \oplus \cdots \oplus X_p)^{\perp} + (X_{p+1} \oplus \cdots \oplus X_{n+2m})^{\perp} = E$.

Proof. Assume (i) and let $\bar{X}_i = \bar{K}_1, \ldots, \bar{M}_m$ be the eigenspaces of $\bar{\varphi}$. Their sum $\Sigma$ splits, $\Sigma = \bar{K}_1 \overset{\perp}{\oplus} \ldots \overset{\perp}{\oplus} \bar{K}_n \overset{\perp}{\oplus} (\bar{L}_1 \oplus \bar{M}_1) \overset{\perp}{\oplus} \ldots \overset{\perp}{\oplus} (\bar{L}_m \oplus \bar{M}_m)$. From the representation of $1$ as g.c.d. of the factors $\frac{\bar{f}}{X-\nu_1}, \ldots, \frac{\bar{f}}{X-\mu_m}$ of $\bar{f}$ we conclude that $\Sigma = E$. Because the $\bar{L}_i$ and $\bar{M}_i$ are totally isotropic we obviously have (ii). It remains to prove the converse implication (ii) $\Rightarrow$ (i).

If (ii) holds we can orthogonally separate $K_1$ and $\Sigma_1 := K_2 \oplus \cdots \oplus M_m$ (by Theorem 2 in Chapter VI) for, by (ii) we have $\Sigma^{\perp\perp} = K_1^{\perp\perp} + \cdots + M_m^{\perp\perp} \subset K_1^{\perp\perp} + \Sigma_1^{\perp\perp} \subset \Sigma^{\perp\perp}$. From this, and by choosing $p = 1$ in (ii), we see that $\Sigma^{\perp\perp} = K_1^{\perp\perp} \oplus \Sigma_1^{\perp\perp}$. Furthermore, we conclude that $\Sigma_1^{\perp\perp} = K_2^{\perp\perp} + \cdots + M_m^{\perp\perp}$ so the step may be repeated $n+m$ times in order to obtain a decomposition $E = E_1 \overset{\perp}{\oplus} \cdots \overset{\perp}{\oplus} E_n \overset{\perp}{\oplus} F_1 \overset{\perp}{\oplus} \cdots \overset{\perp}{\oplus} F_m$ with $K_j \subset E_j$, $L_i \oplus M_i \subset F_i$. In $F_i$ we can use symplectic separation (Theorem 4 in Chapter VI) in order to obtain Witt decompositions $F_i = (L_i^{\perp\perp} \oplus L_i') \overset{\perp}{\oplus} F_{i_0}$ with $M_i \subset L_i' \subset L_i'^{\perp}$. We define $\bar{\varphi}$ as a dilatation with multipliers $\nu_j$ on $E_j$, $\lambda_i$ on $L_i^{\perp\perp}$, $\mu_i$ on $L_i'$,

$\nu_1$ on all $F_{i_o}$ $(i = 1,\dots,m)$ if $n \neq 0$ in (23) and $\nu_1 = 1$ if $n = 0$ . (Incidentally, if $E$ should be alternate, then we can write $F_{i_o}$ as a sum of two totally isotropic subspaces and merge these with $L_i^{\perp\perp}$ and $L_i'$ respectively. We then need not distinguish between $n = 0$ and $n \neq 0$ in (24).) We have thus found an isometry $\bar{\varphi}$ on $E$ which extends $\varphi$ and which has $\bar{f}(\bar{\varphi}) = 0$ on $E$ . Q.E.D.

Corollary. Let $(E,\Phi)$ and $(k,*,\varepsilon)$ be as in Theorem 10 but assume char $k \neq 2$. Let $\varphi : V \longrightarrow V$ be an involutory isometry, $\varphi^2 = 1\!\!1_V$. Then $\varphi$ can be extended to a (metric) involution on all of $E$ if and only if $K_1 := \ker(\varphi - 1\!\!1)$ and $K_2 := \ker(\varphi + 1\!\!1)$ satisfy the conditions $(K_1 + K_2)^{\perp\perp} = K_1^{\perp\perp} + K_2^{\perp\perp}$ and $K_1^{\perp} + K_2^{\perp} = E$ .

If the characteristic of $k$ is allowed to be 2 then the problem in the corollary becomes considerably involved (see Chapter IX) .

*

## References to Chapter X

[1] W. Allenspach, Erweiterung von Isometrien in alternierenden Räumen. Ph. D. Thesis, University of Zurich 1973.

[2] W. Bäni, Linear topologies and sesquilinear forms. Comm. in Algebra 14 (1977), 1561-1587.

[3] L. Brand, Erweiterung von algebraischen Isometrien in sesquilinearen Räumen. Ph. D. Thesis, University of Zurich 1974.

[4] J. Dieudonné, La dualité dans les espaces vectoriels topologiques. Ann. Ecole Norm. Sup. 59 (1942), 108-139.

[5] H. Gross and E. Ogg, Quadratic forms and linear topologies. On completions. Ann. Acad. Sci. Fenn. Ser. A.I, 584 (1974), 1-19.

[6] G. Köthe, Topological Vector Spaces I. Grundlehren Band 159, Springer Berlin, Heidelberg, New York 1969.

[7] G. W. Mackey, On infinite - dimensional linear spaces. Trans. Amer. Math. Soc. 57 (1945), 155-207.

[8] F. Maeda and S. Maeda, Theory of symmetric lattices. Grundlehren Band 173, Springer Berlin, Heidelberg, New York 1970.

[9] S. Maeda, Remarks on the problems in the book: Theory of symmetric lattices. Contained in Colloquia Mathematica Societatis Jańos Bolyai 14, Lattice Theory, Szeged (Hungary) (1974), 227-229.

[10] E. Ogg, Die abzählbare Topologie und die Existenz von Orthogonalbasen in unendlichdimensionalen Räumen. Math. Ann. 188 (1970) 233-250.

CHAPTER ELEVEN

# CLASSIFICATION OF FORMS OVER ORDERED FIELDS

## 1. Introduction

In this chapter we shall show that a certain kind of commutative ordered fields, the so called SAP fields, lend themselves very naturally for the construction of $\aleph_0$-forms which admit a simple classification with respect to isometry. We shall first say a few words about the fields and then describe the type of $\aleph_0$-forms to be studied.

In what follows $k$ is an orderable commutative field. We identify orderings with the corresponding sets $P \subset k$ of positive elements (Thus $P$ is a subgroup of index 2 in the multiplicative group $\dot{k}$ and $P$ is additively closed). $X(k)$ is the set of all orderings on $k$. Each $\alpha \in \dot{k}$ defines a signature $\hat{\alpha}: X(k) \to \{1,-1\}$ by sending $P \in X(k)$ into $\mathrm{sig}_P(\alpha)$ which is $+1$ if $\alpha \in P$ and $-1$ if $\alpha \in -P$. $X(k)$ is endowed with the coarsest topology that renders all signatures $\hat{\alpha}$ continuous. A subbasis for the system of open subsets of $X(k)$ is the system of sets $H(\alpha) = \{P \in X(k) | \alpha \in P\}$ where $\alpha$ runs through $\dot{k}$ ([11], page 208). The field $k$ is called a SAP-field if it has the following "strong approximation property": Any two disjoint closed subsets of $X(k)$ are separated by some signature $\hat{\alpha}$, i.e. one of the two sets belongs to $H(\alpha)$ and the other to $H(-\alpha)$ ([11], page 108).

The reader who is not inclined to become involved in topological considerations may assume $X(k)$ to be finite and discard all of the topology that will occasionally come up in what follows. Examples with finite $X(k)$ are provided by the algebraic number fields which is to show that the discussion will not become shallow if $X(k)$ is finite.

We terminate our introductory remarks on fields by formulating a property on fields which is crucial for the discussion in this chapter.

It reads

(1) Every binary form $\langle 1,-s\rangle$, $s \in \dot{k}_s$, represents some element in each coset of $\dot{k}/\dot{k}_s$ ($\dot{k}_s$ is the multiplicative subgroup of sums of squares).

Examples of such fields will be given in Section 3. We shall now turn to the forms to be classified.

Let $\Phi : E \times E \to k$ be a symmetric, non degenerate bilinear form on the $\aleph_o$-dimensional k-vectorspace $E$. We recall that for $G$ a subspace $\|G\|$ is the set $\{\Phi(g,g) \mid g \in G \setminus \{0\}\}$ and $(E,\Phi)$ is called <u>stable</u> (in itself) if $\|E\| = \bigcap \{\|F^{\perp}\| \mid F \subset E$ and $\dim F < \infty\}$. Notice that this equality cannot hold for <u>finite</u> dimensional non degenerate $E \neq (0)$. Each field $k$ admits stable forms: if for $\beta \in \dot{k}$ we let $\langle\beta,\ldots\rangle$ be the orthogonal sum of $\aleph_o$ copies of $\langle\beta\rangle$ then every orthogonal sum $\oplus^{\perp}_{\beta\in I} \langle\beta,\ldots\rangle$, where card $I \leq \aleph_o$, is a stable space. If a stable space is isotropic then it is an orthogonal sum of hyperbolic planes.

<u>Definition</u>. $(E,\Phi)$ is called weakly stable if $E$ splits off some orthogonal summand which is stable.

It is not difficult to show that $(E,\Phi)$ is weakly stable if and only if the set $\|E\|_\infty := \bigcap \{\|F^{\perp}\| \mid F \subset E$ and $\dim F < \dim E\}$ is not empty. If $(E,\Phi)$ is weakly stable and isotropic it is an orthogonal sum of hyperbolic planes. Thus we shall have to study anisotropic forms only.

The aim of this chapter is to classify the weakly stable $\aleph_o$-forms over SAP fields which satisfy (1). Our results generalize in several directions work on quasistable forms done in [13]. <u>Quasistable</u> forms are weakly stable forms that split off a stable orthogonal summand of <u>finite</u> codimension (The concept is due to MAXWELL). Let us turn to the details now.

## 2. Weakly isotropic forms

All forms are symmetric bilinear forms.

The SAP fields mentioned in the introduction can be characterized by many other properties ([4][11][15]) . One such equivalent property is of particular interest here. Call totally indefinite a form $\varphi = <\alpha_1,\alpha_2,\ldots,\alpha_m>$ if for each $P \in X(k)$ we have $\{sig_p(\alpha_1),\ldots$ $\ldots,sig_p(\alpha_m)\} = \{1,-1\}$ (cf. Cor. 3 in Section 4). We also put down the

Definition. $\varphi$ is called weakly isotropic if $\aleph_o\varphi = \varphi \oplus \varphi \oplus \varphi \oplus \ldots$ is isotropic.

Thus $\varphi = <\alpha_1,\ldots,\alpha_m>$ is weakly isotropic if and only if there are sums of squares, $s_1,s_2,\ldots s_m \in \dot{k}_s$ such that $\sum_1^m \alpha_i s_i = 0$ , i.e. if and only if there is some natural $N$ such that $N\varphi = \varphi\oplus\ldots\oplus\varphi$ is isotropic.

SAP fields may be characterized by the following property (a kind of "HASSE principle") on finite dimensional forms ([4]):

(2) Every totally indefinite form is weakly isotropic.

We wish to compare (2) with the following "axiom" on ordered fields.

(3) For $A,B$ any disjoint closed subsets of $X(k)$ and $\alpha \in \dot{k}$ any element which is positive at all orderings of $A$ (i.e. $\alpha \in P$ for all $P \in A$) there exists $\beta \in \dot{k}$ such that $\alpha-\beta^2$ is positive at all orderings of $A$ and negative at all orderings of $B$ .

Remark. For fields $k$ with finite $X(k)$ property (3) coincides with "axiom 2.1" in [13], i.e., by the next lemma, Maxwell's axiom describes precisely the SAP fields with finite $X(k)$ and with property (1). If $X(k)$ is finite and if all $P \in X(k)$ are archimedian then, as remarked in [13], the weak approximation theorem for real valuations can be used to show that $k$ satisfies (3).

Lemma. (1) & (2) <=> (3) .

Proof. We show first that (1) is equivalent with the following property (called "the shovel")

(4) for all $\alpha,\beta \in \dot{k}$ we have $\{\alpha s + \beta x^2 | x \in k,\ s \in \dot{k}_s\} = \{\alpha x^2 + \beta s | x \in k,\ s \in \dot{k}_s\}$ .

Indeed, if $\alpha,\beta \in \dot{k}$ and $s \in \dot{k}_s$ where $k$ satisfies (1) then $\langle 1,-s\rangle$ represents an element in the coset (mod $\dot{k}_s$) of $-\alpha^{-1}\beta$ . From this we can conclude that $\dot{k}_s \subset \{x^2 + \alpha^{-1}\beta s | x \in k,\ s \in \dot{k}_s\}$ and we see that (4) must hold. Conversely, if in (4) we let $\alpha = 1$ we obtain $\dot{k}_s \subset \{x^2+\beta s | x \in \dot{k}_s\}$ for arbitrary $\beta \in \dot{k}$ hence we have (1). Thus

(5) (1)<=>(4) .

We now turn to reformulating (1) & (2). Let $A,B$ disjoint closed subsets in $X(k)$ where $k$ is any SAP field. Let $\alpha \in \dot{k}$ be a given element with $\alpha$ positive at all orderings in $A$ . Let $D := \{P \in X(k) | -\alpha \in P\}$ . We may pick an $\varepsilon \in \dot{k}$ which is positive at all $P \in A$ and negative at all $P \in B \cup D$ . Then the form $\langle 1,-\alpha,\varepsilon\rangle$ is totally indefinite. By (2) we obtain an equation $s_1 - \alpha s_2 + \varepsilon s_3 = 0$ with $s_i \in \dot{k}_s$ . If now we make the further assumption that $k$ satisfies (4) then we may write the element $-\alpha s_2 s_1$ in the form $-\alpha x^2+\varepsilon s_4$ . Without loss of generality $x \neq 0$ . If we substitute in our equation we obtain $\alpha - s_1^2 x^{-2} = x^{-2}(\varepsilon s_3 s_1+\varepsilon s_4)$ . This shows that the element $\alpha-s_1^2x^{-2}$ is positive at all orderings of $A$ and negative at all orderings of $B$, i.e. (3) holds with $\beta = s_1 x^{-1}$ . Thus (1) & (2) => (3). The converse is easy: Disjoint closed subsets of $X(k)$ can by (3) be separated even by elements of the very special shape $s-\beta^2$ where $s \in \dot{k}_s$ may be picked beforehand; so $k$ must of course be a SAP field. Further, if $\alpha \in \dot{k}_s$ is given then $X(k)$ is the disjoint union of $A := \{P \in X(k) | \alpha \in P\}$

and $B := \{P \in X(k) \mid -\alpha \in P\}$ so that any separating element has the signature $\hat{\alpha}$, i.e. must lie in the same coset modulo $\dot{k}_s$ as the element $\alpha$. Thus by (3) every coset in $\dot{k}/\dot{k}_s$ contains some $s-\beta^2$. Since $s$ is arbitrary in $\dot{k}_s$ this proves (1).

Problem. Describe the orderable fields with property (1) (for non orderable fields the problem is trivial).

## 3. Examples of fields in connection with properties (1) and (2)

We list here a few examples of fields which satisfy some of the conditions (1) and (2) introduced in the two previous sections. Indications as to where proofs may be found are given.

All pythagorean fields $k$ (i.e. $k$ with $\dot{k}^2 = \dot{k}_s$) satisfy (1); $\mathbb{R}((t_1))((t_2))$ is one which violates (2) ([15] Satz (2.2)). On the other hand $\mathbb{Q}((t))$ satisfies (2) ([15], §3) and violates (1) (because (3) is violated when $\alpha = 2+t^2$). $\mathbb{R}$ or any field with just one ordering is an obvious candidate for (3), i.e. for (1) and (2). Any real algebraic numberfield (i.e.formally real finite algebraic extension of $\mathbb{Q}$) satisfies (3) by the Remark in Section 2. Another example which has both (1) and (2) but infinitely many orderings is $k = \mathbb{R}(t)$ by Thm. 8 in [19]. More generally, we may take $k$ to be any finite algebraic orderable extension of a function field $k_o(t)$ where $k_o$ is hereditarily euclidean (this means that not merely in $k_o$ but in each finite algebraic orderable extension field of $k_o$ every element or its negative is a square [16]); by Thm. 1 in [4] these fields are examples with (1) and (2) and, of course, with infinitely many orderings. Finally $k = \mathbb{R}(t_1,t_2)$ violates (1) (since $1\cdot(t_1^2+t_2^2) \notin \{1x^2+t_1 s \mid x \in k, s \in \dot{k}_s\}$ (4) is violated) and it also violates (2) by [15], Satz (2.2).

## 4. A remark on Hilbert ordered skew-fields

In this short section we collect a few simple facts in connection with ordered fields. We think that the statements which we wish to make are more perspicuous if we include the non commutative case at this spot. We shall need only HILBERT's conception of an ordered skew field $k$ which amounts to the specification of an additively closed subgroup of index 2 in the multiplicative group $\dot{k}$ of $k$ ([8] §13, Sätze der Anordnung, furthermore §33, Satz 60 where the famous example of an ordered skew field is given). In particular $P$ is an invariant subgroup of $\dot{k}$. (For other concepts of orderings see [2], Appendix I p. 127-128 and [9]).

Let $M \subset \dot{k}$ be a subset which contains $1$ ; we let $\hat{M}$ be the set of all finite sums of finite products $m_1 x_1^2 m_2 x_2^2 m_3 x_3^2 \ldots$ where $m_i \in M$ and $x_i \in \dot{k}$ .

By an application of Zorn's lemma in the manner of [3] or [18] it is not difficult to prove the following

<u>Lemma</u>: Let $1 \in M \subset \dot{k}$ . There exists an ordering $P$ on $k$ with $M \subset P$ if and only if finite sums of finite products $m_1 x_1^2 m_2 x_2^2 \ldots$ ($m_i \in M$ , $x_i \in \dot{k}$ ) do not vanish or, equivalently, if $-1 \notin \hat{M}$ .

One then deduces the following corollaries

<u>Corollary 1</u>. Let $1 \in M \subset \dot{k}$ and $\alpha \in \dot{k}$ . If $-1, \alpha \notin \hat{M}$ then there is an ordering $P$ on $k$ with $M \subset P$ and $\alpha \in -P$ .

<u>Corollary 2</u>. Let $1 \in M \subset \dot{k}$ . Then $\hat{M} = \bigcap \{P \in X(k) \mid P \supset M\}$ .

<u>Corollary 3</u>. Assume that $k$ is ordered and $(\alpha_\iota)_{\iota \in I}$ a totally indefinite family in $k$ (i.e. for each $P \in X(k)$ there are $\nu, \mu \in I$ with $\alpha_\nu \in P$ and $\alpha_\mu \in -P$). Then there is a finite subfamily which is totally indefinite.

<u>Proof</u>. Set $\bar{\alpha}_\iota := \alpha_\iota \cdot \alpha_o^{-1}$ for some fixed index $o \in I$ so that

$1 \in M := \{\bar{\alpha}_\iota \mid \iota \in I\}$ . Since $M$ is totally indefinite we must have $-1 \in \hat{M}$ by the lemma, $-1 = \Sigma\Pi\bar{\alpha}_i x_i^2$ . Let $M_o$ be the set of the $\bar{\alpha}_i$ occuring in this sum. Since $-1$ is invariably negative there must for each $P \in X(k)$ be an element in $M_o$ which is negative at $P$ . So $\{1\} \cup M_o$ is totally indefinite and so is the finite subfamily $(\{1\} \cup M_o)\alpha_o$ of $(\alpha_\iota)_{\iota\in I}$ .

## 5. Two HASSE Principles

Let $(e_i)_{i\in I}$ be an orthogonal basis in a k-space $E$ equipped with a nondegenerate symmetric form $\Phi$ and $k$ an orderable commutative field. For fixed ordering $P \in X(k)$ we can sort out the $e_i$ according to whether $\Phi(e_i,e_i)$ is positive or negative at $P$ . Call $n^+(P)$ the cardinality of the set of the $e_i$ with positive inner product and $n^-(P)$ the cardinality of the remaining $e_i$ . The two cardinals do not depend on the choice of the basis. The pair $\text{ind}_\Phi(P) = (n^-(P)\ ,\ n^+(P))$ is called <u>the</u> (inertial) <u>index</u> of $(E,\Phi)$ at $P$ . In the theory of finite dimensional spaces one can, of course, do with <u>one</u> of the two cardinals (or, as is customary, with the difference of the two).

Another way of putting it is to introduce the $k_P$-ification $\Phi_P$ of the form $\Phi$ where $k_P$ is the real closure of the ordered field $(k,P)$. If $\Psi$ is another form over $k$ then an equality of indices, $\text{ind}_\Phi(P) = \text{ind}_\Psi(P)$ , is equivalent with an isometry $\Phi_P \cong \Psi_P$ . We also see that $\Phi_P$ is isotropic if and only if $\Phi$ is <u>indefinite</u> at $P$ , i.e. if the product $n^+(P)\cdot n^-(P)$ is not zero.

It will turn out that the forms which are investigated in this chapter are characterized by their indices at all $P$ (Section 6, Theorem). In other words they will satisfy the following "weak HASSE principle"

(WH) $\qquad \Phi \cong \Psi \Longleftrightarrow \Phi_P \cong \Psi_P \quad$ for all $P \in X(k)$ .

We now compare (WH) with the following "strong HASSE principle"

(SH) $\Phi$ is isotropic <=> $\Phi_P$ is isotropic for all $P \in X(k)$ .

Let $\mathcal{C}$ be a class of $\aleph_o$-forms over $k$ with the property

(6) if $\Phi = \varphi \oplus \Phi'$ with $\dim \varphi$ finite then $\Phi \in \mathcal{C} \Leftrightarrow \Phi' \in \mathcal{C}$ .

Examples of $\mathcal{C}$ with (6) are the class of quasi-stable forms, the class of weakly stable forms. We now justify part of the terminology by the following

Lemma. If $\mathcal{C}$ satisfies (6) then we have the implication:

(SH) for all $\Phi \in \mathcal{C}$ => (WH) for all $\Phi, \Psi \in \mathcal{C}$

Proof. Assume that for all $P \in X(k)$ we have $\Phi_P \cong \Psi_P$ . We shall construct an isometry between $(E,\Phi)$ and $(F,\Psi)$ recursively. Assume that $E = G \overset{\perp}{\oplus} E'$ and $F = H \overset{\perp}{\oplus} F'$ with finite dimensional $G$ and $T$ an isometry $G \to H$ . One may start with $G = (0)$ . We show how to extend $T$ to $G \overset{\perp}{\oplus} (e)$ where $e \in E'$ is prescribed. Assume first that $\Phi(e,e) \neq 0$ . Let $\Phi'$ and $\Psi'$ be the restrictions of $\Phi$ and $\Psi$ to $E'$ and $F'$ respectively. By an application of Witt's theorem we conclude that $\Phi'_P \cong \Psi'_P$ $(P \in X(k))$ . Now the form $\langle -\Phi(e,e) \rangle \oplus \Phi'$ is obviously isotropic. Therefore by (SH) we conclude that $\langle -\Phi(e,e) \rangle \oplus \Psi'$ is isotropic. This means that $\Psi'$ must contain a vector $f$ with inner product $\Phi(e,e)$ . It is obvious that we can therefore extend $T$ by sending $e$ into $f$ . If we should have that $\Phi(e,e) = 0$ then $e$ is contained in a plane spanned by an orthogonal pair of non isotropic vectors $e_1, e_2$ . By applying the former argument twice we extend $T$ to $G \overset{\perp}{\oplus} k(e_1, e_2)$ . Alternate application of this procedure to $E$ and $F$ yields the desired isometry $(E,\Phi) \cong (F,\Psi)$ .

Remark. Condition (6) does not hold for the class $\mathcal{C}$ of all stable forms. Nevertheless the implication in the lemma is valid in this case too. To see this one has to modify the above proof in the following way: By (SH) for stable forms we get that $\aleph_o \cdot \langle -\Phi(e,e)\rangle \oplus \Psi'$ is isotropic. But this implies that $\langle -\Phi(e,e)\rangle \oplus \aleph_o\Psi'$ is isotropic, and the stability of $\Psi'$ means that $\aleph_o\Psi' \cong \Psi'$. So $\langle -\Phi(e,e)\rangle \oplus \Psi'$ is isotropic.

## 6. The classification

k is assumed to be an orderable (commutative) field; forms are symmetric and non degenerate and of dimension $\aleph_o$.

In the following theorem we characterize the property SAP and the property (1) and SAP (i.e. (3) by the lemma in Sec. 2) via the behaviour of certain classes of $\aleph_o$-forms.

Theorem. For any k we have

SAP <=> (WH) for stable forms <=> (SH) for stable forms.
(3) <=> (WH) for quasistable forms <=> (SH) for quasistable forms.
(3) <=> (WH) for weakly stable forms <=> (SH) for weakly stable forms.

Proof. We set out by showing that (3) implies (SH) for weakly stable forms. If $\Phi_P$ is isotropic for all $P \in X(k)$ then $(\alpha_i)_{i\in\mathbb{N}}$ is totally indefinite for any diagonalization $\Phi = \overset{\perp}{\underset{\mathbb{N}}{\oplus}} \langle\alpha_i\rangle$. Thus by Corollary 3 in Sec. 5 $\Phi$ splits off an orthogonal summand $\varphi$ which is totally indefinite and of finite dimension. Hence if k has SAP then we may quote (2) and conclude that $\aleph_o\varphi$ is isotropic; so $\aleph_o\Phi$ is isotropic. If $\Phi$ is weakly stable, $\Phi = \Phi_1 \oplus \Phi_2$ with $\Phi_2$ stable then $\aleph_o\Phi \cong \aleph_o\Phi_1 \oplus \Phi_2$ by the stability of $\Phi_2$. If k has (3), then it has (1) and we may quote (4) to conclude from the existence of an isotropic vector of $\aleph_o\Phi_1 \oplus \Phi_2$ that $\Phi_1 \oplus \Phi_2$ must contain a non zero isotropic vector.

Thus we see that the assumptions on k in the first "column" of the theorem individually imply the (SH)-statements at the far end of the corresponding row. The latter imply the corresponding (WH)-statements by the lemma in Section 5.

Assume then (WH) for stable forms. We want to show that $k$ has SAP by showing that (2) holds. If $\varphi = \langle\alpha_1,\ldots,\alpha_n\rangle$ is totally indefinite then $\Phi := \aleph_o\varphi = \oplus_{i=1}^n\langle\alpha_i,\alpha_i,\ldots\rangle$ must be isometric to an orthogonal sum of hyperbolic planes and hence isotropic. This shows that the (WH)-statements in the middle column of the theorem imply (2), i.e. SAP.

It remains to prove that (WH) for quasistable forms implies (1): indeed, if $s \in \dot{k}_s$ and $\alpha \in \dot{k}$ are prescribed then $\langle 1,-s\rangle \oplus \langle -\alpha,-\alpha,\ldots\rangle$ is isotropic because it is isometric with $\langle 1,-1\rangle \oplus \langle -\alpha,-\alpha,\ldots\rangle$ by (WH). Q.e.d.

Remark 1. In the whole chapter we discuss symmetric forms only. It is, however, not difficult to extend the results to hermitean forms over quadratic extensions of $k$ or quaternion division algebras with the usual involutions when the fixed field $k$ has the requisite properties(investigated in the present chapter).

Remark 2. If, for the moment, we let $k$ be not formally real then $X(k) = \emptyset$ and conversely. Therefore, if $(E,\Phi)$ is weakly stable over such a $k$ then it must contain a totally isotropic subspace of infinite dimension. Hence if $\operatorname{char} k \neq 2$ then $E$ is an orthogonal sum of hyperbolic planes. Thus we see that in the case of non orderable fields only the characteristic two situation leaves room for a discussion of weakly stable spaces. This discussion is carried out in Chapter VII.

## 7. Canonical representatives for quasistable forms

$k$ is an ordered commutative field with property (3).

With every non empty finite family $(X_i)$ of non empty disjoint closed subsets $X_i$ of $X(k)$ we associate once and for all a family of elements $\varepsilon(i) \in k$ with $\varepsilon(i)$ negative at all orderings $P \in X_i$ and positive at all $P$ of the other $X_j$ in the family.

Let $(E,\Phi)$ be quasistable, i.e. $E = F \overset{\perp}{\oplus} G$ with $\dim F$ finite and $G$ stable. It is advantageous to choose $F$ minimal in the sense that $\|G\| \cap \|F\| \subset \{0\}$. Furthermore we may assume that the restriction $\Psi$ of $\Phi$ to $G$ is anisotropic for otherwise $G$ and $E$ are both orthogonal sums of hyperbolic planes (which is the canonical form of an isotropic quasistable space whenever the characteristic is not two).

To bring $\Psi$ into canonical shape pick some $\alpha \in \|G\| \setminus \{0\}$ and let $I$ be the set of all $P \in X(k)$ with the property that all elements of $\alpha^{-1}\|G\|$ are positive at $P$. If now $\beta \in \dot{k}$ is any element which happens to be positive at all orderings of $I$ then the quasistable form $\langle -\beta \rangle \oplus \alpha^{-1}\Psi$ is totally indefinite and hence isotropic by the theorem of the last section. Therefore $\beta \in \alpha^{-1}\|G\|$. We see: $\alpha^{-1}\|G\|$ is the set of <u>all</u> $\beta \in k$ which are positive at all $P \in I$. Clearly $I$ is a closed subset of $X(k)$.

Let then $f_1, \ldots, f_m$ be some orthogonal basis for $F$. There is - by what we have just seen - for every $f_i$ some $P \in I$ at which $\alpha^{-1}\Phi(f_i,f_i)$ is negative for otherwise $\Phi(f_i,f_i) \in \|F\| \cap \|G\| \subset \{0\}$ by our normalization. For all $0 \leq j \leq m$ we then consider the sets

$$X_j := \{P \in I \mid \text{precisely } j \text{ among } \alpha^{-1}\Phi(f_1,f_1), \ldots, \alpha^{-1}\Phi(f_m,f_m) \text{ are neg. at } P\}.$$

The sets $X_j$ are closed subsets of $X(k)$ and $I = \cup X_j$. Let $J = \{i_1, \ldots, i_r\}$ be the subscripts $j$ with $X_j \neq \emptyset$ and let $\varepsilon(i_1), \ldots$ $\ldots, \varepsilon(i_r)$ be the elements $\in k$ associated with the family $(X_j)_{j \in J}$. By the classification theorem of Section 6 we obtain the canonical representation

$$(7) \qquad \alpha^{-1}\Phi \cong i_1\langle \varepsilon(i_1)\rangle \oplus \ldots \oplus i_r\langle \varepsilon(i_r)\rangle \oplus \alpha^{-1}\Psi$$

Here one can easily read off indices of the form $\alpha^{-1}\Phi$ : for $P \notin I$

the index is invariably $(\aleph_o,\aleph_o)$ ; for $P \in I$ the index is $(i,\aleph_o)$ if $P \in X_i$ .

We see that the collection of all families $(\varepsilon(i))$ corresponding to the collection of all non empty finite families $(X_i)$ of non empty disjoint closed subsets of $X(k)$ form a complete and irredundant system of invariants for the description of quasistable forms over a field satisfying (3). May the closed sets $X_i$ be chosen completely arbitrary in $X(k)$ ? There is a condition! Recall that $(E,\Phi)$ must admit countably infinite bases $(e_i)$ . Passing to the inner products $\Phi(e_i,e_i)$ we see that $\dot{k}$ must admit subsets $M$ of cardinality $\leq \aleph_o$ such that $\cup X_i$ is the set of all $P \in X(k)$ for which $M$ is positive at $P$ . For certain fields this requires $\cup X_i$ to be a "large" subset of $X(k)$ . On the other hand, if this condition is satisfied then we may set $\Psi := \oplus^{\perp}_{\mu \in M} \langle\mu,\mu,\ldots\rangle$ and define a quasistable $\Phi$ by the right hand side of (7).

In [13] MAXWELL has given canonical representatives when $X(k)$ is finite; they can easily be obtained by our classification theorem. Let $X(k) = \{P_1,\ldots,P_m\}$ and choose (by SAP) once and for all elements $\eta_1,\ldots,\eta_m \in k$ with

$$\eta_i \in P_j \ (i \neq j) \quad \text{and} \quad \eta_i \in -P_i \quad (i,j=1,\ldots,m) .$$

We abbreviate $n_i := n^-(P_i)$ where $(n^-(P_i),\aleph_o)$ is the index of $\alpha^{-1}\Phi$ at $P_i$ . By the classification theorem of the last section we obtain

$$(8) \quad \alpha^{-1}\Phi \cong n_1\langle\eta_1\rangle \oplus \ldots \oplus n_m(\eta_m) \oplus \alpha^{-1}\Psi .$$

For finite $X(k)$ the two normal forms (7) and (8) are of a different nature. We shall illustrate it by an

Example. Let $k = \mathbb{Q}(\sqrt{2})$ . So $X(k) = \{P_1,P_2\}$ and the only possibilities for non empty finite families of non empty disjoint subsets of $X(k)$ are $(\{P_1\})$, $(\{P_2\})$, $(\{P_1,P_2\})$, $(\{P_i\})_{i=1,2}$ .

A natural choice for the corresponding families $(\varepsilon(i))$ is, in turn, $(-1)$, $(-1)$, $(-1)$, $(\sqrt{2},-\sqrt{2})$. Then (7) will always turn out either as $\alpha^{-1}\Phi \cong i\langle -1\rangle \oplus \langle stable\rangle$ or as $\alpha^{-1}\Phi \cong i_1\langle\sqrt{2}\rangle \oplus i_2\langle -\sqrt{2}\rangle \oplus$ (stable) with $i_1 \neq i_2$.

In order to arrive at (8) we need choose the $\eta_i$. It is natural to set $\eta_1 = \sqrt{2}$ and $\eta_2 = -\sqrt{2}$. Then for all forms over $k$ (8) will read as $\alpha^{-1}\Phi \cong n_1\langle\sqrt{2}\rangle + n_2\langle -\sqrt{2}\rangle \oplus$ (stable). Thus the normal form of $E := \langle -3\rangle \oplus \langle 1,1,\ldots\rangle$ in the style of (7) is $\langle -1\rangle \oplus \langle 1,1,\ldots\rangle$ whereas in the manner of (8) it reads as $\langle\sqrt{2}\rangle \oplus \langle -\sqrt{2}\rangle \oplus \langle 1,1,\ldots\rangle$.

Remark 1. $X(k)$ may be infinite but $\|G\| = \bigcap\{\|H^{\perp}\| \mid H \subset E$ and $\dim H < \infty\}$ be so large as to make $I$ turn out finite. Then one can choose elements in $k$ which are negative at any one ordering of $I$ and positive at all the other orderings of $I$ and again arrive at a standard representation in the style of (8).

Remark 2. From (7) we see that the indices of $\alpha^{-1}\Phi$ (and hence the indices of $\Phi$ ) at $P \notin I$ are invariably $(\aleph_0,\aleph_0)$. Therefore, instead of saying that a quasistable $\Phi$ is determined by all indices we may say that a quasistable $(E,\Phi)$ is determined by the collection of the following invariants: the set $\|E\|_\infty$, which is of the kind that there is $I \subset X(k)$ with $\alpha^{-1}\|E\|_\infty = \{\beta \mid \beta \in P$ for all $P \in I\}$ for some $\alpha \in \|E\|_\infty$, and by the indices of $\Phi$ at the orderings in $I$. (cf.[13] Thm. 2.5).

## 8. Fields over which all $\aleph_0$-forms are quasistable

k is an orderable commutative field; forms are symmetric and non degenerate.

Lemma. For any $k$ the following two statements are equivalent

(9) Each $\aleph_0$-form over $k$ is quasistable

10) $X(k)$ is finite and every $\aleph_o$-form $(E,\Phi)$ with $\|E\| \subset \dot{k}_s$ represents 1.

<u>Proof</u>. (10)$\Longrightarrow$(9): Let $\Phi = \overset{\perp}{\underset{\mathbb{N}}{\oplus}}\langle\alpha_i\rangle$ be an arbitrary $\aleph_o$-form over $k$ and assume (10). Let $\beta_1,\ldots,\beta_m$ be a (finite) set of representatives in $\dot{k}/\dot{k}_s$. By expressing all $\alpha_i$ in terms of $\beta_j$ and factors from $\dot{k}_s$ we see that $\Phi$ splits into an orthogonal sum $\Phi = \beta_1\Phi_1 \oplus\ldots\oplus\beta_m\Phi_m$ with $\|\Phi_i\| \subset \dot{k}_s$. At least one $\Phi_i$ is of infinite dimension and each such $\Phi_i$ has $\dot{k}_s \subset \|\Phi_i\|$ by (10), i.e. $\|\Phi_i\| = \dot{k}_s$. Hence $\Phi$ is an orthogonal sum of a finite dimensional form and finitely many stable ones. Therefore $\Phi$ is quasistable.

(9)$\Longrightarrow$(10): Since every stable $\Phi$ with $\|\Phi\| \subset \dot{k}_s$ has $\|\Phi\| = \dot{k}_s$ and thus represents 1 we see that every quasistable form $\Psi$ with $\|\Psi\| \subset \dot{k}_s$ represents 1. This shows one half of (9)$\Longrightarrow$(10). It remains to show that a $k$ with infinite $X(k)$ admits forms that are not quasistable. If $X(k)$ is infinite then - since each $P \in X(k)$ is made up of full equivalence classes in $\dot{k}/\dot{k}_s$ - the group $\dot{k}/\dot{k}_s$ is infinite as well. We also recall that the intersection of a finite number of subgroups in $\dot{k}$, each of finite index in $\dot{k}$, has itself finite index in $\dot{k}$ so that the intersection $\bigcap_1^n P_i$ of finitely many orderings must contain infinitely many equivalence classes of $\dot{k}$. As $\bigcap\{P|P \in X(k)\}$ reduces to $\dot{k}_s$ we can therefore determine a countable sequence $(P_i)_{i\in\mathbb{N}}$ of orderings with $\bigcap_1^n P_i \not\subset P_{n+1}$ for all $n \geq 1$. We pick $\alpha_n \in (\bigcap_1^n P_i) \setminus P_{n+1}$ and define $E := \overset{\perp}{\underset{\mathbb{N}}{\oplus}}\langle\alpha_i\rangle$. No stable subspace of $E$ can contain a line $\langle\alpha_i\rangle$ since $\alpha_i$ is negative at $P_{i+1}$ and $\Phi(x,x)$ positive at $P_{i+1}$ for all $x \in (\langle\alpha_1\rangle +\ldots+\langle\alpha_i\rangle)^{\perp}$. In particular $E$ cannot be quasistable.

From Hasse Theory we know that the orderable finite algebraic extensions $k$ of $\mathbb{Q}$ satisfy (10); hence the

<u>Corollary</u>. Let $k$ be a real algebraic number field. Each non-

degenerate $\aleph_o$-form is determined up to isometry by its positive and negative indices at all real completions of k . (If k is not real then there are no indices and all non degenerate $\aleph_o$- forms are isometric, to wit, they are orthogonal sums of hyperbolic planes. Cf. the remark at the end of Section 6 .)

This corollary was proved in [14] and independently in [12] .

*

## References to Chapter XI

[1] W. Bäni and H. Gross, On SAP fields. Math. Z. 162 (1978) 69-74.

[2] R. Baer, Linear algebra and projective geometry. Academic Press New York, 1952.

[3] N. Bourbaki, Algèbre chap. VI , groupes et corps ordonnés, ASI 1179, Hermann, Paris, 1952.

[4] R. Elman, T.Y. Lam, A. Prestel, On some Hasse Principles over Formally Real Fields. Math. Z. 134 (1973) 291-301.

[5] C.J. Everett and H.J. Ryser, Rational vector spaces. Duke Mathematical Journal, vol. 16 (1949) 553-570.

[6] H. Gross and R.D. Engle, Bilinear forms on k - vectorspaces of denumerable dimension in the case of char (k) = 2 , Commentarii Mathematici Helvetici, vol. 40 (1965) 247-266.

[7] H. Gross and H.R. Fischer, Non - real fields k and infinite dimensional k - vectorspaces. Mathematische Annalen, vol. 159 (1965) 285-308.

[8] D. Hilbert, Grundlagen der Geometrie. Teubner, Stuttgart, 1956.

[9] S.S. Holland, Orderings and Square roots in * - fields. J. Alg. 46 (1977) 207-219.

[10] I. Kaplansky, Forms in infinite - dimensional spaces, Anais da Academia Brasileira de Ciencias, vol. 22 (1950) 1-17.

[11] M. Knebusch, A. Rosenberg, R. Ware, Structure of Witt rings, quotients of abelian group rings, and orderings of fields. Bull. Amer. Math. Soc. 77 (1971) 205-210.

[12] L.E. Mattics, Quadratic forms of countable dimension over algebraic number fields. Comment. Math. Helv. 43 (1968) 31-40.

[13] G. Maxwell, Classification of countably infinite hermitean forms over skewfields. Amer. J. Math. 96 (1974) 145-155.

[14] O.T. O'Meara, Infinite dimensional quadratic forms over algebraic number fields. Proc. Amer. Math. Soc. 10 (1959) 55-58.

[15] A. Prestel, Quadratische Semi - Ordnungen und quadratische Formen. Math. Z. 133 (1973) 319-342.

[16] A. Prestel and M. Ziegler, Erblich euklidische Körper. Journal reine angew. Math. 274/275 (1975) 196-205.

[17] L.J. Savage, The application of vectorial methods to geometry. Duke Mathematical Journal, vol. 13 (1946) 521-528.

[18] T. Szele, On ordered skew fields. Proc. Amer. Math. Soc. 3 (1952) 410-413.

[19] E. Witt, Theorie der quadratischen Formen in beliebigen Körpern. J. reine angew. Math. 176 (1937) 31-44.

*

<u>Postscript</u>. After this chapter had been written we received a preprint of a paper by A. Prestel and R. Ware, entitled "Almost isotropic quadratic forms" ( to appear in J. London Math. Soc. (2), 19 (1979)) In this paper the authors characterize the fields which satisfy SAP & (1 by various other properties, e.g.: The pythagorean closure of $k$ is SAP .

CHAPTER TWELVE

# CLASSIFICATION OF SUBSPACES IN SPACES WITH DEFINITE FORMS

## Introduction

In the whole chapter $(E,\Phi)$ will be a positive definite hermitean space of dimension $\aleph_0$ over the divisionring $k$ with involution $\xi \mapsto \xi^\tau$. If $\tau \neq 1\!\!1$ then it follows from Dieudonné's lemma that $k$ is either a quadratic extension $k = k_0(\gamma)$ over an ordered field $(k_0,<)$ with $0 > \gamma^2 \in k_0$ and $(x+y\gamma)^\tau = x-y\gamma$ for all $x,y \in k_0$; or $k$ is a quaternion algebra $(\frac{\alpha,\beta}{k_0})$ with $k_0$ ordered, $\alpha,\beta < 0$ and $\tau$ being the usual "conjugation". If $\tau = 1\!\!1$, possible only when $k$ is commutative, then $\phi$ is symmetric and $k = k_0$ is ordered.

$(E,\Phi)$ will be assumed strongly universal throughout, i.e. $\|F\| = \|E\|$ for all infinite dimensional non degenerate subspaces $F$ of $E$; furthermore we assume that $1 \in \|E\|$.

*

In Chapter Five we described a complete set of orthogonal invariants for subspaces $V \subset (E,\Phi)$ under the assumption that $E$ abounds in isotropic vectors. Here we shall do the same under the opposite assumption that $(E,\Phi)$ is positive definite. The difference concerning both the difficulty of the problem and the theorems which hold is remarkable.

Let us illustrate the situation by an example. Assume that the symmetric space $(E,\Phi)$ over $\mathbb{R}$ is spanned by an orthogonal basis $(e_i)_{i\in\mathbb{N}}$ with $\|e_{2i}\|=+1$ and $\|e_{2i+1}\| = -1$. Then the cardinal $\dim E/V$ is the only invariant for $\perp$-dense subspaces $V$ of finite codimension; in other words, all $V \subset E$ with $V^\perp = (0)$ and equal $\dim E/V < \infty$ make up one orbit under the orthogonal group of $E$. The same is true for $k$ any subfield of $\mathbb{R}$; in fact the nature of $k$ is entirely irrelevant as is shown in Chapter 5. Assume now that $k = \mathbb{R}$ and $(e_i)$

is orthonormal and let the orthogonal group act on the set of $\perp$-dense subspaces $V \subset E$ with, say, $\dim E/V = 2$. There are 3 orbits. If $k$ is replaced by the real closure of $\mathbb{Q}$ we obtain $2^{\aleph_0}$ orbits. Besides the cardinal $\dim E/V$ we have here certain "arithmetical" invariants. They make apparent a "rigidity" of the definite space $(E,\Phi)$.

In the first seven sections we prove our principal result on $\perp$-dense subspaces; it is the fundament for the whole chapter.

1. Standard bases for $\perp$-dense subspaces and their matrices

1.1. Let $(k_o,<)$ be the ordered subfield of the divisionring $k$. The Cantor completion process with Cauchy systems in $k_o$ leads to an ordered field. Of this field we shall need in the following only the subfield $\bar{k}_o$ of the limits of the denumerable Cauchy sequences in $k_o$.

1.2. In the whole chapter $K$ is the real closure of $k_o$. Notice that $k_o$ need not be dense in $K$ for non archimedean ordered $k_o$ (Satz 1.2 in [1]). For a nice characterization of ordered fields which are dense in their real closure the reader may consult [5] and §3 in [4].

1.3. We let the given involution $\tau$ on $k$ act as the identity on the fields $\bar{k}_o$, $K$. So $\bar{k}_o \otimes k$, $K \otimes k$ become involutorial rings. Since $\alpha,\beta,\gamma^2 < 0$ the rings $\bar{k}_o \otimes k = \bar{k}_o(\gamma)$, $K \otimes k = K(\gamma)$ resp. $\bar{k}_o \otimes k = \left(\frac{\alpha,\beta}{\bar{k}_o}\right)$, $K \otimes k = \left(\frac{\alpha,\beta}{K}\right)$ are in fact still division rings. We think of $\bar{k}_o \otimes k$, $K \otimes k$ as endowed with the usual "euclidean" topology induced by the "norm" $xx^\tau$ $(x \in \bar{k}_o \otimes k$ or $K \otimes k)$.

1.4. Besides the hermitean k-space $(E,\Phi)$ under investigation we shall often make use of the n-tupel spaces $k^n$, $(\bar{k}_o \otimes k)^n$ etc. equipped with the usual hermitean product $\langle , \rangle$; $\langle (\xi_i), (\eta_i) \rangle = \Sigma \xi_i \eta_i^\tau$. On a few occasions in Appendix II we use the same notations also for infinite convergent series, e.g. $\langle (\alpha_i)_{i \in \mathbb{N}}, (\beta_i)_{i \in \mathbb{N}} \rangle = \sum_{\mathbb{N}} \alpha_i \beta_i^\tau$, $\| (\alpha_i)_{i \in \mathbb{N}} \| =$

$= \sum_{\mathbb{N}} \alpha_i \alpha_i^\tau$ and the like.

1.5. Since $(E,\Phi)$ is strongly universal every infinite dimensional subspace of $(E,\Phi)$ is spanned by an orthonormal basis. Hence every $x \in E$ is contained in some finite dimensional subspace spanned by an orthonormal basis. In other words, for every $\alpha \in \|\Phi\|$ there exists $n \in \mathbb{N}$ and suitable $X \in k^n$ such that $\alpha = \langle X,X \rangle$ . From this we conclude

1.5.1. Let $M$ be the $N \times N$ matrix of some positiv definite hermitean form $\chi$ with respect to the involution $\tau$ of $k$ , $\dim \chi = N \leq \aleph_o$ . If $\|\chi\| \subset \|\Phi\|$ then $M$ is a m-Gram matrix for a suitable $m \leq \aleph_o$ . This means that there exists a $m \times N$ matrix $C$ over $k$ such that $M = C^{tr} \cdot C^\tau$ . If the number $n$ in 1.5 may be chosen independently of $\alpha$ then we may pick for $m$ any number exceeding $nN$ .

Indeed, diagonalize: $M = Q^{tr} \cdot D \cdot Q^\tau$ with $D = [d_1, \ldots]$ diagonal. $d_i \in \|\Phi\|$ . Hence we find a set of $N$ pairwise orthogonal m-tuples $Y_1, \ldots$ with $\langle Y_i, Y_i \rangle = d_i$ . Take the $Y_i$ as columns for a matrix $Y$ , $Y^{tr} \cdot Y^\tau = D$ . $C := (Q^{tr} Y^{tr})^{tr}$ will do.

1.6. Let $(\lambda_i)_{i \in \mathbb{N}}$ be a sequence in $k$ and $(\gamma_{ij})_{ij \in \mathbb{N}}$ an invertible row-finite matrix with

$$\sum_j \gamma_{ij} \gamma_{rj}^\tau = \delta_{ir} \qquad \text{(Kronecker)} . \tag{1.6.1}$$

We define a new sequence $(\bar{\lambda}_i)_{i \in \mathbb{N}}$ by

$$\lambda_i = \sum \bar{\lambda}_j \gamma_{ij}^\tau \tag{1.6.2}$$

and claim:

(1.6.3) There exists for every $N \in \mathbb{N}$ a $N' \in \mathbb{N}$ such that $\sum_1^N \lambda_i \lambda_i^\tau \leq \sum_1^{N'} \bar{\lambda}_i \bar{\lambda}_i^\tau$ and symmetrically for every $N \in \mathbb{N}$ there exists $N' \in \mathbb{N}$ such that $\sum_1^N \bar{\lambda}_i \bar{\lambda}_i^\tau \leq \sum_1^{N'} \lambda_i \lambda_i^\tau$ . In particular, if one of the two sequences $(\sum_1^N \lambda_i \lambda_i^\tau)_{N \in \mathbb{N}}$ , $(\sum_1^N \bar{\lambda}_i \bar{\lambda}_i^\tau)_{N \in \mathbb{N}}$ is a Cauchy sequence then so is the other.

Proof. For fixed $i \in \mathbb{N}$ let $j(i)$ be the smallest natural number such that $\gamma_{ij} = 0$ when $j > j(i)$. Let $N_1 = \max\{j(1),\dots,j(N)\}$, so $\lambda_i = \Sigma_1^{j(i)} \bar{\lambda}_j \gamma_{ij}^\tau = \Sigma_1^{N_1} \bar{\lambda}_j \gamma_{ij}^\tau$. $\Sigma_1^N \lambda_i \lambda_i^\tau = \Sigma_{j,r=1}^{N_1} \bar{\lambda}_j (\Sigma_{i=1}^{N} \gamma_{ij}^\tau \gamma_{ir}) \bar{\lambda}_r^\tau \leq$ $\Sigma_{j,r=1}^{N_1} \bar{\lambda}_j (\Sigma_1^{N_2} \gamma_{ij}^\tau \gamma_{ir}) \bar{\lambda}_r^\tau$ for $N_2 \geq N$ since our quantity is obviously monotonic with increasing $N$. We choose $N_2 = \max\{i(1),\dots,i(N_1)\}$ (our matrix $(\gamma)$ is of course also column-finite in view of (1.6.1) and have $\Sigma_1^{N_2} \gamma_{ij}^\tau \gamma_{ir} = \delta_{jr}$ for $1 \leq j,r \leq N_1$. Hence $\Sigma_1^N \lambda_i \lambda_i^\tau \leq \Sigma_1^{N_1} \bar{\lambda}_i \bar{\lambda}_i^\tau$.

1.7. Standard bases. Let $V \subset (E,\Phi)$ be some infinite dimensional subspace. There always (1.5) exist bases $\mathfrak{B} = (v_i)_{i \in \mathbb{N}} \cup (f_\iota)_{\iota \in J}$ of $E$ such that

(1.7.1) $(v_i)_{i \in \mathbb{N}}$ is an orthonormal basis of $V$;

(1.7.2) $(f_\iota)_{\iota \in J}$ is an orthonormal basis of some supplement of $V$ in $E$.

With respect to a fixed basis $\mathfrak{B}$ we set for the whole chapter

(1.7.3) $\alpha_{\iota i} = \Phi(f_\iota, v_i)$ $\quad (\iota \in J,\ i \in \mathbb{N})$

(1.7.4) $A_{\iota\kappa n} = \Sigma_{i=1}^n \alpha_{\iota i} \alpha_{\kappa i}^\tau$ $\quad (\iota,\kappa \in J,\ n \in \mathbb{N})$

Since $0 \leq \|f_\iota - \Sigma_{i=1}^n \alpha_{\iota i} v_i\| = 1 - A_{\iota\iota n}$ we have

(1.7.5) $0 \leqslant A_{\iota\iota n} < 1$ $\quad (\iota \in J,\ n \in \mathbb{N})$.

We call standard basis for the embedding $V \subset E$ a basis $\mathfrak{B}$ which satisfies (1.7.1), (1.7.2) and

(1.7.6) $(A_{\iota\kappa n})_{n \in \mathbb{N}}$ is a Cauchy sequence for all $\iota,\kappa \in J$.

In general there will be no standard basis for an embedding $V \subset E$ but we do have the following important case:

(1.7.7) If $(k_o,<)$ is archimedean then there are always standard bases.

Proof. $(A_{\iota\iota n})_n$ is bounded by 1.7.5 and obviously monotonic, hence Cauchy in the archimedean case. If all $(A_{\iota\iota n})_n$, $\iota \in J$, are Cauchy then so are $(A_{\iota\kappa n})_n$ by the Schwarz inequality (this follows just as in the commutative case).

Furthermore, by 1.6 and by the Schwarz inequality, we have

(1.7.8) If $\mathfrak{B} = (v_i)_{\mathbb{N}} \cup (f_\iota)_J$ is a standard basis for $V \subset E$ and $(\bar{v}_i)_{\mathbb{N}}$ any orthonormal basis of $V$ then $\bar{\mathfrak{B}} = (\bar{v}_i)_{\mathbb{N}} \cup (f_\iota)_J$ is again a standard basis for $V \subset E$ and $\lim_{n\to\infty} \bar{A}_{\iota\kappa n} = \lim_{n\to\infty} A_{\iota\kappa n}$ in $\bar{k}_o \otimes k$ $(\iota,\kappa \in J)$. If $V \subset E$ admits at least one standard basis then the union $\mathfrak{B}$ of any orthonormal families $(v_i)_{\mathbb{N}}$ (spanning $V$) and $(f_\iota)_J$ (spanning a supplement of $V$ in $E$) is a standard basis for $V \subset E$.

(1.7.9) Thus, whenever we do have a standard basis for $V \subset E$ we may contemplate the positive semidefinite matrix $A = (A_{\iota\kappa})_{\iota,\kappa\in J}$ where $A_{\iota\kappa} = \lim_{n\to\infty} A_{\iota\kappa n} \in \bar{k}_o \otimes k$ ; we call $A$ <u>the matrix associated with the standard basis</u>.

## 2. <u>The matrix of a ⊥-dense subspace with standardbasis</u>

Assume that $V \subsetneqq (E,\Phi)$ admits a standard basis $\mathfrak{B} = (v_i)_{i\in\mathbb{N}} \cup (f_\iota)_{\iota\in J}$. We can always arrange for the typical $x \in E$ to be represented in the form $x = \Sigma_1^n \xi_i f_i + \Sigma_1^N \lambda_i v_i$. Setting $S_N :=$ $\Sigma_{j=1}^N (\lambda_j + \Sigma_1^n \xi_i \alpha_{ij})(\lambda_j + \Sigma_1^n \xi_i \alpha_{ij})^\tau$ we have the identity ($\delta_{\iota\kappa}$: Kronecker)

$$(2.1) \qquad N \geq n : 0 \leq S_N = \|x\| + \sum_{\iota,\kappa=1}^{n} \xi_\iota (A_{\iota\kappa} - \delta_{\iota\kappa}) \xi_\kappa^\tau + \sum_{\iota,\kappa=1}^{n} \xi_\iota (A_{\iota\kappa N} - A_{\iota\kappa}) \xi_\kappa^\tau .$$

We assert:

(2.2) The matrix $A_{\iota\kappa} - \delta_{\iota\kappa}$ is negative semidefinite.

<u>Proof</u>. Assume that for some given $\xi_1,\ldots,\xi_n$ we have $\sum_{\iota,\kappa=1}^{n} \xi_\iota (A_{\iota\kappa} - \delta_{\iota\kappa}) \xi_\kappa^\tau \neq 0$. Choose $N$ so large that $\left| \sum_{\iota,\kappa=1}^{n} \xi_\iota (A_{\iota\kappa N} - A_{\iota\kappa}) \xi_\kappa^\tau \right| < \left| \sum_{\iota,\kappa=1}^{n} \xi_\iota (A_{\iota\kappa} - \delta_{\iota\kappa}) \xi_\kappa^\tau \right|$. Consider then the vector $x_N = \sum_1^n \xi_i f_i + \Sigma_1^N \lambda_j v_j$ with $\lambda_j = -\sum_1^n \xi_i \alpha_{ij}$. $S_N = 0$ for this choice of the $\lambda_j$ and (2.1) shows that $\sum_{\iota,\kappa=1}^{n} \xi_\iota (A_{\iota\kappa} - \delta_{\iota\kappa}) \xi_\kappa^\tau < 0$ since $\|x\| > 0$. Q.e.d.

We also see that we cannot have, at the same time, both "$\Sigma\xi_\iota(A_{\iota\kappa}-\delta_{\iota\kappa})\xi_\kappa^\tau=0$" and "$\Sigma\xi_i f_i \in V + V^\perp$" unless all $\xi_i$ be zero. Thus

(2.3) If $A_{\iota\kappa}-\delta_{\iota\kappa}$ is not definite then $V + V^\perp \neq E$.

The converse does not, of course, hold as illustrated by theorem (2.6). From (2.1) we deduce by the same kind of arguments the following

(2.4) Assume that on the k-vectorspace $E$ we introduce a hermitean sesquilinear form $\Psi$ with respect to the involution $\tau$ of $k$ by specifying the products on some basis $\mathfrak{B} = (w_i)_{i\in\mathbb{N}} \cup (f_\iota)_{\iota\in J}$ of $E$ in the following manner. We require firstly $(w_i)_{i\in\mathbb{N}}$ to be orthonormal for $\Psi$, secondly we require $(f_\iota)_{\iota\in J}$ to be orthonormal for $\Psi$, thirdly we choose the $\alpha_{\iota i} = \Psi(f_\iota, w_i)$ in such a way that (1.7.6) and (2.2) hold. Then we have

(2.4.1) $\Psi$ is positive semidefinite on all $E$,

(2.4.2) If $W$ is the span of $(w_i)_{i\in\mathbb{N}}$ then $W^\perp \neq (0)$ iff there exist $(\xi_1,\ldots,\xi_m) \neq (0,\ldots,0)$ such that the sequence $(\sum_{\iota,\kappa=1}^{m} \xi_\iota A_{\iota\kappa n}\xi_\kappa^\tau)_n$ is eventually constant when $n \to \infty$.

(2.4.3) If $W^\perp = (0)$ then the form $\Psi$ is definite on all of $E$ and the matrix $(A_{\iota\kappa})$ is definite.

Notice that $(E,\Psi)$ thus defined need not be strongly universal.

2.5. Corollary. Assume that $V \subset (E,\Phi)$ with $V^\perp = (0)$ admits a standardbasis $\mathfrak{B}$. Then the matrix $(A_{\iota\kappa})$ associated with this basis is positive definite and the matrix $A_{\iota\kappa} - \mathbb{1}$ is negative semidefinite. Furthermore $k$ admits non eventually constant denumerable Cauchy sequences; hence the topological space $\bar{k}_o \otimes k$ has a denumerable basis and is thus a complete space.

2.6. A theorem. As $\Phi$ is assumed to be strongly universal we see that $\|\Phi\|$ coincides with the subset $Q := \{\langle X,X\rangle \mid X \in k^m, m \in \mathbb{N}\}$ of $k_o$. Let $\bar{Q}$ be the closure of $Q$ in $\bar{k}_o$.

In order not to have to interrupt the train of thoughts we have delegated the proof of the following instructive theorem to Appendix II.

Theorem. Assume that $\bar{Q} = \bar{k}_o^+$ (the non negative elements of $\bar{k}_o$) and that $k_o$ possesses a non eventually constant denumerable Cauchy sequence. Let $A$ be any positive definite $J \times J$-matrix over $\bar{k}_o \otimes k$, card $J \leq \aleph_o$, with $A - \mathbb{1}$ negative semidefinite. Then there is a positive definite hermitean k-space $(E,\Psi)$ which contains a standard basis $\mathfrak{B} = (v_i)_{\mathbb{N}} \cup (f_\iota)_J$ with $\perp$-dense span of the $(v_i)_{\mathbb{N}}$ and with $A$ as the associated matrix. Here $(E,\Psi)$ need not be strongly universal.

Remark. We have chosen the assumption $\bar{Q} = \bar{k}_o^+$ for the sake of convenience; the assumption is actually too strong.

Corollary. Let $(E,\Phi)$ be a strongly universal k-space with $1 \in \|\Phi\|$. Consider the situation of the theorem but make the stronger assumption $Q = k_o^+$. Then the space $(E,\Psi)$ constructed in 2.6 is isometric with $(E,\Phi)$.

Proof. We keep the notations of 2.6. The subspace $\bar{E} := k(v_i)_{i\in\mathbb{N}}$ of $(E,\Psi)$ is strongly universal being isometric to $(E,\Phi)$. Furthermore $Q = \|\Phi\| = \|\bar{E}\| \subset \|\Psi\| \subset k_o^+ = Q$ hence we have equality throughout.

2.7. Consider the situation of theorem 2.6 for some fixed $n = \text{card } J = \dim E/V \leq \aleph_o$. We may represent the matrix $A = (A_{\iota\kappa})_{\iota,\kappa\in J}$ as a point with the coordinates $A_{\iota\kappa}$ in a $\frac{1}{2}\, n\cdot(n+1)$ - dimensional space. By our theorem 2.6 we are interested in the convex region $\mathcal{R}$ which is the intersection of the two cones

$$K_1 : (A_{\iota\kappa}) \text{ positive definite},$$

$$K_2 : (A_{\iota\kappa}) - \mathbb{1} \text{ negative semidefinite}.$$

To every point of $\mathcal{R}$ corresponds a dense embedding $V \subset (E,\Phi)$ (and conversely). The invariants for $V \subset E$ which we are going to set up in the next section will enable us to replace the study of orbits in the set of $\perp$-dense $V \subset E$ under the orthogonal group of $(E,\Phi)$ by the study of orbits in the region $\mathcal{R}$ under some more accessible group.

Here is a picture of $\mathcal{R}$ for $n = 2$ and $\Phi$ symmetric. One may take e.g. $k_o = \mathbb{Q}$ so $\bar{k}_o = \mathbb{R}$ and $\mathcal{R} \subset \mathbb{R}^3$ :

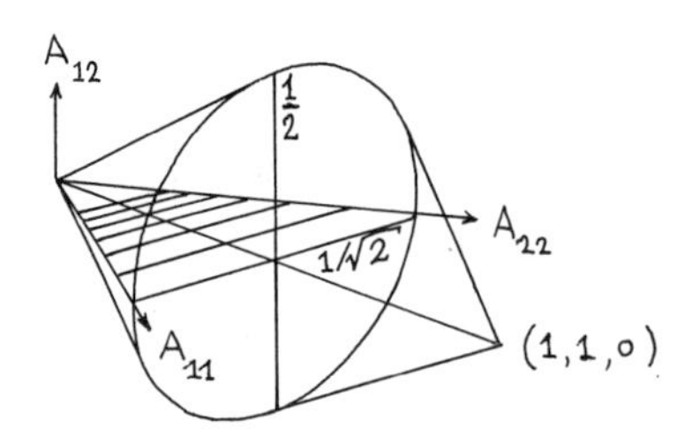


When $n = 2$ and $k = \mathbb{R}$ then $\mathcal{R}$ is the union of 3 orbits by the Main Theorem in Section 4. Where are the orbits!

$\mathcal{R}$ for $n = 2$ ; only the nearer half of the surface (and with the equator ellipse excluded) belongs to $\mathcal{R}$ .

## 3. The $\psi$ – invariant of a $\perp$-dense subspace

3.1. Let $\mathfrak{B} = (v_i)_{i \in \mathbb{N}} \cup (f_\iota)_{\iota \in J}$ , $\bar{\mathfrak{B}} = (\bar{v}_i)_{i \in \mathbb{N}} \cup (\bar{f}_\iota)_{\iota \in J}$ be two standard bases for the embedding $V \subset E$ , $V^\perp = (0)$ . We have

(3.1.1) $$\bar{f}_\iota = \Sigma_{\kappa=1}^{\kappa(\iota)} \gamma_{\iota\kappa} f_\kappa + \Sigma_{i=1}^{i(\iota)} \xi_{\iota i} v_i$$

for certain row-finite matrices $(\gamma_{\iota\kappa})$ , $(\xi_{\iota i})$ over $k$ ; $i(\iota)$ is the smallest natural number such that $\xi_{\iota i} = 0$ for $i > i(\iota)$ ; similarly for $\kappa(\iota)$ . $(\gamma_{\iota\kappa})$ is invertible.

Let $(A_{\iota\kappa})$ , $(\bar{A}_{\iota\kappa})$ be the matrices over $\bar{k}_o \otimes k$ associated with $\mathfrak{B}$ and $\bar{\mathfrak{B}}$ respectively and $\mathbb{1} = (\delta_{\iota\kappa})_{\iota,\kappa \in J}$ the unit matrix. We assert

(3.1.2) $$\bar{A}_{\iota\kappa} - \delta_{\iota\kappa} = \sum_{\nu,\mu=1}^{\nu(\iota),\mu(\kappa)} \gamma_{\iota\nu}(A_{\nu\mu} - \delta_{\nu\mu})\gamma^\tau_{\kappa\mu} \qquad (\gamma_{\iota\nu} \in k) \quad .$$

Proof. By 1.7.8 $\bar{\bar{\mathfrak{B}}} = (\bar{v}_i)_{i \in \mathbb{N}} \cup (f_\iota)_{\iota \in J}$ is yet another standard basis for $V \subset E$ and $\bar{\bar{A}}_{\iota\kappa} = A_{\iota\kappa}$ $(\iota,\kappa \in J)$ . By making the transition from $\mathfrak{B}$ to $\bar{\mathfrak{B}}$ via $\bar{\bar{\mathfrak{B}}}$ it becomes evident that we may assume at the outset that

(3.1.3) $\quad \bar{v}_i = v_i \quad (i \in \mathbb{N}) \quad$ in $\mathfrak{B}$ , $\bar{\mathfrak{B}}$ .

From (3.1), (3.1.3) we get $\bar{\alpha}_{\iota i} = \Phi(\bar{f}_\iota, v_i) = \Sigma_{\kappa=1}^{\kappa(\iota)} \gamma_{\iota\kappa}\alpha_{\kappa i} + \xi_{\iota i}$ and, by making use of the fact that $(v_i)_{\mathbb{N}}$ , $(\bar{f}_\iota)_J$ , $(f_\iota)_J$ are orthonormal families, it follows that

(3.1.4) $$\bar{A}_{\iota\kappa m} - \delta_{\iota\kappa} = \Sigma_{\nu,\mu=1}^{\nu(\iota),\mu(\kappa)} \gamma_{\iota\nu}(A_{\nu\mu m} - \delta_{\nu\mu})\gamma_{\kappa\mu}^{\tau} \quad \text{for every } m \geq i(\iota),\ i(\kappa) .$$

From this (3.1.2) follows since we have convergence for $m \to \infty$.

3.2. The invariant $\psi$ . We see from 3.1.2 that the quantities appropriate for the description of a $\perp$-dense embedding $V \subset E$ are not the matrices $A$ associated with standard bases but rather the matrices $A - \mathbb{1}$ . In the special case when $k_o = \bar{k}_o$ (3.1.2) says that $A - \mathbb{1}$ transforms just like an ordinary hermitean symmetric tensor (i.e. like a hermitean form) when the standard basis describing the embedding $V \subset E$ is changed. If $k_o \subsetneqq \bar{k}_o$ then the embedding $V \subset E$ still determines the equivalence class of the matrix $A - \mathbb{1}$ (over $\bar{k}_o \otimes k$) modulo the transformations 3.1.2 over k induced by the allowable "coordinate transformation (3.1.1)" which describes the transition from one standard basis $\mathfrak{B}$ to another such basis $\bar{\mathfrak{B}}$ .

For the rest of the chapter we shall set ($\delta_{\iota\kappa}$ = Kronecker)

(3.2.1) $$\psi_{\iota\kappa} := A_{\iota\kappa} - \delta_{\iota\kappa} \qquad (\iota,\kappa \in J)$$

and by $\psi$ we shall understand the class of the matrix $(\psi_{\iota\kappa})$ modulo the transformations (3.1.2) which we shall set down here once more

(3.2.2) $$\bar{\psi}_{\iota\kappa} = \Sigma\gamma_{\iota\nu}\psi_{\nu\mu}\gamma_{\kappa\mu}^{\tau} \in \bar{k}_o \otimes k \qquad (\gamma_{\iota\nu} \in k) .$$

Definition (3.2.1) is justified by the fact that the quantity $\psi$ actually characterizes the dense embedding $V \subset E$ ("main theorem"). The easy half of this characterization is contained in the proof of 3.1.2. We state it in the following form

3.2.3. Theorem. Let $V \subset E$, $\bar{V} \subset E$ with $V^{\perp} = \bar{V}^{\perp} = (0)$ be embeddings which admit standard bases. If there exists a metric automorphism $T$ of $(E,\Phi)$ with $TV = \bar{V}$ then $\psi = \bar{\psi}$ for the corresponding psis, i.e. an equation (3.2.2) holds for the components $\psi_{\iota\kappa}$, $\bar{\psi}_{\iota\kappa}$ where $(\gamma_{\iota\kappa})_{\iota,\kappa\in J}$ is a row-finite invertible matrix over $k$.

## 4. The Main Theorem on $\perp$-dense subspaces and the plan of its proof

Our first principal result is a converse of theorem 3.2.3 above. In order to prove it we have to put down an additional condition on the field $k$ (c.f. Remark 5.6 below).

We have set down as a general assumption for the whole chapter that $(E,\Phi)$ be strongly universal and that $1 \in \|\Phi\|$. This means that the set $\|\Phi\|$ is made up of all sums $\langle X,X\rangle$ where $X \in k^m$ and where $m$ ranges in $\mathbb{N}$. For the proof of the main theorem we need a finiteness condition to the effect that we obtain all of $\|\Phi\|$ if we bound $m$ by a suitable bound. Let $Q_s \subset k_0^+$ be the set of all sums $\langle X,X\rangle$ where $X \in k^s$. We shall assume $k$ to satisfy

(4) There exists $s \in \mathbb{N}$ such that $k_0^+ = Q_s$.

Remark. It is not difficult to provide examples of division rings $(k,\tau)$ that admit positive definite forms $\Phi$ and which satisfy (4). Let us consider the case where $\tau = \mathbb{1}$. In this case (4) says that $k_0$ has finite Pythagoras number [7] and admits one ordering only. (The Pythagoras number of an ordered commutative field is the smallest cardinal $s \leq \aleph_0$ such that every sum of squares in the field equals a sum of at most $s$ squares.) Algebraic number fields with unique ordering provide examples with $s = 4$. Of course, there always are the real closed fields with $s = 1$.

We are now able to state

The Main Theorem (on $\perp$-dense subspaces). Let $V,\bar{V} \in E$ with

$V^{\perp} = \bar{V}^{\perp} = (0)$ be embeddings which admit standard bases such that an equation (3.2.2) holds. Provided that $k$ satisfies (4) there exists a metric automorphism $T$ of $(E,\Phi)$ with $TV = \bar{V}$.

The proof of the Main Theorem will be given in the course of the next three sections. In sec. 5 we derive from the assumptions of the main theorem that there are standard bases such that the associated matrices are equal (Lemma 1). In sec. 7 we prove that if the associated matrices are equal one may always introduce bases such that $\alpha_{\iota i} = \bar{\alpha}_{\iota i}$ (1.7); this will of course guarantee the existence of the required automorphism (Lemma 3). For the recursive construction in the proof of Lemma 3 another lemma is needed the proof of which is somewhat lengthy; we could have shortened it to a fraction were it not for the non commutative cases which we are not willing to drop. It will be inserted as lemma 2 in section 6.

## 5. Proof of the Main Theorem: the first lemma

5.1. Lemma 1. Let $V \subset E$, $\bar{V} \subset E$ with $V^{\perp} = \bar{V}^{\perp} = (0)$ be embeddings which admit standard bases $\mathfrak{B}, \bar{\mathfrak{B}}$ such that the corresponding $\psi, \bar{\psi}$ are equal; i.e. (3.2.2) holds for $(\gamma_{\iota\kappa})_{\iota,\kappa\in J}$ a row-finite invertible matrix over $k$. If $k$ satisfies (4) then there exists a standard basis $\mathfrak{B}'$ for $V \subset E$ such that $\psi'_{\iota\kappa} = \bar{\psi}_{\iota\kappa}$ $(\iota,\kappa\in J)$.

Proof of lemma 1:

5.2. If for all $\iota \in J \subset \mathbb{N}$ we succeed in determining natural numbers $i(\iota)$ and $\xi_{\iota i} \in k$, $1 \leq i \leq i(\iota)$ such that the vectors

$$(5.2.1) \qquad f'_{\iota} = \Sigma_{\kappa=1}^{\kappa(\iota)} \gamma_{\iota\kappa} f_{\kappa} + \Sigma_{i=1}^{i(\iota)} \xi_{\iota i} v_i \qquad (\iota\in J)$$

are an orthonormal family then by (3.1.2) $\psi'_{\iota\kappa} = \Sigma\, \gamma_{\iota\nu}\psi_{\nu\mu}\gamma^{\tau}_{\kappa\mu} = \bar{\psi}_{\iota\kappa}$.

5.3. The induction assumptions. We shall assume that we have already determined numbers $i(\iota) \in \mathbb{N}$ for $\iota \leq N$ as well as the corresponding $\xi_{\iota i}$, $1 \leq i \leq i(\iota)$ such that $(f'_\iota)_{\iota \leq N}$ is an orthonormal family. We shall show how to find $i(N+1)$ and the corresponding $\xi_{N+1,i}$, $1 \leq i \leq i(N+1)$ such that $(f'_\iota)_{\iota \leq N+1}$ is an orthonormal family; this will also cover the very first step, i.e. the determination of $\iota(1)$ and $\xi_{1i}$, $1 \leq i \leq \iota(1)$.

5.4. <u>Reformulation of 5.3</u>. Let $m \geq \max\{i(1),\ldots,i(N)\}$ be a fixed natural number. We set

(5.4.0) $$\xi_{\iota i} := 0 \quad (1 \leq \iota \leq N \quad , \quad i(\iota) < i \leq m);$$

hence we may write

(5.4.1) $$f'_\iota = \Sigma_{\kappa=1}^{\kappa(\iota)} \gamma_{\iota\kappa} f_\kappa + \Sigma_{i=1}^{m} \xi_{\iota i} v_i \qquad (\iota \leq N) \quad .$$

If we set

(5.4.2) $$\lambda_{\iota i} := \xi_{\iota i} + \Sigma_{\kappa=1}^{\kappa(\iota)} \gamma_{\iota\kappa} \alpha_{\kappa i} \qquad (1 \leq \iota \leq N,\ 1 \leq i \leq m)$$

and

(5.4.3) $$\Gamma(m)_{\iota\kappa} := \bar{A}_{\iota\kappa} - \Sigma_{i=m+1}^{\infty} (\Sigma_{\nu=1}^{\nu(\iota)} \gamma_{\iota\nu} \alpha_{\nu i})(\Sigma_{\mu=1}^{\mu(\kappa)} \gamma_{\kappa\mu} \alpha_{\mu i})^\tau$$

then orthonormality of the family (5.4.1) is easily verified to be equivalent with

(5.4.4) $$\Sigma_{i=1}^{m} \lambda_{\iota i} \lambda_{\kappa i}^\tau = \Gamma(m)_{\iota\kappa} \qquad (1 \leq \iota, \kappa \leq N) \quad .$$

Simplifying the expression for $\Gamma(m)_{\iota\kappa}$ in (5.4.3) shows that

(5.4.5) $$\Gamma(m)_{\iota\kappa} \in k \quad .$$

The convergence of the infinite sum in (5.4.3) follows directly from the existence of the infinite sum $A_{\nu\mu}$.

5.5. <u>Choice of $i(N+1)$</u>. Consider the matrix $\Gamma = (\Gamma(m)_{\iota,\kappa})_{\iota,\kappa \leq N+1}$. For $m \to \infty$ the matrix $\Gamma$ converges to the positive definite (2.4.3) matrix $(\bar{A}_{\iota\kappa})_{\iota,\kappa \leq N+1}$. We may therefore pick $m$ so large that $\Gamma(m)$ is positive definite (Continuity of the principal minor determinants). In fact, if $s$ is the number in (4) we shall require that $m$ be chosen such that we have both

(5.5.1) $\quad m > N\cdot s$ and $\Gamma$ positive definite .

Such an $m$ we take as our candidate for $i(N+1)$ . By (4) and (1.5.1) we then have

(5.5.2) $\quad \Gamma = C^{tr}\cdot C^{\tau}$ for a matrix $C = (c_{i\iota})_{\substack{1\leq\iota\leq N+1\\ 1\leq i\leq m}}$ over $k$ .

In the hermitean m-tupel space $(k^m;\langle,\rangle)$ we consider the vectors $x_\iota = (\lambda_{\iota 1},\ldots,\lambda_{\iota m})$ , $1 \leq \iota \leq N$ and the $N+1$ columnvectors $y_1,\ldots,y_{N+1}$ of the matrix $C$ in (5.5.2). By (5.4.4) we have $\langle x_\iota,x_\kappa\rangle = \Gamma(m)_{\iota\kappa}$ , $1 \leq \iota,\kappa \leq N$ . By (5.5.2) we have $\langle y_\iota,y_\kappa\rangle = \Gamma(m)_{\iota\kappa}$, $1 \leq \iota,\kappa \leq N+1$ . By Witt's theorem we can extend the isometry $\varphi : y_i \mapsto x_i$ $(1\leq i\leq N)$ to all of $k^m$ . If we set

$$x_{N+1} := \varphi y_{N+1}$$

we have

(5.5.3) $\quad \langle x_\iota,x_\kappa\rangle = \Gamma(m)_{\iota\kappa}$ , $1 \leq \iota,\kappa \leq N+1$ .

Now we should pick $i(N+1)$ and the corresponding $\xi_{N+1,i}$ $(1\leq i\leq i(N+1))$ in such a fashion that the vectors $f'_1,\ldots,f'_N,f'_{N+1}$ of (5.2.1) are orthonormal. Just as in (5.4) this orthonormality amounts to the condition

(5.5.4) $\quad \Sigma\lambda_{\iota i}\lambda^{\tau}_{\kappa i} = \Gamma(m)_{\iota\kappa} \qquad (1\leq\iota,\kappa\leq N+1)$ .

The $\lambda_{\iota i}$ with $1 \leq \iota \leq N$ and $1 \leq i \leq m$ are known by (5.4.2), and the

(5.5.5) $\quad \lambda_{N+1,i} := \xi_{N+1,i} + \Sigma_{\kappa=1}^{\kappa(N+1)}\gamma_{N+1,\kappa}\alpha_{\kappa,i} \qquad (1\leq i\leq m)$

are the unknowns since the $\xi_{N+1,i}$ are unknown. Now (5.5.3) shows that (5.5.4) is solved with $\lambda_{N+1,1},\ldots,\lambda_{N+1,m}$ as the components of $x_{N+1} := \varphi y_{N+1}$ . Hence we can solve for the $\xi_{N+1,1},\ldots,\xi_{N+1,m}$ in (5.5.5). These elements and $i(N+1) := m$ have all the requisite properties. The proof of lemma 1 is thus complete.

5.6. Remark. In retrospect it can be made plausible that we do need the assumption (4) instead of mere strong universality of $(E,\Phi)$

if we want to prove lemma 1. For if the lemma holds then the matrix $\Gamma(m)$ as introduced by (5.4.3) must be a "m-Gram-matrix" for <u>some</u> $m \in \mathbb{N}$, i.e. there must exist a $m \times N$ matrix $C$ over $k$ with $\Gamma(m) = C^{tr} \cdot C^{\tau}$ for <u>some</u> $m \in \mathbb{N}$. Without a finiteness assumption it is conceivable that <u>all</u> matrices $C$ with $C^{tr} \cdot C^{\tau} = \Gamma(m)$ have more than $m$ rows for all $m \in \mathbb{N}$. Now the way in which the $N \times N$ matrix $\Gamma(m)$ is defined, namely as a difference of matrices over $\bar{k}$, makes it necessary to require the following of the field $k$ (recall that $N$ varies in $\mathbb{N}$ in the course of the proof): Every hermitean form $\Gamma$ in finitely many variables which has $\|\Gamma\| \subset \bar{k}_o^+ - \bar{k}_o^+$ satisfies $\|\Gamma\| \subset Q_s$ for suitable $s \in \mathbb{N}$. Hence condition (4).

<u>5.7</u>. The following remark will be useful at some later stage. Let $(v_i)_{i \in \mathbb{N}} \cup (f_\iota)_{\iota \in J}$ be a standard basis for $V \subset E$, $V^\perp = (0)$. If we change finitely many $f_\iota$ modulo $V$ to $f'_\iota$, $f'_\iota = f_\iota + \Sigma \xi_{\iota i} v_i$ such that these $f'_\iota$ are orthonormal, then the $f'_\iota$ can be completed to a standard basis $(v_i)_{i \in \mathbb{N}} \cup (f'_\iota)_{\iota \in J}$ such that $f_\iota - f'_\iota \in V$ and $\psi'_{\iota\kappa} = \psi_{\iota\kappa}$ $(\iota, \kappa \in J)$. Indeed, we only have to apply the proof of 5.1.3 to the case where $(\gamma_{\iota\kappa})_{\iota\kappa \in J}$ of 3.2.2 is the unit matrix.

## 6. Proof of the Main Theorem: the second lemma

<u>6.1</u>. Consider the situation $r = \dim E/V < \infty$. For some fixed standard basis $\{f_1, \ldots, f_r\} \cup (v_i)_{i \in \mathbb{N}}$ we look at the matrix $A_n = (A_{\iota\kappa n})_{1 \leq \iota, \kappa \leq r}$ defined in 1.7. For $n \to \infty$ the matrix $A_n$ converges to the positive definite (2.4.3) matrix $A = (A_{\iota\kappa})_{1 \leq \iota, \kappa \leq r}$. Hence there is $n_o \in \mathbb{N}$ such that $A_n$ is positive definite for $n \geq n_o$ and for these $n$ there is an inverse $B_n = (B_{\iota\kappa n})$ of $A_n$; ${}^{tr}B_n = B_n^\tau$ and $B_n$ is positive definite for $n \geq n_o$. $B_n$ converges with $n \to \infty$ to $B = A^{-1}$.

The matrix $S_n = A_{n+1} - A_n$ is positive semidefinite and the identity

(6.1.1) $\qquad B_n - B_{n+1} = B_{n+1}[S_n + S_n B_n S_n]B_{n+1}$

shows that $B_n - B_{n+1}$ is positive semidefinite. Hence for fixed $X \in k^r$ the sequence $({}^{tr}XB_nX^\tau)_n$ is monotonically decreasing and not eventually constant since $V^\perp = (0)$ (2.4.2).

6.2. For the recursive construction of the automorphism in lemma 7.0 of the next section the maximum of a certain function defined on the sphere $S^{m-1} \subset (K \otimes k)^m$,

(6.2.1) $\qquad S^{m-1} : \langle X,X\rangle = 1$

turns out to be of central importance. We shall now define this function. We set $A := (\alpha_{sj})_{1\le s\le r, 1\le j\le m}$ with the $\alpha_{sj}$ as defined in 1.7; furthermore $C_n := {}^{tr}A^\tau B_n A$, $C := {}^{tr}A^\tau BA$. For $X \in (K\otimes k)^m$ we then define

(6.2.2) $\qquad \pi_{m,n}(X) = {}^{tr}XC_nX^\tau \;,\quad \pi_{m,\infty}(X) = {}^{tr}XCX^\tau .$

6.3. Lemma 2. Let $n \ge n_o$, $n \ge m$. (i) $\pi_{m,n}$ has a maximum $\mu_{m,n}$ on $S^{m-1}$ with $0 \le \mu_{m,n} \le 1$ ($\mu_{m,m} \in \{0,1\}$). (ii) For fixed $m$ there exist $n$ such that $\mu_{m,n} < 1$. (iii) $\pi_{m,\infty}$ has a maximum $\mu_m$ on $S^{m-1}$ and $\mu_m < \mu_{m,n} \le 1$ for all $m \in \mathbb{N}$.

Just as in the case of commutative fields $k$ with archimedean ordered $k_o$ (i.e. $K = \mathbb{R}$) one finds that the maximum problem of lemma 2 is a equivalent to an eigenvalue problem, namely

(6.3.1) $\qquad {}^{tr}X\cdot C_n = \mu\,{}^{tr}X \;, \qquad X \in (K\otimes k)^m .$

Before proving the spectral theorem for the fields admitted here we shall list some consequences in connection with (6.3.1).

6.4. If $X$ is a nonzero eigenvector for the eigenvalue $\mu$ in (6.3.1) then $\pi_{m,n}(X) = \mu\cdot\langle X,X\rangle$. Hence $\mu \in K$. We have $0 \le \mu \le 1$ (and $\mu \in \{0,1\}$ if $m=n$).

<u>Proof</u>. To show that $\mu \leq 1$ we shall prove that $\mu^2 - \mu \leq 0$ . Set $\mathcal{D} := (\alpha_{sj})_{1 \leq s \leq r, m < i \leq n}$ (if $m=n$ $\mathcal{D}$ will not be needed) so $A_m = A_n - \mathcal{D} \cdot {}^{tr}\mathcal{D}^\tau$ . We have $\mu^2 \cdot \langle X,X \rangle = {}^{tr}(\mu X)(\mu X)^\tau = {}^{tr}X \cdot C_n \cdot {}^{tr}C_n^\tau X^\tau = {}^{tr}X \cdot {}^{tr}A^\tau B_n \, A {}^{tr}A^\tau B_n A X^\tau = {}^{tr}X \cdot {}^{tr}A^\tau B_n (A_n - \mathcal{D} \cdot {}^{tr}\mathcal{D}^\tau) B_n A X^\tau = \pi_{m,n}(X) - {}^{tr}({}^{tr}\mathcal{D}^\tau B_n A X^\tau)^\tau \cdot ({}^{tr}\mathcal{D}^\tau B_n A X^\tau)$ . As $\pi_{m,n}(X) = \mu \langle X,X \rangle$ we therefore see that $(\mu^2 - \mu)\langle X,X \rangle \leq 0$ . If we should have $m = n$ then we find $(\mu^2 - \mu)\langle X,X \rangle = 0$ .

<u>6.4.1</u>. There are only finitely many $n \in \mathbb{N}$ such that (6.3.1) possesses a nonzero eigenvector $X_n$ with eigenvalue 1.

<u>Proof</u> (indirect). Assume there were such $X_n$ with $n$ ranging in an infinite subset $P \subset \mathbb{N}$ . We keep the notations of the proof of 6.4 where we have shown that $(\mu^2 - \mu)\langle X,X \rangle = - {}^{tr}({}^{tr}\mathcal{D}^\tau B_n A X^\tau)^\tau \cdot ({}^{tr}\mathcal{D}^\tau B_n A X^\tau)$ . Hence ${}^{tr}\mathcal{D}^\tau B_n A X_n^\tau = 0$ for $n \in P$ . Since $V^\perp = (0)$ we cannot have that the rows of the $r$by$(n-m+1)$ matrix $\mathcal{D}$ are linearly dependent for infinitely many $n \in \mathbb{N}$ . Therefore $B_n A X_n^\tau = 0$ for $n \geq n_1$ and $n \in P$ . $B_n$ is invertible for $n \geq n_o$ (6.1) so $A X_n^\tau = 0 \in (K \otimes k)^m$ for $n \geq n_o + n_1$ and $n \in P$ . Therefore (6.2.2) we find for these $n$ that $0 = {}^{tr}X_n {}^{tr}A^\tau B_n A X_n^\tau = \pi_{m,n}(X_n) = \langle X_n, X_n \rangle$ , a contradiction

<u>6.5</u>. The following statements are equivalent. (i) $X \in S^{m-1}$ is a point of maximum for $\pi_{m,n}$ . (ii) $X \in S^{m-1}$ is an eigenvector of (6.3.1) belonging to an eigenvalue $\mu$ which renders $C_n - \mu \mathbb{1}$ negative semidefinite.

<u>Proof</u>. Assume (ii) and let $X + Z$ be arbitrary in $S^{m-1}$ , $\langle X+Z, X+Z \rangle = 1$ . We find

$$\text{(6.5.1)} \quad \pi(X+Z) - \pi(X) = \mu[\langle X,Z \rangle + \langle Z,X \rangle] + {}^{tr}Z C_n Z^\tau = {}^{tr}Z(C_n - \mu \mathbb{1})Z^\tau$$

which shows that (i) holds. Assume conversely that (i) holds. We then consider the function $H(Y) := \pi_{m,n}(Y) - \mu \langle Y,Y \rangle$ on the vectorspace $(K \otimes k)^m$ for suitable value of the "constant" $\mu$ : We can fix $\mu \in K$ such that the "derivatives"

$$D_T = \lim_{K \ni \Delta \to o} \frac{1}{\Delta} [H(X+\Delta T)-H(X)]$$

vanish in arbitrary directions $T \in (K \otimes k)^m$ . Indeed, if we choose $\mu = \max_{S^{m-1}} \pi_{m,n}(Y) = \pi_{m,n}(X)$ then $C_n - \mu \mathbb{1}$ is negative semidefinite on $S^{m-1}$ hence so on all of $(K \otimes k)^m$ (since $K$ is real closed). Therefore $H(X+\Delta T) - H(X) = H(X+\Delta T) = {}^{tr}(X+\Delta T)(C_n - \mu \mathbb{1})(X+\Delta T)^\tau \leq 0$ for all $\Delta \in K$ . Hence the quantity

$$(6.5.2) \quad \lim_{K \ni \Delta \to o} \frac{1}{\Delta}(H(X+\Delta T) - H(X)) = {}^{tr}XC_n X^\tau + {}^{tr}TC_n X^\tau - \mu(\langle X,T \rangle + \langle T,X \rangle) = D_T$$

is both $\leq 0$ and $\geq 0$ so $D_T = 0$ .

We shall now evaluate (6.5.2) for suitable choices of $T$ . Let $T = T_i$ be the m-tuple $(0,\ldots,\delta,\ldots 0)$ with $\delta$ at the $i^{th}$ place. We shall first discuss the non commutative case, $K \otimes k = (\frac{\alpha,\beta}{K})$ . Assume that $\delta$ is one of the three basis vectors $e_i$ with $\delta^\tau = -\delta$ . A banal calculation shows that $D_{T_i} = 0$ (6.5.2) yields $(X = (\xi_1,\ldots,\xi_m))$:

$$(6.5.3) \quad 0 = \delta\Sigma - \Sigma^\tau\delta + \mu\xi_i\delta - \mu\delta\xi_i^\tau \quad \text{where} \quad \Sigma = \sum_{j=1}^{m} \xi_j C_{nji}$$

for $1 \leq i \leq m$ . In other words $(\Sigma - \mu\xi_i^\tau) = \delta^{-1}(\Sigma^\tau - \mu\xi_i)\delta$ . Letting in turn $\delta$ be $e_1$, $e_2$ shows that $e_1^{-1} \cdot e_2$ commutes with $\Sigma - \mu\xi_i^\tau$ $(1 \leq i \leq m)$ so $\Sigma - \mu\xi_i$ commutes with $e_3 = \alpha e_1^{-1} \cdot e_2$ $(\alpha = e_1^2 \in k_o)$ . Application of (6.5.3) to $\delta = e_3$ thus yields

$$(6.5.4) \quad \Sigma - \Sigma^\tau - \mu(\xi_i^\tau - \xi_i) = 0 \qquad (1 \leq i \leq m)$$

We also have

$$(6.5.5) \quad \Sigma + \Sigma^\tau - \mu(\xi_i^\tau + \xi_i) = 0 \qquad (1 \leq i \leq m)$$

which is read off from $D_{T_i} = 0$ if we let $\delta = 1$ . Addition of the two equations shows that (6.3.1) holds. The same conclusion is obtained in the easier commutative case which we leave to the reader.

6.6. The spectral theorem. Let $C : (K \otimes k)^m \longrightarrow (K \otimes k)^m$ be a hermitean linear map, i.e. a map which satisfies

(6.6.1) $\langle CX,Y\rangle = \langle X,CY\rangle$ for all $X,Y \in (K\otimes k)^m$ .

Then there exists an orthogonal basis of $((K\otimes k)^m ;\langle,\rangle)$ consisting of eigenvectors of $C$ .

The spectral theorem could be proved just as in the well known commutative case with $\tau = 1\!\!1$ by making use of the Dieudonné determinant. In order to discuss the "characteristic polynomial" it is however crucial that the involutorial field $(k,\tau)$ has $\xi\xi^\tau$ in its center for all $\xi \in k$ . In the skew case this leaves us just with the quaternion algebras and, in this case, a straight forward reduction to the commutative case can be given as follows:

<u>Proof of the spectral theorem</u>. The field $L = K(\sqrt{\alpha})$ is algebraically closed (whether $k = k_o(\sqrt{\alpha})$ or $k = (\frac{\alpha,\beta}{k_o}))$ and $L \subset K \otimes k$ . Hence by the commutative case of the spectral theorem the L-linear endomorphism $C$ of the finite dimensional L-space $(K\otimes k)^m$ admits an eigenvector. etc.

6.9. <u>Corollary</u>. Let $\Phi$ be a hermitean form in finitely many variables over the division ring $(\frac{\alpha,\beta}{K})$ (hermitean with respect to "conjugation"). Assume that $K$ is real closed ($K$ may be archimedean or not). Then $\Phi$ can be diagonalized by a unitary transformation of the coordinates. $\Phi$ has therefore a maximum on the unitsphere.

6.10. <u>Proof of Lemma 2</u>. Let $C$ be the map defined by $C_n$ (with respect to the basis $e_i = (0,\ldots,0,1,0\ldots,0) \in (K\otimes k)^m$). By the spectral theorem there is an orthogonal basis $\mathfrak{B}$ of eigenvectors for $C$ ; the matrix $\bar{C}_n$ with respect to $\mathfrak{B}$ is diagonal, $\bar{C}_n = VC_nV^{-1} = [\mu_1,\ldots,\mu_m]$ . $V$ has the eigenvectors as rows and $\mu_i$ are the eigenvalues. Since $K$ is real closed we may choose the eigenvectors $X$ to have $\langle X,X\rangle = 1$ so that $V$ is unitary, $V^{-1} = {}^{tr}V^\tau$. The equation $V(C_n-\mu 1\!\!1)\,{}^{tr}V^\tau = \bar{C}_n - \mu 1\!\!1$ shows that the matrices $C_n - \mu 1\!\!1$ and $[\mu_1,\ldots,\mu_m] - \mu 1\!\!1$ are at the same time negative semidefinite so

$\mu = \max\{\mu_1,\dots,\mu_m\}$ qualifies for (6.5); if $X$ is an eigenvector for $\mu$ then $X$ is a point of maximum for $\pi_{m,n}$. Assertions (i) of the lemma now follow from (6.4). $\pi_{m,\infty}$ has analogous properties. The inequality in (iii) follows from (6.1). Assertion (ii) is taken care of by 6.4.1 (It follows equally from (iii) and the fact that the eigenvalues depend continuously on $C_n$ .).

## 7. End of the proof of the Main Theorem: the third lemma

7.0. Lemma 3. Let $V \subset E$, $\bar{V} \subset E$ with $V^{\perp} = \bar{V}^{\perp} = (0)$ be embeddings which admit standard bases $\mathfrak{B},\bar{\mathfrak{B}}$ such that $\psi_{\iota\kappa} = \bar{\psi}_{\iota\kappa}$ for all $\iota,\kappa \in J$ (card $J = \dim E/V = \dim E/\bar{V}$). If $k$ satisfies condition (4) then there exist standard bases $\mathfrak{B}',\mathfrak{B}''$ for $V \subset E$, $\bar{V} \subset E$ respectively such that $\Phi(f'_\iota,v'_i) = \Phi(f''_\iota,v''_i)$ (for all $\iota \in J$, $i \in \mathbb{N}$), i.e. the map which sends members of $\mathfrak{B}'$ into corresponding members of $\mathfrak{B}''$ induces a metric automorphism of $E$ which maps $V$ onto $\bar{V}$.

7.1. Proof of Lemma 3 when card $J = \dim E/V < \infty$.

7.1.0. Let $\mathfrak{B} = (v_i)_{i\in\mathbb{N}} \cup \{f_1,\dots,f_n\}$, $\bar{\mathfrak{B}} = (\bar{v}_i)_{i\in\mathbb{N}} \cup \{\bar{f}_1,\dots,\bar{f}_n\}$ be the two bases which have $\psi_{\iota\kappa} = \bar{\psi}_{\iota\kappa}$ or, equivalently, $A_{\iota\kappa} = \bar{A}_{\iota\kappa}$ $(1\leq\iota,\kappa\leq n)$. We are going to construct a metric automorphism $T: E \to E$ with $TV = \bar{V}$ in the following manner. Let $(e_s)_{s\in\mathbb{N}}$ be some fixed basis of $E$. Assume that we have already determined finite dimensional subspaces $V_r \subset V$, $\bar{V}_r \subset \bar{V}$ with the properties

(i) there is an isometry $\bar{T}_r: k(\bar{f}_1,\dots,\bar{f}_n) \oplus \bar{V}_r \to k(f_1,\dots,f_n) \oplus V_r$ with $\bar{T}_r\bar{f}_i = f_i$ and $\bar{T}_r\bar{V}_r = V_r$ $(1\leq i\leq n)$, (ii) $V_r,\bar{V}_r$ are spanned by orthonormal bases.

Let $e$ be the first member of our auxiliary basis $(e_s)_{s\in\mathbb{N}}$ of $E$ with $e \notin k(\bar{f}_1,\dots,\bar{f}_n) \oplus \bar{V}_r$. By (1.5) we can find a finite dimensional space $\bar{W} \subset \bar{V}_r^{\perp} \cap \bar{V}$ spanned by an orthonormal basis $(\bar{x}_i)_{i\in I}$ and such that $e \in k(\bar{f}_1,\dots,\bar{f}_n) \oplus (\bar{V}_r \oplus \bar{W})$. We shall determine a space

$W \subset V_r^{\perp} \cap V$ with $\dim W = \dim \bar{W}$ and spanned by an orthonormal basis $(x_i)_{i\in I}$ such that $\Phi(f_\iota, x_i) = \Phi(\bar{f}_\iota, \bar{x}_i)$ $(1 \leq \iota \leq n \; ; \; i \in I)$ . It is then clear that $\bar{T}_r$ can be extended to an isometry $k(\bar{f}_1,\ldots,\bar{f}_n) \oplus (\bar{V}_r \oplus \bar{W}) \to k(f_1,\ldots,f_n) \oplus (V_r \oplus W)$ by sending $\bar{x}_i$ into $x_i$ . The step may then be repeatet with $V_{r+1} = V_r \oplus W$ and $T_{r+1} = \bar{T}_{r+1}^{-1}$ instead of $\bar{V}_r, \bar{T}_r$ . Alternating in this fashion between $V$ and $\bar{V}$ we make sure that $k(f_1,\ldots,f_n) \oplus \cup V_r$, $k(\bar{f}_1,\ldots,\bar{f}_n) \oplus \cup \bar{V}_r$ will both exhaust all of $E$ .

In order to carry through this program we need solve only the following problem whose solution may then be applied to the embeddings $V' = k(v_1',\ldots,v_r')^{\perp} \cap V \subset V' \oplus k(f_\iota)_J$ , $\bar{V}' = k(\bar{v}_1',\ldots,\bar{v}_r')^{\perp} \cap \bar{V} \subset \bar{V}' \oplus k(\bar{f}_\iota)_J$ where the $v_i', \bar{v}_i'$ are the basis vectors which we have already constructed.

<u>7.1.1</u>. (Problem). Given a vector $\bar{x} \in \bar{V}$ with $\|\bar{x}\| = 1$ find a vector $x \in V$ with $\|x\| = 1$ and prescribed $\Phi(f_\iota, x)$ , to wit

$$\Phi(f_\iota, x) = \bar{\eta}_\iota \quad \text{where} \quad \bar{\eta}_\iota = \Phi(\bar{f}_\iota, \bar{x}) \; , \; \iota \in J \; . \tag{7.1.1.1}$$

Notice that by the density of $V$ there exist, of course, vectors $z \in V$ with (7.1.1.1); the problem is whether there are such vectors with $\|z\| \leq 1$ !

7.1.2. <u>Solution of problem 7.1.1</u>. In order to solve this problem we shall look out for a vector $z \in V$ with $\Phi(f_\iota, z) = \bar{\eta}_\iota$ and $\|z\| < 1$. Then $1 - \|z\| \in \|\Phi\|$ by assumption (4) so ( .1.5) the space $V \cap z^{\perp} \cap k(f_\iota)_{\iota\in J}^{\perp}$ contains a vector $v$ with $\|v\| = 1 - \|z\|$ . $x = v+z$ then has the required properties.

Our candidate for such a vector $z$ is (cf. Remark 7.1.3)

$$z_n = \Sigma_1^n \lambda_i v_i \; , \; \text{where} \; \lambda_i = \sum_{\nu,\mu\in J} \bar{\eta}_\nu^{\tau} B_{\nu\mu n} \alpha_{\mu i} \in k \tag{7.1.2.1}$$

and where the natural number $n$ shall be chosen suitably. (7.1.1.1) is satisfied for $x = z_n$ . For $\|z_n\|$ we find $\|z_n\| = \Sigma_1^n \lambda_i \lambda_i^{\tau} =$

$\sum_{\nu,\mu\in J} \bar{\eta}_\nu^{\tau} B_{\nu\mu n} \bar{\eta}_\mu$ . Hence $\lim_{n\to\infty} \|z_n\|$ exists and equals $\sum_{\nu,\mu\in J} \bar{\eta}_\nu^{\tau} B_{\nu\mu} \bar{\eta}_\mu =$ $\sum_{\nu,\mu\in J} \bar{\eta}_\nu^{\tau} \bar{B}_{\nu\mu} \bar{\eta}_\mu$ since $A_{\nu\mu} = \bar{A}_{\nu\mu}$ by the assumption of the lemma $(\psi_{\nu\mu} = A_{\nu\mu} + \delta_{\nu\mu})$ . Thus, if we had $\sum \bar{\eta}_\nu^{\tau} \bar{B}_{\nu\mu} \bar{\eta}_\mu < 1$ then for suitably large $n$ we would indeed have $\|z_n\| < 1$ and the proof of Lemma 3 would be complete (in the $\dim E/V < \infty$ - case).

Now, how large can $\Sigma \bar{\eta}_\nu^{\tau} \bar{B}_{\nu\mu} \bar{\eta}_\mu$ get when (7.1.1.1) $\bar{\eta}_\nu = \Phi(\bar{f}_\nu, \bar{x})$ , $\bar{x} = \Sigma_1^m \bar{\xi}_i \bar{v}_i$ for some $m \in \mathbb{N}$ and $\|\bar{x}\| = \Sigma_1^m \bar{\xi}_i \bar{\xi}_i^{\tau} = 1$ ? Part (iii) of (6.3) tells precisely that $\sum_{\nu,\mu\in J} \bar{\eta}_\nu^{\tau} \bar{B}_{\nu\mu} \bar{\eta}_\mu < 1$ on the sphere "$\Sigma \bar{\xi}_i \bar{\xi}_i^{\tau} = 1$" .

7.1.3. Remark. One may ask how the choice of the candidate 7.1.2.1 is motivated. How does one chance upon the formula for the components $\lambda_i$ ? The answer is: work out the problem for symmetric forms in the case of archimedean $k_o$ , $\bar{k}_o = \mathbb{R}$ by looking for the shortest $z \in V$ satisfying 7.1.1.1. An obvious tool is Lagranges "method of undetermined multipliers". Straight forward discussion of Lagrange's function $H = \|z\| - \sum_{\iota\in J} \mu_\iota (\Phi(f_\iota, z) - \bar{\eta}_\iota) = \Sigma \lambda_i^2 - \sum_{\iota\in J} \mu_\iota (\alpha_{\iota i} \lambda_i - \bar{\eta}_\iota)$ yields $2\lambda_i = \sum_{\sigma\in J} \mu_\sigma \alpha_{\sigma i}$ and the multipliers $\mu_\sigma$ turn out to be $\mu_\sigma = 2 \sum_{\nu\in J} \bar{\eta}_\nu B_{\nu\sigma n}$ ; hence our formula for the $\lambda_i$ in this case.

If one is interested in the mere existence of a $z \in V$ with $\|z\| \leq 1$ and (7.1.1.1), i.e. if one doesn't need a formula for $z$ which lends itself for generalization then one can also argue differently in the case of commutative $k$ with archimedean $k_o$ ; see [8], Lemma III.9.

7.2. Proof of the Lemma 3 when card $J = \dim E/V = \aleph_o$. Let $\mathfrak{B} = (v_i)_{i\in\mathbb{N}} \cup (f_\iota)_{\iota\in J}$ , $\bar{\mathfrak{B}} = (\bar{v}_i)_{i\in\mathbb{N}} \cup (\bar{f}_\iota)_{\iota\in J}$ , $(J=\mathbb{N})$, be the standard bases with $\psi_{\iota\kappa} = \bar{\psi}_{\iota\kappa}$ , i.e. $A_{\iota\kappa} = \bar{A}_{\iota\kappa}$ for $\iota,\kappa \in J$ . We shall build up new standard bases $\mathfrak{B}'$, $\mathfrak{B}''$ with $\Phi(f_\iota', v_i') = \Phi(f_\iota'', v_i'')$ $(\iota\in J,\ i\in\mathbb{N})$ step by step. Let $(w_i)_{i\in\mathbb{N}}$, $(\bar{w}_i)_{i\in\mathbb{N}}$ be "auxiliary" bases of $V$, $\bar{V}$ respectively. We put down the following

7.2.1. Induction Assumptions. Assume that we have already determined families $\mathfrak{B}'_{n,m} = (v'_i)_{1\le i\le n} \cup (f'_\iota)_{1\le\iota\le m}$ , $\mathfrak{B}''_{n,m} = (v''_i)_{1\le i\le n} \cup (f''_\iota)_{1\le\iota\le m}$ such that $(v'_i)$, $(v''_i)$, $(f'_\iota)$, $(f''_\iota)$ are orthonormal families with

$$\Phi(f'_\iota, v'_i) = \Phi(f''_\iota, v''_i) \qquad 1 \le \iota \le m \,,\ 1 \le i \le n \tag{7.2.1.1}$$

and $k(v'_i)_{1\le i\le n} = V \cap \operatorname{span} \mathfrak{B}'_{n,m}$ , $k(v''_i)_{1\le i\le n} = \bar{V} \cap \operatorname{span} \mathfrak{B}''_{n,m}$ ; furthermore

$$f'_\iota - f_\iota \in V \,,\ f''_\iota - \bar{f}_\iota \in \bar{V} \,,\quad 1 \le \iota \le m \,. \tag{7.2.1.2}$$

We are going to show how to extend the fragments $\mathfrak{B}'_{n,m}$ , $\mathfrak{B}''_{n,m}$ such that the induction assumptions just mentioned continue to hold for the extended families.

Let $V'_o = V \cap k(v'_i)^\perp_{i\le n}$ , $V''_o = \bar{V} \cap k(v''_i)^\perp_{i\le n}$ . We consider the following subspaces of $(E,\Phi)$:

$$E' = k(f'_1,\ldots,f'_m) \oplus V'_o \,,\ E'' = k(f''_1,\ldots,f''_m) \oplus V''_o \,. \tag{7.2.1.3}$$

We assert that not only do these spaces $E',E''$ admit standard bases but there are such bases with

$$A'_{\iota\kappa} = A''_{\iota\kappa} \tag{7.2.1.4}$$

for the associated matrices. Indeed, since $A_{\iota\kappa} = \bar{A}_{\iota\kappa}$ for all $\iota,\kappa$ in particular so for $1 \le \iota,\kappa \le m$ . Replacing the $f_\iota, \bar{f}_\iota$ by $f'_\iota, f''_\iota$ does not by (3.1) affect the associated matrices in view of (7.2.1.2). Without affecting these matrices we may furthermore switch from $(v_i)_{i\in\mathbb{N}}$ , $(\bar{v}_i)_{i\in\mathbb{N}}$ to any other orthonormal bases of $V,\bar{V}$ respectively; this permits us to chop off $k(v'_1,\ldots,v'_n)$ , $k(v''_1,\ldots,v''_n)$ as orthogonal summands in $V$, $\bar{V}$ respectively: because of (7.2.1.1) equality of the associated matrices is thereby inherited by $E'$ and $E''$ . This proves (7.2.1.4).

7.2.2. The adjunction of further $v'_i$, $v''_i$ to $\mathfrak{B}'_{n,m}$, $\mathfrak{B}''_{n,m}$ . Let $w$ be the first member of our auxiliary basis $(w_i)_{i\in\mathbb{N}}$ of $V$ not contained in $k(v'_1,\ldots,v'_n)$ . $w = x + y$ , $y \in k(v'_1,\ldots,v'_n)$ , $x \in V'_0$ . We are going to enlarge $\mathfrak{B}'_{n,m}$ , $\mathfrak{B}''_{n,m}$ in such a way that $w \in \text{span}\,\mathfrak{B}'_{n',m'}$ .

Because of (7.2.1.4) we may apply 7.1 to the spaces (7.2.1.3): there are orthonormal bases $(w'_i)_{i\in\mathbb{N}}$ , $(w''_i)_{i\in\mathbb{N}}$ of $V'_o$, $V''_o$ respectively such that

$$\Phi(f'_\iota,w'_i) = \Phi(f''_\iota,w''_i) \;; \qquad 1 \le \iota \le m \;,\; i \in \mathbb{N} \,.$$

There is $s \in \mathbb{N}$ such that $x \in k(w'_1,\ldots,w'_s)$ . We now pass from $\mathfrak{B}'_{n,m}$, $\mathfrak{B}''_{n,m}$ to $\mathfrak{B}'_{n+s,m}$, $\mathfrak{B}''_{n+s,m}$ by setting $v'_{n+i} = w'_i$ , $v''_{n+i} = w''_i$ $(1\le i\le s)$ . In a later step we adjoin enough new $v''_i$ to $\mathfrak{B}''_{n,m}$ so that the extended family contains the first member $\bar{w}$ of $(\bar{w}_i)_{i\in\mathbb{N}}$ not contained in the span of $\mathfrak{B}''_{n,m}$ . Alternating between $\mathfrak{B}'_{n,m}$ and $\mathfrak{B}''_{n,m}$ is necessary to ensure that the growing bases $(v'_i)$, $(v''_i)$ will both exhaust $V$ and $\bar{V}$ respectively. Induction assumptions 7.2.1 are trivially preserved by this step. There remains

7.2.3. <u>The adjunction of further</u> $f', f''$ <u>to</u> $\mathfrak{B}'_{n,m}, \mathfrak{B}''_{n,m}$ . Let $B'_{n,m}$, $\mathfrak{B}''_{n,m}$ be as in 7.2.1. For purposes of calculations we complete $(v'_i)_{i\le n}$, $(v''_i)_{i\le n}$ to orthonormal bases $(v'_i)_{i\in\mathbb{N}}$, $(v''_i)_{i\in\mathbb{N}}$ of $V, \bar{V}$ respectively. We also complete the $f'_1,\ldots,f'_m$ and the $f''_1,\ldots,f''_m$ as indicated in 5.7. Setting $\alpha'_{\iota i} := \Phi(f'_\iota,v'_i)$ , $\alpha''_{\iota i} := \Phi(f''_\iota,v''_i)$ $(\iota\in J,$ $i\in\mathbb{N})$ we shall therefore have $\sum_{\mathbb{N}} \alpha'_{\iota i}\alpha'^{\tau}_{\kappa i} = A_{\iota\kappa}$, $\sum_{\mathbb{N}} \alpha''_{\iota i}\alpha''^{\tau}_{\kappa i} = \bar{A}_{\iota\kappa}$ . We then try to determine $N \in \mathbb{N}$ and $\xi_i \in k$ for $1 \le i \le n+N$ such that

(7.2.3.1) $$f^*_{m+1} := f'_{m+1} + \Sigma_1^{n+N} \xi_i v'_i$$

satisfies

(7.2.3.2) $$\|f^*_{m+1}\| = 1 \;,\quad \Phi(f^*_{m+1},f'_\iota) = 0 \qquad (1\le\iota\le m) \;,$$
$$\Phi(f^*_{m+1},v'_i) = \Phi(f''_{m+1},v''_i) \qquad (1\le i\le n)$$

or, equivalently,

(7.2.3.3) $$\Sigma_1^{n+N} \xi_i\xi_i^\tau + \Sigma_1^{n+N} \alpha'_{m+1,i}\xi_i^\tau + \Sigma_1^{n+N} \xi_i\alpha'^{\tau}_{m+1,i} = 0$$

(7.2.3.4) $$\Sigma_1^{n+N} \xi_i\alpha'^{\tau}_{\iota i} = 0 \qquad (1\le\iota\le m)$$

(7.2.3.5) $$\alpha'_{m+1,i} + \xi_i = \alpha''_{m+1,i} \qquad (1\le i\le n)$$

We set

(7.2.3.6) $\quad X_i = \xi_i + \alpha'_{m+1,i} \qquad (1 \leq i \leq n+N)$

We see that (7.2.3.5) settles the values of $X_i$ for $1 \leq i \leq n$. The other two conditions now read ($\alpha'_{\iota i} = \alpha''_{\iota i}$ $(1\leq i\leq n,\ 1\leq\iota\leq m)$ by (7.2.1.1)):

(7.2.3.7) $\quad \Sigma_{n+1}^{n+N} X_i X_i^{\tau} = \Sigma_1^{n+N} \alpha'_{m+1,i}\alpha'^{\tau}_{m+1,i} - \Sigma_1^{n} \alpha''_{m+1,i}\alpha''^{\tau}_{m+1,i}$

(7.2.3.8) $\quad \Sigma_{n+1}^{n+N} X_i \alpha'^{\tau}_{\iota i} = \Sigma_1^{n+N} \alpha'_{m+1,i}\alpha'^{\tau}_{\iota i} - \Sigma_1^{n} \alpha''_{m+1,i}\alpha''^{\tau}_{\iota i} \qquad (1\leq\iota\leq m)$ .

Or if we finally make use of the fact that $\sum_{i=1}^{\infty} \alpha'_{\iota i}\alpha'^{\tau}_{\kappa i} = A_{\iota\kappa} = \bar{A}_{\iota\kappa} = \sum_{i=1}^{\infty} \alpha''_{\iota i}\alpha''^{\tau}_{\kappa i}$

(7.2.3.7') $\quad \Sigma_{n+1}^{n+N} X_i X_i^{\tau} = \Sigma_{n+1}^{\infty}\alpha''_{m+1,i}\alpha''^{\tau}_{m+1,i} - \Sigma_{n+N+1}^{\infty}\alpha'_{m+1,i}\alpha'^{\tau}_{m+1,i}$

(7.2.3.8') $\quad \Sigma_{n+1}^{n+N} X_i \alpha'^{\tau}_{\iota i} = \Sigma_{n+1}^{\infty}\alpha''_{m+1,i}\alpha''^{\tau}_{\iota i} - \Sigma_{n+N+1}^{\infty}\alpha'_{m+1,i}\alpha'^{\tau}_{\iota i} \qquad (1\leq\iota\leq m)$

Call $L_{m+1,m+1}(N)$ the righthand side in (7.2.3.7') and $L_{m+1,\iota}(N)$ $(1\leq\iota\leq m)$ the righthand side in (7.2.3.8'), furthermore $L_{\nu\mu}(N) := \Sigma_{n+1}^{n+N}\alpha'_{\nu i}\alpha'^{\tau}_{\mu i}$ $(1\leq\nu,\mu\leq m)$ . From (7.2.3.7), (7.2.3.8) it is clear that $L(N)$ is a matrix over $k$ . In the hermitean space $(k^N;\langle,\rangle)$ we consider the vectors $X = (X_{n+1},\ldots,X_{n+N})$ , $Y_\iota = (\alpha'_{\iota,n+1},\ldots,\alpha'_{\iota,n+N})$ where $1\leq\iota\leq m$ . Hence our last two equations read

(7.2.3.7") $\quad \langle X,X\rangle = L_{m+1,m+1}(N)$

(7.2.3.8") $\quad \langle X,Y_\iota\rangle = L_{m+1,\iota}(N) \qquad (1 \leq \iota \leq m)$

For $N \to \infty$ the matrix $L(N)$ converges to the positiv definite matrix $(V^{\perp} = (0))$ $(\Sigma_{n+1}^{\infty}\alpha''_{\nu i}\alpha''^{\tau}_{\mu i})_{\nu,\mu\leq m+1}$ so $L(N)$ is positive definite for sufficiently large $N \in \mathbb{N}$ . Hence by assumption (4) and (1.5.1) we have $L_{\iota,\kappa}(N) = \langle Z_\iota, Z_\kappa\rangle$ $(1\leq\iota,\kappa\leq m+1)$ for $Z_\iota$ suitable vectors of $k^N$ $(N>1 + ms)$. Since $\langle Y_\iota, Y_\kappa\rangle = L_{\iota\kappa}(N)$ for $1 \leq \iota,\kappa \leq m$ we can by Witts theorem extend the isometry $\varphi : Z_\iota \mapsto Y_\iota$ $(1\leq\iota\leq m)$ to all of $k^N$ . Then $X := \varphi Z_{m+1} = (X_{n+1},\ldots,X_{n+N})$ solves our conditions (7.2.3.7") and (7.2.3.8"). From the $X_i$ $(n+1\leq i\leq N+n)$ we solve for the $\xi_i$ and thus find our vector $f^*_{m+1}$ of (7.2.3.1).

From $\mathfrak{B}'_{n,m}$, $\mathfrak{B}''_{n,m}$ we now pass to $\mathfrak{B}'_{n,m+1}, \mathfrak{B}''_{n,m+1}$ by adjoining $f^*_{m+1}$ and $f''_{m+1}$ respectively. All induction assumptions continue to hold. Shall we exhaust all of $E$ by iterating steps 7.2.2 and 7.2.3 alternately (and alternating in each step in turn between the roles of $V$ and $\bar{V}$)? The point to keep in mind is that we keep changing the bases $(f'_\iota)_{\iota\in J}$, $(f''_\iota)_{\iota\in J}$ in each application of step (7.2.3) (see the beginning of (7.2.3)) so that $f^*$ is - in each application - the match to a member $f''$ in a basis which is continually being changed (and of course the same if the roles of $V$ and $\bar{V}$ are interchanged). However, in each change of the basis $(f'_\iota)$, $(f''_\iota)$ the members are merely modulated modulo $V$, $\bar{V}$ respectively.Since we do exhaust $V$, $\bar{V}$ by repeated application of (7.2.2) we shall therefore also exhaust supplements of $V$, $\bar{V}$ in $E$ by repeated application of (7.2.3). This completes 7.2.

## 8. An important special case: $\perp$-dense hyperplanes

Embeddings $V \subset E$ with $\dim E/V = 1$ are remarkably simpler to deal with than those of arbitrary finite codimension. Therefore, and because hyperplanes are distinguished objects we state here a generalized version of our main theorem for this special case.

8.1. If $S = (s_i)_{i\in\mathbb{N}}$ is a sequence in an ordered set we let $A[S]$ be the set of all elements larger than all $s_i$. Thus an equality $A[S] = A[S']$ for two sequences $S,S'$ says that each $s_n$ is overtaken by a $s'_{n'}$, and vice-versa. If $A[S] = A[S']$ we also write $S \overset{\circ}{\sim} S'$. $\overset{\circ}{\sim}$ is an equivalence relation.

8.2. Given a sequence $X = (\xi_i)_{i\in\mathbb{N}}$ in the divisionring $(k,\tau)$ we associate with it the sequence $X^* = (-1+\sum_1^n \xi_i\xi_i^\tau)_{n\in\mathbb{N}}$ in the ordered field $k_o \subset k$. We define:

(8.2.1) If $X$ and $Y$ are sequences in $(k,\tau)$ we set $X \sim Y$ if and only if there exists $0 \neq \gamma \in k$ such that $A[X^*] = A[\gamma\gamma^\tau \cdot Y^*]$.

The relation $\sim$ is an equivalence relation and we let $[X]$ be the equivalence class of the sequence $X$ .

8.3. Let $V \subset E$ have $V^{\perp} = (0)$ and $\dim E/V = 1$ . A basis for the embedding will here be a basis $\mathfrak{B}$ satisfying 1.7.1 and 1.7.2, $\mathfrak{B} = (v_i)_{i\in\mathbb{N}} \cup \{f\}$ . As in 1.7 we set

$$(8.3.1) \qquad \alpha_i = \Phi(f,v_i) \quad , \quad i \in \mathbb{N} \quad ,$$

and furthermore we associate with the basis $\mathfrak{B}$ the sequence

$$(8.3.2) \qquad X = (\alpha_i)_{i\in\mathbb{N}} \; .$$

It is trivial to say that $X$ determines $V \subset E$ to within an automorphism of $E$ . We shall prove that $[X]$ is an orthogonal invariant which characterizes $V \subset E$ up to automorphisms of $E$ . By our main theorem this is evident if $\mathfrak{B}$ is a standard basis for then "$X^* \overset{\circ}{\sim} \gamma\gamma^{\tau} Y^*$" is obviously equivalent with "$\lim X^* = \gamma\gamma^{\tau} \lim Y^*$". However $\mathfrak{B}$ need not be a standard basis.

Our theorem now reads

8.4. Theorem. Let $V,\bar{V}$ be $\perp$-dense hyperplanes in $(E,\Phi)$. If there exists a metric automorphism $T$ of $(E,\Phi)$ with $TV = \bar{V}$ then $[X] = [\bar{X}]$ for all bases of the embedding. Conversely, if $[X] = [\bar{X}]$ for at least some bases $\mathfrak{B},\bar{\mathfrak{B}}$ and if (4) holds then there exists a metric automorphism $T$ of $E$ with $TV = \bar{V}$ .

8.5. Proof of 8.4. The first half of the theorem is a routine verification (One makes use of 1.6). Assume then that $[X] = [\bar{X}]$ with respect to bases $\mathfrak{B} = (v_i)_{\mathbb{N}} \cup (f)$ , $\bar{\mathfrak{B}} = (\bar{v}_i)_{\mathbb{N}} \cup (\bar{f})$ . We first remark that a basis $\mathfrak{B}'$ can be introduced such that $A[X'^*] = A[\bar{X}^*]$ : By assumption $A[\bar{X}^*] = A[\gamma\gamma^{\tau}X^*]$ for suitable $\gamma \neq 0$ . Now if $\gamma\gamma^{\tau} \neq 1$ we may assume that $\gamma\gamma^{\tau} < 1$ (for otherwise we interchange the roles of $X$ and $\bar{X}$ since $A[X^*] = A[\frac{1}{\gamma\gamma^{\tau}} \cdot \bar{X}^*]$). By (1.5) and (4) there are $\lambda_1,\dots,\lambda_m \in k$ such that $\Sigma_1^m \lambda_i \lambda_i^{\tau} = 1-\gamma\gamma^{\tau}$ . In $V \cap f^{\perp}$ there are $m$

orthonormal vectors $v_1',\dots,v_m'$ . If we set $f' = \gamma f + \sum_1^m \lambda_i v_i'$ we have $\|f'\| = 1$ . $\{v_1',\dots,v_m'\}$ can be completed to an orthonormal basis $(v_i')_{\mathbb{N}}$ of $V$ . With respect to the basis $\mathfrak{B}' = (v_i')_{\mathbb{N}} \cup (f')$ we now have

$$(8.5.1) \qquad (\sum_1^n \bar{\alpha}_i \bar{\alpha}_i^\tau)_n \approx (\sum_1^n \alpha_i' \alpha_i'^\tau)_n \, .$$

We claim that bases can be introduced such that $\alpha_i = \bar{\alpha}_i$ for all $i \in \mathbb{N}$ . The program is the same as in 7.1 and it is only 7.1.1 which needs to be considered. The solution of problem 7.1.1 is much simpler here as we do not need lemma 6.3. The given vector $\bar{x}$ in 7.1.1 can be completed to an orthonormal basis which we call again $(\bar{v}_i)_{i\in\mathbb{N}}$ ; (8.5.1) will be inherited (see (1.6)). Since $V^\perp = (0)$ there is an infinity of $\alpha_i' \neq 0$ so that we find $N \in \mathbb{N}$ with

$$S_N := \sum_1^N \alpha_i' \alpha_i'^\tau > \bar{\alpha}_1 \bar{\alpha}_1^\tau \qquad (\bar{v}_1 = \bar{x}) \, .$$

The vector $x := S_N^{-1} \cdot \bar{\alpha}_1^\tau \cdot \Sigma_{i=1}^N \alpha_i' v_i'$ satisfies $\|x\| < 1$ and $\Phi(f',x) = \Phi(\bar{f},\bar{x})$ . This terminates the proof of 8.4.

8.5.2. Remark. There is, in principle, no obstacle to treat the case of arbitrary $V \subset E$ in the same fashion as we have done with hyperplanes. For practical purposes, however, the emerging complications would be prohibitive.

## Appendix I. An interpretation of the invariant $\psi$ in Section 3

In this section we shall interpret $\psi$ in a natural fashion as a hermitean form on the space $\bar{k}_o \otimes (E,\Phi)$ .

1. The set $H := \{(\xi_i)_{i\in\mathbb{N}} \mid \xi_i \in \bar{k}_o \otimes k$ and $\sum_1^\infty \xi_i \xi_i^\tau$ konvergent$\}$ is a $\bar{k}_o \otimes k$-vector space with an obvious hermitean form $\langle (\xi_i)_{\mathbb{N}}, (\eta_i)_{\mathbb{N}} \rangle = \sum_{\mathbb{N}} \xi_i \eta_i^\tau$ (recall that for the fields $k$ of chap. XII we do have the Schwarz inequality $\langle x,y\rangle\langle x,y\rangle^\tau \leq \langle x,x\rangle\langle y,y\rangle$ for all $x,y \in H$). $H$ is complete under the topology $\mu$ induced by the positive definite form $\langle\, ,\rangle$ (Notice that $\langle x,x\rangle^{1/2}$ need not be an element of $\bar{k}_o \otimes k$ . If, however, $k_o$

happens to be dense in its real closure then by the Hauschild characterization of these fields ([5] Satz 13) the completion $\tilde{k}_o$ by means of Cauchy systems is real closed; furthermore in the situation of interest to us $k_o$ will contain (2.5) a non eventually constant denumerable Cauchy sequence which implies that $\tilde{k}_o = \bar{k}_o$. Thus in these cases $H$ will be a complete normed vector space over $\bar{k}_o \otimes k$ with norm $\langle x,x\rangle^{1/2} \in \bar{k}_o$ for all $x \in H$).

2. Let $V \subset (E,\Phi)$ with $V^\perp = (0)$ be an embedding which admits a standard basis $\mathfrak{B} = (v_i)_{i\in\mathbb{N}} \cup (f_\iota)_{\iota\in J}$. We fix some orthonormal basis $(e_i)_{i\in\mathbb{N}}$ of $E$ and indentify the k-space $(E,\Phi)$ with a subset of the $\bar{k}_o\otimes$ k-space $(H,\langle,\rangle)$ by letting $x = \sum_{\mathbb{N}}\xi_i e_i \in E$ ($\xi_i \neq 0$ for finitely many $i$ only) correspond to $(1 \otimes \xi_i)_{i\in\mathbb{N}} \in H$. With no risk of confusion we shall sometimes not distinguish notationally between $E$ and $1 \otimes E$ in $H$.

Now for $(z_i)_{i\in\mathbb{N}}$ any orthogonal set in $H$ the sum $\sum_1^\infty z_i$ will converge in $H$ if and only if $\sum_1^\infty \langle z_i,z_i\rangle$ converges (for, if $n > m$ we see by $\|\sum_1^n z_i - \sum_1^m z_i\| = \|\sum_{m+1}^n z_i\| = \sum_{m+1}^n \langle z_i,z_i\rangle = \sum_1^n\langle z_i,z_i\rangle - \sum_1^m\langle z_i,z_i\rangle$ that $(\sum_1^n z_i)_n$ is Cauchy iff $(\sum_1^n\langle z_i,z_i\rangle)_n$ is Cauchy). Hence $H$ is precisely the completion of $E$. We may apply this argument furthermore to the orthogonal set $(v_i)_{i\in\mathbb{N}}$ and have (since $\mathfrak{B}$ is a standard basis)

(A1) for $x \in E$ the sum $\sum_{\mathbb{N}}\Phi(x,v_i)v_i$ converges to an element $x^* \in \tilde{V} \subset H$ ($\tilde{V}$ the completion of $V$ in $H$). We have

(A2) $x^* - 1 \otimes x \perp V$.

Proof. For arbitrarily fixed $j \in \mathbb{N}$ and $j < n$ we set $x_n = \sum_1^n\phi(x,v_i)\cdot v_i$ and have $\langle x^*-x,v_j\rangle = \langle x_n-x,v_j\rangle + \langle x^*-x_n,v_j\rangle = 0 + \langle x^*-x_n,v_j\rangle$ so, by the Schwarz inequality, $\langle x^*-x_n,v_j\rangle\langle x^*-x_n,v_j\rangle^\tau \leq \|x^*-x_n\|\cdot 1$. Since $\|x^*-x_n\| \to 0$ for $n \to \infty$ the assertion is now evident.

3. By (A1), (A2) we now have

(A3) $E \subset \tilde{V} \oplus \tilde{V}^\perp \subset H$.

If $k_o$ is archimedean ordered then we have of course $\tilde{V} \oplus \tilde{V}^{\perp} = H$ by standard arguments (For each non-void finite orthonormal set $D$ in $H$ one invariably has $\sum_{z \in D} \langle x,z\rangle\langle x,z\rangle^{\tau} \leq \|x\|$ for all $x \in H$. Hence if $k_o$ is archimedean we may call for sup and see that $\sum_{B} \langle x,z\rangle\langle x,z\rangle^{\tau}$ exists for $B$ an arbitrary orthonormal set in $H$).

We set

$$\text{(A4)} \qquad E_V := \tilde{V} \oplus \tilde{V}^{\perp}$$

and let $\pi_{\tilde{V}}$, $\pi_{\tilde{V}^{\perp}}$ be the projections corresponding to this decomposition.

4. The invariant $\psi$ qua hermitean form. Besides the form $\langle,\rangle$ we may introduce the hermitean form $\chi$ on $E_V$ by setting $\chi(x,y) = \sum_{\mathbb{N}} \langle x,v_i\rangle\langle y,v_i\rangle^{\tau}$ $(x,y \in E_V)$. Obviously $\chi(x,y) = \langle \pi_{\tilde{V}}x, \pi_{\tilde{V}}y\rangle$. Another form may be introduced on $\bar{k}_o \otimes E$ by setting

$$\text{(A5)} \qquad \Psi(f_\iota,f_\kappa) = \psi_{\iota\kappa}\ ,\ \Psi(f_\iota,v_i) = \Psi(v_j,v_i) = 0 \quad (\iota,\kappa \in J;\ i,j \in \mathbb{N})\ .$$

This gives the intended interpretation of the invariant $\Psi = \{\psi_{\iota\kappa}\}$ of 3.2.1:

$$\text{(A6)} \qquad \text{for } x,y \in E \text{ we have } \Psi(1\otimes x, 1\otimes y) = -\langle \pi_{\tilde{V}^{\perp}}x,\ \pi_{\tilde{V}^{\perp}}y\rangle\ .$$

Restricting all forms $\langle,\rangle$, $\chi,\Psi$ to $E = 1 \otimes E$ yields

$$\text{(A7)} \qquad \Phi(x,y) = \chi(x,y) - \Psi(x,y) \qquad (x,y \in E)\ .$$

If $\Psi$ vanishes on $E$ then we have Parseval's identity: $\Phi(x,y) = \chi(x,y)$.

For later reference we state explicitely

$$\text{(A8)} \qquad \psi_{\iota,\kappa} = 0 \quad (\iota,\kappa \in J) \text{ if and only if } \tilde{V}^{\perp} = (0)\ ,\ \text{i.e. } E_V = \tilde{V}\ .$$

and prove the following

5. Lemma. Let $V = W \oplus W^{\perp}$ ($W$ some subspace of $(E,\Phi)$) be such that $V \subset E$ admits a standard basis. For $e \in E \setminus V$ set $W_1 = W \oplus k(e)$ and $V_1 = W_1 \oplus W_1^{\perp}$. If the embedding $V \subset E$ satisfies one of the equivalent properties in (A8) then so does $V_1 \subset E$. In particular, $V_1 \subset E$ admits a standard basis.

Proof. Case 1: $e \notin W^{\perp\perp}$. Let $w \in W^{\perp} \setminus W_1^{\perp}$ so $W^{\perp} = W_1^{\perp} \oplus k(w)$. On one hand we have $1 \otimes e \in \tilde{V} = \widetilde{W + W^{\perp}} = \tilde{W} + \widetilde{W^{\perp}} = \tilde{W} \oplus \widetilde{W_1^{\perp}} \oplus \bar{k}_o \otimes kw$ (the 2nd equality because every subsequence of a Cauchy-sequence $(\sum_1^n \xi_i \xi_i^{\tau})_n$ is Cauchy); on the other hand we have $1 \otimes e \notin \tilde{W} + \widetilde{W_1^{\perp}}$ (for otherwise $1 \otimes e \in \tilde{W}$ since $e \perp W_1^{\perp}$, which gives $e \perp W^{\perp}$ contradicting $e \notin W^{\perp\perp}$). Therefore $1 \otimes w \in \tilde{W} \oplus \widetilde{W_1^{\perp}} \oplus \bar{k}_o \otimes ke = \tilde{W}_1 \oplus \widetilde{W_1^{\perp}}$. Ergo $\tilde{W} \oplus \widetilde{W^{\perp}} \subset \tilde{W}_1 \oplus \widetilde{W_1^{\perp}}$ i.e. $\tilde{V} \subset \tilde{V}_1$. Hence $\tilde{V}_1^{\perp} \subset \tilde{V}^{\perp} = (0)$. From $E \subset \tilde{V}_1$ one concludes that $V_1 \subset E$ admits a standard basis. Case 2: $e \in W^{\perp\perp}$ is trivial since then $W + W^{\perp} \subset W_1 + W_1^{\perp}$.

## 6. A characterization of the special case $\{\psi_{\iota\kappa}\} = \{0\}$

The situation in which the semidefinite matrix $\psi_{\iota\kappa} = A_{\iota\kappa} - \mathbb{1}$ is the zero matrix for a $\perp$-dense embedding $V \subset E$ can be characterized by a locally convex topology associated with the embedding $V \subset E$.

If $V \subset E$ is $\perp$-dense then $\Phi$ sponsors a non degenerate dual pairing and a weak topology $\sigma$ which is hausdorff: Let here $\sigma = \sigma(E,V)$ be the coarsest locally convex topology which renders all linear forms $f_v: x \mapsto \Phi(x,v)$ $(v \in V)$ continuous (1.3); a subbasis of closed neighbourhoods is formed by the sets $\{x \in E | \Phi(x,v)\Phi(x,v)^{\tau} \leq 1\}$ where $v$ ranges in $V$. Next to this topology we have on $E$ the topology $\mu$ induced by the positive definite form $\Phi$; $\mu \geq \sigma$.

Theorem. Let $V \subset E$ with $V^{\perp} = (0)$ be an embedding which admits a standard basis. Then the following statements are equivalent.
(i) $\psi_{\iota\kappa} = 0$ for all $\iota, \kappa \in J$; (ii) $V$ is dense in $(E,\mu)$; (iii) $\mu$-Cauchy sequences in $E$ which converge weakly to $0$ converge to $0$.

This theorem is of particular interest in the light of theorem 2.6 (in section 2) which says that one can have "total" orthonormal families with $\psi_{\iota\kappa}$ arbitrarily prescribed. Only for $\psi \equiv 0$ is such a family topologically dense!

Proof. (iii) $\Rightarrow$ (ii). We approximate $x \in E$. Let $(v_i)_{i \in \mathbb{N}}$ be an orthonormal basis of $V$. Set $x_j = x - \Sigma_1^j \Phi(x,v_n)v_n$ $(j \in \mathbb{N})$. $(x_j)_{\mathbb{N}}$ converges weakly to $0$, i.e. we have

(A9) for all $n \in \mathbb{N}$: $\lim\limits_{j \to \infty} \Phi(x_j, v_n) = 0$.

Since there is a standard basis the sequence $(\Sigma_1^j \Phi(x,v_n)\Phi(x,v_n)^\tau)_{j \in \mathbb{N}}$ is Cauchy and this is the same as to say that $(x_j)_{j \in \mathbb{N}}$ is Cauchy, $(x_j)_{\mathbb{N}}$ thus qualifies for (iii) and converges to $0$: $x = \Sigma_{\mathbb{N}} \Phi(x,v_n)v_n$ which means that $V$ is dense in $E$ as $x$ was arbitrary.

(ii $\Rightarrow$ (i). From (ii) we obtain $x = \Sigma_{\mathbb{N}} \Phi(x,v_n)v_n$ and then $\|x\| = \Sigma_{\mathbb{N}} \Phi(x,v_n)\Phi(x,v_n)^\tau$ by standard arguments. Thus $\Phi(x,x) = \chi(x,x)$; in other words $\Psi(x,x) = 0$ for all $x \in E$ by (A8). This means $\psi_{\iota\kappa} = 0$ for all $\iota, \kappa \in J$.

(i) $\Rightarrow$ (iii). Let be given an $\varepsilon > 0$ and $(x_j)_{\mathbb{N}}$ a $\mu$-Cauchy sequence which satisfies (A9). We have to show that $\|x_j\| \to 0$ for $j \to \infty$. Since we assume (i) we have by (A7) Parseval's identity in $E$, particularly

$$\|x_j - \Sigma_1^r \Phi(x_j,v_n)v_n\| = \Sigma_{r+1}^\infty \Phi(x_j,v_n)\Phi(x_j,v_n)^\tau .$$

We write $x_j = (x_j - x_N) + x_N$; since $(x_j)_{\mathbb{N}}$ is Cauchy we may pick $N$ so large that $\|x_j - x_N\| < \varepsilon^2$ (all $j > N$). We may furthermore choose $r$ so large that $\Sigma_{r+1}^\infty \Phi(x_N,v_n)\Phi(x_N,v_n)^\tau < \varepsilon^2$. We then have $\|x_j - \Sigma_1^r \Phi(x_j,v_n)v_n\| \leq 4\varepsilon^2$. Since $(x_j)_{\mathbb{N}}$ satisfies (A9) by assumption we may finally choose $j$ so large that $\|\Sigma_1^r \Phi(x_j,v_n)v_n\| < \varepsilon^2$. Then $\|x_j\| < 9\varepsilon^2$. This shows that $\|x_j\| \to 0$.

## Appendix II. The proof of a theorem in Section 2

We shall prove a theorem of Sec. 2 which we set down here once more:

Theorem. Assume that $\bar{Q} = \bar{k}_0^+$ (the nonnegative elements of $\bar{k}_0$) and that $k_0$ possesses a noneventually constant denumerable Cauchy sequence. Let $A$ be any positive definite $J \times J$ matrix over $\bar{k}_0 \otimes k$, card $J \leq \aleph_0$, with $A - \mathbb{1}$ negative semidifinite. Then there is a positive definite hermitean k-space $(E,\Psi)$ which contains a standard basis $\mathfrak{B} = (v_i)_{\mathbb{N}} \cup$ $\cup (f_\iota)_J$ with $\perp$-dense span of the $(v_i)_{\mathbb{N}}$ and with $A$ as its associate matrix.

Remarks. The assumption on the existence of at least one non eventually constant Cauchy sequence is clearly necessary; it makes sure that the elements of $k_o$, $\bar{k}_o$ can be written as denumerably infinite sums without zero terms.

The following situation will often occur: We are given a convergen series $t = \Sigma_1^\infty t_i$ in $\bar{k}_o \otimes k$ and a sequence of $\delta_i > 0$ in $k_o$ and we should represent $t$ as a sum over $k$, $t = \Sigma_1^\infty r_i$, $r_i \in k$, $|r_i - t_i| < \delta$ This is indeed easy: Pick some fixed sequence $(\varepsilon_i)_{i\in\mathbb{N}}$ with $\varepsilon_i \to 0$, $0 < \varepsilon_i \in k_o$. Choose $r_1 \in k$ with $|t_1 - r_1| < \min(\varepsilon_1, \delta_1, \delta_2)$. Assume we have already chosen $r_1, \ldots, r_n$ such that $|t_i - r_i| < \delta_i$ and $|\Sigma_1^n t_1 - \Sigma_1^n r_i| < \min(\varepsilon_n, \delta_{n+1})$. $t_{n+1} + (\Sigma_1^n t_i - \Sigma_1^n r_i)$ lies in the $\delta_{n+1}$-disk of $t_{n+1}$ and we may pick $r_{n+1}$ such that $|t_{n+1} - r_{n+1}| < \delta_{n+1}$ and $|\Sigma_1^{n+1} t_i - \Sigma_1^{n+1} r_i|$ is arbitrarily small, in particular $< \min(\varepsilon_{n+1}, \delta_{n+2}$ The step may now be continued. $\sum\limits_1^\infty r_i$ converges with $t$ as its limit.

Proof of the theorem

1. We shall show how to pick sequences $\alpha_i = (\alpha_{i\sigma})_{\sigma\in\mathbb{N}}$, $\alpha_{i\sigma} \in k$ such that all series $\sum\limits_{\sigma\in\mathbb{N}} \alpha_{\iota\sigma} \alpha_{\kappa\sigma}^\tau$ converge and equal $A_{\iota\kappa}$. We shall do this in such a fashion that for any $n, p \in \mathbb{N}$ the "tails" of the first $n$ sequences $\alpha_1, \ldots, \alpha_n$

$$\text{(B1)} \quad \left\{ \begin{array}{l} (\alpha_{1,p+1}, \alpha_{1p+2}, \ldots), \\ (\alpha_{2,p+1}, \ldots) \qquad , \\ (\alpha_{n,p+1}, \ldots) \qquad , \end{array} \right.$$

are linearly independent. This guarantees that we are going to have the situation of (2.4.3); in view of (2.4) our proof will be complete if we can exhibit such sequences $\alpha_i$ for $i \in J$. It will be a step by step construction so we start by discussing the

Induction assumptions. Assume that we have already found sequences $\alpha_1,\ldots,\alpha_n$ such that $\sum_{\sigma\in\mathbb{N}} \alpha_{\iota\sigma}\alpha^{\tau}_{\kappa\sigma} = A_{\iota\kappa}$ $(\iota,\kappa \leq n)$, $\alpha_{\iota\sigma} \in k$ and such that all tails (B1) are linearly independent.

We put down as a purely technical device the further induction assumption that for an infinity of $\sigma \in \mathbb{N}$ all $\alpha_1,\ldots,\alpha_n$ have their $\alpha_{i\sigma} = 0$.

The subsequent proof takes also care of the case when $\alpha_{n+1}$ is the first sequence to be picked, i.e. where no sequences $\alpha_1,\ldots,\alpha_n$ have been picked yet.

$\alpha_{n+1}$ will be found in three steps. First step: We pick a sequence $\bar{\alpha}_{n+1}$ which behaves as it should according to (2.4) except that it has its coordinates $\bar{\alpha}_{n+1\sigma}$ in $\bar{k}_o \otimes k$. Actually we shall require that we have

(B2) $\qquad \Sigma\bar{\alpha}_{n+1\sigma}\alpha^{\tau}_{i\sigma} = A_{n+1,i} \quad (i=1,\ldots n)$

(B3) $\qquad \Sigma\bar{\alpha}_{n+1\sigma}\bar{\alpha}^{\tau}_{n+1\sigma} < A_{n+1,n+1}$ .

Second step: We approximate $\bar{\alpha}_{n+1}$ by a sequence $\bar{\bar{\alpha}}_{n+1}$ which still behaves correctly and which now has its coordinates $\bar{\bar{\alpha}}_{n+1\sigma}$ in $k$. Third step : To get the desired $\alpha_{n+1}$ we add a correction to $\bar{\bar{\alpha}}_{n+1}$ in such a manner that (B2) is saved and such that $\Sigma\alpha_{n+1\sigma}\alpha^{\tau}_{n+1\sigma} = A_{n+1,n+1}$.

Remark. Steps 2 and 3 cannot be accomplished in one sweep: when approximating $\bar{\alpha}_{n+1}$ by $\bar{\bar{\alpha}}_{n+1}$ one can save the linear (B 2) but not at the same time equality in the "quadratic" (B 3) ; since $\bar{\bar{\alpha}}_{n+1}$ is close to $\bar{\alpha}_{n+1}$ it is easy to save "<" in (B 3).

Step 1: choice of $\bar{\alpha}_{n+1}$. Let $A$ be the positive definite matrix $(A_{\iota\kappa})_{1\leq\iota,\kappa\leq n+1}$. There is a matrix

$$C = \begin{pmatrix} \boxed{\mathbb{1}_n} & \begin{matrix}\lambda_1\\ \vdots \\ \lambda_n\end{matrix} \\ 0 \dots 0 & 1 \end{pmatrix}$$ such that $C^{tr}AC^{\tau}$ is of the form

$$\begin{pmatrix} \boxed{A_{ik}} & \begin{matrix}0\\ \vdots \\ 0\end{matrix} \\ 0 \dots 0 & A'_{n+1,n+1} \end{pmatrix} .$$

$A'_{n+1,n+1}$ is in $\bar{k}_o$ and positive. Represent $A'_{n+1,n+1}$ as the limit of a convergent sum $\sum_{\sigma\in\mathbb{N}} \gamma_\sigma\gamma_\sigma^\tau$ with $\gamma_\sigma \neq 0$ . Form a sequence $\alpha'_{n+1} = (\alpha'_{n+1\sigma})_{\sigma\in\mathbb{N}}$ in the following way: At the infinitely many $\sigma \in \mathbb{N}$ where $\alpha_1,\dots,\alpha_n$ have their coordinates zero, we insert the $\gamma_\sigma$ but take care to leave an infinity of these $\alpha'_{n+1}$ to be zero so as to save the induction assumption. For all the other $\sigma$ we set $\alpha'_{n+1\sigma} = 0$ . We have $\langle\alpha'_{n+1},\alpha_i\rangle = 0$ $(i=1,\dots,n)$ and $\langle\alpha'_{n+1},\alpha'_{n+1}\rangle = A'_{n+1,n+1}$ . The sequence

$$\bar{\alpha}_{n+1} = \frac{1}{2}\,\alpha'_{n+1} - \lambda_1\alpha_1 - \dots - \lambda_n\alpha_n$$

satisfies $\langle\bar{\alpha}_{n+1},\alpha_i\rangle = A_{n+1i}$ $(i=1,\dots,n)$ and $\langle\bar{\alpha}_{n+1},\bar{\alpha}_{n+1}\rangle = A_{n+1,n+1} - \frac{3}{4}A'_{n+1,n+1} > 0$ . In other words $\bar{\alpha}_{n+1}$ satisfies (B 2) and (B 3) . Before going over to step 2 we need some preparations.

2. We may pick positive $\delta,\ \delta_1 \in k_o$ with $\langle\bar{\alpha}_{n+1},\bar{\alpha}_{n+1}\rangle \equiv \|\bar{\alpha}_{n+1}\| \leq \delta_1^2$ and $\delta^2 + 2\delta\delta_1 < A_{n+1,n+1} - \|\bar{\alpha}_{n+1}\|$ . We choose now a convergent series $\sum_{\sigma\in\mathbb{N}} \delta^{(\sigma)} = \delta^2$ with $0 < \delta^{(\sigma)} \in k_o$ .

Pick a sequence of natural numbers $N(1) < N(2) < \dots$ such that

$$\{(\alpha_{11},\dots,\alpha_{1N(1)}),\ \dots\ ,(\alpha_{n1},\dots,\alpha_{nN(1)})\}$$
$$\{(\alpha_{1N(1)+1},\dots,\alpha_{1N(2)}),\ \dots\ ,(\alpha_{nN(1)+1},\dots,\alpha_{nN(2)})\}$$
$$\vdots \qquad\qquad\qquad \vdots$$
$$\{(\alpha_{1N(\sigma)+1},\dots,\alpha_{1N(\sigma+1)}),\ \dots\ ,(\alpha_{nN(\sigma)+1},\dots,\alpha_{nN(\sigma+1)})\}$$

are all linearly independent sets of tuples. Notation: $\alpha_m^{(\sigma)} := (\alpha_{mN(\sigma-1)+1},\dots,\alpha_{mN(\sigma)})$ . Accordingly we divide up the sequence $\bar{\alpha}_{n+1}$ into adjacent tuples $\bar{\alpha}_{n+1}^{(\sigma)}$ of lengths $N(1),\ N(2)-N(1),\ N(3)-N(2),\dots$

There is a uniquely determined linear isomorphism

$\Delta: k^n \to k(\alpha_1^{(\sigma)},\ldots,\alpha_n^{(\sigma)})$ which assigns to an n-tuple $\varepsilon^{(\sigma)} = (\varepsilon_1^{(\sigma)},\ldots,\varepsilon_n^{(\sigma)})$ a vector $\alpha^{(\sigma)}$ such that

(B4) $$<\alpha^{(\sigma)},\alpha_i^{(\sigma)}> = \varepsilon_i^{(\sigma)} \qquad i = 1,\ldots,n \quad .$$

We keep the notation "$\Delta$" for the extension ($\bar{k}_o$-ification) of $\Delta$ to $(\bar{k}_o \otimes k)^n \to \bar{k}_o \otimes k(\alpha_1^{(\sigma)},\ldots,\alpha_n^{(\sigma)})$ . We shall make our choice for the $\varepsilon^{(\sigma)}$ .

Let $\bar{\varepsilon}_i^{(\sigma)} = <\bar{\alpha}_{n+1}^{(\sigma)},\alpha_i^{(\sigma)}>$; this can also be expressed by saying $\Delta\bar{\varepsilon}^{(\sigma)} = \bar{\alpha}_{n+1}^{(\sigma)}$ . "$\sum_{\sigma\in\mathbb{N}} \bar{\varepsilon}_i^{(\sigma)} = A_{n+1i}$ , (i=1,...n)" is a reformulation of (B 2) . $\Delta$ is continuous (1.3); there is a $\mu^{(\sigma)} > 0$ in $k_o$ such that $\|\alpha^{(\sigma)} - \bar{\alpha}_{n+1}^{(\sigma)}\| < \delta^{(\sigma)}$ when $\|\varepsilon^{(\sigma)} - \bar{\varepsilon}^{(\sigma)}\| < \mu^{(\sigma)}$ .

Notice that by the convergence of $\sum_{\sigma\in\mathbb{N}} \|\bar{\alpha}_{n+1}^{(\sigma)}\|$ (B 3) and the convergence of $\sum_{\sigma\in\mathbb{N}} \delta^{(\sigma)}$ the sum $\sum_{\sigma\in\mathbb{N}} \|\alpha^{(\sigma)}\|$ will be convergent (a triviality if $k_o$ is archimedean).

<u>Step 2: choice of</u> $\bar{\bar{\alpha}}_{n+1}$. Choose $n$ convergent series $\sum_{\sigma\in\mathbb{N}} \varepsilon_1^{(\sigma)},\ldots$ $\ldots,\sum_{\sigma\in\mathbb{N}} \varepsilon_n^{(\sigma)}$ in $k$ with limits $A_{n+1,1},\ldots,A_{n+1,n}$ such that $\varepsilon^{(\sigma)} := (\varepsilon_1^{(\sigma)},\ldots,\varepsilon_n^{(\sigma)})$ satisfies $\|\varepsilon^{(\sigma)} - \bar{\varepsilon}^{(\sigma)}\| < \mu^{(\sigma)}$ . These are our choices for the $\varepsilon_i^{(\sigma)}$ in (B4). We define $\bar{\bar{\alpha}}_{n+1}$ by juxtaposition of the $\alpha^{(\sigma)}$ which solve (B4). We have $<\bar{\bar{\alpha}}_{n+1}\alpha_i> = \sum_{\sigma\in\mathbb{N}} <\alpha_{n+1}^{(\sigma)},\alpha_i^{(\sigma)}> = \sum_{\sigma\in\mathbb{N}} \varepsilon_i^{(\sigma)} = A_{n+1,i}$ (i=1,...,n) ; $\|\bar{\bar{\alpha}}_{n+1} - \bar{\alpha}_{n+1}\| = \sum_{\sigma\in\mathbb{N}} \|\alpha_{n+1}^{(\sigma)} - \bar{\alpha}_{n+1}^{(\sigma)}\| \le \sum_{\sigma\in\mathbb{N}} \delta^{(\sigma)} = \delta^2$ . Expressing $\bar{\bar{\alpha}}_{n+1}$ as $\bar{\alpha}_{n+1} + (\bar{\bar{\alpha}}_{n+1} - \bar{\alpha}_{n+1})$ we find (by making use of the Schwartz inequality) $\|\bar{\bar{\alpha}}_{n+1}\| < \|\bar{\alpha}_{n+1}\| + \delta^2 + 2\delta\delta_1 < A_{n+1,n+1}$ . So $\bar{\bar{\alpha}}_{n+1}$ satisfies (B2), (B3) in place of $\bar{\alpha}_{n+1}$ (and has zeros at the right places $\sigma \in \mathbb{N}$).

<u>Step 3: Choice of</u> $\alpha_{n+1}$ . There remains to adjust the "length" $\|\bar{\bar{\alpha}}_{n+1}\|$ . Choose a convergent series $\sum_{i\in\mathbb{N}} \kappa_i = A_{n+1,n+1} - \|\bar{\bar{\alpha}}_{n+1}\| > 0$ in $k_o$ with $\kappa_i$ positive and in $Q \cap k_o$ ($\bar{Q} = \bar{k}_o^{(+)}$ by the assumption of the theorem). Then there is a $n_i \in \mathbb{N}$ such $k^{n_i}$ contains a $n_i$-tuple $\beta_i$ with $\|\beta_i\| = \kappa_i$ .

We fill into the "empty holes" of $\overset{=}{\alpha}_{n+1}$ the components of these vectors $\beta_i$ but take care to leave an infinity of zeros. Call $\alpha_{n+1}$ the new sequence thus obtained from $\overset{=}{\alpha}_{n+1}$. $\|\alpha_{n+1}\|$ exists, i.e. the infinite sum is convergent, and equals $\|\overset{=}{\alpha}_{n+1}\| + \sum_{\mathbb{N}}\kappa_i = A_{n+1,n+1}$; $\langle\alpha_{n+1},\alpha_i\rangle = \langle\overset{=}{\alpha}_{n+1},\alpha_i\rangle = A_{n+1,i}$ $(i=1,..,n)$. It is evident that the "tails" of $\alpha_{n+1}$ are independent of those in (B1). This terminates the proof of the theorem stated at the beginning of the Appendix.

*

9. Standard bases for arbitrary subspaces (Definitions and existence)

In this short section we shall introduce the concept of a standard basis for an embedding $V \subset (E,\Phi)$ where $V$ is an arbitrary subspace of $E$. It will enable us to discuss the embedding entirely in terms of $\perp$-dense embeddings; the results of sections 1-7 will make it possible to give a complete set of invariants for the orbit of $V$ under the orthogonal group of $E$. In this connection see the postscript on page 327.

If $\dim V^{\perp} < \infty$ we shall have $E = V^{\perp\perp} \oplus V^{\perp}$ and the embedding $V \subset E$ is completely described by the $\perp$-dense embedding $V \subset V^{\perp\perp}$ and by the isometry class of $V^{\perp}$. Thus in order to exclude banalities in the sequel we shall put down the general assumption

(9.0) $$\dim V^{\perp} \in \{0,\aleph_0\} \quad .$$

Standard bases for $V \subset E$ (Definition). Let $(E,\Phi)$ be a $\aleph_0$-dimensional positive definite k-space where $k$ is as indicated under the caption of chap. XII. Let $V \subset E$ be a subspace satisfying (9.0). By a standard basis for the embedding $V \subset E$ we mean a Hamel basis

(9.1) $$\mathfrak{B} = (v_i)_{i\in\mathbb{N}} \cup (w_i)_{i\in\mathbb{N}} \cup (f_\iota)_{\iota\in J} \cup (e_\iota)_{\iota\in I}$$

of the space $E$ such that we have

(9.2) $(v_i)_{\mathbb{N}}$ is an orthonormal basis of $V$

(9.3) $(w_i)_{\mathbb{N}}$ is an orthonormal basis of $V^{\perp}$ if $V^{\perp} \neq (0)$ (and $\emptyset$ otherwise)

(9.4) $(f_\iota)_J$ is an orthonormal basis of some supplement of $V$ in $V^{\perp\perp}$ if $V \neq V^{\perp\perp}$ (and $\emptyset$ otherwise)

(9.5) $(e_\iota)_I$ is an orthonormal basis of some supplement of $V^{\perp\perp}+V^\perp$ in $E$ if $V^{\perp\perp}+V^\perp \neq E$ (and $\emptyset$ otherwise)

(9.6) $e_\iota \perp f_\kappa$ $(\iota\in J,\kappa\in I)$ and furthermore, setting

(9.7) $\alpha_{\iota i} = \Phi(f_\iota,v_i)$ , $\beta_{\iota i} = \Phi(e_\iota,v_i)$ , $\gamma_{\iota i} = \Phi(e_\iota,w_i)$

(9.8) $(\sum_{i=1}^n \alpha_{\iota i}\alpha^\tau_{\kappa i})_n$ , $(\sum_{i=1}^n \beta_{\iota i}\beta^\tau_{\kappa i})_n$ , $(\sum_{i=1}^n \gamma_{\iota i}\gamma^\tau_{\kappa i})_n$ are Cauchy sequences.

Some of these conditions may of course be vacuous, e.g. if $E= V \oplus V^\perp$ and so $\mathfrak{B} = (v_i)_{\mathbb{N}} \cup (w_i)_{\mathbb{N}}$ . If $V^\perp = (0)$ then $\mathfrak{B}$ reduces to $\mathfrak{B} = (v_i)_{\mathbb{N}} \cup (f_\iota)_J$ and is a standard basis in the sense of 1.7. There always are bases $\mathfrak{B}$ satisfying (9.2) through (9.6).

Existence of standard bases

Assume that a basis $\mathfrak{B}$ as in (9.1) is a standard basis for $V \subset E$ . $\mathfrak{B}$ gives rise to $\perp$-dense embeddings, in fact (1.7):

(9.9) $(v_i)_{\mathbb{N}} \cup (f_\iota)_J$ is a standardbasis for the $\perp$-dense embedding $V \subset V^{\perp\perp}$ ,

(9.10) $(v_i,w_i)_{\mathbb{N}} \cup (f_\iota,e_\kappa)_{\iota\in J,\kappa\in I}$ is a standard basis for the $\perp$-dense embedding $V \oplus V^\perp \subset E$ ,

(9.11) $(v_i)_{\mathbb{N}} \cup (f_\iota,e_\kappa)_{\iota\in J,\kappa\in I}$ is a standard basis for the $\perp$-dense embedding $V \subset V \oplus k(f_\iota)_J \oplus k(e_\iota)_I$ ,

(9.12) $(w_i)_{\mathbb{N}} \cup (e_\iota)_I$ is a standard basis for the $\perp$-dense embedding $V^\perp \subset V^\perp \oplus k(e_\iota)_I$ .

<u>Theorem</u>. Let $(E,\Phi)$ be as in 9.1 and $S$ some arbitrarily fixed supplement of $V^{\perp\perp}+V^\perp$ in $E$ . The embedding $V \subset E$ admits a standard basis if and only if the $\perp$-dense embeddings $V \oplus V^\perp \subset E$ and $V \subset V^{\perp\perp}\oplus S$ admit standard bases.

<u>Proof</u>. Assume that $(r_i)_{\mathbb{N}} \cup (g_\lambda)_L$ is a standard basis for the

$\perp$-dense embedding $V \oplus V^{\perp} \subset E$. Let $(v_i)_{\mathbb{N}}$, $(w_i)_{\mathbb{N}}$ be orthonormal bases of $V, V^{\perp}$ respectively. By (1.7.8) we have that $(v_i, w_i)_{\mathbb{N}} \cup (g_{\lambda})_L$ is still a standard basis for $V \oplus V^{\perp} \subset E$.
Consider the embedding $V \subset V^{\perp\perp} \oplus S$. Since the orthogonal of $V$ in $V^{\perp\perp} \oplus S$ is $(0)$ we can find an orthonormal family $(f_{\iota})_J \cup (e_{\iota})_I$ such that the span of the $f_{\iota}$ is a supplement of $V$ in $V^{\perp\perp}$ and the span of the $e_{\iota}$ is a supplement of $V^{\perp\perp}$ in $V^{\perp\perp} \oplus S$. By (1.7.8) $(v_i, w_i)_{\mathbb{N}} \cup (f_{\iota}, e_{\kappa})_{\iota \in J, \kappa \in I}$ is again a standard basis for the $\perp$-dense embedding $V + V^{\perp} \subset E$. It follows that $\mathfrak{B} = (v_i)_{\mathbb{N}} \cup (w_i)_{\mathbb{N}} \cup (f_{\iota})_J \cup (e_{\iota})_I$ is a standard basis for $V \subset E$ ((9.8) is seen to hold since subsequences of denumerable Cauchy systems are Cauchy).

Conversely, if $V \subset E$ admits a standard basis then by (9.8) it follows directly that $V + V^{\perp} \subset E$ and $V \subset V^{\perp\perp} \oplus S$ must admit standard bases.

Corollary. If there exists some supplement $S$ of $V^{\perp\perp} + V^{\perp}$ in $E$ such that the $\perp$-dense embeddings $V + V^{\perp} \subset E$ and $V \subset V^{\perp\perp} + S$ admit standard bases then each of the (infinitely many) bases $\mathfrak{B}$ of $E$ satisfying (9.1) through (9.6) is a standard basis for $V \subset E$.

## 10. The matrices associated with a standard basis

Let the standard basis $\mathfrak{B}$ of $V \subset E$ be as in (9.1). We define a $J$ by $J$ matrix $A$, two $I$ by $I$ matrices $B$, $C$, and a $I$ by $J$ matrix $D$ over $\bar{k}_0 \otimes k$ as follows:

$$(10.1) \qquad A := ((\Sigma_{i=1}^{\infty} \alpha_{\iota i} \alpha_{\kappa i}^{\tau})_{\iota,\kappa \in J})$$

$$(10.2) \qquad B := ((\Sigma_{i=1}^{\infty} \beta_{\iota i} \beta_{\kappa i}^{\tau})_{\iota,\kappa \in I})$$

$$(10.3) \qquad C := ((\Sigma_{i=1}^{\infty} \gamma_{\iota i} \gamma_{\kappa i}^{\tau})_{\iota,\kappa \in I})$$

$$(10.4) \qquad D := ((\Sigma_{i=1}^{\infty} \beta_{\iota i} \alpha_{\kappa i}^{\tau})_{\iota \in I,\kappa \in J}) .$$

$A,B,C,D$ are called the <u>matrices associated with $\mathfrak{B}$</u>. Clearly, some of the definitions may be vacuous. E.g. if $V^{\perp} = 0$ then $\mathfrak{B}$ reduces to $(v_i)_{\mathbb{N}} \cup (f_\iota)_J$ and $A$ is the only matrix defined; if $V^{\perp\perp} = V$ then $\mathfrak{B}$ reduces to $(v_i)_{\mathbb{N}} \cup (w_i)_{\mathbb{N}} \cup (e_\iota)_I$ and $B,C$ are the only matrices defined. Notice that $D$ exists if and only if $A$ and $B$ exist, i.e, if and only if $J \neq \emptyset$ and $I \neq \emptyset$ .

<u>Necessary and sufficient conditions for four</u> $\bar{k}_o \otimes k$-<u>matrices to be the matrices associated with a standard basis</u>.

In the Theorem of section 9 we have seen that standard bases $\mathfrak{B}$ of $V \subset E$ can be constructed by "superposition" of standard bases $\mathfrak{B}_1$, $\mathfrak{B}_2$ of the $\perp$-dense embeddings $V \oplus V^{\perp} \subset E$ and $V \subset V^{\perp\perp} \oplus S$ respectively where $S$ is an arbitrarly fixed supplement of $V^{\perp\perp} + V^{\perp}$ in $E$ .

Let us start out with an arbitrary standard basis $\mathfrak{B}$ of $V \subset E$ as in (9.1). Set $S = k(e_\iota)_I$ . We shall write down the matrices for the two dense embeddings with respect to the bases $\mathfrak{B}_1 = (v_i,w_i)_{\mathbb{N}} \cup (e_\iota,f_\kappa)_{\iota\in J,\kappa\in J}$, $\mathfrak{B}_2 = (v_i)_{\mathbb{N}} \cup (e_\iota,f_\kappa)_{\iota\in I,\kappa\in J}$ (obviously arranged):

(10.5) $\qquad \mathfrak{B}_1$ has the matrix $A_1 = \begin{pmatrix} B + C & D \\ tr_D\tau & A \end{pmatrix}$ (embedding $V \oplus V^{\perp} \subset E$)

(10.6) $\qquad \mathfrak{B}_2$ has the matrix $A_2 = \begin{pmatrix} B & D \\ tr_D\tau & A \end{pmatrix}$ (embedding $V \subset V^{\perp\perp} \oplus S$)

<u>Theorem</u>. Assume that the field $k$ satisfies $\bar{k}_o^+ = \bar{Q}$ (cf. (4)) and that $k_o$ possesses a non eventually constant denumerable Cauchy sequence. For $I,J$ at most denumerable sets let four matrices $A,B,C,D$ over $\bar{k}_o \otimes k$ in turn be $J$ by $J$, $I$ by $I$, $I$ by $I$, $I$ by $J$ . Under these assumptions there exists a positive definite hermitean k-space $(E,\Psi)$ which contains a subspace $V$ such that $A,B,C,D$ are the matrices associated with a suitable standard basis of the embedding $V \subset E$ if and only if the following holds:

(10.7) $\qquad C$ is positive definite and hermitean

(10.8) $\qquad \mathbb{1} - A_1 := \mathbb{1} - \begin{pmatrix} B+C & D \\ tr_D\tau & A \end{pmatrix}$ is positive semidefinite and hermitean

(10.9) $A_2 := \begin{pmatrix} B & D \\ tr_D\tau & A \end{pmatrix}$ is positive definite and hermitean

Proof of the Theorem. The necessity of the three conditions follows from (9.4), (10.5), (10.6) and the Corollary in 2.5.

Assume conversely that the three conditions are satisfied. $(\mathbb{1}-A_1)+A_2$ is positive definite so $\mathbb{1}-C$ is positive definite. $\mathbb{1}-A_2 = (\mathbb{1}-A_1) + \begin{pmatrix} C & 0 \\ 0 & 0 \end{pmatrix}$ is positive semidefinite. $A_1 = A_2 + \begin{pmatrix} C & 0 \\ 0 & 0 \end{pmatrix}$ is positive definite. Hence $C, A_1, A_2$ qualify for the Theorem in 2.6.

On a k-vector space $E$ spanned by a basis $\mathfrak{B} = (v_i)_{\mathbb{N}} \cup (w_i)_{\mathbb{N}} \cup (f_\iota)_J \cup (e_\iota)_I$ we are going to define a hermitean form $\Psi$. We declare $(v_i)_{\mathbb{N}} \cup (w_i)_{\mathbb{N}}$ and $(f_\iota)_J \cup (e_\iota)_I$ to be orthonormal; furthermore $\Phi(f_\iota, w_i) = 0$ $(\iota \in J,\ i \in \mathbb{N})$. By the corollary in 2.6 we can define products $\Phi(e_\iota, v_i)$, $\Phi(f_\kappa, v_i) \in k$ such that $(v_i)_{\mathbb{N}} \cup (e_\iota, f_\kappa)_{\iota \in I, \kappa \in J}$ is a standard basis for the dense embedding $k(v_i)_{\mathbb{N}} \subset k(v_i)_{\mathbb{N}} \oplus k(f_\iota)_J \oplus k(e_\iota)_I$ and has $A_2$ as its associated matrix. It follows already that $k(v_i)_{\mathbb{N}}^{\perp} = k(w_i)_{i \in \mathbb{N}}$. By the same corollary we can furthermore define products $\Phi(e_\iota, w_i) \in k$ such that $(w_i)_{\mathbb{N}} \cup (e_\iota)_I$ is a standard basis for the dense embedding $k(w_i)_{\mathbb{N}} \subset k(w_i)_{\mathbb{N}} \oplus k(e_\iota)_I$ and has $C$ as its associated matrix. It follows that $k(w_i)_{\mathbb{N}}^{\perp} = k(v_i)_{\mathbb{N}} \oplus k(f_\iota)_J$. We may therefore set $k(v_i)_{\mathbb{N}} = V$, $k(w_i)_{\mathbb{N}} = V^{\perp}$, $k(v_i)_{\mathbb{N}} \oplus k(f_\iota)_J = V^{\perp\perp}$. From 2.4.3 applied to the dense embedding $V + V^{\perp} \subset E$ with matrix $A_1$ we learn that the form $\Psi$ thus defined is positive definite on the whole space.

Remark 1. Notice that the space constructed in the above proof need not be strongly universal. For this to be the case it is sufficient to assume that $Q = k_o^+$.

It is worthwhile to consider the special case where $V = V^{\perp\perp}$ in the above theorem (or equivalently, $J = \emptyset$ and $I \neq \emptyset$, i.e. only the matrices $B$ and $C$ are defined). We then obtain the

Corollary. Assume that the field $k$ satisfies $\bar{k}_o^+ = \bar{Q}$ and that $k_o$ possesses a non eventually constant denumerable Cauchy sequence. Let

$B,C$ be two $I$ by $I$ matrices over $\bar{k}_o \otimes k$, card $I \leq \aleph_o$. There exists a positive definite hermitean k-space $(E,\Psi)$ which contains a $\perp$-closed subspace $V$ such that $B,C$ are the matrices associated with a suitable standard basis of the embedding $V \subset E$ if and only if the following holds:

(10.10) $\quad B,C$ are positive definite and hermitean,

(10.11) $\quad \mathbb{1} - (B+C)$ is positive semidefinite.

If furthermore $k_o^+ = \mathbb{Q}$ then $(E,\Psi)$ is strongly universal.

Remark 2. Let (as in section 2.7) $K_1$ be the (open) convex cone of positive definite hermitean matrices $A \in (\bar{k}_o \otimes k)^{I\times I}$ and $K_2$ the (closed) convex cone of positive semidefinite matrices $\mathbb{1}-A$, $A \in K_1$. $K_2 = \overline{\mathbb{1}-K_1}$. $\frac{1}{2}(K_1 \cap K_2)$ is then the largest subset $S$ of $K_1$ such that arbitrary pairs $A,B \in S$ will define a closed embedding with $A,B$ its associated matrices. For given $A \in K_1$ we may pick $B$ anywhere in the convex set $[-A + (K_1 \cap K_2)] \cap K_1$.

## 11. The main theorem on arbitrary subspaces (Statement)

Theorem. Let $(E,\Phi)$ be a positive definite strongly universal k-space with $\dim E = \aleph_o$, $1 \in \|\Phi\|$ and where the field $k$ is as described under the caption of Chap. XII. Assume furthermore that $k$ satisfies (4). Let $V \subset E$ be a subspace with $\dim V^\perp \in \{0,\aleph_o\}$ and $I,J$ sets with card $I = \dim E/V^\perp + V^{\perp\perp}$, card $J = \dim V^{\perp\perp}/V$. Assume that the embedding $V \subset E$ admits a standard basis; then the associated matrices $A,B,C,D$ are in turn $J$ by $J$, $I$ by $I$, $I$ by $I$, $I$ by $J$. Some of these matrices may not be defined depending on whether $I = \emptyset$ or $J = \emptyset$ or both.

The equations $\mathcal{A}\cdot{}^{tr}\mathcal{A}^\tau = A$, $\mathcal{B}\cdot{}^{tr}\mathcal{B}^\tau = B$, $\mathcal{C}\cdot{}^{tr}\mathcal{C}^\tau = C$ (if defined) possess solutions $\mathcal{A},\mathcal{B},\mathcal{C}$ which share (B1) p. 300 and which are in turn $J$ by $\mathbb{N}$, $I$ by $\mathbb{N}$, $I$ by $\mathbb{N}$. Let in the sequel $\mathcal{A},\mathcal{B},\mathcal{C}$ be any such solutions.

Consider a second embedding $\bar{V} \subset E$ with analogous objects $\bar{A}, \bar{B}, \bar{C}, \bar{D}$ .

We assert: There exists a metric automorphism $T$ of $E$ with $TV = \bar{V}$ if and only if card $I$ = card $\bar{I}$ , card $J$ = card $\bar{J}$ and the following holds.

If $I \neq \emptyset$ and $J \neq \emptyset$ (equivalently: all matrices $A,B,C,D$ are defined) then there exist column finite matrices

(11.0) $\quad \Gamma,\Omega,\Delta,\Xi$ with $\Gamma,\Omega$ bijective

such that

(Ge)

$$
\begin{cases}
(11.1) & \bar{A}-\mathbb{1} = {}^{tr}\Omega(A-\mathbb{1})\Omega^{\tau} \\
(11.2) & \bar{B}-\mathbb{1} = {}^{tr}\Gamma(B-\mathbb{1})\Gamma^{\tau} + {}^{tr}\Delta(A-\mathbb{1})\Delta^{\tau} + {}^{tr}\Gamma D\Delta^{\tau} + {}^{tr}\Delta\cdot{}^{tr}D^{\tau}\cdot\Gamma \\
 & \qquad - {}^{tr}\Xi\Xi^{\tau} - {}^{tr}\Gamma C\Xi^{\tau} - {}^{tr}\Xi\cdot{}^{tr}C^{\tau}\cdot\Gamma^{\tau} \\
(11.3) & \bar{C} = {}^{tr}\Gamma C\Gamma^{\tau} + {}^{tr}\Gamma C\Xi^{\tau} + {}^{tr}\Xi\cdot{}^{tr}C^{\tau}\cdot\Gamma^{\tau} + {}^{tr}\Xi\Xi^{\tau} \\
(11.4) & \bar{D} = {}^{tr}\Gamma D\Omega^{\tau} + {}^{tr}\Delta(A-\mathbb{1})\Omega^{\tau} \;.
\end{cases}
$$

If $J = \emptyset$ and $I \neq 0$ (equivalently: only $B$ and $C$ are define[d]) then there exist column finite matrices

(11.5) $\quad \Gamma,\Xi$ with $\Gamma$ bijective

such that

(Cl)

$$
\begin{cases}
(11.6) & \bar{B}-\mathbb{1} = {}^{tr}\Gamma(B-\mathbb{1})\Gamma^{\tau} - {}^{tr}\Xi\Xi^{\tau} - {}^{tr}\Gamma C\Xi^{\tau} - {}^{tr}\Xi\cdot{}^{tr}C^{\tau}\Gamma^{\tau} \\
(11.7) & \bar{C} = {}^{tr}\Gamma C\Gamma^{\tau} + {}^{tr}\Gamma C\Xi^{\tau} + {}^{tr}\Xi\cdot{}^{tr}C^{\tau}\cdot\Gamma^{\tau} + {}^{tr}\Xi\Xi^{\tau} \;,
\end{cases}
$$

or equivalently,

(Cl')

$$
\begin{cases}
(11.6') & \bar{B} + \bar{C} - \mathbb{1} = {}^{tr}\Gamma(B+C-\mathbb{1})\Gamma^{\tau} \\
(11.7') & \bar{C} = ({}^{tr}\Gamma C + {}^{tr}\Xi)\cdot{}^{tr}({}^{tr}\Gamma C + {}^{tr}\Xi)^{\tau} \;.
\end{cases}
$$

(Cl) is obtained from (Ge) by deleting all terms which contain $A,\bar{A},D,\bar{D},\Omega,\Delta$ .

If $J \neq \emptyset$ and $I = 0$ (equivalently: only $A$ is defined) there exists a column finite bijective matrix $\Omega$ such that

(De)

$$
\begin{cases}
(11.8) & \bar{A} - \mathbb{1} = {}^{tr}\Omega(A-\mathbb{1})\Omega^{\tau} \\
(11.9) & \dim \bar{V}^{\perp} = \dim V^{\perp} \;.
\end{cases}
$$

(De) is obtained from (Ge) by suppressing all terms which contain $B,\bar{B},C,\bar{C},D,\bar{D},\Gamma,\Delta,\Xi$ and by adding (11.9).

If $J = I = \emptyset$ (equivalently: no matrices are defined) then

$$\text{(11.10)} \qquad \dim \bar{V}^{\perp} = \dim V^{\perp}$$

*

Table listing all cases of the theorem

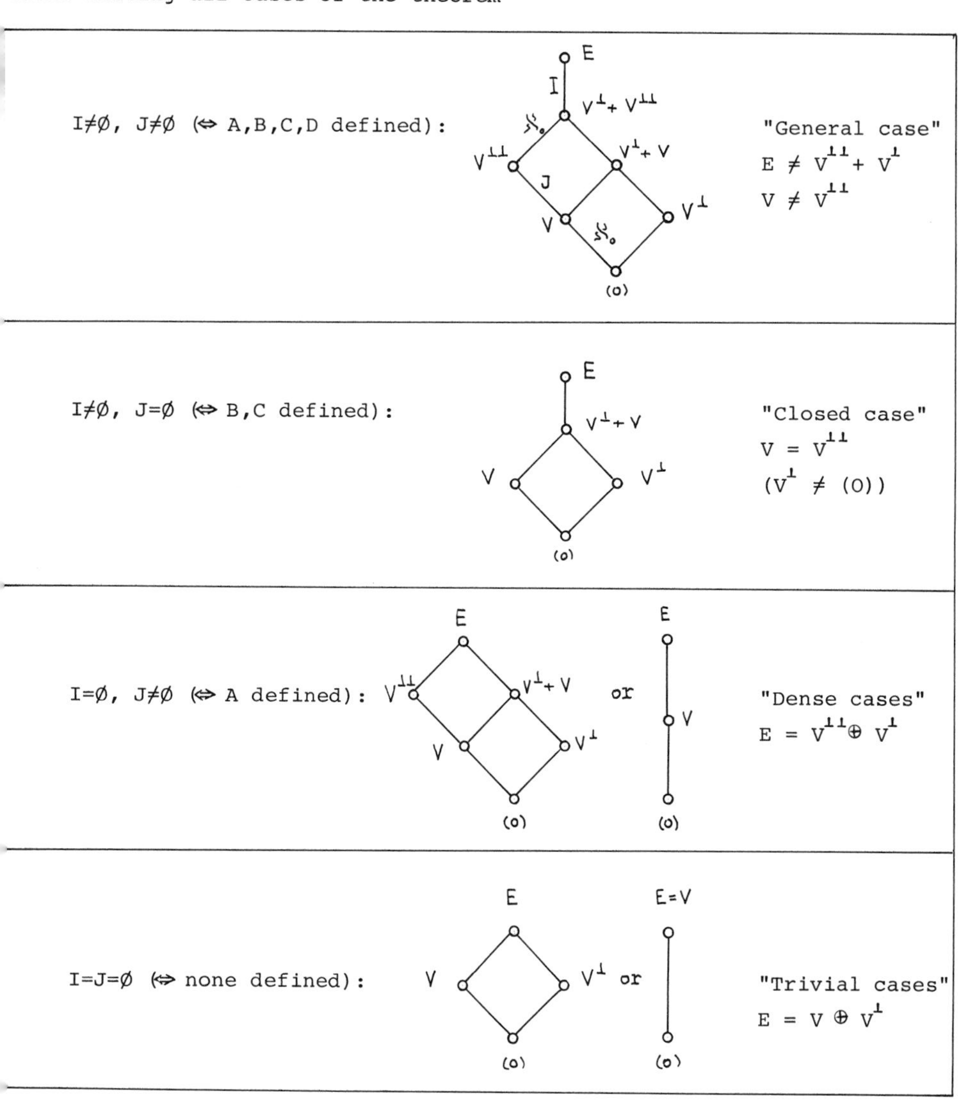

## 12. The proof

We shall indicate the steps to be taken; details can be supplied by the reader who has gone through Sections 1 - 7.

Necessity of the conditions in the theorem. If we switch from a standard basis $\mathfrak{B}$ of some embedding $V \subset E$ to another standard basis $\tilde{\mathfrak{B}}$ of $V \subset E$ one has

$$(12.1) \qquad \tilde{v}_i = \sum_j \Phi_{ji} v_j \, , \quad \tilde{w}_i = \sum_j \Psi_{ji} w_j \, , \quad \tilde{f}_\iota = \sum_j \Omega_{j\iota} f_j + \sum_j \Lambda_{j\iota} v_j \, ,$$

$$\tilde{e}_\iota = \sum_j \Gamma_{j\iota} e_j + \sum_j \Delta_{j\iota} f_j + \sum_j \Xi_{j\iota} w_j + \sum_j \Theta_{j\iota} v_j \, ,$$

where

(12.2) $\Phi,\Psi,\Omega,\Lambda,\Gamma,\Delta,\Xi,\Theta$ are column finite and $\Phi,\Psi,\Omega,\Gamma$ bijective.

$$(12.3) \qquad \Phi \cdot {}^{tr}\Phi^{\tau} = \mathbb{1} \, , \quad \Psi \cdot {}^{tr}\Psi^{\tau} = \mathbb{1}$$

$$(12.4) \qquad E_f := (\Phi(\tilde{f}_\iota, \tilde{f}_\kappa)) = {}^{tr}\Omega\Omega^{\tau} + {}^{tr}\Lambda\Lambda^{\tau} + {}^{tr}\Omega A \Lambda^{\tau} + {}^{tr}\Lambda \cdot {}^{tr}A^{\tau} \cdot \Omega^{\tau} = \mathbb{1}$$

$$(12.5) \qquad E_e := (\Phi(\tilde{e}_\iota, \tilde{e}_\kappa)) = {}^{tr}\Gamma\Gamma^{\tau} + {}^{tr}\Delta\Delta^{\tau} + {}^{tr}\Theta\Theta^{\tau} + {}^{tr}\Xi\Xi^{\tau} + {}^{tr}\Gamma B \Theta^{\tau} +$$

$${}^{tr}\Theta \cdot {}^{tr}B^{\tau} \cdot \Gamma^{\tau} + {}^{tr}\Gamma C \Xi^{\tau} + {}^{tr}\Xi \cdot {}^{tr}C^{\tau}\Gamma^{\tau} + {}^{tr}\Delta A \Theta^{\tau} + {}^{tr}\Theta \cdot {}^{tr}A^{\tau} \cdot \Delta^{\tau} = \mathbb{1} .$$

$$(12.6) \qquad E_{ef} := (\Phi(\tilde{e}_\iota, \tilde{f}_\kappa)) = {}^{tr}\Gamma B \Lambda^{\tau} + {}^{tr}\Delta \cdot {}^{tr}A^{\tau} \cdot \Lambda^{\tau} + {}^{tr}\Theta \cdot {}^{tr}A^{\tau} \cdot \Omega^{\tau} + $$
$$+ {}^{tr}\Delta\Omega^{\tau} + {}^{tr}\Theta\Lambda^{\tau} = 0 .$$

Computation of the matrices $\tilde{A},\tilde{B},\tilde{C},\tilde{D}$ associated with $\tilde{\mathfrak{B}}$ from (12.1) and elimination of $\Theta$ and $\Lambda$ by means of (12.4) - (12.6) gives formulas (Ge) in section 11. Thus, if there is an automorphism $T$ of $E$ with $T\bar{V} = V$ then $\mathfrak{B}$ and $\tilde{\mathfrak{B}} := \bar{\mathfrak{B}}^{T}$ are standard bases for $V \subset E$ and hence there are matrices (11.0) satisfying (Ge) (or (Cl), (De) respectively according to which of $A,B,C,D$ are defined). Furthermore $\dim E/V^{\perp\perp} + V^{\perp} = \dim E/\bar{V}^{\perp\perp} + \bar{V}^{\perp}$, $\dim V^{\perp\perp}/V = \dim \bar{V}^{\perp\perp}/\bar{V}$, $\dim V^{\perp} = \dim \bar{V}^{\perp}$

Sufficiency of the conditions in the theorem. Let $\mathfrak{B} = (v_i)_{\mathbb{N}} \cup (w_i)_{\mathbb{N}} \cup (f_\iota)_J \cup (e_\iota)_I$ be the standard basis in the definition of $A,B,C,D$. The first step to take is to change $\mathfrak{B}$ to a standard basis $\tilde{\mathfrak{B}} = (v_i)_{\mathbb{N}} \cup (w_i)_{\mathbb{N}} \cup (\tilde{f}_\iota)_J \cup (\tilde{e}_\iota)_I$ such that the matrices $\tilde{A},\tilde{B},\tilde{C},\tilde{D}$

associated with $\tilde{\mathfrak{B}}$ satisfy

(12.7) $\bar{A} = \tilde{A}$ , $\bar{B} = \tilde{B}$ , $\bar{C} = \tilde{C}$ , $\bar{D} = \tilde{D}$ .

This is accomplished just as in the proof of lemma 1 in sec. 5. Define $M_{\iota\kappa} = \sum_{i=1}^{\infty} \Phi(\bar{g}_\iota,\bar{v}_i)\Phi(\bar{g}_\kappa,\bar{v}_i)^\tau$ where the $\bar{g}_\iota$ with even subscript are the $\bar{e}_\kappa$ , those with odd subscript the $\bar{f}_\kappa$ in the standard basis $\bar{\mathfrak{B}}$ which defines $\bar{A},\bar{B},\bar{C},\bar{D}$ . $M_{\iota\kappa}$ is the positive definite matrix over $\bar{k}_o \otimes k$ associated with the standard basis $(\bar{v}_i)_{\mathbb{N}} \cup (\bar{e}_\iota,\bar{f}_\kappa)_{\iota\in I,\kappa\in J}$ for the dense embedding $\bar{V} \subset \bar{V}^{\perp\perp} \oplus k(\bar{e}_\iota)_I$ . A routine verification shows that the difference $\Gamma_{\iota\kappa} := \sum_{i=1}^{\infty} \Phi(g_\iota,v_i)\Phi(g_\kappa,v_i)^\tau - M_{\iota\kappa}$ lies in $k$ (the $g_\iota$ are analogous to $\bar{g}_\iota$ ). Cf. 5.4.5. Now the reasoning of 5.5 may be applied and we may assume that (12.7) holds, or rather, we may assume without loss of generality that we have from the beginning

(12.8) $\bar{A} = A$ , $\bar{B} = B$ , $\bar{C} = C$ , $\bar{D} = D$ .

Let then $\mathfrak{B},\bar{\mathfrak{B}}$ be standard bases of $V \subset E$ , $\bar{V} \subset E$ which have (12.8). We are going to change the bases in two successive steps. First we contemplate the dense embeddings $V^\perp \subset V^\perp \oplus k(e_\iota)_I$ , $\bar{V}^\perp \subset \bar{V}^\perp \oplus k(\bar{e}_\iota)_I$ with associated matrices $B = \bar{B}$ . Lemma 3 in sec. 7 shows that there are standard bases $\mathfrak{B}',\bar{\mathfrak{B}}'$, $\mathfrak{B}' = (v_i)_{\mathbb{N}} \cup (w_i')_{\mathbb{N}} \cup (f_\iota)_J \cup (e_\iota')_I$ and similarly for $\bar{\mathfrak{B}}'$ , such that

(12.9) $\Phi(e_\iota',w_i') = \Phi(\bar{e}_\iota',\bar{w}_i')$ $i \in \mathbb{N}$ , $\iota \in I$ .

We then consider the embeddings $V \subset V^{\perp\perp} \oplus k(e_\iota')_I$ , $\bar{V} \subset \bar{V}^{\perp\perp} \oplus k(\bar{e}_\iota')_I$ . We see (lemma 3 in sec. 7) that we may change $\mathfrak{B}',\bar{\mathfrak{B}}'$ to bases $\mathfrak{B}'',\bar{\mathfrak{B}}''$ where $\mathfrak{B}'' = (v_i'')_{\mathbb{N}} \cup (w_i')_{\mathbb{N}} \cup (f_\iota'')_J \cup (e_\iota'')_I$ , and similarly for $\bar{\mathfrak{B}}''$ , such that

(12.10) $\Phi(f_\iota'',v_i'') = \Phi(\bar{f}_\iota'',\bar{v}_i'')$ , $\Phi(e_\kappa'',v_i'') = \Phi(\bar{e}_\kappa'',\bar{v}_i'')$ $(i\in\mathbb{N},\ \iota\in J,\ \kappa\in I)$

Since all changes in this last step are made modulo $V^{\perp\perp}$ , the equalities (12.9) are preserved, i.e. we shall have

(12.11) $\Phi(e''_\iota, w'_i) = \Phi(\bar{e}''_\iota, \bar{w}'_i) \qquad (\iota \in I,\ i \in \mathbb{N})$ .

Thus, when B,C are both defined then (12.10), (12.11) show that the obvious map $\mathfrak{B}'' \to \bar{\mathfrak{B}}''$ induces an isometry T of E with $TV = \bar{V}$ . Assume that B or C is not defined. Then (12.10) shows that $\mathfrak{B}'' \to \bar{\mathfrak{B}}''$ defines an isometry $T_0 : V \to \bar{V}$ . By assumption $\dim V^\perp = \dim \bar{V}^\perp \in \{0, \aleph_0$ so (1.5) $V^\perp$ and $\bar{V}^\perp$ are isometric. Therefore and since $E = V \oplus V^\perp = \bar{V} \oplus \bar{V}^\perp$ we see that T can be extended to all of E. Q. E. D.

## 13. Embeddings that split

We keep the notations of sections 9 and 10 as well as assumption (9.0). Forms and the field are as described under the caption of chap. XII.

*

The simplest case of a $\perp$-dense embedding $V \subset (X,\Phi)$ , $\dim X = \aleph_0$ , has

$$\dim X/V = 1 .$$

By taking orthogonal sums of at most countably many, say $a_1$ embeddings $V_\iota \subset (X_\iota, \Phi_\iota)$ of this type and setting $E_1 = \overset{\perp}{\bigoplus_\iota} X_\iota$ , $V_1 = \bigoplus_\iota V_\iota$ we obtain a dense embedding

$$V_1 \subset (E_1, \Phi_1) \ ; \quad \dim E_1/V_1 = a_1 .$$

The simplest kind of a "non trivial" $\perp$-closed embedding $V \subset (X,\Phi)$ has

$$\dim X/V+V^\perp = 1 .$$

By taking an orthogonal sum of at most countably many, say $a_2$ embeddings $V_\kappa \subset (X_\kappa, \Phi_\kappa)$ of this sort and setting $E_2 = \overset{\perp}{\bigoplus_\kappa} X_\kappa$ , $V_2 = \bigoplus_\kappa V_\kappa$ we obtain a closed embedding

$$V_2 \subset (E_2, \Phi_2) \ ; \quad \dim E_2/V_2+V_2^\perp = a_2 .$$

We then pass to an orthogonal sum $E = E_1 \overset{\perp}{\oplus} E_2 \overset{\perp}{\oplus} E_3$ , $V = V_1 \oplus V_2$

where $V_1 \subset E_1$ and $V_2 \subset E_2$ are either embeddings as above or else $E_1 = (0)$ or $E_2 = (0)$ but not both $E_1 = E_2 = (0)$, and where $(E_3, \Phi_3)$ is non degenerate and arbitrary up to $a_3 := \dim E_3 \leq \aleph_0$ and $a_3$ infinite when $a_2 = 0$. We then have an embedding $V \subset E$ with

$$V^\perp = (V_2^\perp \cap E_2) \overset{\perp}{\oplus} E_3 , \quad V^{\perp\perp} = E_1 \oplus V_2 \tag{13.1}$$

$$\dim V = (a_1 + a_2)\aleph_0 = \aleph_0 \tag{13.2}$$

$$\dim V^\perp = a_2\aleph_0 + a_3 \in \{0, \aleph_0\} \tag{13.3}$$

$$\dim V^{\perp\perp}/V = a_1 \leq \aleph_0 \tag{13.4}$$

$$\dim E/V^{\perp\perp} + V^\perp = a_2 \leq \aleph_0 . \tag{13.5}$$

Remark. If $a_2 = 0$ we have $a_3 = \dim V^\perp$ so $a_3$ is an invariant of the embedding; on the other hand if $a_2 \neq 0$ we may replace the embedding $V_2 \subset E_2$ by $V_2 \subset E_2 \overset{\perp}{\oplus} E_3$, i.e. we may assume that $a_3 = 0$ ( $E_2$ and $E_2 \overset{\perp}{\oplus} E_3$ are isometric) in the above decomposition of $E$.

Lemma 1. Assume that $V \subset E$ is an arbitrary embedding but that we are not in the trivial case where $E = V + V^\perp$. Then the following statements are equivalent: (i) there is an orthogonal decomposition of the space $E$,

$$E = \bigoplus^\perp X_\iota \quad \text{with} \quad V = \bigoplus (V \cap X_\iota) \quad \text{and}$$
$$\dim X_\iota / (X_\iota \cap V) + ((X_\iota \cap V)^\perp \cap X_\iota) = 1 , \tag{13.6}$$

(ii) there is an orthogonal decomposition of $E$,

$$E = E_1 \overset{\perp}{\oplus} E_2 \overset{\perp}{\oplus} E_3 , \quad V = V_1 \oplus V_2 \quad \text{with} \quad V_1 \subset E_1 , \quad V_2 \subset E_2 \tag{13.7}$$

where "$V_1 \subset E_1$", "$V_2 \subset E_2$" are $\perp$-dense resp. $\perp$-closed embeddings of the kind exhibited above or else $E_1 = (0)$ or $E_2 = (0)$ but not both.

The proof rests on some crude combinatorial arguments and will not be written out.

<u>Definition</u>. An embedding $V \subset E$ with $V + V^{\perp} \neq E$ is <u>split</u> if it satisfies one of the conditions (i), (ii) in lemma 1.

<u>Lemma 2</u>. If the embedding $V \subset E$ splits and admits a standard basis then there is a decomposition (13.6) where all embeddings $V \cap X_i \subset X_i$ admit standard bases.

<u>Proof</u>. Consider a decomposition of $E$ as in (13.7). By the remark after (13.5) we may assume that $\dim E_3 \in \{0, \aleph_0\}$ so that $E_3$ is spanned by an orthogonal basis if $E_3 \neq (0)$. The assertion now follows by making systematic use of the corollary in section 9.

## 14. Conditions for a ⊥-dense embedding to split

<u>Theorem</u>. Assume that the field satisfies condition (4) and that $V \subset (E,\Phi)$ is a ⊥-dense embedding which admits a standard basis. The following are equivalent. (i) The embedding splits. (ii) A suitable standard basis has a diagonal associated matrix A. (iii) The matrix $A - \mathbb{1}$ over $\bar{k}_0 \otimes k$, where $A$ is the matrix associated with any standard basis, can be diagonalised over $k$ (i.e. there exists a columnfinite bijective matrix $\Omega$ over $k$ with ${}^{tr}\Omega(A-\mathbb{1})\Omega^{\tau}$ diagonal).

<u>Proof</u>. (i) ⇒ (ii) follows by using lemma 2 in sec. 13 and the definition of $A$. (ii) ⇒ (i): If $A$ is diagonal then by the corollary in 2.6 there is a (strongly universal) orthogonal sum of dense embeddings $V_i \subset X_i$, $\dim X_i/V_i = 1$ such that its associated matrix $\bar{A}$ is $A$. Since then (11.8) holds for $\Omega = \mathbb{1}$ we conclude from the main theorem that there is an isometry $T : \oplus^{\perp} X_i \to E$ with $T \oplus^{\perp} V_i = V$. So $V \subset E$ splits.

(i) ⇒ (iii): Since we have proved (i) ⇒ (ii) we obtain from (i) that there is a diagonal $\bar{A} - \mathbb{1}$ for suitable standard basis. Thus

$\bar{A}-\mathbb{1} = {}^{tr}\Gamma(A-\mathbb{1})\Gamma^{\tau}$ for some columnfinite bijective $\Gamma$ over $k$ by (3.1.2).

(iii) $\Longrightarrow$ (i): Assume conversely that ${}^{tr}\Gamma(A-\mathbb{1})\Gamma^{\tau}$ is a diagonal matrix $D$ for a columnfinite bijective $\Gamma$ over $k$. $D$ is negative semidefinite just as $A-\mathbb{1}$. Pick $\lambda_i \in Q \setminus \{0\}$ such that $-1 < \lambda_i^2 D_{ii} \leq 0$. The $\lambda_i$ define a diagonal matrix $\Lambda$ so that $-1 < ({}^{tr}\Lambda D \Lambda^{\tau})_{ii} \leq 0$. Define $\bar{A}-\mathbb{1} := {}^{tr}\Lambda D\Lambda^{\tau} = {}^{tr}(\Gamma^{\tau}\Lambda^{\tau})^{\tau}\cdot(A-\mathbb{1})\cdot(\Gamma^{\tau}\Lambda^{\tau})$. Again by the corollary in 2.6 there is a (strongly universal) $(\bar{E},\bar{\Phi})$ and a splitting embedding $\bar{V} \subset \bar{E}$ which has $\bar{A}$ as its associated matrix for suitable standard basis. By the main theorem $V \subset E$ must split as well.

Corollary 1. Assume that $k_0 = K$ (i.e. $k_0$ real closed). Every $\perp$-dense embedding admitting a standard basis splits.

Proof. (iii) $\Longrightarrow$ (i) of the theorem.

Corollary 2. The dense subspaces $V \subset E$ of Corollary 1 which have $\dim E/V = n$ for fixed $n \leq \aleph_0$ form $n+1$ orbits under the orthogonal group; the nullity of the semidefinite matrix $A-\mathbb{1}$ is the only invariant.

Example 1. Assume that in the commutative field $k$ every positive element possesses a square root. Let $V \subset (E,\Phi)$ be a dense embedding with standard basis and with $\Phi$ symmetric and $\dim E/V = 2$. If the embedding splits then by (iii) of the theorem there is a bijective $2\times 2$ matrix $\Gamma$ which diagonalizes $\psi = A-\mathbb{1} \in \bar{k}^{2\times 2}$,

$$\gamma_{11}\gamma_{12}\psi_{11} + (\gamma_{21}\gamma_{12} + \gamma_{11}\gamma_{22})\psi_{12} + \gamma_{21}\gamma_{22}\psi_{22} = 0 .$$

Not all coefficients of the $\psi_{ij}$ are zero. In other words, the $\psi_{ij}$ are linearly dependent over $k$. Assume conversely the latter to be the case:

$$\alpha\psi_{11} + \gamma\psi_{12} + \beta\psi_{22} = 0 \quad (\alpha, \beta, \gamma \in k \text{ and not all zero}) .$$

We assert that $V \subset E$ splits. If $\psi_{12} = 0$ we quote (ii) of the theorem . Assume then that $\psi_{12} \neq 0$ . If $\gamma = 0$ then both $\alpha,\beta$ are nonzero ($\psi_{11}\psi_{22} - \psi_{12}^2 \geq 0$ as $\psi$ is semidefinite by 2.4.1) and we set $\gamma_{11} = -(-\alpha\beta)^{1/2}\beta^{-1}$ , $\gamma_{12} = (-\alpha\beta)^{1/2}\alpha^{-1}$ , $\gamma_{21} = 1$ , $\gamma_{22} = \beta\alpha^{-1}$ . We have $\det(\gamma_{ij}) \neq 0$ , $\sum\gamma_{r1}\psi_{rs}\gamma_{s2} = 0$ and may quote (iii) of the theorem. If on the other hand $\gamma \neq 0$ , say $\gamma = 1$ , we set $\gamma_{11} = \sigma$ , $\gamma_{12} = \alpha\sigma^{-1}$ , $\gamma_{21} = 1$ , $\gamma_{22} = \beta$ where $\sigma$ is a nonzero root of $X^2\beta - X + \alpha = 0$ (the discriminant $1-4\alpha\beta$ is positive since $\psi_{11}\psi_{22} > \psi_{12}^2 = (\alpha\psi_{11} + \beta\psi_{22})^2$) . $\operatorname{Det}(\gamma_{ij}) = \sigma\beta - \alpha\sigma^{-1} = \sigma^{-1}(\sigma^2\beta-\alpha) = \sigma^{-1}(\sigma-2\alpha) \neq 0$ and again $\sum \gamma_{r1}\psi_{rs}\gamma_{s2} = 0$ .

Example 2. Let $k$ be a proper subfield of $\mathbb{R}$ . Then $[\mathbb{R}:k] \geq \aleph_0$ by Thm. 17, p. 316 of [6]. In the k-vector space $\mathbb{R}$ we can find at least $2^{\aleph_0}$ 3-dimensional "disjoint" subspaces $R_\iota$ ($R_\iota \cap R_\kappa = (0)$ for $\iota \neq \kappa$) . In each $R_\iota$ we pick three linearly independent vectors $\psi_{11}$ , $\psi_{12}$ , $\psi_{22}$ which define a dense enbedding $V_\iota \subset E$ . These $2^{\aleph_0}$ embeddings do not split and lie in different orbits.

## 15. Conditions for $\perp$-closed embeddings to split

Theorem. Assume that the field satisfies $k_0^+ = \mathbb{Q}$ (cf. 2.6) and that $V \subset (E,\Phi)$ is a $\perp$-closed embedding which admits a standard basis. The following are equivalent. (i) The embedding splits. (ii) A suitable standard basis has diagonal associated matrices $B$ , $C$ . (iii) There exist columnfinite matrices $\Gamma$ , $\Xi$ with $\Gamma$ bijective such that

$$(15.1) \qquad D_1 := {}^{tr}\Gamma B\Gamma^\tau - {}^{tr}\Xi\Xi^\tau - {}^{tr}\Gamma\mathcal{C}\Xi^\tau - {}^{tr}\Xi\cdot{}^{tr}\mathcal{C}^\tau\cdot\Gamma^\tau + \mathbb{1} - {}^{tr}\Gamma\Gamma^\tau$$

$$(15.2) \qquad D_2 := ({}^{tr}\Gamma\mathcal{C} + {}^{tr}\Xi)\cdot{}^{tr}({}^{tr}\Gamma\mathcal{C} + {}^{tr}\Xi)^\tau$$

are diagonal ( $C$ is some fixed solution of $C\cdot{}^{tr}C^{\tau} = C$ which shares (B1) on p. 300 ); cf. (C1) in the theorem of 11.

Corollary. Let $k_0$ be real closed. Every $\perp$-closed embedding (with standard basis) $V \subset (E,\Phi)$ with $\dim E/V+V^{\perp} < \infty$ splits.

Proof. By the spectral theorem in 6.6 it follows just as in the commutative case that the hermitean matrices $B$ , $C$ can simultaneously be diagonalized. Hence we are in the situation (iii) of the theorem with $\Xi = 0$ .

Proof of the theorem. (i) $\Rightarrow$ (ii) follows by using lemma 2 in 13 and the definition of $B$ , $C$ .

(ii) $\Rightarrow$ (i): One constructs a $\perp$-closed split embedding $\bar{V} \subset (\bar{E},\bar{\Phi})$ with $B$ , $C$ as associated matrices (use 2.6) and $E$ , $\bar{E}$ isometric. By Sec. 11, $V \subset E$ will split.

(i) $\Rightarrow$ (iii): We apply (i) $\Rightarrow$ (ii) and (11.6'), (11.7').

(iii) $\Rightarrow$ (i): Assume $D_1$ , $D_2$ in (15.1), (15.2) to be diagonal. $D_2$ is positive simidefinite and so is $\mathbb{1} - (D_1+D_2) = {}^{tr}\Gamma(\mathbb{1}-B-C)\Gamma^{\tau}$ ($\mathbb{1} - B - C$ is positive semidefinite by (10.8)). Hence $D_1 - \mathbb{1}$ is negative semidefinite. In fact we have that $D_2$ is positive definite as $\Gamma$ is bijective. Pick a diagonal matrix $\Lambda$ such that ${}^{tr}\Lambda D_2\Lambda^{\tau}$ and ${}^{tr}\Lambda(D_1-\mathbb{1})\Lambda^{\tau}$ have diagonal elements in $]0,1]$ resp. $]-1,0]$ . There is such a $\Lambda$ over $\mathbb{Q}$ . Define $\bar{C} := {}^{tr}\Lambda D_2\Lambda^{\tau}$ , $\bar{B} - \mathbb{1} := {}^{tr}\Lambda(D_1-\mathbb{1})\Lambda^{\tau}$ . We find

$$\bar{C} = ({}^{tr}(\Gamma^{\tau}\Lambda^{\tau})^{\tau}C + {}^{tr}(\Xi^{\tau}\Lambda^{\tau})^{\tau})\cdot{}^{tr}({}^{tr}(\Gamma^{\tau}\Lambda^{\tau})^{\tau}\cdot C + {}^{tr}(\Xi^{\tau}\Lambda^{\tau})^{\tau})^{\tau}$$

$$\mathbb{1} - \bar{B} - \bar{C} = {}^{tr}(\Gamma^{\tau}\Lambda^{\tau})^{\tau}(\mathbb{1}-B-C)(\Gamma^{\tau}\Lambda^{\tau}) \ .$$

In particular, $\bar{B} + \bar{C} - \mathbb{1}$ is negative semidefinite. $\bar{B}$ and $\bar{C}$ are positive definite by our normalization of diagonal elements via $\Lambda$ . Hence $\bar{B}$ , $\bar{C}$ qualify for the corollary in Sec. 10: there is a closed embedding with the matrices $\bar{B}$ , $\bar{C}$ , and it splits by (ii) $\Rightarrow$ (i) of our theorem. Hence $V \subset E$ splits by Section 11.

## 16. Parseval embeddings

We consider the situation of Appendix I and shall make use of the terminology introduced on that occasion. Forms and the field are as described under the caption of Chapter XII .

*

Let $V$ be an arbitrary subspace of $(E,\Phi)$ . If $\tilde{V}$ is its (norm-topology) closure in $\tilde{E}$ we set

(16.1) $V^{\circ} := \tilde{V} \cap E$ ("normtopology closure of $V$ in $E$")

$V^{\circ}$ is obviously an invariant of the orbit of $V$ (under the orthogonal group of $E$ ). Let

$$\mathcal{V}(V;\ +,\ \cap,\ \perp,\ \circ)$$

be the smallest sublattice of the lattice $L(E)$ of all subspaces of $E$ which contains $V$ and which is stable under the operations $\perp$ and $\circ$ . The lattice looks as follows

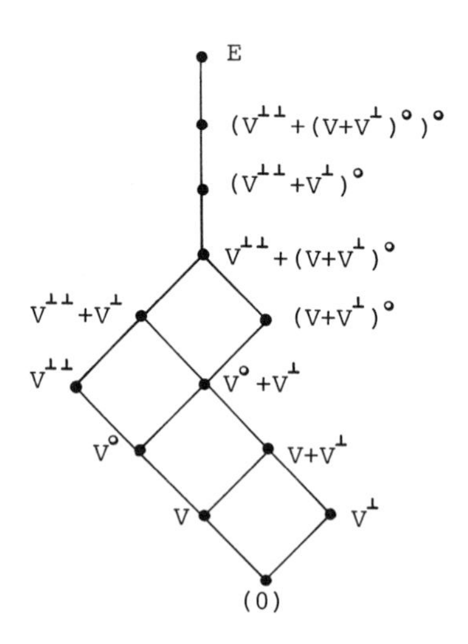

The verification is straight forward: $\perp$-closed hyperplans in $E$ are closed so $\perp$-closed subspaces of $E$ are closed; the inclusions given by the diagram are therefore correct and stability under $^\circ$ is evident. Stability under $\perp$ and $+$ is evident too. The only non trivial point with $\cap$ is the intersection $V^{\perp\perp} \cap (V+V^\perp)^\circ$. Now - just as in Hilbert spaces - we have that the sum of two orthogonal and closed subspaces in $\tilde{E}$ is closed, in particular $V^{\perp\perp} \cap (V+V^\perp)^\circ = V^{\perp\perp} \cap (\tilde{V}+\widetilde{V^\perp})$. Hence if $d = x+y$ with $d \in V^{\perp\perp}$, $x \in \tilde{V}$, $y \in \widetilde{V^\perp}$ then $y = d-x \in \widetilde{V^\perp} \cap \tilde{V}^{\perp\perp} \subset \tilde{V}^\perp \cap \tilde{V}^{\perp\perp} = (0)$. This gives $V^{\perp\perp} \cap (V+V^\perp)^\circ = V^{\perp\perp} \cap \tilde{V} = V^\circ$ as indicated in the diagram. Finally we remark that $(V^{\perp\perp}+V^\perp) \cap (V+V^\perp)^\circ = V^\perp + V^\circ$ by modularity and by what we just proved. The remaining intersections pose no problems.

Remark. It is clear that the cardinal numbers defined by the above lattice $\mathcal{V}$ - we mean the dimensions of quotient spaces of neighbouring elements in $\mathcal{V}$ - are invariants of the orbit of $V$ under the orthogonal group. We shall now discuss special cases where these cardinal numbers turn out to be a complete set of invariants.

Definition. Let $V \subset E$ be an arbitrary embedding which admits a standard basis. It is called parseval if $(V+V^\perp)^\circ = E$ (cf. 16.1) ($V$ is then also called a parseval subspace). To motivate the terminology we remark that for any standard basis $(v_i)_{i\in\mathbb{N}} \cup (f_j)_J$ we have

(16.2) $$\text{if } V^\perp = 0 \text{ then } V^\circ = \{x \in E \mid \|x\| = \sum_{i=1}^{\infty} \Phi(x,v_i)\Phi(x,v_i)^\tau\} = \text{Rad}\,\Psi .$$

Proof. If $x \in V^\circ$ then (A 2) $x = \sum_1^\infty \Phi(x,v_i)v_i$ and so $\|x\| = \sum_1^\infty \Phi(x,v_i)\Phi(x,v_i)^\tau$. Conversely, if the latter is the case then (A 7) we have $\psi(x,x) = 0$ which by (A 6) means that $\pi_{\tilde{V}^\perp}x = 0$, i.e. $x \in \tilde{V}$. Finally $V^\circ = \text{Rad}\,\Psi$ follows from (A 7).

Thus if $V \subset E$ is $\perp$-dense and parseval then $(v_i)_{\mathbb{N}}$ is a topologically dense orthonormal set and we have Parseval's identity on all of $E$.

16.3. Theorem. Let $V \subset E$ be an arbitrary embedding with a standard basis. If $V \subset E$ is parseval then $\mathcal{V}(V; +, \cap, \perp, \circ)$ has a diagram as follows

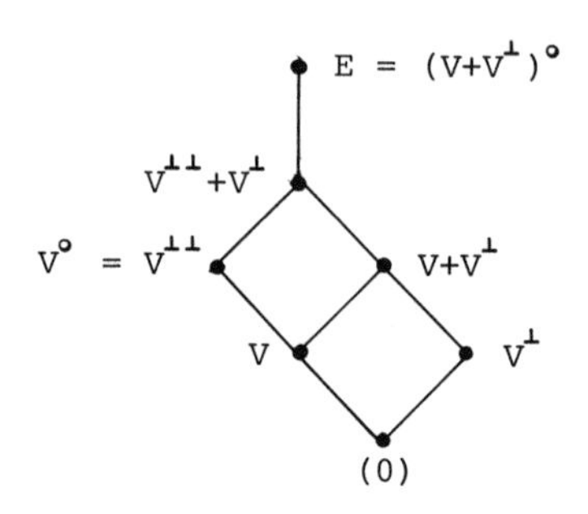


Furthermore, if the field has complete $k_0$, $k_0 = \bar{k}_0$ (cf. 2.5), then the two cardinals

$$(16.4) \qquad \dim V^{\perp\perp}/V \quad , \quad \dim E/V^{\perp\perp}+V^\perp$$

are a complete set of invariants for the orbit of $V$ under the orthogonal group of $E$. In particular the embedding splits. The embedding $V \subset V^{\perp\perp}$ is parseval.

Proof. Consider the matrices $A$, $B$, $C$, $D$ associated with a standard basis for $V \subset E$. By (A 7) we have $(V+V^\perp)^\circ = E$ if and only if the form $\Psi$ belonging to the dense embedding $V + V^\perp \subset E$ is the zero form. Hence by the theorem in sec. 10 we have $(V+V^\perp)^\circ = E$ if and only if either $A$, $B$, $C$, $D$ are all defined and

$$A = \mathbb{1} \quad , \quad B+C = \mathbb{1} \quad , \quad D = 0$$

or only $B$ and $C$ are defined and

$$B + C = \mathbb{1}$$

or only $A$ is defined and

$$A = \mathbb{1}$$

or $E = V + V^{\perp}$ (in which case nothing remains to be shown). In all cases we may quote the theorem in Sec. 10: Set $\Delta = \Xi = 0$, $\Omega = \mathbb{1}$ and pick $\Gamma$ such that $\bar{C} = {}^{tr}\Gamma C \Gamma^{\tau}$ where $\bar{C}$ is an analogous object. There always is such a $\Gamma$ as we have only one isometry class over $k$. Hence any two embeddings with equal invariants (16.4) can be transformed into each other. Finally by Sec. 13 we see that for prescribed $\dim V^{\perp\perp}/V$, $\dim E/V^{\perp\perp}+V^{\perp}$ there is a parseval embedding $V \subset E$ that splits.

(16.5) Corollary ([8]). Let $H$ be a real or complex separable Hilbert space and let $B_1$, $B_2$; $\bar{B}_1$, $\bar{B}_2$ be complete orthogonal sets such that their spans $V_1$, $V_2$; $\bar{V}_1$, $\bar{V}_2$ in $H$ satisfy $V_1 \subset V_2$, $\bar{V}_1 \subset \bar{V}_2$, $\dim V_2/V_1 = \dim \bar{V}_2/\bar{V}_1$. Then there is a metric automorphism of $H$ with $TV_i = \bar{V}_i$ $(i = 1,2)$.

Proof. Apply (16.3) to the $\perp$-dense parseval embeddings $V_1 \subset V_2$, $\bar{V}_1 \subset \bar{V}_2$ to get an isometry $T : V_2 \to \bar{V}_2$ with $TV_1 = \bar{V}_1$. $T$ can be extended to all of $H$.

There is another instance where the cardinal invariants of $\mathcal{V}(V; +, \cap, \perp, \circ)$ are a complete set of invariants, to wit,

(16.6) Theorem. Let $V \subset (E,\Phi)$ be a $\perp$-dense embedding with a standard basis. $(V; +, \cap, \perp, \circ)$ has the diagram:

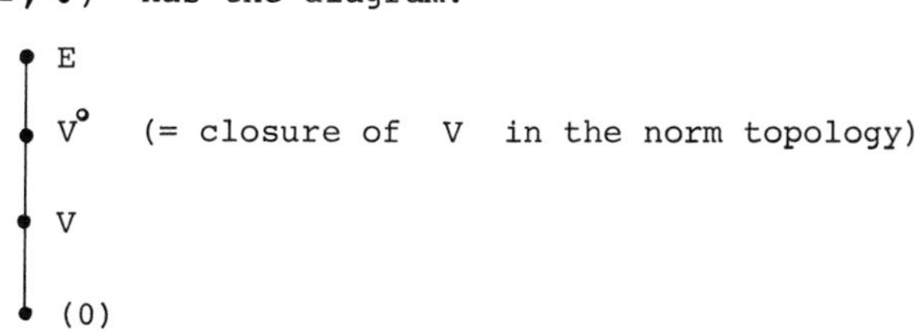

If $k$ has complete $k_0$, $k_0 = \bar{k}_0$ (cf. **2.5**) then the two cardinal invariants

(16.7) $\dim E/V^\circ$ , $\dim V^\circ/V$

are a complete set of invariants for the orbit of $V$ under the orthogonal group of $E$ .

Proof. (16.3)

(16.8) Extension Theorem. Let $V$ , $\bar{V}$ be isometric parseval subspaces of $(E,\Phi)$ . For $T : V \to \bar{V}$ a fixed isometry the following are equivalent. (i) $T$ is a homeomorphism with respect to the two linear topologies $\sigma_1 = \sigma(\Phi)|_V$ , $\sigma_2 = \sigma(\Phi|_{V^{\perp\perp}})|_V$ resp. $\bar{\sigma}_1 = \sigma(\Phi)|_{\bar{V}}$ , $\bar{\sigma}_2 = \sigma(\Phi|_{\bar{V}^{\perp\perp}})|_{\bar{V}}$ and $\dim V^\perp = \dim \bar{V}^\perp \in \{0,\aleph_0\}$ . (ii) $T$ can be extended to a metric automorphism of $E$ .

Proof of (16.8) when $V^\perp = 0$ . Assume (i). It is easy to deduce that $\dim E/V = \dim E/\bar{V}$ . Let $e$ be a fixed element of $E \setminus V$ . $\bar{H} = T(e^\perp \cap V)$ is a $\bar{\sigma}_1$-closed hyperplane of $\bar{V}$ , $\bar{H}^{\perp\perp} \cap V = \bar{H}$ . So $\bar{H}^\perp \neq (0)$ and thus $T(e^\perp \cap V) = \bar{e}_0^\perp \cap \bar{V}$ for suitable $\bar{e}_0 \in E$ . Pick $v \in V \setminus (e^\perp)$ and $\lambda \in k$ such that $\Phi(e,v) = \Phi(\lambda\bar{e}_0, Tv)$ ; set $\bar{e} = \lambda e_0$ . We now have $\Phi(e,v) = \Phi(\bar{e}, Tv)$ for all $v \in V$ . In particular, if $(v_i)_{i\in\mathbb{N}}$ is an orthonormal basis of $V$ and $\bar{v}_i = Tv_i$ then

$$\Phi(e,v_i) = \Phi(\bar{e},\bar{v}_i) \quad , \quad i \in \mathbb{N} .$$

We have thus shown that for arbitrary $x \in E$ there is some $\bar{x} \in E$ with $\Phi(x,v_i) = \Phi(\bar{x},\bar{v}_i)$ , $i \in \mathbb{N}$ . The map $\bar{T} : x \mapsto \bar{x}$ is well defined as $\bar{V}^\perp = (0)$ ; it is injective as $V^\perp = (0)$ and it is linear; the embeddings are parseval, $\bar{T}$ also preserves the form:
$\Phi(x,y) = \sum_1^\infty \Phi(x,v_i)\Phi(y,v_i)^\tau = \sum_1^\infty \Phi(\bar{x},\bar{v}_i)\Phi(\bar{y},\bar{v}_i)^\tau = \Phi(\bar{x},\bar{y})$ by (16.2). $\bar{T}$ is surjective as $T^{-1}$ gives raise to the inverse assignment.

Proof of (16.8) when $V^{\perp\perp} = V$. Assume (i). It follows that $\dim E/V+V^{\perp} = \dim E/\bar{V}+\bar{V}^{\perp}$. In contrast to the previous case we shall here have a recursive construction of the desired extension of $T$. We shall enlarge $V$ by adjoining a prescribed vector $e$ outside of $V$. Let $(v_i)_{\mathbb{N}}$, $(w_i)_{\mathbb{N}}$ be orthonormal bases of $V$, $V^{\perp}$ respectively $(\bar{v}_i)_{\mathbb{N}}$, $(\bar{w}_i)_{\mathbb{N}}$ similar objects with $\bar{v}_i = Tv_i$ $(i \in \mathbb{N})$.

Case 1 : $e \in E \setminus V+V^{\perp}$. We find (by the dense case already treated) a vector $\bar{e}_0 \in E$ with $\Phi(e,v_i) = \Phi(\bar{e}_0,\bar{v}_i)$ $(i \in \mathbb{N})$. $\bar{e}_0$ is unique modulo $\bar{V}^{\perp}$. We look for some $w \in \bar{V}^{\perp}$ with $\|\bar{e}_0+w\| = \|e\|$. Since (16.2) $\|e\| = \sum_1^{\infty} \Phi(e,v_i)\Phi(e,v_i)^{\tau} + \sum_1^{\infty} \Phi(e,w_i)\Phi(e,w_i)^{\tau}$ with both sums positive we have $0 < \sum_1^{\infty} \Phi(e,v_i)\Phi(e,v_i)^{\tau} = \sum_1^{\infty} \Phi(\bar{e}_0,\bar{v}_i)\Phi(\bar{e}_0,\bar{v}_i)^{\tau} < \|e\|$. There is $N \in \mathbb{N}$ such that $\sum_{N+1}^{\infty} \Phi(\bar{e}_0,\bar{w}_i)\Phi(\bar{e}_0,\bar{w}_i)^{\tau} < \sum_1^{\infty} \Phi(e,w_i)\Phi(e,w_i)^{\tau}$. Thus if $\alpha := \|\bar{e}_0 - \sum_1^N \Phi(\bar{e}_0,\bar{w}_i)\bar{w}_i\|$ we now have $\alpha < \|e\|$. Since $(E,\Phi)$ is strongly universal there exists in $\bar{V}^{\perp} \cap (\bar{e}_0 - \sum_1^N \Phi(\bar{e}_0,\bar{w}_i)\bar{w}_i)^{\perp}$ a vector $\bar{w}$ of length $\|\bar{w}\| = \|e\| - \alpha$. We define $Te := \bar{e}_0 + \bar{w} - \sum_1^N \Phi(\bar{e}_0,\bar{w}_i)\bar{w}_i$. Thereby we have extended $T$ from $V$ to $V_1 = V \oplus k(e)$.

By the lemma in Appendix I the embeddings $V_1 \subset E$, $TV_1 \subset E$ are again parseval. The isometriy $T: V_1 \to TV_1$ is still homeomorphic with respect to $\sigma(\Phi)$ : a map which is continuous on a $\sigma(\Phi)$-closed hyperplane $V$ of $V_1$ is continuous on $V_1$ since closed hyperplanes are neighbourhoods. $V_1^{\perp\perp} = V_1$. Therefore the embeddings $V_1 \subset E$, $TV_1 \subset E$ satisfy again (i) of the theorems and the step by step construction may be continued.

Case 2 : $e \in V + V^{\perp}$. It is sufficient to take care of the component of $e$ in $V^{\perp}$, so we assume $e \in V^{\perp}$. As $\bar{V}^{\perp}$ is strongly universal $(\dim \bar{V}^{\perp} = \aleph_0$ as $\dim \bar{V}^{\perp} = \dim V^{\perp} \neq 0)$ we find $\bar{e} \in \bar{V}^{\perp}$ with $\|\bar{e}\| = \|e\|$. $T$ extends to $V_1 := V \oplus k(e)$ by sending $e$ into $\bar{e}$ ; the extension is still weakly homeomorphic and the embeddings $V_1 \subset E$,

$TV_1 \subset E$ satisfy (i) of the theorem.

Since $\dim V^{\perp} = \dim \bar{V}^{\perp}$ and $\dim E/V + V^{\perp} = \dim E/\bar{V} + \bar{V}^{\perp}$ the steps described in the two cases may be repeated. Alternating between the roles of $V$, $\bar{V}$ one constructs ascending chains of $\perp$-closed parseval spaces $V \subset V_1 \subset V_2 \subset \ldots$ ; $\bar{V} \subset \bar{V}_1 \subset \bar{V}_2 \subset \ldots$ together with a compatib sequence of isometries $T_i : V_i \to \bar{V}_i$ such that $\cup V_i \supset E$, $\cup \bar{V}_i \supset E$.

<u>Proof of (16.8).</u> If $\dim V^{\perp} = 0$ we have a proof. If $V^{\perp} \neq 0$ then $\dim V^{\perp} = \aleph_0$. We may by (i) of the theorem and by the dense case already treated (parseval embeddings $V \subset V^{\perp\perp}$, $\bar{V} \subset \bar{V}^{\perp\perp}$) extend $T$ to an isometry $T_1 : V^{\perp\perp} \to \bar{V}^{\perp\perp}$. If we succeed in showing that $T_1$ is weakly homeomorphic we may quote the closed case already treated and extend $T_1$ to all of $E$.

Continuity of $T_1$, $T_1^{-1}$ : Let be given the (subbasis-) neighbourhoo $\bar{e}^{\perp} \cap \bar{V}^{\perp\perp}$. We are going to construct a neighbourhood $e^{\perp} \cap V^{\perp\perp}$ with $T_1(e^{\perp} \cap V^{\perp\perp}) = \bar{e}^{\perp} \cap \bar{V}^{\perp\perp}$. It suffices to consider the cases $\bar{e} \in \bar{V}^{\perp\perp}$, $\bar{e} \notin \bar{V}^{\perp\perp} + \bar{V}^{\perp}$. The first is trivial. Let $\bar{e} \notin \bar{V}^{\perp\perp} + \bar{V}^{\perp}$. It follows that there is $e \in E$ with $\Phi(e,v_i) = \Phi(\bar{e},\bar{v}_i)$ for $v_i$, $Tv_i$ the members of orthonormal bases of $V$, $\bar{V}$ respectively. We test $e^{\perp} \cap V^{\perp\perp}$ ; let $y \in e^{\perp} \cap V^{\perp\perp}$ and set $\bar{y} = T_1 y$. As the embeddings $V + V^{\perp} \subset E$, $\bar{V} + \bar{V}^{\perp} \subset E$ are parseval we have $0 = \Phi(e,y) = \sum_{i=1}^{\infty} \Phi(e,v_i)\Phi(y,v_i)^{\tau} + {}$
$+ \sum_{i=1}^{\infty} \Phi(e,w_i)\Phi(y,w_i)^{\tau} = \sum_{i=1}^{\infty} \Phi(e,v_i)\Phi(y,v_i)^{\tau} + 0 = \sum_{i=1}^{\infty} \Phi(\bar{e},\bar{v}_i)\Phi(\bar{y},\bar{v}_i)^{\tau} =$
$= \sum_{i=1}^{\infty} \Phi(\bar{e},\bar{v}_i)\Phi(\bar{y},\bar{v}_i)^{\tau} + \sum_{i=1}^{\infty} \Phi(\bar{e},\bar{w}_i)\Phi(\bar{y},\bar{w}_i)^{\tau} = \Phi(\bar{e},\bar{y})$. So $T_1(e^{\perp} \cap V^{\perp\perp}) \subset$ $\subset \bar{e}^{\perp} \cap \bar{V}^{\perp\perp}$. But both are hyperplanes in $\bar{V}^{\perp\perp}$ and therefore we even have equality. The proof of the extension theorem is thus complete.

*

## References to Chapter XII

[1] R. Baer, Dichte, Archimedizität und Starrheit geordneter Körper. Math. Ann. 188 (1970) 165-205.

[2] H. Gross, Ueber isometrische Abbildungen in abzählbar dimensionalen Räumen über reellen Körpern. Comment. Math. Helv. 43 (1968) 348-357.

[3] -, Eine Bemerkung zu dichten Unterräumen reeller quadratischer Räume. Comment. Math. Helv. 45 (1970) 472-493.

[4] P. Hafner and G. Mazzola, The cofinal character of uniform spaces and ordered fields. Zeitschrift f. math. Logik u. Grundl. d. Math. 17 (1971) 377-384.

[5] K. Hauschild, Cauchyfolgen höheren Typus in angeordneten Körpern. Zeitschr. f. math. Logik u. Grundl. d. Math. 13 (1967) 55-66.

[6] N. Jacobson, Lectures in Abstract Algebra, vol. III. van Nostrand New York (1964).

[7] A. Prestel, Remarks on the Pythagoras and Hasse Number of Real Fields. J. reine angew. Math. 303/304 (1978) 284-294.

[8] U. Schneider, Beiträge zur Theorie der sesquilinearen Räume unendlicher Dimension. Ph. D. Thesis University of Zurich 1975.

*

<u>Postscript</u> (concerning Section 11). In an interesting paper, "Subspaces of positive definite inner product spaces of countable dimension" (to appear in Pac. J. Math.), W. Bäni gives a classification in a different vein. Only the groundfield $\mathbb{R}$ (or $\mathbb{C}$, $\mathbb{H}$) is admitted. The upshot is that, on the one hand, a large class of subspaces can be characterized by cardinal number invariants and that, on the other hand, an attempt to classify <u>all</u> subspaces is - in a very precise sense - futile.

CHAPTER THIRTEEN

# CLASSIFICATION OF $\perp$-DENSE SUBSPACES WITH DEFINITE FORMS

## 1. Introduction

The fields $k$ admitted in this chapter are the same as those of Chapter Twelve but with the additional proviso that $k_o$ is archimedean ordered. $(E,\Phi)$ will be a non degenerate hermitean space of dimension $\aleph_o$ which is weakly universal and has $1 \in \|\Phi\|$. In contrast to Chapter Twelve the space $(E,\Phi)$ is not assumed to be positive definite.

We refer to Section 1 of Chapter Twelve for notational conventions.

*

As in Chapter Twelve we are concerned with the problem of finding a complete set of orthogonal invariants for a subspace $V$ of $E$. As told in the caption only $\perp$-dense subspaces will be discussed. We shall first reduce the problem by dismissing several easy cases that can arise.

Case 1. Assume that dense subspaces $V,\bar{V}$ contain both infinite dimensional totally isotropic subspaces. Provided that $\dim E/V = \dim E/\bar{V}$ it is not difficult to give a recursive construction for an isometry $T : E \to E$ that maps $V$ onto $\bar{V}$. The nature of the underlying field is entirely irrelevant. This is a special case of results proved in Chapter Five, it will not be discussed any further.

Case 2. Assume that dense subspaces $V,\bar{V}$ contain only finite dimensional totally isotropic subspaces. Let $E_o$ be an orthogonal sum of a maximal number of hyperbolic planes in $V$ and $E = E_o \oplus E_o^{\perp}$. $V \cap E_o^{\perp}$ is a dense subspace of $E_o^{\perp}$ and anisotropic. It is obvious that the problem of mapping $V$ onto $\bar{V}$ isometrically is reduced to the situation where $V$ and $\bar{V}$ carry anisotropic forms (restrictions of $\Phi$). To this case we now turn.

Case 3. $\Phi|_{V\times V}$ is anisotropic. By Dieudonné's lemma it follows that $\Phi$ is definite on all lines in $E$. Since $\Phi$ is weakly universal so is $\Phi|_{V\times V}$. Hence we conclude that $\Phi|_{V\times V}$ must be definite when it is anisotropic. Without loss of generality we may and we shall assume that, with the exception of section 10, we have

(1.1) $\Phi|_{V\times V}$ is positive definite

$V$ will then admit an orthonormal basis $(v_i)_{i\in\mathbb{N}}$. It is not difficult to see that there are orthonormal bases $(f_\iota)_{\iota\in J}$ that span supplements of $V$ in $E$. Thus

$$\mathfrak{B} = (v_i)_{i\in\mathbb{N}} \cup (f_\iota)_{\iota\in J}$$

satisfies the requirements for standard bases in XII.1.7. We shall keep the definitions of $\alpha_{\iota i}$ and $A_{\iota\kappa n}$ given there as well as the notion "standardbasis" described in (XII.1.7.6). (XII.1.7.8) will then remain valid here too.

If there exists a standard basis for our $V \subset E$ then (XII.1.7.5) does not, of course, hold any more; in fact we have the

(1.2) Lemma. $A_{\iota\kappa} - \delta_{\iota\kappa}$ is negative semi-definite if and only if $\Phi$ is positive definite on all of $E$.

(1.3) If standard bases exist then the principal results on $\perp$-dense subspaces in Chapter Twelve continue to hold. These are (i): If $TV = \bar{V}$ for an isometry $T : E \to E$ then the associated hermitean tensors $A_{\iota\kappa} - \delta_{\iota\kappa}$, $\bar{A}_{\iota\kappa} - \delta_{\iota\kappa}$ (which are over $\bar{k}_o \otimes k$) are equivalent over $k$ and, conversely, if we impose condition (4) of Chapter Twelve on the field $k_o$ we have (ii) that equivalence over $k$ of these tensors guarantees that $V$ and $\bar{V}$ belong to the same orbit under the orthogonal group of $E$ (XII.3.2.3 and XII.4).

(1.4) Our discussion shows that we are left with the situation where $\Phi|_{V\times V}$ is positive definite but admits no standard basis (for the embedding $V \subset E$). An instance of particular interest is that of an <u>archimedean ordered</u> $k_o$. $\mathfrak{B}$ not being a standard basis then means that for at least one $\iota \in J$ the monotonic sequence $(A_{\iota\iota n})_{n\in\mathbb{N}}$ is not bounded, i.e. "$\lim_{n\to\infty} \Delta_{\iota\iota n} = \infty$". In order to illustrate what can be done in such cases we shall start out with the case

$$\dim E/V = 2 .$$

As an application we shall derive from it general results for the situation where $k_o = \mathbb{R}$ (section 9).

In order that we need not interrupt the train of thoughts later on we start with an elementary section on an identity.

## 2. <u>Digression on Lagranges Identity</u>

In a commutative field one has the identity

$$(2.1) \qquad (x_1^2+x_2^2)(y_1^2+y_2^2) - (x_1y_1+x_2y_2)^2 = (x_1y_2-x_2y_1)^2 .$$

It is a special case of the identity which expresses multiplicativity of the norm for Caley numbers ("octaves") (See [2]; it may also be conceived as an instance of the Lagrange Identity in exterior algebra (cf. [3] page 155). Furthermore it is a rather special case of Pfisters theorem on products of sums of squares in commutative fields $F$ : Let $m = 2^n$ and $x_1,\dots,x_m, y_1,\dots,y_m \in F$ ; then there exist $z_2,\dots,z_m \in F$ such that $(x_1^2+\dots+x_m^2)(y_1^2+\dots+y_m^2) - (x_1y_1+\dots+x_my_m)^2 = z_2^2+\dots+z_m^2$ ([5] p. 297). We shall give here a generalization of (2.1) in a different direction (see 2.9 below).

Let $A = \begin{pmatrix} a_{11} & a_{12} \\ a_{21} & a_{22} \end{pmatrix}$ be hermitean over $(k,\tau)$, $a_{12}^\tau = a_{21}$. Set

$$(2.2) \qquad D := a_{11}a_{22} - a_{12}a_{12}^\tau \quad , \quad A^* = \begin{pmatrix} a_{22} & -a_{12} \\ -a_{21} & a_{11} \end{pmatrix} \quad (\text{"adjoint of } A\text{"}).$$

For the non commutative fields $k$ admitted in the present chapter we see that $D$ belongs to the center of $k$ and $A^*A = AA^* = D\mathbb{1}$ .

Consider arbitrary elements ${}^{tr}X = (x_1,x_2)$ , ${}^{tr}Y = (y_1,y_2) \in k^2$ ($X,Y$ as column vectors). We shall investigate the quantity

$$(2.3) \qquad \Delta := {}^{tr}XAX^{\tau}\cdot{}^{tr}YA^*Y^{\tau} - \langle X,Y\rangle\langle X,Y\rangle^{\tau}\cdot D$$

where $\langle,\rangle$ is the usual hermitean product in $k^2$ , $\langle X,Y\rangle = \Sigma x_i y_i^{\tau}$ .

Define

$$(2.4) \qquad A_1 = A + \begin{pmatrix} y_1^{\tau}y_1 & y_1^{\tau}y_2 \\ y_2^{\tau}y_1 & y_2^{\tau}y_2 \end{pmatrix}$$

and its determinant $D_1$ in analogy to (2.2). A straight forward verification shows that

$$(2.5) \qquad \Delta = D_1\cdot{}^{tr}XAX^{\tau} - D\cdot{}^{tr}XA_1X^{\tau}$$

(notice that $xyz + (xyz)^{\tau} = zxy + (zxy)^{\tau}$ for $x,y,z$ elements of the fields admitted in this chapter).

On the other hand we have identically

$$(2.6) \qquad D_1\cdot(D_1A - DA_1) = D\cdot A_1SA_1 + A_1SA\,SA_1 \quad \text{where} \quad S := A_1^* - A^* .$$

Now the right end of (2.5) may be spliced with the left end of (2.6) .

Remark. $S$ is positive semidefinite because for arbitrary $Z = (z_1,z_2)$ we have

$$(2.7) \qquad {}^{tr}ZSZ^{\tau} = \langle Z,Z\rangle\langle Y,Y\rangle - \langle Z,Y\rangle\langle Z,Y\rangle^{\tau} \geq 0 .$$

Summarizing we have

(2.8) Let $A$ be an arbitrary hermitean $2\times 2$ matrix and $D$ its "determinant" as defined in (2.2). Let ${}^{tr}X = (x_1,x_2)$, ${}^{tr}Y = (y_1,y_2) \in k^2$ be arbitrary. Assume that $D \geq 0$ . Define the matrix $A_1$ by (2.4) and let $D_1$ be its determinant. Abbreviate $S := A_1^*-A^*$ , $U := {}^{tr}X\cdot A_1$ , $V := {}^{tr}XA_1\cdot S$ . Then we have $D_1\{{}^{tr}XAX^{\tau}\cdot{}^{tr}YA^*Y^{\tau} - \langle X,Y\rangle\langle X,Y\rangle^{\tau}\cdot D\} = D_1\{{}^{tr}X\,(D_1A - DA_1)X^{\tau}\} = D(U\cdot S\cdot{}^{tr}U^{\tau}) + V\cdot A\cdot{}^{tr}V^{\tau} \geq V\cdot A\cdot{}^{tr}V^{\tau}$ (we have the inequality since $S$ is positive semidefinite).

Let now $A$ be a positive definite matrix. In particular $a_{11} \neq 0$ and since $\Sigma x_i a_{ij} x_j^\tau = a_{11}(x_1 + x_2 \cdot \frac{a_{12}^\tau}{a_{11}})(x_1 + x_2 \cdot \frac{a_{12}^\tau}{a_{11}})^\tau + \frac{D}{a_{11}} x_2 x_2^\tau$ we see that $D > 0$ since the form is definite. Thus $A$ will be invertible, $A^{-1} = \frac{A^*}{D}$ and $A^{-1}$ is positive definite as well. Since

$D_1 = D + {}^{tr}YA^*Y^\tau = D(1 + {}^{tr}YA^{-1}Y^\tau) \geq D > 0$ we see that $D_1$ is non zero and $A_1$ invertible. We may thus divide in (2.8) by $D_1 \cdot D$ and obtain the announced generalization of (2.1):

(2.9) Let $A$ be a positive definite hermitean $2\times 2$ matrix and $X, Y \in k^2$ arbitrary. With the notations of (2.8) we have proved

$${}^{tr}XAX^\tau \cdot {}^{tr}YA^{-1}Y^\tau - \langle X,Y\rangle\langle X,Y\rangle^\tau = D_1 {}^{tr}X\left(\frac{A}{D} - \frac{A_1}{D_1}\right)X^\tau = U \cdot \frac{S}{D_1} \cdot {}^{tr}U^\tau + V \frac{A^{-1}}{D_1} \cdot {}^{tr}V^\tau \geq$$

and the left hand side is zero if and only if ${}^{tr}X \cdot A_1 \cdot S = (0,0)$ .

In case that $\tau$ is the identity we can give a less involved formula. First of all we remark that

(2.10) if $(k,\tau)$ is commutative and $(a_{ij})$ an arbitrary hermitean $2\times 2$ matrix we have

$$\Sigma x_i a_{ij} x_j^\tau \cdot \Sigma y_i a_{ij} y_j^\tau - \Sigma x_i a_{ij} y_j^\tau \cdot (\Sigma x_i a_{ij} y_j^\tau)^\tau =$$

$$(x_1 y_2 - x_2 y_1)(x_1 y_2 - x_2 y_1)^\tau \cdot (a_{11} a_{22} - a_{12} a_{12}^\tau) \ .$$

From this identity the verification of which is straight forward we get

(2.11) Let $\tau$ be the identity (possible only when $k$ is commutative). If $A = (a_{ij})$ is an invertible symmetric $2\times 2$ matrix we have

$${}^{tr}XAX \cdot {}^{tr}ZA^{-1}Z - \langle X,Z\rangle^2 = \frac{1}{D}({}^{tr}XA\hat{Z})^2$$

where $D = \det A$ , $X$ and $Z = (z_1, z_2)$ in $k^2$ are arbitrary and $\hat{Z} = (z_2, -z_1)$ . Thus the left hand side vanishes if and only if ${}^{tr}XA$ and $Z$ are linarly dependent. If $A$ is positive definite then $D > 0$ and the left hand side is non negative.

For $A = \mathbb{1}$ we get back to (2.1).

Remark. If one merely wishes to show that the left hand side of the equation in (2.9) is non-negative one may proceed as follows. From (XII 6.9) it follows that an arbitrary hermitean $m\times m$ matrix $A$ can be diagonalized over $\bar{k}_o \otimes k$; i.e. there exists an invertible $m\times m$ matrix $B$ over $\bar{k}_o \otimes k$ with $BA\cdot{}^{tr}B^{\tau} = D$ (diagonal). Assume that $A$ is positive definite, then $A^{-1}$ exists and $B\cdot A^{-1}\cdot{}^{tr}B^{\tau} = D^{-1}$ since $B\cdot{}^{tr}B = \mathbb{1}\cdot D = \Lambda^2$ for a diagonal matrix over $\bar{k}_o = \mathbb{R}$. We now have ${}^{tr}XAX^{\tau}\cdot{}^{tr}YA^{-1}Y^{\tau} - \langle X,Y\rangle\langle X,Y\rangle^{\tau} = {}^{tr}UU^{\tau}\cdot{}^{tr}VV^{\tau} - \langle U,V\rangle\langle U,V\rangle^{\tau}$ (where ${}^{tr}U := {}^{tr}X\cdot{}^{tr}B^{\tau}\cdot\Lambda$, ${}^{tr}V := {}^{tr}Y\cdot{}^{tr}B^{\tau}\cdot\Lambda^{-1}$) and this is non negative since for all positive definite hermitean forms $\chi$ over the fields in this chapter we have the Schwarz inequality $\chi(u,u)\cdot\chi(v,v)-\chi(u,v)\chi(u,v)^{\tau} \geq 0$.

## 3. The example "dim E/V = 2". The invariant $\{\Omega_{ij}\}$

We keep the notations of section 1 in chap. XII and assume that

(3.1) $\quad (k_o,<)$ is archimedean ordered, so $\mathbb{R} = \bar{k}_o$.

Let $V \subset (E,\Phi)$ be a $\perp$-dense subspace with

(3.2) $\quad \dim E/V = 2$.

3.3. Let $(v_i)_{i\in\mathbb{N}}$ be an orthonormal basis of $V$ and $\{f_1,f_2\}$ an orthonormal basis of some supplement of $V$ in $E$,

(3.3.1) $$\mathfrak{B} = (v_i)_{\mathbb{N}} \cup \{f_1,f_2\}.$$

We recall that for each $n \in \mathbb{N}$ the Gram matrix $(A_{\iota\kappa n})$ is positive semidefinite. Since $V^{\perp} = (0)$ there is $n_o \in \mathbb{N}$ such that the center element of $k$,

(3.3.2) $$D_n := A_{11n}A_{22n} - A_{12n}A_{12n}^{\tau}$$

is non zero. Hence the hermitean matrix $A_n = (A_{\iota\kappa n})_{1\leq\iota,\kappa\leq 2}$ is positive definite and invertible for $n \geq n_o$ (cf. 2.2). $A_n^{-1}$ is positive definite for $n \geq n_o$; therefore we find (cf. the remark after 2.8) that (for $n \geq n_o$) $D_{n+1} - D_n = D_n\,{}^{tr}YA_n^{-1}Y^{\tau} \geq 0$ where

${}^{tr}Y = (\alpha^{\tau}_{1,n+1}, \alpha^{\tau}_{2,n+1})$ (cf. 2.4) . In other words $D_n$ is monotonically increasing with $n$ .

3.4. From the previous paragraph we conclude that the following are equivalent: (i) $\mathfrak{B}$ is no standard basis, (ii) One at least among $\lim_{n\to\infty} A_{11n}$ , $\lim_{n\to\infty} A_{22n}$ is $\infty$ , (iii) $\lim_{n\to\infty} D_n = \infty$ .

3.5. From the identity (XII.6.1.1) we conclude that for $X \in k^2$ an arbitrary "columnvector" the sequence ${}^{tr}XA_n^{-1}X^{\tau}$ is monotonically decreasing as $n$ tends to $\infty$ . Therefore (3.1) the limit

$${}^{tr}XBX^{\tau} := {}^{tr}X\cdot\lim_{n\to\infty} A_n^{-1}X^{\tau} \text{ exists in } \bar{k}_o \otimes k = \mathbb{R} \otimes k .$$

For suitable choices of $X$ we see that $B := \lim_n A_n^{-1}$ exists in $\bar{k}_o \otimes k$ . For $B = (B_{ij})$ we have

$$B = \lim_{n\to\infty} \frac{A_n^*}{D_n} .$$

3.6. Consider now two $\perp$-dense embeddings $V \subset E$ , $\bar{V} \subset E$ with bases $\mathfrak{B}, \bar{\mathfrak{B}}$ as in (3.3.1). We assume that $\mathfrak{B}$ and hence $\bar{\mathfrak{B}}$ possess one (and hence all) properties of (3.4). We recall that if $V \subset E$ should admit at least one standard basis then all bases $\mathfrak{B}$ are standard bases. In our example we shall generally assume that there are no standard bases. If there is a metric automorphism $T$ of $E$ with $TV = \bar{V}$ then one finds an invertible $2 \times 2$ matrix $N = (\nu_{ij})_{1\le i,j\le 2}$ and a natural number $n_1$ such that with respect to the bases $\bar{\mathfrak{B}}$, $\mathfrak{B}^o = (\bar{v}_i)_{\mathbb{N}} \cup \{Tf_1, Tf_2\}$ we have

$$(3.6.1) \qquad (\bar{A}_{ijn}) - \mathbb{1} = N\cdot((\mathring{A}_{ijn}) - \mathbb{1})\cdot{}^{tr}N^{\tau} \qquad (n \ge n_1)$$

(cf.(XII.3.1)).

In Sec. 3 of Chapter XII where we did have standard bases one could let $n$ go to infinity and from (3.6.1) derive a transformation for the $\psi_{ij} := \lim_{n\to\infty} (A_{ijn}) - \mathbb{1}$ . In the present situation this makes no sense however, by (3.5) we do have the existence of the limits

$$(3.6.2) \qquad \Omega_{ij} := \lim_{n\to\infty} \frac{A_{ijn}}{D_n} \in \bar{k}_o \otimes k \quad .$$

We shall see in Sec. 6 below that (3.6.1) does indeed yield a transformation law for the quantities $\Omega_{ij}$ ; it will be akin to the transformation laws of tensor densities (as arise e.g. in differential geometry) in that the $\Omega_{ij}$ transform like a tensor but with an extra factor depending on the transformation matrix.

3.7. The invariants. Consider the situation of the previous paragraph. We decide for the $\Omega_{ij}$ $(1\le i,j\le 2)$ as the characteristic quantities of the subspace $V \subset E$ , or rather - as the $\Omega_{ij}$ do depend on the bases $\mathfrak{B}$ - for the class of $(\Omega_{ij})$ modulo the appropriate transformations.

There is an important argument in favor of the quantities $\Omega_{ij}$ . In the proof of the main theorem in chapter XII a crucial rôle was played by the function $\pi_{m,n}$ (see XII.6.2); we had $\pi_{m,n}(X) = ({}^{tr}X{}^{tr}A^{\tau})\cdot B_n^{tr}({}^{tr}X{}^{tr}A^{\tau})^{\tau}$ , in other words $\pi_{m,n}$ is essentially defined by means of $B$ . Thus the functions $\pi_{m,n}$ of Lemma XII.6.3 - associated with bases $\mathfrak{B}$ of embeddings - are also defined when $\mathfrak{B}$ is no standard basis; in fact $\pi_{m,n}$ will continue to be of major importance . By ( 3.5) we have

$$(3.7.1) \qquad \Omega_{11} = B_{22} \;,\quad \Omega_{22} = B_{11} \;,\quad \Omega_{12} = -B_{12}$$

(3.8) If $V \subset E$ admits no standard basis then $\Omega_{11}\Omega_{22} - \Omega_{12}\Omega_{12}^{\tau} = 0$ . Indeed, by (3.6.2) $\Omega_{11}\Omega_{22} - \Omega_{12}\Omega_{12}{}^{\tau} = \lim\limits_{n\to\infty} \frac{1}{D_n} = 0$ .

## 4. A change of the basis in $V$ does not affect the $\Omega_{ij}$

We keep the notations of the previous section and assert:

(4.1) If in (3.3.1) the basis $(v_i)_{\mathbb{N}}$ of $V$ is replaced by another orthogonal basis $(\bar{v}_i)_{\mathbb{N}}$ of $V$ then the $\Omega_{ij}$ of (3.6.2) are not changed.

It will be sufficient to show that for arbitrarily fixed $X \in k^2$ the number ${}^{tr}X\cdot\Omega\cdot X^{\tau}$ will remain unaffected, i.e. that we shall have

(4.2) $\qquad {}^{tr}X\cdot\Omega\, X^{\tau} = {}^{tr}X\bar{\Omega}\cdot X^{\tau}$

where $\bar{\Omega}$ is defined with respect to $\bar{\mathfrak{B}} = (\bar{v}_i)_{\mathbb{N}} \cup \{f_1,f_2\}$ , and $\bar{v}_i = \sum_j \lambda_{ij} v_j \quad (i \in \mathbb{N})$ .

4.3. We have

(4.3.1) $\qquad \bar{\alpha}_{\iota i} = \phi(f_\iota,\bar{v}_i) = \sum_j \alpha_{\iota j}\lambda_{ij}^{\tau} \qquad (\iota=1,2)$

and, since $(\bar{v}_i)_{\mathbb{N}}$ is orthonormal, furthermore

(4.3.2) $\qquad \Sigma\lambda_{rj}\lambda_{sj}^{\tau} = \delta_{rs}$ (Kronecker) .

We abbreviate as follows

(4.3.3) $\qquad [\iota,i,n] := \Sigma_{j=1}^{n}\alpha_{\iota j}\lambda_{ij}^{\tau}$

(4.3.4) $\qquad$ for $m,n \in \mathbb{N}$ $A(m,n)_{\iota\kappa} := (\Sigma_{i=1}^{m}[\iota,i,n][\kappa,i,n]^{\tau})_{1\le\iota,\kappa\le 2}$ .

4.4. $A(m,n)$ is a Gram matrix , thus whenever its "determinant"

$$D(m,n) := A(m,n)_{11}A(m,n)_{22} - A(m,n)_{12}A(m,n)_{12}^{\tau}$$

is non zero then $D(m,n) > 0$ and $A(m,n)$ is positive definite, $A(m,n)^{-1}$ exists and is positive definite as well. Since the matrix $(\lambda_{ij})$ is row-finite each sum $[\iota,i,\infty] := \Sigma_{j=1}^{\infty}\alpha_{\iota j}\lambda_{ij}^{\tau}$ has only a finite number of non zero summands and thus

(4.4.1) $\qquad D(m,\infty) := \lim_{n\to\infty} D(m,n)$

trivially exists.

Now if $x \in E$ satisfies $x \perp \bar{v}_i$ for all $i \in \mathbb{N}$ then $x = 0$ as $V^{\perp} = (0)$ ; from this fact one concludes that $D(m,\infty)$ cannot vanish for all $m$ . Hence there is $m_o \in \mathbb{N}$ such that $D(m,\infty) \neq 0$ for $m \geq m_o$ ; therefore there exists $n_o(m) \in \mathbb{N}$ such that

(4.4.2) $\qquad D(m,n) \neq 0$ for $m \geq m_o$ and $n \geq n_o(m)$ .

4.5. From (4.4.2) we obtain that $A(m,n)^{-1}$ exists and is positive definite for $m \geq m_o$ and $n \geq n_o(m)$. By the argument following (3.3.2) we have that $D(m+1,n) - D(m,n)$ is nonnegative ; therefore

(4.5.1) $\quad D(t,n) > 0$ for all $t \geq m \geq m_o$ and $n \geq n_o(m)$.

We may therefore consider the function

$$L(t,n) := {}^{tr}X \cdot \frac{A(t,n)}{D(t,n)} \cdot X^{\tau} \quad \text{for all } t \geq m \geq m_o \text{ and } n \geq n_o(m)$$

Let us look, for some fixed $t$, at the matrices $A(t+1,n)$, $A(t,n)$. They are related just as $A_1$ and $A$ in (2.4) (with $[\iota,t+1,n_o(m)]^{\tau}$ in the rôle of $y_\iota$). Hence we may argue as in (2.9) to conclude that $L(t,n) - L(t+1,n) \geq 0$ for $t \geq m \geq m_o$ and $n \geq n_o(m)$. Thus the nonnegative value $L(t,n)$ decreases monotonically as $t$ tends to $\infty$ ; so by (3.1) there exists the limit (in $\bar{k}_o \otimes k$)

(4.5.2) $\quad L(\infty,n) := \lim_{t\to\infty} L(t,n) \leq L(t,n)$ for all $t \geq m \geq m_o$, $n \geq n_o(m)$.

4.6. If in $A(t,n)$ one expands the products $[\iota,i,n][\kappa,i,n]^{\tau}$ we see by (4.3.2) that

(4.6.1) $\quad L(\infty,n) = {}^{tr}X \dfrac{A_n}{D_n} X^{\tau}$

where $A_n$ is, as usual, the matrix $(\Sigma_{j=1}^{n} \alpha_{\iota j} \alpha^{\tau}_{\kappa j})_{1\leq\iota,\kappa\leq 2}$ and $D_n$ its "determinant". If $n \to \infty$ then by (3.5) the right hand side in (4.6.1) has a limit which is $\leq L(\infty,n)$ ; the limit clearly is ${}^{tr}X\Omega X^{\tau}$. Thus ${}^{tr}X\Omega X^{\tau} \leq L(\infty,n)$ and by (4.5.2) we obtain

(4.6.2) $\quad {}^{tr}X\Omega X^{\tau} \leq L(t,n)$ for all $t \geq m \geq m_o$ and $n \geq n_o(m)$.

4.7. If $n \to \infty$ then $L(t,n)$ converges to some limit $L(t,\infty)$ by (4.4.1). ${}^{tr}X\Omega X^{\tau} \leq L(t,\infty)$ by (4.6.2). Now $L(t,\infty)$ is by the very definitions (4.3.1), (4.3) equal to ${}^{tr}X \cdot \bar{\Omega}_t \cdot X^{\tau}$ where $\bar{\Omega}_t$ is the matrix $(\frac{\bar{A}_{\iota\kappa t}}{\bar{D}_t})$. Hence ${}^{tr}X\Omega X^{\tau} \leq {}^{tr}X\bar{\Omega}_t X^{\tau}$ for all $t \geq m_o$. If we let $t$ tend to $\infty$ then $\bar{\Omega}_t$ converges to $\bar{\Omega}$ so

$${}^{tr}X\Omega X^{\tau} \leq {}^{tr}X\bar{\Omega}X^{\tau} .$$

The opposite inequality follows by a symmetric argument. This proves (4.2).

## 5. Transformation formula for $D_n$

5.1. Let $\mathfrak{B}, \bar{\mathfrak{B}}$ be two bases (3.3.1) for the embedding $V \subset E$ with

(5.1.1) $v_i = \bar{v}_i \quad (i \in \mathbb{N})$ , $\bar{f}_j - \Sigma \nu_{jr} f_r \in V \quad (j=1,2)$ where $N = (\nu_{jr})$

is invertible. By using the fact that $(v_i)_{\mathbb{N}}$ , $\{f_1, f_2\}$ , $\{\bar{f}_1, \bar{f}_2\}$ are orthonormal it follows that

(5.1.2) $\bar{A}_n - \mathbb{1} = N(A_n - \mathbb{1}) \cdot {}^{tr}N^{\tau} \qquad (m \geq n_o)$

where $n_o$ is sufficiently large (cf. XII 3.1.4) . $A_n$ is the Gram matrix of 3.3. In order to express $\bar{D}_n$ (3.3.2) in terms of $D_n$ we need

5.2. Lemma on noncommutative $2 \times 2$ determinants. Let $M = \begin{pmatrix} ab \\ cd \end{pmatrix}$ be a $2 \times 2$ matrix. We define

(5.2.1) $$|M| := \begin{cases} -cb & \text{if } a = o \\ ad - aca^{-1}b & \text{if } a \neq o \end{cases}$$

(5.2.2) If $A$ is a hermitean $2 \times 2$ matrix then $|M \cdot A \cdot {}^{tr}M^{\tau}| - |A| \Delta_M = 0$ where $\Delta_M = aa^{\tau} \cdot dd^{\tau} + bb^{\tau} \cdot cc^{\tau} - ca^{\tau} \cdot bd^{\tau} - db^{\tau} \cdot ac^{\tau}$ and $\Delta_M \neq o$ for $M$ invertible.

The proof of this formula is elementary but somewhat tedious; it relies essentially on the fact that the field (if not commutative) is of type $(\frac{\alpha, \beta}{k_o})$ . Let $A = \begin{pmatrix} u & w \\ w^{\tau} & v \end{pmatrix}$ . Substituting the definitions for $|A|$ and $|M \cdot A \cdot {}^{tr}M^{\tau}|$ most of the terms on the left hand side of (5.2.2) will cancel by systematically observing that $\mathrm{Tr}(xyz) = \mathrm{Tr}(zxy)$ where $\mathrm{Tr}(n)$ means $n + n^{\tau}$ ; in this way the left hand side of (5.2.2) is seen to reduce to

$(bw^{\tau}a^{\tau}dw^{\tau}c^{\tau} + awb^{\tau}cwd^{\tau} + awb^{\tau}dw^{\tau}c^{\tau} + bw^{\tau}a^{\tau}cwd^{\tau}) - (dw^{\tau}a^{\tau}bw^{\tau}c^{\tau} + cwb^{\tau}awd^{\tau})$ $- (ww^{\tau}ca^{\tau}bd^{\tau} + ww^{\tau}db^{\tau}ac^{\tau}) = \mathrm{Tr}p \cdot \mathrm{Tr}q - \mathrm{Tr}(pq) - \mathrm{Tr}(pq^{\tau}) = 0$ ; $p = b^{\tau}aw$ and $q = d^{\tau}cw$ . Q.E.D.

(5.2.3) Remark. 1) One can almost get the formula (5.2.2) without computations by making use of the Dieudonné determinant (for a nice exposition of this determinant see[1] p. 151):
$|MA\cdot{}^{tr}M^{\tau}| = |M||A||{}^{tr}M^{\tau}|\cdot r_1 = |A|\cdot|M|\cdot|{}^{tr}M^{\tau}|\cdot r_2 = |A||M\cdot{}^{tr}M^{\tau}|\cdot r_3 =$
$|A|\cdot\Delta\cdot r_4$ where $r_1,r_2,r_3,r_4 \in [k^*,k^*]$, the commutator subgroup of $k^* = k\setminus\{0\}$. Since $A$ is hermitean the two determinants are in $k_o$, so $r_4 \in k_o \cap [k^*,k^*]$; therefore $r_4 = \pm 1$ since quaternions that are products of commutators are obviously of "norm" 1.

2) The same conclusion as in the previous remark may be obtained as follows. One makes use of the fact that the algebra $(\frac{\alpha,\beta}{k_o})$ is isomorphic to an algebra of $2\times 2$ matrices over the commutative field $k(\sqrt{\alpha})$ by associating with a quaternion $q = a_o + ia_1 + j(a_2 + ia_3) = z_1 + jz_2$ the matrix

$$\omega(q) = \begin{pmatrix} z_1 & z_2 \\ \beta z_2^{\tau} & z_1^{\tau} \end{pmatrix} \qquad \text{where } z_1^{\tau} = a_o - ia_1 \quad \text{etc.}$$

If $(q_{rs})$ is a hermitean $2\times 2$ matrix over $(\frac{\alpha,\beta}{k_o})$ and $|(q_{rs})|$ as defined in 5.2.1 then $|(q_{rs})| \in k_o$ and for the usual determinant of the $4\times 4$ matrix $(\omega(q_{rs}))$ one finds

$$\det(\omega(q_{rs})) = |(q_{rs})|^2 .$$

Therefore $|MA{}^{tr}M^{\tau}|^2 = \det(\omega(M)\omega(A)\omega({}^{tr}M^{\tau})) = \det\omega(A)\cdot\det\omega(M{}^{tr}M^{\tau}) =$
$|A|^2\cdot|M{}^{tr}M^{\tau}|^2 = |A|^2\Delta_M^2$.

5.3. Consider now the situation of 5.1. By formula (5.2.2) we have

$$(5.3.1) \qquad |\bar{A}_n - \mathbb{1}| = |A_n - \mathbb{1}|\cdot\Delta_N$$

By the definition (5.2.1) we find $|\bar{A}_n - \mathbb{1}| = \bar{D}_n - \bar{A}_{11n} - \bar{A}_{22n} + 1$ and similarly for $|A_n - \mathbb{1}|$; expressing here $\bar{A}_{iin}$ by means of (5.1.2) we finally have from (5.3.1)

$$(5.3.2) \qquad \bar{D}_n = 1 + \sum_{j,r=1}^{2} \nu_{1j}(A_{jrn} - \delta_{jr})\nu_{1r}^{\tau} + \sum_{j,r=1}^{2} \nu_{2j}(A_{jrn} - \delta_{jr})\nu_{2r}^{\tau} + (D_n - A_{11n} - A_{22n} + 1)\Delta_N .$$

5.4. Assume that the basis $\mathfrak{B}$ for the $\perp$-dense embedding $V \subset E$ is no standard basis (cf. 3.4). Let $\bar{\mathfrak{B}}$ be a second basis for $V \subset E$ satisfying (5.1.1).

(5.4.1) The limit $\rho(\mathfrak{B},\bar{\mathfrak{B}}) := \lim_{n\to\infty} \frac{\bar{D}_n}{D_n} \in \bar{k}_o \otimes k (= \mathbb{R}\otimes k)$ exists, and

(5.4.2) $$\rho(\mathfrak{B},\bar{\mathfrak{B}}) = \Sigma \nu_{1j}\Omega_{jr}\nu_{1r}^{\tau} + \Sigma \nu_{2j}\Omega_{jr}\nu_{2r}^{\tau} + (1-\Omega_{11}-\Omega_{22})\Delta_N$$

$$\text{(where } \Delta_N = \nu_{11}\nu_{11}^{\tau}\nu_{22}\nu_{22}^{\tau} + \nu_{12}\nu_{12}^{\tau}\nu_{21}\nu_{21}^{\tau} - \nu_{21}\nu_{11}^{\tau}\nu_{12}\nu_{22}^{\tau} - \nu_{22}\nu_{12}^{\tau}\nu_{11}\nu_{21}^{\tau}\text{)},$$

(5.4.3) $$\rho(\mathfrak{B},\bar{\mathfrak{B}})\,\rho(\bar{\mathfrak{B}},\mathfrak{B}) = 1 .$$

These assertions follow directly from (5.3.2) and the definition (3.6.2).

## 6. Transformation law for the quantities $\Omega_{ij}$

As before we assume that $V \subset E$ admits no standardbasis.

Let $\mathfrak{B},\bar{\mathfrak{B}}$ two arbitrary bases for $V \subset E$, $V^{\perp} = (0)$, $\dim E/V = \dim E/\bar{V} = 2$. There is a uniquely determined, invertible $2 \times 2$ matrix $N = (\nu_{ij})$ over $k$ such that

$$\bar{f}_i - \Sigma \nu_{ij} f_j \in V .$$

We assert that

$$\bar{\Omega} = \rho \cdot N \cdot \Omega \cdot {}^{tr}N^{\tau} \quad \text{where (5.4)} \quad \rho = \rho(\bar{\mathfrak{B}},\mathfrak{B}) .$$

Indeed, by (5.1.2) we have for $n \geq n_o$ (sufficiently large) that $\bar{\Omega}_n - \frac{\mathbb{1}}{\bar{D}_n} = \frac{D_n}{\bar{D}_n} \cdot N \cdot \Omega_n \cdot {}^{tr}N^{\tau} - \frac{1}{\bar{D}_n} N \cdot {}^{tr}N^{\tau}$. As $n \to \infty$ both $D_n, \bar{D}_n$ tend to infinity, hence the assertion.

## 7. The main theorem ($\dim E/V = 2$; $D_n \to \infty$)

7.1. Theorem. Let $(E,\Phi)$ be a weakly universal k-space with $\dim E = \aleph_o$, $1 \in \|\Phi\|$ and where the field $k$ is as described under the caption of Chapter Thirteen. Let $V,\bar{V} \subset E$ be positive definite subspaces with $\dim E/V = \dim E/\bar{V} = 2$ , $V^{\perp} = \bar{V}^{\perp} = (0)$ and such that there are no standard bases for the embeddings $V \subset E$, $\bar{V} \subset E$ . If there exists a metric automorphism $T$ of $E$ with $TV = \bar{V}$ then there exists an invertible $2 \times 2$ matrix $N = (\nu_{ij})$ over $k$ such that

$$(7.1.1) \quad \rho := \Sigma\nu_{1j}\Omega_{jr}\nu_{1r}^{\tau} + \Sigma\nu_{2j}\Omega_{jr}\nu_{2r}^{\tau} + (1-\Omega_{11}-\Omega_{22})\Delta_N > 0$$

and

$$(7.1.2) \quad \bar{\Omega}_{ij} = \rho^{-1}\cdot\Sigma\nu_{ir}\Omega_{rs}\nu_{js}^{\tau}$$

($\Delta_N$ as in 5.4.2). Conversely, if the field $k$ satisfies condition (4) in chap. XII then the two conditions will be sufficient for the existence of a metric automorphism $T$ of $E$ with $TV = \bar{V}$ .

7.2. Proof of necessity. Let $\mathfrak{B},\bar{\mathfrak{B}}$ be the two arbitrary bases with respect to which $\Omega$ and $\bar{\Omega}$ are defined. Let $\mathfrak{B}'$ be the "intermediate" basis $(\bar{v}_i)_{i\in\mathbb{N}} \cup \{Tf_1,Tf_2\}$ . $\Omega_{ij} = \Omega'_{ij}$ by (4.1). Expressing $\bar{\mathfrak{B}}$ in terms of $\mathfrak{B}'$ (Sec.6) yields $\bar{\Omega} = \rho\cdot N\cdot\Omega\cdot{}^{tr}N^{\tau}$ with $\rho = \rho(\bar{\mathfrak{B}},\mathfrak{B}')$ . Trivially $\rho > 0$ and by (5.4.2) $\rho$ is of the requisite shape. Since $\Omega'_{ij} = \Omega_{ij}$ we obtain (7.1.1) and (7.1.2).

7.3. Proof of sufficiency. 1. In order to construct $T$ we first show that we can introduce a basis $\mathfrak{B}'$ for $V \subset E$ such that $\Omega'_{ij} = \bar{\Omega}_{ij}$. In order to find $\mathfrak{B}' = (v'_i)_{i\in\mathbb{N}} \cup (f'_1,f'_2)$ we set $v'_i = v_i$ ($v_i$ in the basis $\mathfrak{B}$ defining $\Omega$) and $f'_i = \Sigma\nu_{ij}f_j + \Sigma_1^n\zeta_{ir}v_r$ $(i=1,2)$ and seek to determine $n$ and the $\zeta_{1r},\zeta_{2r}$ $(1\leq r\leq n)$ such that $\{f'_1,f'_2\}$ is orthonormal. The "orthogonality conditions" read $\mathbb{1} = +NA_n\cdot{}^{tr}X^{\tau} + X^{tr}A_n^{\tau}\cdot{}^{tr}N^{\tau} + X\cdot{}^{tr}X^{\tau} + N\cdot{}^{tr}N^{\tau}$ , where $A_n$ is the $2 \times n$ matrix $(\alpha_{ij})$ and $X$ is the $2 \times n$ matrix $(\zeta_{ir})$ . With the substitution $Y := NA_n + X$ the orthogonality conditions become

$$Y\cdot{}^{tr}Y^{\tau} = \mathbb{1} + N(A_n-\mathbb{1})\cdot{}^{tr}N^{\tau} \quad .$$

In other words, the given hermitean matrix $M_n := \mathbb{1} + N(A_n-\mathbb{1})\cdot{}^{tr}N^{\tau}$ should, for suitable $n$ , be expressible as a n-Gram matrix. In view of the assumption (4) of chap. XII put down on $k$ by the theorem this is the case (XII 1.5.1) if $M_n$ is positive definite for some n . $M_n$ is positive definite if and only if $\delta_n := M_{n11}M_{n22} - M_{n12}M^{\tau}_{n12} > 0$ . Now $M_n - \mathbb{1} = N(A_n - \mathbb{1}) \cdot {}^{tr}N^{\tau}$ gives $\delta_n - M_{n11} - M_{n22} + 1 = |N(A_n - \mathbb{1})\,{}^{tr}N^{\tau}|$ $= |A_n - \mathbb{1}|\Delta_N = (D_n - A_{11n} - A_{22n} + 1)\Delta_N$ by (5.2.2). Therefore we see that $\lim_{n\to\infty} \delta_n/D_n$ exists and equals $\rho$ of (7.1.1). As $D_n$ is nonnegative $(n \geq n_o)$ and $\rho > 0$ we have $D_n > 0$ for $n \geq n_o$ . Hence we can find an orthonormal basis $\{f_1', f_2'\}$ (and hence $\mathfrak{B}'$). By (6) we have $\Omega' = \rho N \cdot \Omega \cdot {}^{tr}N^{\tau}$ with $\rho = \rho(\mathfrak{B}', \mathfrak{B})$ . Therefore $\bar{\Omega} = \Omega'$ .

2. For the rest of the proof we may (by what we just proved) assume that
$$\bar{\Omega} = \Omega \ .$$
The construction of $T$ is practically identical with the one given in XII 7.1; the only difference being that instead of "$A_{\iota\kappa} = \bar{A}_{\iota\kappa}$" - which is meaningless here - we must work with the induction assumption "$\Omega_{\iota\kappa} = \bar{\Omega}_{\iota\kappa}$" . We needed the assumption in order to conclude that $B_{\iota\kappa} = \bar{B}_{\iota\kappa}$ where $B_{\iota\kappa} = (A^{-1})_{\iota\kappa}$ . By (3.6.1) we get this more directly from "$\Omega_{\iota\kappa} = \bar{\Omega}_{\iota\kappa}$" . The only thing which remains to be done here is the non trivial verification that the induction assumption can be saved from one construction step to the next. Technically speaking, this amounts to the following: By definition $\Omega_{ij} = \lim_{n\to\infty} \frac{A_{ijn}}{D_n}$ . Now, after the first step described in XII 7.1.1 the vector $v_1$ is chopped off. Setting $\alpha_i = \phi(f_i, v_1)$ $(i=1,2)$ one is therefore left with the associated matrix $A'_{ijn} = A_{ijn} - \alpha_i\alpha_j^{\tau}$ . The question is: Do we after the first step have $\Omega'_{ij} = \lim \frac{A'_{ijn}}{D'_n} = \bar{\Omega}'_{ij}$ ? Here $\bar{\Omega}'_{ij}$ and $\bar{\alpha}_i$ are of course analogously defined and $\alpha_i = \bar{\alpha}_i$ by the very construction of XII 7.1.1 Expanding $D'_n$ we find $(n \geq n_o)$ $\frac{D'_n}{D_n} = 1 - {}^{tr}X \cdot A_n^{-1} X^{\tau}$ where $X = \begin{pmatrix} \alpha_1 \\ \alpha_2 \end{pmatrix}$ .

In other words $\frac{D'_n}{D_n} = 1 - \pi_{1n}(1) > 0$ by Lemma XII.6.3 . We obtain thus the interesting formula $(n\to\infty)$

(7.3.1) $\Omega'_{ij} = \Omega_{ij}(1-\pi_{1,\infty}(1))^{-1}$ where $\Omega'_{ij}$ belongs to the embedding $k(v_i)_{i\geq 2} \subset k(v_i)_{i\geq 2} \oplus k(f_1,f_2)$ and $\pi_{m,\infty}$ is the function in Lemma XII.6.3.

It is now evident that after the first construction step we shall have $\Omega'_{ij} = \bar{\Omega}'_{ij}$ and hence the proof of the sufficiency is complete.

7.4. Corollary 1. Let $V,\bar{V} \subset$ be as in theorem (7.1) and $TV = \bar{V}$ for a metric automorphism of $E$ then $\rho(1\text{-trace }\bar{\Omega}) = (1\text{-trace }\Omega)\Delta_N$

Proof. We have (Sec. 6) $|\bar{A}_n - 1\!\!1| = |A_n - 1\!\!1|\Delta_N$ ; expanding this gives $\bar{D}_n - \bar{A}_{11n} - \bar{A}_{22n} + 1 = (D_n - A_{11n} - A_{22n} + 1)\Delta_n$ . Divide by $D_n$ and factor out $\frac{\bar{D}_n}{D_n}$ on the left. For $n \to \infty$ the assertion follows. Another possibility is to divide (7.1.1) by $\rho$ and use (7.1.2).

7.5. Corollary 2. Let $V \subset (E,\Phi)$ be as in the theorem. The statements "trace $\Omega = 0$", "$0 <$ trace $\Omega < 1$", "trace $\Omega = 1$", "trace $\Omega > 1$" are properties of the orbits (under the orthogonal group of E) of $V$ .

Proof. If trace $\Omega = 0$ then $\Omega_{11} = \Omega_{22} = 0$ , hence $\Omega = 0$ by (3.8). By (7.1.2) $\bar{\Omega} = 0$ for any $\bar{V} \subset E$ in the same orbit as $V$ . If trace $\Omega \neq 0$ the assertions follow by checking the signature of both sides in the equation (7.4).

## 8. Embeddings (dim $E/V = 2$, $D_n \to \infty$) that split

For the definition of "splitting embedding" see XII 13. Corresponding to XII 14 we have here

8.1. Theorem. Assume that $k$ satisfies (4) of Chapter Twelve and that $V \subset E$ is a $\perp$-dense embedding which admits no standard basis. Let $\dim E/V = 2$ . Then the embedding splits if and only if there is some invertible $2 \times 2$ matrix over $k$ such that (7.1.1) holds and $\bar{\Omega}$ of (7.1.2) turns out diagonal.

Proof. If $V \subset E$ splits then we shall have $\rho > 0$ and $\bar{\Omega}$ diagonal for suitable $N$ . Conversely, if $\Omega_{12} = 0$ , then $\Omega_{11}\Omega_{22} = 0$ by

(3.8), say $\Omega_{22} = 0$. We define an embedding $\bar{V} \subset \bar{E}$ with $\bar{\alpha}_{1,2i} = \bar{\alpha}_{2,2i-1} = 0$, $\sum_i \bar{\alpha}_{1,2i-1}\bar{\alpha}^{\tau}_{1,2i-1} = \infty$ and $\sum_i \bar{\alpha}_{2,2i}\bar{\alpha}^{\tau}_{2,2i} = \Omega_{11}^{-1}$ if $\Omega_{11} \neq 0$ and $\infty$ otherwise. We have $\Omega_{ij} = \bar{\Omega}_{ij}$ (i=1,2). We now quote Theorem 7.1.

The following theorem has no analogue when standard bases are present.

8.2. <u>Theorem</u>. Let $V \subset (E,\Phi)$ be as in 8.1 $V \subset E$ splits iff and only if $\Omega_{12} = 0$ or if $\Omega_{11}\Omega_{12}^{-1} \in k$ ( [4] , Satz 8 ) .

<u>Proof</u>. If $V \subset E$ splits then by (8.1) we have $\Omega_{12} = \rho\Sigma\nu_{1j}\Omega_{jr}\nu^{\tau}_{2r} = 0$ for some invertible $N = (\nu_{ij})$. Multiplying this equation by $\Omega_{11}$ yields (3.8) $(\nu_{11}\Omega_{11} + \nu_{12}\Omega^{\tau}_{12})(\Omega_{11}\nu^{\tau}_{21} + \Omega_{12}\nu^{\tau}_{22}) = 0$ hence either $\Omega_{12} = 0$ or $\Omega_{11}\Omega_{12}^{-1} \in k$. Conversely, if $\Omega_{12} = 0$ we quote 8.1; if $\Omega_{11}\Omega_{12}^{-1} = \varepsilon \in k$ we set $N = \begin{pmatrix} 1 & \varepsilon^{\tau} \\ \lambda & -\lambda\varepsilon^{\tau} \end{pmatrix}$ for some $\lambda \in k_o \setminus \{0\}$. We have $\Delta_N = 2\lambda^2\varepsilon\varepsilon^{\tau} \neq 0$ and for $\rho$ (7.1.1) we find $\rho > 0$ if $\lambda$ is chosen sufficiently small. We may thus quote 7.1 to conclude that $\bar{\Omega}_{12} = 0$ for a suitable embedding. Hence $V \subset E$ splits by 8.1.

8.3. <u>Corollary</u>. Assume that $\tau$ is the identity and that every positive element in $k$ is a square. If $V \subset E$ is as in (8.1) then $V \subset E$ splits if and only if $\Omega_{11}, \Omega_{22}, \Omega_{12}$ are linearly dependent over $k$ ( [4] , p. 484 ) .

<u>Proof</u>. If $V \subset E$ splits then the assertion follows from (8.2). Let conversely $\alpha\Omega_{11} + \beta\Omega_{22} + \gamma\Omega_{12} = 0$ ($\alpha, \beta, \gamma$ in $k$ and not all zero). The case where $\Omega_{12} = 0$ is taken care of by (8.1) so let $\Omega_{12} \neq 0$ and hence $\Omega_{11}\Omega_{22} \neq 0$. Since $\Omega_{11}\Omega_{22} - \Omega_{12}^2 = 0$ it follows from our assumption that $\gamma^2 - 4\alpha\beta$ is a square; on the other hand division of our relation by $\Omega_{11}$ yields $\alpha + \beta(\Omega_{12}\Omega_{11}^{-1})^2 + \gamma(\Omega_{12}\Omega_{11}^{-1}) = 0$ ; in other words $\Omega_{12}\Omega_{11}^{-1}$ solves the equation $\alpha + \gamma\zeta + \beta\zeta^2 = 0$. As its discriminant is non negative its solutions are in $k$. In particular $\Omega_{12}\Omega_{11}^{-1} \in k$ which brings us back to 8.2.

## 9. An application to dense embeddings when $k_o = \mathbb{R}$

In this section we assume that $k_o = \bar{k}_o = \mathbb{R}$, in other words, $\Phi$ is symmetric and $k = \mathbb{R}$ or $\Phi$ is hermitean and $k$ is either $\mathbb{C}$ or $\mathbb{H} = (\frac{-1,-1}{\mathbb{R}})$ with $\tau$ the usual conjugation. Every k-space $(E,\Phi)$ is then weakly universal. As we assume that $1 \in \|\Phi\|$ we have that $E$ is of the form $E_{(+1)} \overset{\perp}{\oplus} E_{(-1)}$ where the spaces $E_{(+1)}$ and $-E_{(-1)}$ admit orthonormal bases. Without loss of generality we assume that $\dim E_{(+1)} = \aleph_o$. $n_- := \dim E_{(-1)}$ is a uniquely determined cardinal.

Our results on the case "dim E/V = 2" enable us to completely discuss the $\perp$-dense case where $\dim E/V$ is finite. This is accomplished by making use of splittings and of "combinatorial" considerations. Our first principal result is Thm. 9.2.

9.1. The second part of 7.3 shows the following. If $(\Omega_{ij})_{1\leq i,j\leq 2}$ is diagonal for a certain basis $\mathfrak{B} = (v_i)_{\mathbb{N}} \cup \{f_1,f_2\}$ then $E$ splits as follows:

$$E = (V_1 \oplus k(f_1)) \overset{\perp}{\oplus} (V_2 \oplus k(f_2)) \ , \quad V_1 \oplus V_2 = V \ ;$$

we say that $V \subset E$ splits with respect to $f_1$ and $f_2$. If $\Omega_{ij} = 0$ for $i,j = 1,2$ then $V \subset E$ splits with respect to each orthonormal basis $\{f_1,f_2\}$ which spans a supplement of $V$ in $E$.

The following situation will occur frequently below: $V \subsetneq E$ is a $\perp$-dense embedding which admits no standard basis and there is an $f_1 \in E \setminus V$ such that $A_{11} = \sum_1^\infty \Phi(f_1,v_i)\Phi(f_2,v_i)^\tau$ turns out to be finite for some (and hence each) orthonormal basis $(v_i)_{\mathbb{N}}$ of $V$. If then $\{f_1,f_2\}$ is orthonormal and spans a supplement of $V$ in $E$ then $V \subset E$ splits with respect to $\{f_1,f_2\}$. Indeed, if $A_{11} < \infty$ then $\Omega_{11} = 0$ so $\Omega_{12} = 0$ (3.8). (Incidentally, it is an exercise to directly show that $\lim_{n\to\infty} A_{12n}A_{22n}^{-1/2} = 0$ and therefore $\Omega_{22} = A_{11}^{-1}$ in this case.)

9.2. <u>Theorem</u>. Every $\perp$-dense embedding $V \subset E$ with $\dim E/V$ finite and $(V,\phi)$ positive definite splits ($k$ one of $\mathbb{R},\mathbb{C},\mathbb{H} := (\frac{-1,-1}{\mathbb{R}})$) .

The proof will be divided into three cases.

9.2.1. <u>Case I</u> : $\Sigma\phi(x,v_i)\phi(x,v_i)^\tau < \infty$ for all $x \in E \setminus V$ . We quote (1.3).

9.2.2. <u>Case II</u>: $\Sigma\phi(x,v_i)\phi(x,v_i)^\tau = \infty$ for all $x \in E \setminus V$ . We proceed by induction on $\dim E/V$ . For $m = 1,2$ nothing remains to be shown (8.2).

Let $\dim E/V = m+1$ and $F$ some m-dimensional subspace of $E$ contained in $E \setminus V$ . Upon induction assumption there is a splitting $V \oplus F = \overset{m_\perp}{\underset{1}{\oplus}} (V_j \oplus (f_j))$ , $V_j$ $\perp$-dense in $V_j \oplus (f_j)$ . Let $f_{n+1} \in E \setminus (V+F)$, $\phi(f_{m+1},f_i) = \delta_{m+1,i}$ (Kronecker). Consider in turn the $m$ embeddings $V_j \subset V_j \oplus k(f_j,f_{m+1})$ .

We claim that we can always arrange it so that all of these $m$ embeddings are $\perp$-dense embeddings. Firstly, it cannot happen that <u>all</u> of them fail to be $\perp$-dense embeddings, for otherwise there exists $f_{m+1} + \alpha_i f_i + v_i \in V_i^\perp$ with $v_i \in V_i$ . Hence the contradiction $f_{m+1} + \sum_i(\alpha_i f_i + v_i) \in V^\perp$ . Secondly, assume therefore that $V_1 \subset V_1 \oplus (f_1,f_{m+1})$ is $\perp$-dense. This embedding is either of type $\Omega = 0$ and splits with respect to $\{f_1,f_{m+1}\}$ or else $\Sigma\phi(f_{m+1},w_i)\phi(f_{m+1},w_i)^\tau$ is finite for some orthonormal basis $(w_i)_{\mathbb{N}}$ of $V_1$ and hence by (9.1) it splits again with respect to $\{f_1,f_{m+1}\}$. In either case $V_1 = V' \overset{\perp}{\oplus} V''$ with $f_{m+1} \perp V'$ , $f_1 \perp V''$ and $V'' \subset V'' \oplus k(f_{m+1})$ is a $\perp$-dense embedding. Thus if $(z_i)_{\mathbb{N}}$ is any orthogonal basis of $V''$ we shall have $\phi(f_{m+1},z_{i'}) \neq 0$ for infinitely many $i'$ . Partition $(z_i)_{i\in\mathbb{N}}$ into $m$ blocks with each block containing infinitely many of these $i'$ . Let $V^{(1)},\dots,V^{(m)}$ be the spans of these blocks, $V'' = V^{(1)} \overset{\perp}{\oplus}\dots\overset{\perp}{\oplus} V^{(m)}$ . If we now, in the original orthogonal decomposition of $V$ into $V_1,\dots,V_m$ , replace the $V_1,\dots,V_m$ in turn by $\bar{V}_1 := V' \oplus V^{(1)}$ , $\bar{V}_2 := V_2 \overset{\perp}{\oplus} V^{(2)},\dots,\bar{V}_m := V_m \oplus V^{(m)}$ then all embeddings $\bar{V}_i \subset \bar{V}_i + k(f_i,f_{m+1})$ will be $\perp$-dense and $V \oplus F = \overset{m\perp}{\underset{1}{\oplus}} (\bar{V}_j \oplus (f_j))$ .

Writing again $V_j$ instead of $\bar{V}_j$ we can now repeat the above combinatorial trick: All embeddings $V_j \subset V_j \oplus k(f_j,f_{m+1})$ are either of type $\Omega = 0$ and split with respect to $\{f_j,f_{m+1}\}$ or else $\Sigma\Phi(f_{m+1},w_i)\Phi(f_{m+1},w_i)^{\perp}$ is finite for some orthonormal basis $(w_i)_{\mathbb{N}}$ of $V_j$ and hence (9.1) split again with respect to $\{f_j,f_{m+1}\}$ . In any case $V_j = V_{j^o} \overset{\perp}{\oplus} V_{j^1}$ with $f_{m+1} \perp V_{j^o}$, $f_j \perp V_{j^1}$ . We set $V_{m+1} = \sum_{j=1}^{m} V_{j^1}$ and obtain the splitting $E = \overset{m\perp}{\underset{1}{\oplus}} (V_{j^o} \oplus (f_j)) \overset{\perp}{\oplus} (V_{m+1} \oplus (f_{m+1}))$.

We are thus left with the general case:

9.2.3. Case III: Let $F$ be an arbitrary supplement of $V$ in $E$ . The set $\{f \in F \mid \Sigma\Phi(f,v_i)\Phi(f,v_i)^{\tau} < \infty\}$ is a linear subspace $F_o$ of $F$ . Let $F_1 := F_0^{\perp} \cap F$ . We assert that there is a decomposition as follows.

(9.2.3.1) $E = (V_o \oplus F_o) \overset{\perp}{\oplus} (V_1 \oplus F_1)$ , $V = V_1 \oplus V_2$ , $V_i$ $\perp$-dense in $V_i \oplus F_i$ .

If this holds then we may apply cases I,II to the embeddings $V_i \subset V_i \oplus F_i$ respectively and obtain the assertion of the theorem.

For the proof of (9.2.3.1) we assume that $F_o \neq (0)$, $F_1 \neq (0)$ . Pick $f \in (V+F_o) \setminus V$ with $\Phi(f,f) = 1$ and set $F_{oo} = f^{\perp} \cap F_o$ . By an induction assumption there is a decomposition of the required kind for the embedding $V \subset V \oplus (F_{oo} \oplus F_1)$ , $V \oplus F_{oo} \oplus F_1 = (V_{oo} \oplus F_{oo}) \overset{\perp}{\oplus} (V_1 \oplus F_1)$ and, by case II, $V_1 \subset V_1 \oplus F_1$ has a splitting $V_1 \oplus F_1 = \oplus^{\perp} (V_{1j} \oplus (f_j))$. Precisely as in case II one now looks at the embeddings $V_{1j} \subset V_{1j} \oplus k(f_j,f)$ .

Are these embeddings $\perp$-dense? We can always modulate them in the manner explained above in Case II and achieve density. Notice that for this one also has to take into consideration a splitting of

$V_{oo} \subset V_{oo} \oplus F_{oo}$ (which exists by case I). We may therefore assume without loss of generality that our embeddings are $\perp$-dense and that they therefore split (by the results of Sec. 8) with respect to $\{f_j, f\}$ respectively. The desired decomposition for $V \oplus F_o \oplus F_1$ is now obtained just as in Case II.

Remark. In the Introduction to the chapter we gave a reduction of our investigation to the case where (1.1) holds. The discussion given there shows that we have splittings in the cases dismissed. Hence Theorem 2.9 actually holds whether $(V,\Phi)$ is definite or not. We formulate this important result in the following

9.3. Corollary. Let $k$ be one of $\mathbb{R}, \mathbb{C}, \mathbb{H}$ with obvious involution and $(E,\Phi)$ any non degenerate $\aleph_o$-dimensional hermitean space over $k$. If $V \subset E$ is a subspace with $m = \dim E/V \leq \infty$ and $V^\perp = (0)$ then the embedding $V \subset E$ splits, i.e. there is an orthogonal decomposition (Cf. Sec. 14 in Chap. XII.)

$$E = \overset{m}{\underset{i=1}{\oplus^\perp}} (V_i \oplus k(f_i)), \; V = \overset{m}{\underset{i=1}{\oplus}} V_i \quad .$$

Having established the existence of splittings we shall now be in a position to introduce and discuss orthogonal invariants for embeddings:

9.4. Definition. Let $V \subset E$ have $V^\perp = (0)$ and $\dim E/V \leqq \aleph_o$ . Let $(v_i)_{i\in\mathbb{N}}$ be some fixed orthonormal basis of $V$ . Then the set $V_3 := \{x \in E \mid \Sigma_1^\infty \Phi(x,v_i)\Phi(x,v_i)^\tau < \infty\}$ is a linear subspace of $E$ . On $V_3 \times V_3$ we consider the hermitean form $\Psi$ given by

$$\Psi(x,y) = \sum_{i=1}^{\infty} \Phi(x,v_i)\Phi(y,v_i)^\tau - \Phi(x,y) \quad .$$

We set $V_1 := \operatorname{rad} \Psi$ . Obviously $V \subset V_1$ . We let furthermore $V_2$ be some arbitrarly fixed maximal subspace of $V_3$ on which $\Psi$ is negative semidefinite. We put $n_i = \dim V_i/V_{i-1}$ where $V_o = V$ and $i = 1,2,3$ ; furthermore we set $n_4 := \dim E/V_3$ .

By XII 1.6 $V_1$ does not depend on the basis $(v_i)_{i\in\mathbb{N}}$, neither does $\Psi$ by the same token. Therefore $n_1, n_2, n_3$ are well defined and depend only on the embedding $V \subset E$. Obviously

$$(9.5) \qquad n_1 + n_2 + n_3 + n_4 = \dim E/V .$$

$V_2$ is a maximal positive definite overspace of $V$ in $E$, hence a norm topology can be introduced in $V_2$. $V_1$ is the closure of $V$ in this topology (as we have shown in Appendix I of chapter XII). $V_3$ is <u>the</u> maximal overspace of $V$ such that the embedding $V \subset V_3$ admits a standard basis (cf. XII 1.7.8).

<u>9.6</u>. Let us now look at a $\perp$-dense embedding $V \subset E$ with $\dim E/V = 1$. We introduce the element

$$S = \Sigma\Phi(f,v_i)\Phi(f,v_i)^\tau \in \mathbb{R} \cup \{\infty\}$$

where $f$ is a vector with $\phi(f,f) = 1$ and not in $V$ and where $(v_i)_{\mathbb{N}}$ is an orthonormal basis in $V$. It is not difficult to see that the square class of $(S-1)$ in $\mathbb{R} \cup \{\infty\}$ does not depend on the basis $(v_i)_{\mathbb{N}} \cup (f)$ chosen to describe the embedding. It is easy to prove that this square class constitutes a complete set of orthogonal invariants for the orbits under the orthogonal group in the set of positive definite $\perp$-dense hyperplanes in $E$. Thus <u>there are four orbits</u>, corresponding to the square classes (in $\mathbb{R} \cup \{\infty\}$) of

$$-1,\ 0,\ 1,\ \infty .$$

Representatives in the orbits are embeddings $V \subset E$ with $S$ in turn, say, $\frac{1}{2}$, $1$, $2$, $\infty$. The corresponding 4-tuples $(n_1,n_2,n_3,n_4)$ introduced in definition 9.4 above are in turn $(0,1,0,0)$, $(1,0,0,0)$, $(0,0,1,0)$, $(0,0,0,1)$. Hence we see that in a splitting $E = \overset{m}{\underset{i=1}{\oplus^\perp}} (V_i \oplus k(f_i))$ of a $\perp$-dense embedding $V \subset E$ with finite $\dim E/V$ the cardinal $n_1$ counts how many summands will be in the first orbit above, $n_2$ tells how many summands fall in the second orbit above etc.

We are now ready to show

9.7. Theorem. Let $k$ be one of $\mathbb{R}, \mathbb{C}, \mathbb{H}$ with obvious involution and $(E,\Phi)$ a non degenerate $\aleph_0$-dimensional hermitean k-space. The quadruple $\langle n_1, n_2, n_3, n_4 \rangle$ introduced in 9.4 characterizes the orbit of V (under the orthogonal group of E) in the set of positive definite subspaces $V$ of $E$ with $V^\perp = (0)$ and $0 < \dim E/V < \infty$.

Proof. If $V$ and $\bar{V}$ are such spaces and in the same orbit then any isometry $T$ of $E$ maps $V_3$ of definition 9.4 onto $(\bar{V})_3$. Hence $n_4 = \bar{n}_4$. Sylvester's law of inertia applied to the form $\Psi$ of 9.4 gives $n_i = \bar{n}_i$ $(i=1,2,3)$.

Assume conversely that $n_i = \bar{n}_i$ $(1 \leq i \leq 4)$ for the two embeddings $V \subset E$, $\bar{V} \subset E$ respectively. As the latter split by 9.2. we can match summands of splittings in such a way that an automorphism $T$ of $E$ can be defined by sending summands onto each other (cf. the end of 9.6). The proof is thus complete.

9.8. Remark. The dimensions $(n_i)_{1 \leq i \leq 4}$ of 9.4 can of course be introduced for $k$ an arbitrary field as admitted in Chapter thirteen. They are however not immune to extensions of the field. We shall describe here an example of an embedding $V \subset E$ with $n_4 = 2$ such that after a transition of the ground field $k \subset \mathbb{R}$ to $\mathbb{R}$ the $\mathbb{R}$-ification has $n_4 = 1$. On a $k$ space $E = V \oplus k(f_1, f_2)$ where $k \subset \mathbb{R}$, $k \neq \mathbb{R}$ define a symmetric $\Phi$ by declaring a basis $\mathfrak{B} = (v_i)_{\mathbb{N}} \cup \{f_1, f_2\}$ to be a basis for the embedding $V \subset E$ and $\Phi(f_1, v_i) = 1$ $(i \in \mathbb{N})$ and $\Phi(f_2, v_i) = \beta_i$ where we pick the sequence $(\beta_i)_{\mathbb{N}}$ with limit $\beta \in \mathbb{R} \setminus k$ and such that $\Sigma|\beta - \beta_i| < \infty$ in $\mathbb{R}$. We have $V^\perp = (0)$, $\Omega_{11} = \alpha$, $\Omega_{12} = \alpha\beta$, $\Omega_{22} = \alpha\beta^2$ where $\alpha^{-1} = \Sigma(\beta - \beta_i)^2$. In view of 8.2 the embedding does not split. Hence by the remarks in 9.1 we cannot have $n_4 = 1$. As obviously $n_4 \neq 0$ we have $n_4 = 2$. But over $\mathbb{R}$ the embedding does split, so then we shall have $n_4 = 1$.

## 10. Counting orbits of $\perp$-dense subspaces in arbitrary hermitean spaces over $\mathbb{R}$, $\mathbb{C}$ or $\mathbb{H}$

Let the field $k$ be one of $\mathbb{R}$, $\mathbb{C}$, $\mathbb{H}$ with obvious involution. For a change let $(E,\Phi)$ be an arbitrary non degenerate $\aleph_o$-dimensional hermitean space over $k$. So $E = E_{(+1)} \overset{\perp}{\oplus} E_{(-1)}$ as remarked at the beginning of Section 9. Without loss of generality we assume that $\dim E_{(+1)}$ is infinite. $n_- := \dim E_{(-1)}$ ("the index of $E$") is a well determined cardinal between $0$ and $\aleph_o$.

We then have

Theorem. We consider $\perp$-dense subspaces $V \subset E$ with finite $m = \dim E/V$. $(V,\Phi)$ may be definite or not. If the index $n_-$ of $E$ is $\aleph_o$ the orbit of $V$ under the orthogonal group of $(E,\Phi)$ is characterized by $m$ : all $\perp$-dense subspaces of $E$ of equal codimension form one single orbit. If $n_- < \infty$ then the index of $V$ is finite and there are decompositions

$$E = E_o \overset{\perp}{\oplus} E_1 \quad \text{with} \quad V = E_o \oplus V \cap E_1 \, ,$$

where $E_o$ is a sum of $i$ hyperpolic planes ($i$ = index $V$), $V \cap E_1 \subset E_1$ is a $\perp$-dense embedding with $(V \cap E_1,\Phi)$ positive definite. The numbers $n_1,n_2,n_3,n_4$ of Def.9.4 are defined for $V \cap E_1 \subset E_1$ and they do not depend on the particular decomposition of $E$ into $E_o \oplus E_1$. The orbit of $V$ is characterized by the index $i$ of $V$ and the numbers $n_1,n_2,$ $n_3,n_4$. Furthermore, if $r := \min\{m,n_-\}$ then there are precisely

$$\frac{1}{6}(r+1)(r+2)(3m-2r+3)$$

different orbits of $\perp$-dense subspaces $V \subset E$ with $\dim E/V = m$.

Proof. If the index $n_-$ of $E$ is $\aleph_o$ then $V$ contains an infinite dimensional totally isotropic subspace and there is precisely one orbit for fixed $m = \dim E/V$ by Cor. 1 to Thm. 1 in Chap. V.

We are left with the case $n_- < \infty$. Consider an instance with

$i = n_-$ , in other words $E_1$ is positive definite. Each summand $S_j = V_j \oplus (f_j)$ of a splitting of $V \cap E_1 \subset E_1$ is therefore positive definite and there are 2 possibilities for each $S_j$ , viz. $n_1 = 1$ or $n_2 = 1$ . Hence we find $m+1$ orbits when $i = n_-$ . Consider next the case with $i = n_- - 1$ . All $S_j$ with the exception of precisely one, say $S_1$ , are positive definite. For $S_1$ there are two possibilities, namely $n_3 = 1$ or $n_4 = 1$ . Hence we have $2 \cdot m$ orbits when $i = n_- - 1$ . In this fashion we find the number of all orbits,
$1(m+1) + 2 \cdot m + \cdots + (r+1)(m+1-r) = \Sigma_1^{r+1} \nu(m+2-\nu) = (m+2)\Sigma_1^{r+1}\nu - \Sigma_1^{r+1}\nu^2$ .

## 11. Applications to the theory of divergent series

It would lead us too far astray if we wanted to delve into the theory of divergent sequences and series here. Suffice it to say that our results of this chapter yield results on divergent series. As an illustration we may consider Theorem 8.1 which may be rendered as follows:

If $(\zeta_i^1)_{\mathbb{N}}, \ldots, (\zeta_i^m)_{\mathbb{N}}$ are arbitrary sequences of reals such that the sum $\sum_{i=1}^{\infty} (\lambda^1\zeta_i^1 + \cdots + \lambda^m\zeta_i^m)^2$ diverges for arbitrary $\lambda_1, \ldots, \lambda_m \in \mathbb{R}$ then there is a row-finite orthogonal matrix $(\alpha_{ij}) = A$, ${}^{tr}AA = \mathbb{1}$ , such that the transformed sequences $(\eta_i^1)_{\mathbb{N}} = (\Sigma_{j=1}^{\infty}\alpha_{ij}\zeta_j^1)_{i\in\mathbb{N}}, \ldots$
$\ldots, (\eta_i^m)_{\mathbb{N}} = (\Sigma_{j=1}^{\infty}\alpha_{ij}\zeta_j^m)_{i\in\mathbb{N}}$ are pairwise orthogonal in the sense that $\eta_i^r \cdot \eta_i^s = 0$ for all $i$ and $r \neq s$ .

Indeed, we may define a $\perp$-dense embedding $V \subset E$ with $\dim E/V = m$ via a standard basis $(v_i)_{i\in\mathbb{N}} \cup (f_j)_{1\le j\le m}$ by setting $\phi(f_j, v_i) := \zeta_i^j$ . It has $n_4 = m$ so that it splits with respect to any $m$-tuple spanning a supplement of $V$ in $E$ . In particular it will split with respect to $f_1, \ldots, f_m$ . This translates as stated above.

## References to Chapter XIII

[1] E. Artin, Geometric Algebra, Interscience Publ. NY (1957).

[2] F. van der Blij, History of the octaves in Simon Slevin, Wis- en Natuurkundig Tijdschrift (Groningen) 34$^{e}$ Jaargang Avlevering III Februari 1961.

[3] W. Greub, Multilinear Algebra, Springer Verlag NY (1967).

[4] H. Gross, Eine Bemerkung zu dichten Unterräumen reeller quadratischer Räume. Comment. Math. Helv. 45, 472-493 (1970).

[5] T.Y. Lam, The Algebraic Theory of Quadratic Forms, Benjamin, Inc. Reading (Mass) 1973 .

*

Postscript. Our "lazy" treatment of subspaces in the indefinite case, namely via codimension 2 plus combinatorial arguments prohibits a discussion of subspaces of infinite codimension. However, according to W. Bäni, Theorem 9.7 holds for infinite $\dim E/V$ as well. See a forthcoming paper where he also includes the discussion of subspaces that are not necessarily $\perp$-dense.

CHAPTER FOURTEEN

# QUADRATIC FORMS

## Introduction

Quadratic forms are closely related to orthosymmetric sesquilinear forms and, to a large extent, they behave very similarly. In fact, the two concepts partly overlap (cf. Example 2 in Section 3 below). For the purpose of illustration we start with the classical notion of a quadratic form

$$Q: \; E \to k$$

on a k-vector space $E$ over a commutative field $k$ of arbitrary characteristic. The map $Q$ is called a quadratic form if 1) we have $Q(\lambda x) = \lambda^2 Q(x)$ for all $\lambda \in k$, $x \in E$, and 2) the assignment $\Psi$: $(x,y) \mapsto Q(x+y) - Q(x) - Q(y)$ from $E \times E$ into $k$ is bilinear ( $\Psi$ is called the <u>bilinear form associated to $Q$</u> ; it is, by necessity, a symmetric form). Thus, by definition, we have the formula $Q(x+y) = Q(x) + Q(y) + \Psi(x,y)$ . It is easily generalized to finite sums $\sum_{i=1}^{n} \xi_i x_i$ ,

$$Q(\sum_{i=1}^{n} \xi_i x_i) = \sum_{i=1}^{n} \xi_i^2 Q(x_i) + \sum_{1 \le i < j \le n} \xi_i \xi_j \Psi(x_i, y_j) \; .$$

Let us give some examples.

<u>Example 1</u>. Let $(e_\iota)_{\iota \in I}$ be a basis of the k-vector space $E$ . Put some ordering $<$ on the index set I. Choose any matrix $(\alpha_{\iota\kappa})_{\iota,\kappa \in I}$ over $k$ where the entries $\alpha_{\iota\kappa}$ with $\iota > \kappa$ are zero. A typical vector $x \in E$ is a linear combination $\sum_\iota \xi_\iota e_\iota$ with only finitely many $\xi_\iota \neq 0$ ; hence we may define

$$Q(\sum_\iota \xi_\iota e_\iota) := \sum_{\iota \le \kappa} \alpha_{\iota\kappa} \xi_\iota \xi_\kappa$$

and verify that 1) and 2) are satisfied. From the definition we obtain $Q(e_\iota) = \alpha_{\iota\iota}$ and, for $\iota < \kappa$, $\Psi(e_\iota, e_\kappa) = \Psi(e_\kappa, e_\iota) = \alpha_{\iota\kappa}$. For example, in a plane $E = k(e_1, e_2)$ we may choose $(\alpha_{\iota\kappa}) = \begin{pmatrix} \alpha & 1 \\ 0 & \beta \end{pmatrix}$ and we have for $x = \xi_1 e_1 + \xi_2 e_2 \in E$

$$(0) \qquad Q(x) = \alpha\xi_1^2 + \beta\xi_2^2 + \xi_1\xi_2 \quad .$$

Example 2. Take any symmetric bilinear form $\Phi$ and define $Q(x) := \Phi(x,x)$. Condition 1) obviously holds and $\Psi(x,y) := Q(x+y) - Q(x) - Q(y)$ is indeed bilinear because $\Psi(x,y) = 2\Phi(x,y)$. If char $k \neq 2$ then the process is reversible. We can recapture from $Q$ the original form $\Phi$, $\Phi(x,y) = \frac{1}{2}\Psi(x,y)$. Here we gain nothing new by the introduction of quadratic forms. If, on the other hand, the characteristic of $k$ is 2 then the associated form $\Psi$ is identically zero. Hence the manufacture of quadratic forms by applying the "squaring process" to symmetric bilinear forms will never produce, e.g. a quadratic form of the kind (0) (with associated form $\Psi(x,y) = \xi_1\eta_2 + \xi_2\eta_1$). Of course, we could apply the squaring process to forms $\Phi$ which are not symmetric (say the form with matrix $(\alpha_{\iota\kappa})_{\iota\in I}$ in the first example). Artin says in his book on geometric algebra that this is not desirable since such a $\Phi$ would not be uniquely determined by the quadratic form ([1], p. 110). Yet, this is exactly the course on which we shall embark in the following sections.

In this introductory chapter we shall present the concept of quadratic form as advanced by J. Tits (in Sections 2.2, 2.3 of [3]) and C.T.C. Wall [4] and discuss a few related concepts as far as needed in subsequent chapters. We are aware of the fact that the definition of quadratic form presented is not the most general, even for vector spaces; yet the concept is general enough for the purposes of this book.

Assumptions. In the whole chapter $k$ is a division ring with an antiautomorphism $\xi \mapsto \xi^*$ whose square is inner, $\xi^{**} = \varepsilon^{-1}\xi\varepsilon$ and, furthermore, $\varepsilon\varepsilon^* = \varepsilon^*\varepsilon = 1$ for some $\varepsilon \in k$. $\mathrm{Sesq}_*(E)$ is the additive group of all *-sesquilinear forms on the k-vector space $E$ - non-orthosymmetric forms included. Quadratic forms will be defined in terms of the structure $(k,*,\varepsilon)$.

*

## 1. Symmetrization

We define a map

$$\pi : \mathrm{Sesq}_*(E) \to \mathrm{Sesq}_*(E)$$

by setting

$$(1) \qquad (\pi\Phi)(x,y) := \varepsilon\Phi(y,x)^* .$$

Together with $\Phi$ the function $\pi\Phi$ belongs to $\mathrm{Sesq}_*(E)$. We shall verify linearity inthe first argument: $(\pi\Phi)(\lambda x,y) = \varepsilon(\Phi(y,x)\lambda^*)^* = \varepsilon\lambda^{**}\Phi(y,x)^* = \varepsilon\cdot\varepsilon^{-1}\lambda\varepsilon\Phi(y,x)^* = \lambda(\pi\Phi)(x,y)$. The other defining properties are equally obvious as is the fact that $\pi$ is involutorial,

$$(2) \qquad \pi \circ \pi = \mathbb{1} \text{ (identity)} .$$

Definition 1. $\pi + \mathbb{1}$ is called symmetrization operator.

We observe that $\mathrm{im}(\pi+\mathbb{1})$ consists solely of $\varepsilon$-hermitean forms. Indeed, if $\Psi = (\pi+\mathbb{1})\Phi$ then

$$\Psi(x,y) = (\pi\Phi)(x,y) + \Phi(x,y) = \varepsilon\Phi(y,x)^* + \Phi(x,y)$$
$$\Psi(y,x) = (\pi\Phi)(y,x) + \Phi(y,x) = \varepsilon\Phi(x,y)^* + \Phi(y,x)$$

and therefore $\varepsilon\Psi(x,y)^* = \varepsilon(\Phi(y,x)^{**}\varepsilon^* + \Phi(x,y)^*) =$
$= \varepsilon(\varepsilon^{-1}\Phi(y,x)\varepsilon\varepsilon^* + \Phi(x,y)^*) = \Psi(y,x)$ .

Furthermore, all forms $\Psi = \pi\Phi + \Phi$ are trace-valued, $\Psi(x,x) = \varepsilon\Phi(x,x)^* + \Phi(x,x)$, and it is not difficult to prove that $\pi + \mathbb{1}$ applied to $Sesq_*(E)$ produces all $\varepsilon$-hermitean trace-valued forms (cf. Step 1 in the proof of Thm.1 below).

## 2. The process of squaring

In the additive group of $(k,*,\mathbb{1})$ we consider the subgroup

$$(3) \qquad P := \{\varepsilon\xi^*-\xi \mid \xi\in k\} .$$

Let $k/P$ be the factor group and $\alpha \mapsto [\alpha]$ the canonical map. With given $\Phi \in Sesq_*(E)$ we associate a map

$$Q : E \to k/P$$

by setting

$$(4) \qquad Q(x) := [\Phi(x,x)] .$$

We find $Q(x+y) = [\Phi(x,x) + \Phi(y,y) + \Phi(x,y) + \Phi(y,x)] = Q(x) + Q(y) + [\Phi(x,y) + \varepsilon\Phi(y,x)^* - \varepsilon\Phi(y,x)^* + \Phi(y,x)] = Q(x) + Q(y) + [\Psi(x,y)]$ for $\Psi := (\pi+\mathbb{1})\Phi$ the symmetrized $\Phi$.

We next observe that for all $\rho \in P$ and all $\lambda \in k$ we have $\lambda\rho\lambda^* \in P$. Indeed, if $\rho = \varepsilon\xi^* - \xi$ then $\lambda\rho\lambda^* = \varepsilon\cdot\varepsilon^{-1}\lambda\varepsilon\cdot\xi^*\lambda^* - \lambda\xi\lambda^* = \varepsilon\lambda^{**}\xi^*\lambda^* - \lambda\xi\lambda^* = \varepsilon(\lambda\xi\lambda^*)^* - (\lambda\xi\lambda^*) \in P$. Hence, for nonzero $\lambda$, we obtain an automorphism of the group $k/P$ by $[\alpha] \mapsto \lambda[\alpha]\lambda^* := [\lambda\alpha\lambda^*]$. Therefore, we obtain by (4) that $Q(\lambda x) = [\lambda\Phi(x,x)\lambda^*] = \lambda[\Phi(x,x)]\lambda^* = \lambda Q(x)\lambda^*$.

<u>Summary</u>. With each $\Phi \in Sesq_*(E)$ we can associate the pair $(\Psi,Q)$ where $\Psi := (\pi+\mathbb{1})\Phi$ is the symmetrized of $\Phi$ in the sense of the previous section and where the map $Q$ is defined by (4). $\Psi$ and $Q$ satisfy the following relations

$$(5) \qquad Q(x+y) = Q(x) + Q(y) + [\Psi(x,y)] ,$$

$$(6) \qquad Q(\lambda x) = \lambda Q(x)\lambda^* ,$$

$$(7) \qquad \Psi(x,x) = Q(x) + \varepsilon Q(x)^* ,$$

where (7) is short for "$\Psi(x,x) = \mu + \varepsilon\mu^*$ for all $\mu \in Q(x)$" . (Notice that $P$ consists of antisymmetric elements $\rho$ in the sense that $\varepsilon\rho^* = -\rho$ .)

It is to be expected that the surprisingly natural formulae (5), (6), (7) will pass the trial of the pyx as "axioms" for the concept of a quadratic form.

## 3. The concept of quadratic form

Let $\mathcal{Q}$ be the subset of $\mathrm{im}(\pi+\mathbb{1}) \times (k/P)^E$ of all pairs $(\Psi,Q)$ which satisfy (5), (6), (7). We shall now have a look at the assignment

$$\mathrm{Sesq}_*(E) \to \mathcal{Q}$$

defined by

$$\Phi \mapsto (\Psi,Q)$$

where $\Psi := (\pi+\mathbb{1})\Phi$ and $Q$ is defined by $Q(x) := [\Phi(x,x)]$ . $\mathcal{Q}$ is an additive group and the assignment is a group homomorphism. We have

Theorem 1([4] p. 246).

$$\mathrm{Sesq}_*(E)/\mathrm{im}(\pi-\mathbb{1}) \cong \mathcal{Q} .$$

Proof. Since $(\pi+\mathbb{1})\circ(\pi-\mathbb{1}) = 0$ by (2) the kernel of the assignment contains $\mathrm{im}(\pi-\mathbb{1})$ and we obtain a homomorphism $\hat{\kappa}: \mathrm{Sesq}_*(E)/\mathrm{im}(\pi-\mathbb{1}) \to \mathcal{Q}$ .

1) $\hat{\kappa}$ is epimorphic. Let $(e_\iota)_{\iota\in I}$ be a basis of $E$ . Order the index set $I$ . Since $\Psi$ satisfies (7) we have $\Psi(e_\iota,e_\iota) = \lambda_\iota + \varepsilon\lambda_\iota^*$

for certain $\lambda_\iota \in Q(e_\iota)$ . Define a sesquilinear form $\Phi$ by

$$\Phi(e_\iota,e_\kappa) = \begin{cases} \lambda_\iota & \text{if } \iota = \kappa \\ \Psi(e_\iota,e_\kappa) & \text{if } \iota < \kappa \\ 0 & \text{if } \iota > \kappa \end{cases}$$

It is readily verified that $\Psi = (\pi+\mathbb{1})\Phi$ . Furthermore, by the definition of $\Phi$ we find for a typical element $x = \sum_\iota \xi_\iota e_\iota$ that $\Phi(x,x) = \sum_\iota \xi_\iota \Phi(e_\iota,e_\iota)\xi_\iota^* + \sum_{\iota<\kappa} \xi_\iota \Phi(e_\iota,e_\kappa)\xi_\kappa^* = \sum_\iota \xi_\iota \lambda_\iota \xi_\iota^* + \sum_{\iota<\kappa} \xi_\iota \Psi(e_\iota,e_\kappa)\xi_\kappa^*$ .
In other words, by using (5) and (6),
$[\Phi(x,x)] = \sum_\iota Q(\xi_\iota e_\iota) + [\sum_{\iota<\kappa} \Psi(\xi_\iota e_\iota,\xi_\kappa e_\kappa)] = Q(\sum_\iota \xi_\iota e_\iota) = Q(x)$ . Thus we have shown that $(\Psi,Q)$ is the image of $\Phi$ under $\hat{\kappa}$ .

2) $\hat{\kappa}$ is injective. Assume that $\Psi = \pi\Phi + \Phi$ and $Q$ are both identically zero. We have to show that there exists $\chi \in \mathrm{Sesq}_*(E)$ with $\Phi = \pi\chi - \chi \in \mathrm{im}(\pi-\mathbb{1})$ . Since $Q = 0$ we have $\Phi(e_\iota,e_\iota) \in P$ , i.e. $\Phi(e_\iota,e_\iota) = \varepsilon\lambda_\iota^* - \lambda_\iota$ for certain $\lambda_\iota \in k$ . If we define $\chi$ by $\chi(e_\iota,e_\kappa) := \lambda_\iota$ $(\iota=\kappa)$ , $\chi(e_\iota,e_\kappa) := 0$ $(\iota<\kappa)$ , $\chi(e_\iota,e_\kappa) := -\Phi(e_\iota,e_\kappa)$ $(\iota>\kappa)$ then we have $(\pi\chi-\chi)(e_\iota,e_\kappa) = \Phi(e_\iota,e_\kappa)$ for all pairs $\iota$ , $\kappa$ (e.g., if $\iota < \kappa$ then $(\pi\chi-\chi)(e_\iota,e_\kappa) = \varepsilon\chi(e_\kappa,e_\iota)^* = -\varepsilon\Phi(e_\kappa,e_\iota)^* = \Phi(e_\iota,e_\kappa)$ as $\pi\Phi + \Phi = 0$ ). Thus $\Phi \in \mathrm{im}(\pi-\mathbb{1})$ and $\hat{\kappa}$ is shown to be a group isomorphism. This proves Theorem 1.

Definition 2 ([4] p. 245, [3] p. 23). An element $\Phi + \mathrm{im}(\pi-\mathbb{1})$ in $\mathrm{Sesq}_*(E)/\mathrm{im}(\pi-\mathbb{1})$ or - equivalently by Theorem 1 - a pair $(\Psi,Q)$ consisting of a trace-valued $\varepsilon$-hermitean form $\Psi$ and a map $Q\colon E \to k/P$ which satisfy (5), (6), (7), is termed quadratic form on $E$ ( $P$ is the additive subgroup $\{\varepsilon\xi^*-\xi \mid \xi\in k\}$ of $k$ ). $(E,\Psi,Q)$ is called a quadratic space if $(\Psi,Q)$ is a quadratic form on $E$ .

In the following we shall rarely conceive of a quadratic form as being an element in the cokernel of $\pi - 1\!\!1$ ; we give preference to the $(\Psi,Q)$ - interpretation and we think of $\Psi$ as measuring angles between vectors and of $Q$ as measuring lengths of vectors (cf. Lemma 1 below).

Let us look at some special cases.

Example 1. If $(k,*,\varepsilon) = (k,id,1)$ then $k$ is commutative and we are back in the classical situation described in the introduction: $P$ is $\{0\}$ so $Q$ maps into $k$ ; the axioms (5) and (6) are the usual ones and (7) is superfluous as it follows from (5) in this case.

Example 2. If char $k \neq 2$ or if the center of $k$ is not left pointwise fixed by $*$ then there exists a $\gamma$ with $\gamma + \gamma^* = 1$ . Take $\gamma = \frac{1}{2}$ or, if the characteristic is 2, $\gamma = \alpha\sigma^{-1}$ where $\sigma = \alpha + \alpha^*$ and $\alpha \neq \alpha^*$ is an element of the center ( $\sigma^* = \alpha^* + \alpha^{**} = \alpha^* + \varepsilon^{-1}\alpha\varepsilon = \sigma$ ). For any such $\gamma$ each $\varepsilon$-hermitean form $\Psi$ is the symmetrized of $\gamma\Psi$ , $\Psi = (\pi+1\!\!1)\gamma\Psi$ . Therefore, the map $Q\colon E \to k/P$ defined by $Q(x) := \gamma\Psi(x,x) + P$ turns the pair $(\Psi,Q)$ into a quadratic form. Further, $\ker(\pi-1\!\!1) \subset \mathrm{im}(\pi+1\!\!1)$ and hence equality. In particular, if char $k \neq 2$ we may choose $\gamma = \frac{1}{2}$ and from the decomposition $\Phi = (\pi+1\!\!1)\frac{1}{2}\Phi - (\pi-1)\frac{1}{2}\Phi$ we see that $\mathrm{Sesq}_*(E) = \mathrm{im}(\pi+1\!\!1) + \mathrm{im}(\pi-1\!\!1)$ and therefore

$$\mathrm{Sesq}_*(E) = \ker(\pi-1\!\!1) \oplus \ker(\pi+1\!\!1) = \ker(\pi-1\!\!1) \oplus \mathrm{im}(\pi-1\!\!1) .$$

Thus, in any case, <u>if there is such a</u> $\gamma$ <u>then symmetrization is a group isomorphism between</u> $\mathcal{Q} = \mathrm{Sesq}_*(E)/\mathrm{im}(\pi-1\!\!1)$ <u>and</u> $\ker(\pi-1\!\!1)$ . Furthermore, each endomorphism $E \to E$ which preserves $\Psi$ preserves $(\Psi,Q)$ .

Example 3. If char $k = 2$ then the inclusion $\mathrm{im}(\pi+1\!\!1) \subset \ker(\pi-1\!\!1)$ can be proper. The first set consists of all trace-valued $\varepsilon$-hermitean forms, the second set is made up by all $\varepsilon$-hermitean forms.

In particular, if $(k,*,\varepsilon) = (k,\mathrm{id},1)$ then $\mathrm{im}(\pi+\mathbb{1})$ is the set of all alternate forms and $\ker(\pi-\mathbb{1})$ is the set of all symmetric forms. We furthermore call to attention that $P$, as defined by (3), coincides with the subgroup $T := \{\xi+\varepsilon\xi^* \mid \xi\in k\}$ of traces when the characteristic is two. Hence, whenever $X$ is a subspace in a quadratic space $(E,\Psi,Q)$ such that $\Psi$ is identically zero on $X$ then $Q$ maps $X$ homomorphically into the k-vector space $S/T$, $S/T \subset k/T = k/P$.

We finish this section with a formal definition.

Definition 3 ([2] p. 54). A quadratic form $(\Psi,Q)$ on $E$ is called degenerate if and only if the sesquilinear form $\Psi$ is degenerate; we also say that the quadratic space $(E,\Psi,Q)$ is degenerate if the form $\Psi$ is degenerate. Orthogonality with respect to the quadratic form $(\Psi,Q)$ is the same as orthogonality with respect to $\Psi$. Thus $(E,\Psi,Q)$ is degenerate if and only if $E^{\perp} \neq (0)$. (Notice that $\perp$ is a symmetric relation because $\Psi$ is $\varepsilon$-hermitean by Definition 1). If $(E_\iota,\Psi_\iota,Q_\iota)$, $\iota \in I$, is a family of quadratic spaces then its (external) orthogonal sum is the quadratic space $(E,\Psi,Q)$ with $(E,\Psi)$ the (external) orthogonal sum of the sesquilinear spaces $(E_\iota,\Psi_\iota)$, $\iota \in I$, and $Q(\sum_\iota x_\iota) := \sum_\iota Q_\iota(x_\iota)$ for $x_\iota \in E_\iota$ $(\iota\in I)$.

## 4. Isometries between quadratic spaces

A map $D: E \to \bar{E}$ is called isometry between the quadratic spaces $(E,\Psi,Q)$, $(\bar{E},\bar{\Psi},\bar{Q})$ if and only if $D$ is vector space isomorphism which satisfies (8) and (9):

(8) $\qquad \bar{\Psi}(Dx,Dy) = \Psi(x,y) \qquad$ for all $x, y \in E$,

(9) $\qquad \bar{Q}(Dx) = Q(x) \qquad$ for all $x \in E$.

Sometimes we speak of isometries relative to $(\Psi,Q)$ if it is desirable to distinguish them from isometries between the underlying sesquilinear spaces $(E,\Psi)$ , $(\bar{E},\bar{\Psi})$ .

In the situation of Example 1 in the previous section we have (9) ⇒ (8) ; in Example 2 we have (8) ⇒ (9). In the general case we need both conditions. The following easy lemma shows the salient features of an isometry.

Lemma 1. $(E,\Psi,Q)$ and $(\bar{E},\bar{\Psi},\bar{Q})$ are isometric if and only if there are bases $(e_\iota)_{\iota\in I}$ , $(\bar{e}_\iota)_{\iota\in I}$ of $E$ and $\bar{E}$ respectively which are "congruent" in the sense that the following two conditions hold:

$$(8') \qquad \bar{\Psi}(\bar{e}_\iota,\bar{e}_\kappa) = \Psi(e_\iota,e_\kappa) \quad \text{for all} \quad \iota \neq \kappa ,$$

$$(9') \qquad \bar{Q}(\bar{e}_\iota) = Q(e_\iota) \quad \text{for all} \quad \iota .$$

Proof. By the axioms (5) , (6) , (7) it follows that $D : \Sigma\, \xi_\iota e_\iota \mapsto \Sigma\, \xi_\iota \bar{e}_\iota$ satisfies (8) and (9).

Definition 4. A vector $x$ in a quadratic space $(E,\Psi,Q)$ is called singular if $Q(x) = [0]$ (hence a singular element is necessarily isotropic relative to $\Psi$ in view of axiom (7) ). A subspace $X \subset E$ is called totally singular if $Q$ vanishes identically on $X$ (if not char $k \neq 2$ & $\Psi$ alternate then we have the implication "totally singular ⇒ totally isotropic"). A plane $E$ is called hyperbolic for $(\Psi,Q)$ if it admits a basis $\{e_1,e_2\}$ of singular vectors with $\Psi(e_1,e_2) = 1$ .

The proof of the following Lemma 2 is left to the reader.

Lemma 2. A quadratic plane $(F,\Psi,Q)$ is hyperbolic iff it is nondegenerate and contains a nonzero singular vector. If the quadratic space $(E,\Phi,Q)$ is nondegenerate and contains a nonzero singular element then $E$ is spanned by a basis of singular vectors.

Lemma 3. Assume that the nondegenerate quadratic space $(E,\Psi,Q)$ has $\dim E = \aleph_0$ and contains an infinite dimensional totally singular subspace. Then $(E,\Psi,Q)$ is an orthogonal sum of hyperbolic planes (relative to $(\Psi,Q)$).

Proof. Assume that finitely many pairwise orthogonal hyperbolic planes $F_1, \ldots, F_m \subset E$ have been constructed. In order to establish the lemma it suffices to show that there is a further hyperbolic plane $F_{m+1} \subset E_1 := (\overset{m}{\underset{1}{\oplus}} F_i)^\perp$ which contains a previously fixed nonzero vector $a \in E_1$. Because $E_1$ must contain singular vectors different from zero and because it is nondegenerate, $E_1$ is spanned by singular vectors (Lemma 2). As $E_1$ is nondegenerate $a$ cannot be orthogonal to all singular vectors in $E_1$. Hence $a$ can be completed to a nondegenerate plane $F_{m+1} \subset E_1$ that contains a singular vector. $F_{m+1}$ is hyperbolic by Lemma 2. Q.E.D.

Remark. By Lemma 3 all $\aleph_0$-dimensional nondegenerate spaces $(E,\Psi,Q)$ which contain an $\aleph_0$-dimensional totally singular subspace are isometric. There are division rings $(k,*,\varepsilon)$ such that each quadratic k-space satisfies the assumption of Lemma 3. The simplest examples are the commutative $(k,\mathrm{id},1)$ with $\dim_k S/T = [k : k^2]$ finite. The proof is straightforward. Another example is obtained if we let $k$ be the quaternion division algebra $k = \left(\frac{1,X}{k_0}\right)$ where the center $k_0$ is the rational function field $\mathbb{Z}_2(X)$, $\varepsilon = 1$ and $*$ the usual conjugation in $k$. It is straightforward to prove that $\lambda\lambda^*$ ranges over all of $k_0$ if $\lambda$ varies in $k$. Hence it is clear that any hermitean form over $(k,*,1)$ in at least two variables has a nontrivial zero. Hence an $\aleph_0$-dimensional space $(E,\Psi,Q)$ over $(k,*,1)$ will contain infinite dimensional totally isotropic subspaces $V$. As $\dim_k S/T = 1 < \infty$ it follows that, in turn, each such $V$ contains infinite dimensional totally singular subspaces. Hence we may quote Lemma 3 to conclude that there is only 1 isometry class of nondegenerate $\aleph_0$-dimensional quadratic spaces over this quaternion division ring $k$ .

Definition 5. The isometries of a quadratic space $(E,\Psi,Q)$ onto itself form a group under composition. This group is called the orthogonal group of $(E,\Psi,Q)$ .

We terminate this section by describing certain isometries which are found in the orthogonal group of every quadratic space $(E,\Psi,Q)$ . Let $0 \neq a \in E$ . If we should be in the situation that the subgroup $P = \{\varepsilon\xi^*-\xi \mid \xi\in k\}$ reduces to $\{0\}$ , i.e. $(k,*,\varepsilon) = (k,id,1)$ , assume furthermore that $a$ is nonsingular. Hence we may pick

$$(10) \qquad \alpha \in Q(a) \setminus \{0\}$$

and define a map $\Omega_\alpha : E \to E$ by

$$(11) \qquad \Omega_\alpha x = x - \Psi(x,a)\alpha^{-1}a .$$

The maps defined by (11) are called reflexions. We shall now verify that they are elements of the orthogonal group of $(E,\Psi,Q)$. (Notice that $\Psi(a,a) = \alpha + \varepsilon\alpha^*$ by (7) and $Q(a) = [\alpha]$ .)

First: $Q(\Omega_\alpha x) = Q(x) + Q(\Psi(x,a)\alpha^{-1}a) - [\Psi(x,\Psi(x,a)\alpha^{-1}a)] =$ $Q(x) + \Psi(x,a)\alpha^{-1}[\alpha]\alpha^{*-1}\Psi(x,a)^* - [\Psi(x,a)\alpha^{*-1}\Psi(x,a)^*] = Q(x)$ .

Second: $\Psi(\Omega_\alpha x,\Omega_\alpha y) = \Psi(x,y) - \Psi(x,a)\alpha^{-1}\Psi(a,y) - \Psi(x,a)\alpha^{*-1}\Psi(y,a)^*$ $\Psi(x,a)\alpha^{-1}\Psi(a,a)\alpha^{*-1}\Psi(y,a)^*$ . By using $\Psi(a,a) = \alpha + \varepsilon\alpha^*$ we obtain for the last term the expression $\Psi(x,a)\alpha^{*-1}\Psi(y,a)^* + \Psi(x,a)\alpha^{-1}\varepsilon\Psi(y,a)^* =$ $\Psi(x,a)\alpha^{*-1}\Psi(y,a)^* + \Psi(x,a)\alpha^{-1}\Psi(a,y)$ since $\Psi$ is $\varepsilon$-hermitean. Ergo $\Psi(\Omega_\alpha x,\Omega_\alpha y) = \Psi(x,y)$ .

Third: $\Omega_\alpha$ is linear, obviously. Assume $\Omega_\alpha x = 0$ . Hence $x = \lambda a$ with $\lambda = \Psi(x,a)\alpha^{-1}$ . Therefore $\lambda = \Psi(\lambda a,a)\alpha^{-1} = \lambda(\alpha+\varepsilon\alpha^*)\alpha^{-1} =$

$\lambda + \lambda\varepsilon\alpha^*\alpha^{-1}$ whence $\lambda = 0$ , i.e. $x = 0$ . Thus $\Omega_\alpha$ is injective. For given $y \in E$ put $x = y + \mu a$ with $\mu = \Psi(y,a)\alpha^{-1}(1-\Psi(a,a)\alpha^{-1})^{-1}$ and have $\Omega_\alpha x = y$ . Therefore $\Omega_\alpha$ is epimorphic. Hence $\Omega_\alpha$ is an isometry.

The hyperplane $H := a^\perp$ is the set of fixed points of $\Omega_\alpha$ ; so $\Omega_\alpha$ induces an automorphism on the one-dimensional quotient space $E/H$ . We distinguish between $a$ being isotropic and $a$ being anisotropic.

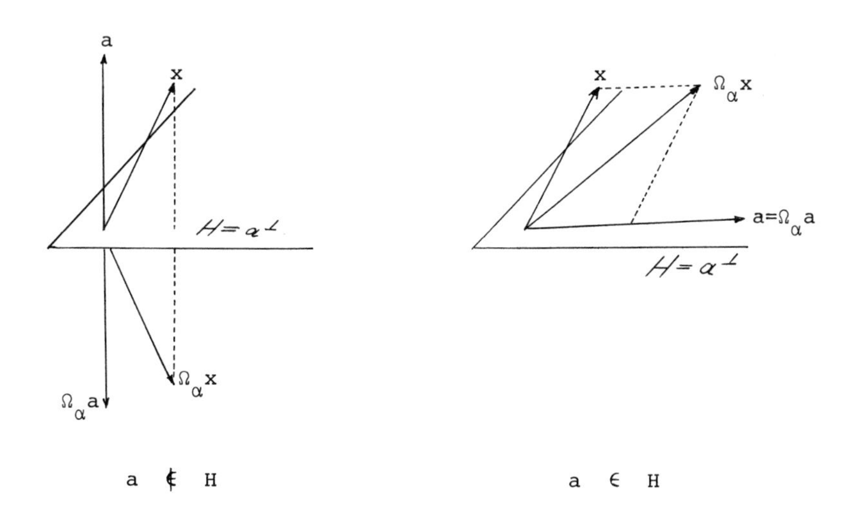


$a \notin H$ $a \in H$

If $a$ is isotropic, $a \in H$ , then $\Omega_\alpha$ induces the identity on $E/H$ ; we speak of a (orthogonal) <u>transvection</u>. If $a$ is anisotropic, $a \notin H$ , then $\Omega_\alpha$ induces a <u>dilatation</u> $\hat{x} \to \mu\hat{x}$ on $E/H$ with $\mu = -\varepsilon\alpha^*\alpha^{-1}$ . We see that $\Omega_\alpha$ is not, in general, an involution in the orthogonal group. However, if $\alpha \in Q(a)$ then $\varepsilon\alpha^* \in Q(a)$ , because $\alpha \equiv \varepsilon\alpha^* \bmod P$ , and we verify that

$$\Omega_\alpha \circ \Omega_{\varepsilon\alpha^*} = \mathbb{1} .$$

## 5. A remark on forms in characteristic two

By now we have become acquainted with three kinds of "forms" in spaces $E$ over division rings $(k,*,\varepsilon)$ of characteristic 2, to wit,

1. trace-valued $\varepsilon$-hermitean forms $\Phi$ ,
2. quadratic forms $(\Psi,Q)$ ,
3. non-trace-valued $\varepsilon$-hermitean forms $\chi$ .

Although the three types do often behave similarly they are marked-ly different objects of study. This becomes particularly manifest if we pass to infinite dimensions. Let us consider here the problem of classi-fying subspaces modulo the action of the associated orthogonal group. In order to enhance the fundamental divergence of one case from another we choose the classical setting of a commutative field, say $k$ an algebraic function field of finitely many variables over an algebraica-ly closed field of characteristic 2, and we let the identity of $k$ be our involution $*$ and $\varepsilon = 1$ so that all forms $\Phi$ , $\Psi$ , $\chi$ are sym-metric bilinear. Let furthermore, in each of the three cases, $E$ be nondegenerate and of dimension $\aleph_0$ and $F \subset E$ a subspace.

In the first case the orbit of $F$ is characterized by seven cardinal number invariants (Cor. 2 in V.2). We refute the possible objection that alternate forms have a "trivial" theory by the remark that we can switch to other stable hermitean forms without essentially upsetting the result and its proof.

A rather more complicated result holds in the second case. There is a complete set of invariants of the orbit of $F$ (Thm. 2 in XVI.4). Besides a finite number of isometry classes which inevitably must make their appearance, and a finite number of obvious cardinal number in-variants, there emerges, as a final invariant, a lattice of linear sub-spaces in the $k^2$-vector space $Q(F^{\perp} \cap F^{\perp\perp})$ of the shape

There are no relations between this lattice and the other invariants in the set besides obvious ones; the lattice can be chosen freely in the $k^2$-vector space $Q(F^{\perp}\cap F^{\perp\perp})$ . One should compare the nondistributive 52-element lattice in XVI.2 with the distributive 14-element lattice in V.2 (a sublattice of the former). These lattices closely reflect the intricacies of the proofs that go with the classification problem. The two diagrams manifest, to our opinion, the increase in complexity in the transition from trace-valued hermitean forms to quadratic forms. The trend is not apparent in finite dimensions; it may be that our experiences run athwart to some expectations.

As is to be expected, on the other hand, the third case of non-trace-valued hermitean forms is the least tame among the three. For no fields do we have a general classification of subspaces. In Chapter VIII we have treated the case of totally isotropic $F$ . The complexity of this special case is roughly the same as that encountered in the classification problem in quadratic spaces. Yet the study of non-trace-valued forms is by no means an academic one. If, in characteristic two, we investigate a trace-valued hermitean form and a self-adjoint endomorphism $U$ then the hermitean form $\Phi_1(x,y) := \Phi(Ux,y)$ is usually not trace-valued any more (see IX.4 for such a situation).

*

## References to Chapter XIV

[1] Artin, Geometric algebra. Interscience Publ., New York 1957.

[2] N. Bourbaki, Formes sesquilinéaires et formes quadratiques. ASI 1272, Hermann Paris, 1959.

[3] J. Tits, Formes quadratiques, groupes orthogonaux, et algèbres de Clifford. Invent. Math. 5 (1968) 19-41.

[4] C.T.C. Wall, On the axiomatic foundations of the theory of Hermitean forms. Proc. Camb. Phil. Soc. 67 (1970) 243-250.

*

We have not touched upon the definition of the Arf invariant of finite dimensional quadratic spaces. Besides [3], which brings a definition of the invariant taylored to the concept of quadratic form as discussed here, we add below a list of further references bearing on the topic (cf. Appendix I to Chapter XVI).

[5] J. Dieudonné, Pseudo-discriminant and Dickson invariant. Pacific J. Math. 5 (1955) 907-910.

[6] R.H. Dye, On the Arf Invariant. J. Algebra 53 (1978) 36-39.

[7] N. Jacobson, Clifford algebras for algebras with involution of type D. J. Algebra 1 (1964) 288-300.

[8] I. Kaplansky, Linear Algebra and Geometry. Allyn & Bacon, Boston 1969, page 29.

[9] - and R.J. Shaker, Abstract Quadratic Forms. Canad. J. Math. 21 (1969) 1218-1233.

[10] W. Klingenberg and E. Witt, Ueber die Arfsche Invariante quadratischer Formen mod 2. J. reine angew. Math. 193 (1954) 121-122.

[11] M. Kneser, Bestimmung des Zentrums der Cliffordschen Algebra. J. reine angew. Math. 193 (1954) 123-124.

[12] T.A. Springer, Note on quadratic forms in characteristic 2. Nieuw archief voor wiskunde (3), X (1962) 1-10.

[13] U. Tietze, Zur Theorie quadratischer Formen über Körpern der Charakteristik 2. J. reine angew. Math. 268-269 (1974) 388-390.

[14] E. Witt, Ueber eine Invariante quadratischer Formen mod 2. J. reine angew. Math. 193 (1954) 119-120.

# APPENDIX I

## THE DIMENSION OF $S/T$ AND A THEOREM ON DIVISION ALGEBRAS IN CHARACTERISTIC 2

1. Let $\xi \mapsto \xi^*$ be an antiautomorphism of the division ring $k$ whose square is inner, $\xi^{**} = \varepsilon^{-1}\xi\varepsilon$, and assume that $\varepsilon\varepsilon^* = \varepsilon^*\varepsilon = 1$. Let $S := \{\xi\in k \mid \varepsilon\xi^*=\xi\}$, $T := \{\xi+\varepsilon\xi^* \mid \xi\in k\}$ be the additive subgroups in $k$ of "symmetric" elements and "traces" respectively. We have $T \subset S$ and the factor group turns into a vector space under the composition $\lambda(\sigma+T) = \lambda\sigma\lambda^* + T$ $(\lambda\in k,\ \sigma\in S)$. The dimension of this vector space is an invariant of the "similarity class" of the structure $(k,*,\varepsilon)$. Indeed, if $(k,{}^\circ,\varepsilon_1)$ satisfies

$$\xi^\circ = \mu^{-1}\xi^*\mu\ ,\quad \varepsilon_1 = \varepsilon\mu^{*-1}\mu$$

for all $\xi \in k$ and some fixed nonzero "multiplyer" $\mu$ then we observe that for the subgroups $S_1, T_1$ relative to $(k,{}^\circ,\varepsilon_1)$ we have

$$S_1 = S\mu\ ,\quad T_1 = T\mu\ ;$$

furthermore, the map $S/T \to S_1/T_1$ defined by

$$\sigma + T \mapsto \sigma\mu + T_1$$

is a $k$-vector space isomorphism (the only nonobvious point is linearity: $(\lambda(\sigma+T))\mu = \lambda\sigma\lambda^*\mu + T_1 = \lambda\sigma\mu\lambda^\circ + T_1 = \lambda(\sigma\mu+T_1)$. We want to show that $\dim_k S/T$ is a power of $2$ if it is finite and not zero. By the previous considerations we may, for this purpose, assume without loss of generality that $(k,*,\varepsilon) = (k,{}^\circ,1)$ by letting $\mu = \varepsilon^*\sigma$ for some fixed $\sigma \in S\setminus\{0\}$.

2. We need a lemma. Recall that a central simple algebra $A$ of finite dimension over its center $k_o$ has $[A:k_o]$ a square (this follows from the Wedderburn-Artin theorem).

Lemma. Let $A$ be an algebra which is central simple and of finite rank $r^2$ over its center $k_o$. Let $*$ be an involution of $A$ which leaves $k_o$ pointwise fixed. Then the dimension of the $k_o$-linear subspace $S := \{\xi \in A \mid \xi^* = \xi\}$ has dimension $\frac{1}{2}r(r+1)$ or $\frac{1}{2}r(r-1)$; only the first case is realized if $r$ is odd or $\mathrm{char}\, k_o = 2$.

Proof. Let $\bar{k}_o$ be the algebraic closure of $k_o$. The algebra $\bar{A} = \bar{k}_o \otimes_{k_o} A$ is central simple and of dimension $r^2$ over $\bar{k}_o$. Hence $A = \mathrm{Mat}_r(\bar{k}_o)$. The involution $*$ extends to an involution $\bar{*}$ on $\bar{A}$ such that the center of $\bar{A}$ is left pointwise fixed under $\bar{*}$ and we have that $\bar{S} := \{\bar{\xi} \in \bar{A} \mid \bar{\xi}^{\bar{*}} = \bar{\xi}\}$ is equal to $\bar{k}_o \otimes S$. In sum, it suffices to prove the assertion of the lemma for the matrix ring $A = \mathrm{Mat}_r(k_o)$ where $k_o$ is assumed algebraically closed.

Let tr be the antiautomorphism of $A$ which sends each $x \in A$ into its transpose. By the theorem of Skolem-Noether the assignment $x \mapsto {}^{tr}(x^*)$ of $A$ into $A$ must be inner; there exists $a \in A$ such that

$$(1) \qquad x^* = a \cdot {}^{tr}x \cdot a^{-1} \quad \text{for all } x \in A .$$

Since $*$ is involutorial, $x = a \cdot {}^{tr}(a^{-1}) \cdot x \cdot {}^{tr}a a^{-1}$ for all $x \in A$, hence $\gamma := {}^{tr}a a^{-1}$ belongs to the center. From $tr^2 = \mathbb{1}$ we obtain furthermore that $\gamma = \pm 1$ : the matrix $a$ is symmetric or skew.

We can make yet some further normalizations. Let $c$ be an invertible matrix in $A$. Then $\varphi: x \mapsto cxc^{-1}$ is an automorphism of $A$ and $\beta := \varphi \circ * \circ \varphi^{-1}$ an involution. For $S' := \{x \in A \mid x^\beta = x\}$ we find

$S' = \varphi(S)$ , hence $\dim_{k_o} S = \dim_{k_o} S'$ . We may replace $*$ by $\beta$ ; instead of (1) we find

$$(2) \qquad x = b \cdot {}^{tr}x \cdot b^{-1} \quad \text{(for all } x \in A \text{)} , \quad b = c \cdot a \cdot {}^{tr}c .$$

Suppose that $\operatorname{char} k_o \neq 2$ . If $a$ is symmetric then we can pick $c$ such that $b$ turns out the unit matrix (because $k_o$ is algebraically closed). In that case $\beta = tr$ and $S'$ consists of all symmetric matrices; therefore $\dim S' = \frac{1}{2} r(r+1)$ . If, on the other hand, $a$ is skew then $c$ can be picked such that $b$ takes the form $\begin{pmatrix} 0 & \mathbb{1} \\ -\mathbb{1} & 0 \end{pmatrix}$ where $\mathbb{1}$ is the unit matrix of size $\frac{r}{2}$ by $\frac{r}{2}$ . In that case $b^{-1} = -b$ . For $x = \begin{pmatrix} t & u \\ v & w \end{pmatrix}$ we find

$$x^{\beta} = -b \cdot {}^{tr}x \cdot b = \begin{pmatrix} {}^{tr}w & -{}^{tr}u \\ -{}^{tr}v & {}^{tr}t \end{pmatrix} .$$

Thus $x \in S'$ if and only if $w = {}^{tr}t$ , $u = -{}^{tr}u$ , $v = -{}^{tr}v$ . As a consequence we obtain that $\dim S' = \frac{1}{2} r(r-1)$ .

Suppose that $\operatorname{char} k_o = 2$ . We can pick $c$ in (2) such that $b$ is either the unit matrix or $b = \begin{pmatrix} 0 & \mathbb{1} \\ \mathbb{1} & 0 \end{pmatrix}$ . In either case we get $\dim S' = \frac{1}{2} r(r+1)$ . Q. E. D.

3. We now return to the division ring $k$ in the first section. We shall assume that $k$ is of finite dimension over its center $k_o$ ;

$$[k:k_o] = r^2 < \infty .$$

Remember that $\dim S/T \neq 0$ implies that $\operatorname{char} k = 2$ and that $*$ is of the so-called first kind, i.e. $*$ leaves the elements of the center $k_o$ fixed. For this situation we establish the following product formula.

Theorem. If $[k:k_o] = r^2 < \infty$ and $0 < \dim S/T < \infty$ then

$$(3) \qquad r \cdot \dim_k S/T = [k_o : k_o^2] ;$$

in particular, both $r$ and $\dim S/T$ are powers of 2.

Proof. We shall consider three vector space structures on the factor group $S/T$, to wit,

| over | operation | dimension |
|---|---|---|
| $k$ | $\lambda(\sigma+T) = \lambda\sigma\lambda^* + T$ | $n$ |
| $k_o$ | $\lambda(\sigma+T) = \lambda\sigma + T$ | $m$ |
| $k_o$ | $\lambda(\sigma+T) = \lambda^2\sigma + T$ | $s$ |

We have $s = m\cdot[k_o:k_o^2]$, $s = n\cdot[k:k_o] = n\cdot r^2$. Thus $n = \frac{1}{r^2}[k_o:k_o^2]\cdot m$. Now $m = \dim_{k_o} S - \dim_{k_o} T$. By the lemma in Section 2 we have $\dim_{k_o} S = \frac{1}{2}r(r+1)$. The map $\xi \mapsto \xi + \xi^*$ is $k_o$-linear, its kernel is $S$ and its image is $T$, ergo $\dim_{k_o} T = \dim_{k_o} k - \dim_{k_o} S = \frac{1}{2}r(r-1)$. Thus $m = r$. This proves (3). Since $[k_o:k_o^2]$ is obviously a power of 2 when finite the same must hold of $r$ and $\dim S/T$ by (3).

Remark. The immediate conclusion from (3) that $r$ is a power of 2 is well known from the theory of the Brauer group $BR(k_o)$ : each prime divisor of the "index" $r$ divides the exponent of $k$ ( = order of the class of $k$ in $Br(k_o)$) (cf. e.g. [1],Theorem 17, p. 76 or [2], Sätze 1 and 2, p. 58, 59 or [3], Lemma 4.4.5, p. 120). Thus if $k$ admits an antiautomorphism that leaves the elements of $k_o$ fixed then its exponent is of course $\leq 2$ and consequently $r$ must be a power of 2.

Example. Let $X_1, \ldots, X_n, Y$ be algebraically independent over $\mathbb{Z}_2$; set $k_o = \mathbb{Z}_2(X_1,\ldots,X_n,Y)$ and let the quaternion division algebra $k = \left(\frac{1,Y}{k_o}\right)$ be equipped with its "conjugation". For this $k$ we have $\dim_k S/T = 2^n$ by (3).

*

## References to Appendix I

[1] A.A. Albert, Structure of Algebras.(3rd print. of rev. ed.) AMS Coll. Publ. XXIV, NY 1968.

[2] M. Deuring, Algebren. (Zweite, korrigierte Auflage), Ergebnisse der Math. vol 41, Springer Berlin 1968.

[3] I.N. Herstein, Noncommutative Rings. Carus Math. Monographs 15, Math. Assoc. of America 1968, distr. by J. Wiley & Sons, Inc.

CHAPTER FIFTEEN

# WITTs THEOREM IN FINITE DIMENSIONS

## 1. Introduction

Witt's Theorem tells that any isometry between subspaces in a finite dimensional space $E$ can be extended to an element of the orthogonal group of $E$. Geometric algebra in finite dimensions pivots on this theorem. Much of the effort put in this book has been aimed at discovering and proving analogous theorems in countable dimension. In this chapter we discuss the finite dimensional case.

As "space" in this book always means vector space the reader who is interested in generalizations of this important theorem to quadratic modules is advised to consult [2, 6, 9, 11] where he may find further references.

We shall first give a proof for quadratic forms in the sense of J. Tits and C.T.C. Wall as explained in the foregoing chapter (Definition 2 in XIV.3). This means that we cover besides the classical results in [1] and [13] the situation of quadratic forms over fields which are allowed to be skew. However, we need hardly retouch Chevalley's proof in [4] (also reproduced in [3]) in order that it cover this far reaching generalization. Furthermore, we do need the result in situations where the form is degenerate on $E$ (e.g. at the end XVI.7); in characteristic 2 it cannot be deduced from the result in the nondegenerate case. Again, it is very easy to adapt Chevalley's proof to the degenerate case. As is well known, it also covers the analogous situation for trace-valued sesquilinear forms.

Thus there remain the sesquilinear forms in characteristic 2 which are not trace-valued. These are harder to deal with but they do not, to us, seem less appropriate than trace-valued forms. Theorem 3 below is on par with its classical trace-valued counterpart; it appears for the first time in [8] where the proof is founded on work in [12]. We shall present Bäni's version of Chevalley's proof.

## 2. Witt's Theorem for finite dimensional quadratic forms and for trace-valued sesquilinear forms

In this section $(E,\Psi,Q)$ is a quadratic space over the division ring $(k,*,\varepsilon)$ as defined in Chapter XIV (Definition 2 in Section 3). In the following theorem $\Psi$ may be a degenerate form. Notice that in characteristic 2 two supplements of $E^{\perp}$ in $E$ need not be isometric relative to $Q$. Therefore we cannot chop off the radical $E^{\perp}$ and in this fashion reduce the mapping problem to a problem between non-degenerate spaces.

Theorem 1 ("Witt"). Let $D: F \to \bar{F}$ be an isometry of the subspaces $F$, $\bar{F}$ in the finite dimensional quadratic space $(E,\Psi,Q)$. In order that $D$ admits an extension to an operation of the orthogonal group of $(E,\Psi,Q)$ it is necessary and sufficient that $D$ maps $E^{\perp} \cap F$ onto $E^{\perp} \cap \bar{F}$.

Proof ([4]). We start with the observation that $D$ can certainly be extended to $F + E^{\perp}$. Hence we may assume without loss of generality that $E^{\perp} \subset F$ and $E^{\perp} \subset \bar{F}$. Then we proceed by induction on the dimension $n$ of $F/E^{\perp}$. If $n = 0$ the assertion is trivial. Let $n \geq 1$ and $F_1$ be a hyperplane of $F$ with $E^{\perp} \subset F_1$. Set $\bar{F}_1 := D(F_1)$. The restriction $D_1$ of $D$ to $F_1$ is an isometry $F_1 \to \bar{F}_1$ which maps $E^{\perp}$ onto itself. By induction $D_1$ extends to an isometry $D_2$ of all of $E$. If $D_2|_F = D$ we are done. Let $D_2|_F \neq D$ and consider $D_2^{-1} \circ D$

This application leaves $F_1$ pointwise fixed. Suppose that it can be extended to an isometry $D_3\colon E \to E$ . Then $D_2 \circ D_3$ extends $D$ . Our argument reduces the problem to the case where $D|_{F_1}$ is the identity on $F_1$ ; in other words, $U := \ker(D-\mathbb{1})$ is a hyperplane of $F$ .

Consider then the 1-dimensional space $G := \{Dx-x \mid x\in F\}$ . Assume that $F' \subset E$ is any subspace with $F' \perp G$ and $F' \cap F = (0) = F' \cap \bar{F}$ . Then

$$(1) \qquad \Psi(Dx,y) = \Psi(x,y) \quad \text{for all} \quad x \in F , \quad y \in F' .$$

We assert that $D$ can be extended to an isometry $\hat{D}\colon F \oplus F' \to \bar{F} \oplus F'$ by leaving $F'$ pointwise fixed. By (1) it is obvious that $\hat{D}$ respects $\Psi$ . Concerning $Q$ we find for all $x \in F$ , $x' \in F'$ that $Q(x+x') = Q(x) + Q(x') + [\Psi(x,x')] = Q(Dx) + Q(x') + [\Psi(Dx,x')] = Q(\hat{D}x) + Q(\hat{D}x') + [\Psi(\hat{D}x,\hat{D}x')] = Q(\hat{D}(x+x'))$ . Notice that after any such transition from $D$ to $\hat{D}$ we have $G = \{\hat{D}z-z \mid z\in F\oplus F'\}$ .

We next convince ourselves that

$$(2) \qquad \Psi(Dx,Dy-y) = -\Psi(Dx-x,y) \quad \text{for all} \quad x , y \in F .$$

Indeed, $\Psi(Dx,Dy-y) = \Psi(Dx,Dy) - \Psi(Dx,y) = \Psi(x,y) - \Psi(Dx,y) = -\Psi(Dx-x,y)$ . In particular $U \perp G$ . We shall now distinguish two cases.

<u>Case 1</u>: $F \not\subset G^{\perp}$ . By (2) therefore $\bar{F} \not\subset G^{\perp}$ . Let $F'$ be a supplement of $U$ in $G^{\perp}$ . Then $F' \cap F \subset G^{\perp} \cap F = U$ ; hence $F' \cap F = F' \cap F \cap U = (0)$ . Similarly one sees that $F' \cap \bar{F} = (0)$ . As we have seen above we may extend $D$ to an isometry $\hat{D}\colon F \oplus F' \to \bar{F} \oplus F'$ . Now $F \oplus F'$ contains the proper subspace $G^{\perp} = U \oplus F'$ and $G^{\perp}$ is a hyperplane. Ergo $F \oplus F' = E$ and we are through.

<u>Case 2</u>: $F \subset G^{\perp}$ . Hence $\bar{F} \subset G^{\perp}$ by (2). As $G \subset F + \bar{F}$ we obtain $G \subset G^{\perp}$ ; in fact $G$ is even singular: for $Dx-x$ a typical element of

$G$ we find $Q(Dx-x) = Q(Dx) + Q(x) + [\Psi(Dx,-x)]$ and, because $\Psi(Dx,-x) = \Psi(x+(Dx-x),-x) = \Psi(x,-x)$ , $Q(Dx-x) = Q(Dx) + Q(x) + [\Psi(x,-x)] = Q(x) + Q(x) + [\Psi(x,-x)] = Q(x-x) = 0$ .

We now show that $F$ and $\bar{F}$ possess a common supplement $F'$ in $G^{\perp}$ . This is trivial when $F = \bar{F}$ . Let $F \neq \bar{F}$ . Then $F = U \oplus (x)$ , $\bar{F} = U \oplus (\bar{x})$ . Obviously $x + \bar{x} \notin F$ , $x + \bar{x} \notin \bar{F}$ so $F \oplus (x+\bar{x}) = \bar{F} \oplus (x+\bar{x})$ . Adding to $(x+\bar{x})$ a supplement of $F + (x+\bar{x})$ in $G^{\perp}$ gives what we are looking for. By the argumentation following (1) we can extend $D$ to an isometry $\hat{D}: F \oplus F' \cong \bar{F} \oplus F'$ . If $G^{\perp}$ $(= F\oplus F' = \bar{F}\oplus F')$ is all of $E$ we are done. This shows that we have reduced the proof of the theorem to the following problem:

Given is a metric automorphism $D$ of the hyperplane $F = G^{\perp}$ which maps $E^{\perp}$ onto $E^{\perp}$ . Prolong $D$ to an isometry of $E$ .

Let $z \in E$ , $z \notin F$ . Assume that we can find $z' \in E$ such that

(3) $$\Psi(x,z) = \Psi(Dx,z') \quad \text{for all} \quad x \in F$$

and

(4) $$Q(z) = Q(z') .$$

Any vector $z'$ with (3) is in $E \setminus F$ . Otherwise $z' = Dw$ for some $w \in F$ , hence $z - w \in F^{\perp}$ by (3). Because $F^{\perp} = G \subset G^{\perp} = F$ we obtain $z \in F$ , contradiction. Thus $E = F \oplus (z) = F \oplus (z')$. By (3) and (4) we can extend $D$ to all of $E$ by sending $z$ into $z'$ . We are left to show the existence of such a $z'$ . The map $F \to k$ defined by $x \mapsto \Psi(D^{-1}x,z)$ is linear and vanishes on $E^{\perp}$ since $D^{-1}$ maps $E^{\perp}$ onto $E^{\perp}$ . Hence it is the restriction of a linear map $f: E \to k$ with $E^{\perp} \subset \ker f$ and therefore of the kind $x \mapsto \Psi(x,z'')$ , i.e. $\Psi(D^{-1}x,z) = \Psi(x,z'')$ for all $x \in F$ . This yields $\Psi(x,z) = \Psi(Dx,z'')$ for all $x \in F$ . For $0 \neq g \in G$ we set $z' = z'' + \lambda g$ . Since $g \perp F$ the vector

$z'$ satisfies (3). Therefore $z'' \notin F$ so $\Psi(z'',g) \neq 0$. Because $g$ is singular as we have seen it follows that

$$Q(z') = Q(z'') + [\Psi(z'',g)\lambda^*]$$

and it is clear that $\lambda$ can be specified such that $Q(z')$ has a prescribed value in $k^+/T$. Thus we can satisfy both (3) and (4) and the proof of the theorem is complete.

We shall list some of the celebrated consequences of Theorem 1.

Corollary 1 ("Cancellation Theorem"). Let $(E,\Psi,Q)$ be of finite dimension and nondegenerate. If $E = F + F^{\perp} = G + G^{\perp}$ and $F \cong G$ then $F^{\perp} \cong G^{\perp}$. (Here $\cong$ means isometry relative to $(\Psi,Q)$.)

Corollary 2. Let $(E,\Psi,Q)$ be as in Corollary 1. All maximal totally singular subspaces of $E$ have the same dimension and are permuted transitively among themselves by the operations of the orthogonal group of $(E,\Psi,Q)$.

Corollary 3. Let $E$ be finite dimensional and $\Psi: E \times E \to k$ a nondegenerate $\varepsilon$-hermitean form; assume that $\operatorname{char} k \neq 2$. If $D: F \to \bar{F}$ is an isometry relative to $\Psi$ between the subspaces $F$, $\bar{F} \subset E$ then $D$ can be extended to an isometry of the whole space $(E,\Psi)$. In particular $F^{\perp}$ and $\bar{F}^{\perp}$ are isometric (isometry with respect to $\Psi$).

Indeed, if $\Psi$ is $\varepsilon$-hermitean then the pair $(\Psi,Q)$ with $Q(x) := \frac{1}{2}\Psi(x,x) + P$ is a quadratic form in the sense of Chapter XIV ($P$ is the additive subgroup $\{\varepsilon\xi^*-\xi \mid \xi \in k\}$ of $k$). Furthermore, an isometry relative to $\Psi$ is an isometry with respect to $(\Psi,Q)$. Hence we may apply Theorem 1 to $F$, $\bar{F}$ in $(E,\Psi,Q)$.

A theorem analogous to Corollary 3 holds in characteristic 2 provided the $\varepsilon$-hermitean form $\Psi$ is assumed to be trace-valued.

Indeed the proof of Theorem 1 goes through if we just forget about $Q$ and take isometry with respect to the sesquilinear form $\Psi$. We then have to replace (4) by the condition $\Psi(z,z) = \Psi(z',z')$. Since $\Psi(z",g) \neq 0$ and $\Psi(g,g) = 0$ we can again set $z' = z" + \lambda g$ and, by trace-valuedness of $\Psi$, solve for $\lambda$ from $\Psi(z"+\lambda g, z"+\lambda g) = \Psi(z,z)$. This establishes

<u>Theorem 2</u>. Let $E$ be finite dimensional and $\Psi: E \times E \to k$ a nondegenerate $\varepsilon$-hermitean form; assume that $\operatorname{char} k = 2$ and $\Psi$ is trace-valued. If $D: F \to \bar{F}$ is an isometry between subspaces of $(E,\Psi)$ then $D$ can be extended to an isometry of the whole space. (In particular $F^{\perp}$ and $\bar{F}^{\perp}$ are isometric.)

In all results of this section we can leave the dimension of $E$ completely arbitrary as long as the dimension of the subspaces $F$, $\bar{F}$ is kept finite. This is entirely obvious since we can chop off from $E$ a finite dimensional orthogonal summand which contains both $F$ and $\bar{F}$.

We remark furthermore that (in this general setting) Theorems 1 and 2 are still equivalent with their corresponding Cancellation Theorems. The proof is the same as that given in the Introduction to Chapter V.

## 3. A Witt type theorem for finite dimensional non-trace-valued sesquilinear forms

We place ourselves in the situation of Theorem 2 above and we keep $k$ of characteristic 2 but allow the form $\Psi$ to be non-trace-valued. Let, as usual, $T$ be the additive subgroup in $(k,*,\varepsilon)$ of all traces $\xi + \varepsilon\xi^*$. For any subspace $X$ in $(E,\Psi)$ we let $X^* = \{x \in X \mid \Psi(x,x) \in T\}$; $\dim X/X^* \leq \dim S/T$. We recall that $E^{*\perp}$ is left pointwise fixed under each operation of the orthogonal group of $(E,\Psi)$. Thus, if $D: F \to \bar{F}$

is to be extended to an element of the orthogonal group we must obviously have that

(5) $F \cap E^{*\perp} = \bar{F} \cap E^{*\perp}$ and $D: F \to \bar{F}$ leaves $F \cap E^{*\perp}$ pointwise fixed.

We shall prove the remarkable

<u>Theorem 3</u> ( [8] ). Let $(E,\Psi)$ be a nondegenerate $\varepsilon$-hermitean form of finite dimension and not necessarily trace-valued. An isometry $D: F \to \bar{F}$ between subspaces $F$ , $\bar{F} \subset E$ can be extended to an isometry of the whole space if and only if (5) is satisfied.

<u>Corollary 4</u>. Let $(E,\Psi)$ in the theorem be of arbitrary dimension but keep $\dim F$ finite. Then the conclusion in the theorem is still valid.

<u>Proof of the corollary</u> (Bäni). The basic idea of the proof is obvious. In order to realize it, let $\mathcal{H}$ be the set $\{H \subset E \mid F + \bar{F} \subset H$ , $H + H^{\perp} = E$ , $\dim H < \infty\}$ . Obviously $E^* = \bigcup_{\mathcal{H}} H^*$ so $E^{*\perp} \cap F = \bigcap \{H^{*\perp} \cap F \mid H \in \mathcal{H}\}$ . As $\dim F$ is finite and the system $H^{*\perp} \cap F$ $(H \in \mathcal{H})$ is directed there is a smallest element, $H_o^{*\perp} \cap F = E^{*\perp} \cap F$ . There is an analogous object $H_1$ for $\bar{F}$ , $H_1^{*\perp} \cap \bar{F} = E^{*\perp} \cap \bar{F}$ . Now each $H$ with $H_o + H_1 \subset H \in \mathcal{H}$ has the requisite property: $E = H \oplus H^{\perp}$ , $\dim H < \infty$ , $H^{*\perp} \cap F = E^{*\perp} \cap F = E^{*\perp} \cap \bar{F} = H^{*\perp} \cap \bar{F}$ because of (5). We can now apply Theorem 3 to $F, \bar{F} \subset (H,\Psi)$ and join the extension of $D$ (to the space $H$ ) with the identity on $H^{\perp}$. Q. E. D.

Let $S$ be the subgroup $\{\xi \in k \mid \varepsilon\xi^* = \xi\}$ of symmetric elements in $k$ ; the factor group $S/T$ is a k-vector space under the composition $\lambda(s+T) = \lambda s \lambda^* + T$ $(\lambda \in k)$ . Each subspace $X$ in $(E,\Psi)$ has its "value space" $\{\Phi(x,x)+T \mid x \in X\}$ in the k-space $S/T$ . Set $R := E^* \cap E^{*\perp}$

and consider any metabolic decomposition of $E$ with respect to $R$,

$$(6) \qquad E = (R \oplus R') \overset{\perp}{\oplus} E_1 \overset{\perp}{\oplus} E_2 \ , \quad E^* = R \oplus E_1 \ , \quad E^{*\perp} = R \oplus E_2 \ .$$

It is very easy to show that the isometry class of $(E,\Psi)$ is characterized by the value space of $E$ in $S/T$ and the isometry classes of $E_1$ and $E_2$ (endowed with the forms induced by $\Psi$). This elementary fact will not be used in the proof of Theorem 3; we shall use it in the proof of the Corollary 5. Here we may point out that if it is assumed that $F$, $\bar{F}$ belong to $E^*$ ([12], Theorem 1.2.1) then the assertion of Theorem 3 reduces immediately to the trace-valued situation of Thm. 2 by using decompositions (6) and the remark that follows (6). Unaware of [7,8] we had realized that the assertion of Theorem 3 invariably holds for <u>arbitrary</u> $F$, $\bar{F}$ provided that $\dim E/E^* = 1$ by doing computational work on the lattice $V(F,E^*)$ $\perp$-stably generated by $F$ and $E^*$ (cf. [5]). Still unaware of the work of Pless, Bäni found the general theorem and observed that, once more, it flows from Chevalley's proof of Theorem 1. (Incidentally, if $F$ is assumed trace-valued then $V(F,E^*)$ is of course the "free" modular lattice generated by the chains $F \subset E^*$ and $E^{*\perp} \subset F^{\perp}$ and hence <u>distributive</u>. This accounts for the easy reduction possible in the special case $F, \bar{F} \subset E^*$ mentioned before.) M. Studer finally drew our attention to [7] and [8].

<u>Proof of Theorem 3</u> (Bäni). The first remark is that $D: F \to \bar{F}$ can certainly be extended to an isometry $F + E^{*\perp} \to \bar{F} + E^{*\perp}$ by leaving $E^{*\perp}$ pointwise fixed. Indeed, if $f \in F$ then $f + Df \in E^*$ which means that $\Phi(f,e) = \Phi(Df,e)$ for all $e \in E^{*\perp}$. By (5) this extends $D$. Thus we can start the argumentation in the proof of Theorem 1 by assuming that $E^{*\perp}$ is in $F$ and left pointwise fixed under $D$. We proceed by induction on the dimension $n$ of $F/E^{*\perp}$. For $n = 0$ the assertion is trivial.

Assume then that $n \geq 1$ and let $F_1$ be a hyperplane of $F$ with $E^{*\perp} \subset F_1$ and proceed as in the proof of Theorem 1. Everything remains unchanged except for discarding $Q$ and having to satisfy

$$\Psi(z,z) = \Psi(z',z') \tag{7}$$

instead of (4). As earlier we find $z'' \in E$ with (3) and have $z'' \in E \setminus F$ by the argumentation that follows (4). Ergo $\Psi(z'',g) \neq 0$. Since $z + z'' \in U^{\perp}$ by (3) ($U$ is the hyperplane in $F$ of points fixed under $D$ ) we have $z + z'' \in E^*$ as $E^{*\perp} \subset U$ by the opening remark in this proof and the choice of $F_1$ ; in other words,

$$\Psi(z,z) \equiv \Psi(z'',z'') \bmod T . \tag{8}$$

Hence, as in the proof of Theorem 2, we set $z' = z'' + \lambda g$ and can solve for $\lambda$ in condition (7). Q. E. D.

Corollary 5 ([5]). Let $k$ be a perfect commutative field and $\Phi: E \times E \to k$ a nondegenerate, not necessarily trace-valued, symmetric bilinear form on the finite dimensional space $E$ . Two subspaces $F$ , $\bar{F} \subset E$ belong to the same orbit under the action of the orthogonal group if and only if there exists a lattice isomorphism $\tau: V(F,E^*) \to V(\bar{F},E^*)$ which in turn sends $F$ , $E^*$ , $F^{\perp}$ , $E^{*\perp}$ into $\bar{F}$ , $E^*$ , $\bar{F}^{\perp}$ , $E^{*\perp}$ and which respects the two indices $\dim(\operatorname{rad} F)$ , $\dim(F^*/\operatorname{rad} F^*)$ .

Proof. If $F$ , $\bar{F}$ fall into the same orbit, $DF = \bar{F}$ for an element of the orthogonal group of $(E,\Psi)$ , then $D$ induces a lattice isomorphism $\tau$ with the properties mentioned. Assume conversely that there is such a $\tau$ . We have a decomposition

$$\begin{aligned} F &= \operatorname{rad} F \overset{\perp}{\oplus} (R \oplus R') \overset{\perp}{\oplus} F_1 \overset{\perp}{\oplus} F_2 \quad \text{with} \\ F^* &= \operatorname{rad} F \oplus R \oplus F_1 \quad \text{and} \\ F \cap F^{*\perp} &= \operatorname{rad} F \oplus R \oplus F_2 . \end{aligned} \tag{9}$$

It follows that the two indices mentioned fix the isometry class of $F$ because $\dim R$, $\dim F_2 \in \{0,1\}$ can be told from the lattice. Hence, by $\tau$, there is an isometry $D: F \to \bar{F}$. How does $D$ operate on $E^{*\perp} \cap F$? If it should happen that $E^{*\perp} \subset \operatorname{rad} F$ then $E^{*\perp} \subset \operatorname{rad} \bar{F}$ by $\tau$ and we can pick $D$ such that it leaves $E^{*\perp}$ pointwise fixed. If $E^{*\perp}$ - which is (0) or 1-dimensional - is in $F \setminus \operatorname{rad} F$ then $E^{*\perp} \subset \bar{F} \setminus \operatorname{rad} \bar{F}$ by $\tau$ and we can arrange the decomposition in (9) such that $E^{*\perp} \subset F_o := (R \oplus R') \overset{\perp}{\oplus} F_1 \overset{\perp}{\oplus} F_2$, $E^{*\perp} \subset \bar{F}_o$ ( $\bar{F}_o$ a supplement of $\operatorname{rad} \bar{F}$ in $\bar{F}$ which contains $E^{*\perp}$ ). $D$ induces an isometry $D_o$: $F_o \to \bar{F}_o$; it maps the space $E^{*\perp}$ onto itself because $E^{*\perp} = F_o^{*\perp} \cap F_o$, $E^{*\perp} = \bar{F}_o^{*\perp} \cap \bar{F}_o$. If $F_o^{*\perp} \cap F_o$ is anisotropic, then $D_o$ must leave it pointwise fixed. If $F_o^{*\perp} \cap F_o$ is isotropic, then we decompose $E = (R \oplus R') \overset{\perp}{\oplus} E_1$, $E = (D_oR \oplus D_oR') \overset{\perp}{\oplus} \bar{E}_1$. Since $R = D_oR = E^{*\perp}$ we have $E_1$, $\bar{E}_1 \subset E^*$ so there is an isometry $D_1: E_1 \to \bar{E}_1$. Joining $D_1$ and $D_o|_{R \oplus R'}$ yields an isometry $E \to E$ which necessarily leaves $E^{*\perp}$ pointwise fixed. Hence $D_o$ leaves $E^{*\perp}$ pointwise fixed. In sum, it follows from $\tau$ that there is an isometry $D: F \to \bar{F}$ which qualifies for Theorem 3.

Remarks. 1. The often mentioned "counter example" to Witt's statement in the non-trace-valued case, namely $\langle \alpha \rangle \cong \langle \alpha \rangle$ & $\langle \alpha \rangle \overset{\perp}{\oplus} \langle \alpha, \alpha \rangle \cong \langle \alpha \rangle \overset{\perp}{\oplus} P$, $P$ a hyperbolic plane and $\alpha \in S \setminus T$, simply ignores (5).

2. The lattice $V(F,E^*)$ in the corollary is always finite for $[k:k^2] = 1$. If $F \not\subset E^*$ & $F^\perp \not\subset E^*$ & $\operatorname{rad} F = \operatorname{rad} F^*$ & $\operatorname{rad}(F^\perp) = \operatorname{rad}(F^{\perp *})$ then it has 28 elements; in all other instances it has less elements. $V(F,E^*)$ is distributive if and only if either $F \subset E^*$ or $F^\perp \subset E^*$. Proofs can be performed "synthetically" or "analytically" (i.e. by making use of appropriate decompositions of $E$ relative to $F$ ).

## References to Chapter XV

[1] C. Arf, Untersuchungen über quadratische Formen in Körpern der Charakteristik 2. J. reine angew. Math. 183 (1941) 148-167.

[2] A. Bak, On modules with quadratic forms. Contained in Springer Lecture Notes, vol. 108 (1969) 55-66. Springer Verlag, Heidelberg.

[3] N. Bourbaki, Formes sesquilinéaires et formes quadratiques. Hermann Paris 1959.

[4] C. Chevalley, The algebraic theory of spinors. Columbia University Press, New York 1954.

[5] H. Gross, Formes quadratiques et formes non traciques sur les espaces de dimension dénombrable. Bull. Soc. Math. de France Mémoire 59 (1979).

[6] M. Kneser, Witts Satz für quadratische Formen über lokalen Ringen. Nachrichten der Akademie der Wissenschaften, Göttingen (1972) 195-203.

[7] V. Pless, On Witt's theorem for nonalternating symmetric bilinear forms over a field of characteristic 2. Proc. Amer. Math. Soc. 15 (1964) 979-983.

[8] - , On the invariants of a vector subspace of a vector space over a field of characteristic two. Proc. Amer. Math. Soc. 16 (1965) 1062-1067.

[9] H.G. Quebbemann, R. Scharlau, W. Scharlau, M. Schulte, Quadratische Formen in additiven Kategorien. Bull. Soc. Math. de France 48 (1976) 93-101.

[10] H. Reiter, Erzeugbarkeit dyadischer orthogonaler Gruppen durch Spiegelungen. J. Algebra 47 (1977) 313-322.

[11] H. Reiter, Witt's theorem for noncommutative semilocal rings. J. Algebra 35 (1975) 483-499.

[12] G.E. Wall, On the conjugacy classes in the unitary, symplectic, and orthogonal groups. J. Australian Math. Soc. 3 (1963) 1-62.

[13] E. Witt, Theorie der quadratischen Formen in beliebigen Körpern. J. reine angew. Math. 176 (1937) 31-44.

## CHAPTER SIXTEEN

## ARFs THEOREM IN DIMENSION $\aleph_0$

### 1. Introduction

In the whole chapter $k$ is a field of characteristic 2 and $\xi \to \xi^*$ an antiautomorphism of the field whose square is inner, $\xi^{**} = \varepsilon^{-1}\xi\varepsilon$ and, furthermore, $\varepsilon\varepsilon^* = 1$ for some $\varepsilon \in k$. Let as usual $S := \{\xi \in k \mid \xi = \varepsilon\xi^*\}$ and $T := \{\xi + \varepsilon\xi^* \mid \xi \in k\}$ be the additive subgroups in $k$ of symmetric elements and traces respectively. The factor group $S/T$ is a k-left vector space under the composition $\lambda(\sigma+T) = \lambda\sigma\lambda^* + T$ $(\sigma \in S,\ \lambda \in k)$. Let $(Q,\Psi)$ be a quadratic form on the k-vector space $E$, i.e. $Q$ is a map from $E$ into the factor group $k/T$ and $\Psi$: $E \times E \to k$ its associated $\varepsilon$-hermitean form. For $X \subset E$ an arbitrary subspace we define the "kernel"

$$K(X) := \{x \in X \cap X^\perp \mid Q(x) = 0\} .$$

$K(X)$ is a linear subspace with

$$\dim \operatorname{rad} X/K(X) \leq \dim S/T .$$

We shall assume in the whole chapter that $\dim S/T < \infty$. Hence $\dim \operatorname{rad} X/K(X)$ will always be finite, in other words, totally isotropic subspaces are "almost" totally singular. This assumption is also essential for the computation of the lattice in Section 2 below. From Section 3 onward we shall assume in addition that $(k, *, \varepsilon)$ is such that

(0) there is only one isometry class in dimension $\aleph_0$ of nondegenerate trace-valued $\varepsilon$-hermitean forms.

Condition (0) is tantamount to assuming that each trace-valued $\aleph_0$-form admits infinite dimensional totally isotropic subspaces. Since $S/T$ is assumed finite it will follow from (0) that in dimension $\aleph_0$ there is only one isometry class of nondegenerate quadratic spaces $(E, \Psi, Q)$. An immediate but important consequence is the following stability property: if $X \subset E$ is infinite dimensional and nondegenerate and $F \subset E$ finite dimensional then $X \cap F^{\perp}$ will contain nondegenerate planes spanned by singular vectors.

The topic of this chapter is the characterization of a single subspace $V \subset E$ - modulo the action of the orthogonal group of $E$ - as given by Glauser in [3] *). The orbit of a subspace $V \subset E$ is, in essence, characterized by the isometry classes of $V$ and $V^{\perp}$, a family of four cardinal numbers and a distributive lattice of height 7 consisting of subspaces in the value space $S/T$. In case of a "perfect" division ring $(\dim S/T = 1)$ we have the two isometry classes and a collection of eleven cardinals.

## 2. Glauser's Lattice

In what follows $\mathcal{V}(V)$ is the smallest sublattice in $\mathcal{L}(E)$ (the lattice of all subspaces of $E$ ) that contains the subspace $V \subset E$ and is closed under the operations $\perp$ and $K$ (as defined in the introduction).

Theorem 1. If $\dim S/T < \infty$ then $\mathcal{V}(V)$ is finite; it is given by the diagram below. If $\dim S/T \geqq 7$ (hence $\geqq 8$) then each nondegenerate quadratic k-space $(E, \Psi, Q)$ of dimension $\aleph_0$ possesses subspaces $V$ such that all 52 elements of $\mathcal{V}(V)$ are different spaces in $E$.

The theorem as well as the following diagram can be found in [3].

---

*) This is a Witt type problem. Whenever we wish to stress the fact that the underlying space is a quadratic space proper and not a sesquilinear space we refer to "Arf".

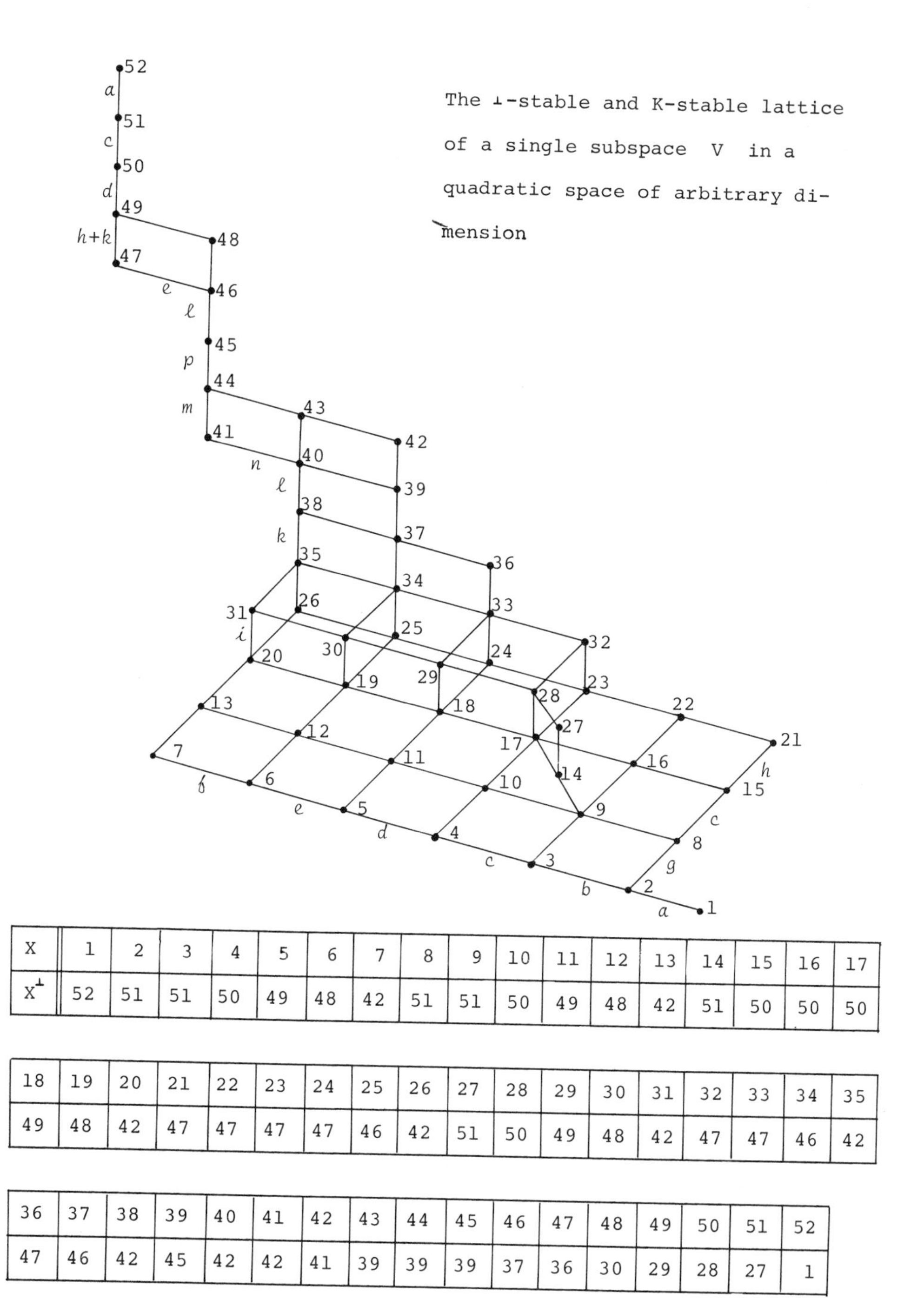


The ⊥-stable and K-stable lattice of a single subspace V in a quadratic space of arbitrary dimension

| X | 1 | 2 | 3 | 4 | 5 | 6 | 7 | 8 | 9 | 10 | 11 | 12 | 13 | 14 | 15 | 16 | 17 |
|---|---|---|---|---|---|---|---|---|---|---|---|---|---|---|---|---|---|
| $X^{\perp}$ | 52 | 51 | 51 | 50 | 49 | 48 | 42 | 51 | 51 | 50 | 49 | 48 | 42 | 51 | 50 | 50 | 50 |

| 18 | 19 | 20 | 21 | 22 | 23 | 24 | 25 | 26 | 27 | 28 | 29 | 30 | 31 | 32 | 33 | 34 | 35 |
|---|---|---|---|---|---|---|---|---|---|---|---|---|---|---|---|---|---|
| 49 | 48 | 42 | 47 | 47 | 47 | 47 | 46 | 42 | 51 | 50 | 49 | 48 | 42 | 47 | 47 | 46 | 42 |

| 36 | 37 | 38 | 39 | 40 | 41 | 42 | 43 | 44 | 45 | 46 | 47 | 48 | 49 | 50 | 51 | 52 |
|---|---|---|---|---|---|---|---|---|---|---|---|---|---|---|---|---|
| 47 | 46 | 42 | 45 | 42 | 42 | 41 | 39 | 39 | 39 | 37 | 36 | 30 | 29 | 28 | 27 | 1 |

$1 = (0)$

$2 = K(V)$

$3 = radV \cap K(V)^{\perp\perp}$

$4 = radV \cap K((radV)^{\perp\perp})^{\perp\perp}$

$5 = radV \cap K(V^{\perp})^{\perp\perp}$

$6 = radV$

$7 = V$

$8 = K(V)^{\perp\perp} \cap K(V^{\perp})$

$9 = radV \cap K(V)^{\perp\perp} + K(V)^{\perp\perp} \cap K(V^{\perp})$

$10 = radV \cap K((radV)^{\perp\perp})^{\perp\perp} + K(V)^{\perp\perp} \cap K(V^{\perp})$

$11 = radV \cap K(V^{\perp})^{\perp\perp} + K(V)^{\perp\perp} \cap K(V^{\perp})$

$12 = radV + K(V)^{\perp\perp} \cap K(V^{\perp})$

$13 = V + K(V)^{\perp\perp} \cap K(V^{\perp})$

$14 = K(V)^{\perp\perp} \cap (radV + K(V^{\perp}))$

$15 = K((radV)^{\perp\perp})$

$16 = radV \cap K(V)^{\perp\perp} + K((radV)^{\perp\perp})$

$17 = radV \cap K((radV)^{\perp\perp})^{\perp\perp} + K((radV)^{\perp\perp})$

$18 = radV \cap K(V^{\perp})^{\perp\perp} + K((radV)^{\perp\perp})$

$19 = radV + K((radV)^{\perp\perp})$

$20 = V + K((radV)^{\perp\perp})$

$21 = K(V^{\perp})$

$22 = radV \cap K(V)^{\perp\perp} + K(V^{\perp})$

$23 = radV \cap K((radV)^{\perp\perp})^{\perp\perp} + K(V^{\perp})$

$24 = radV \cap K(V^{\perp})^{\perp\perp} + K(V^{\perp})$

$25 = radV + K(V^{\perp})$

$26 = V + K(V^{\perp})$

$27 = K(V)^{\perp\perp}$

$28 = K((radV)^{\perp\perp})^{\perp\perp}$

$29 = (radV)^{\perp\perp} \cap K(V^{\perp})^{\perp\perp}$

$30 = (radV)^{\perp\perp}$

$31 = V + (radV)^{\perp\perp}$

$32 = radV \cap K((radV)^{\perp\perp})^{\perp\perp} + K(V^{\perp}) + K(V)^{\perp\perp}$

$33 = radV \cap K(V^{\perp})^{\perp\perp} + K(V^{\perp}) + K(V)^{\perp\perp}$

$34 = radV + K(V^{\perp}) + K(V)^{\perp\perp}$

$35 = V + K(V^{\perp}) + K(V)^{\perp\perp}$

$36 = K(V^{\perp})^{\perp\perp}$

$37 = (radV + K(V^{\perp}))^{\perp\perp}$

$38 = V + K(V^{\perp})^{\perp\perp}$

$39 = rad(V^{\perp})$

$40 = V + rad(V^{\perp})$

$41 = V^{\perp\perp}$

$42 = V^{\perp}$

$43 = V + V^{\perp}$

$44 = V^{\perp} + V^{\perp\perp}$

$45 = rad(V^{\perp})^{\perp}$

$46 = (radV)^{\perp} \cap K(V^{\perp})^{\perp}$

$47 = K(V^{\perp})^{\perp}$

$48 = (radV)^{\perp}$

$49 = K(V^{\perp})^{\perp} + (radV)^{\perp}$

$50 = K((radV)^{\perp\perp})^{\perp}$

$51 = K(V)^{\perp}$

$52 = E$

A table for the operation $K$ is not needed; as it turns out, all totally isotropic elements of the lattice are contained in $rad(V^{\perp})$, hence

$$K(X) = X \cap X^{\perp} \cap K(V^{\perp}) .$$

The proof of Theorem 1 is left to the reader; it is not difficult but time-consuming (cf. VIII.3). We shall, however, make the following comment which helps getting down to the pith of the matter. We say that $\dim x/y < \infty$ for two elements $x$, $y$ in a (arbitrary) modular lattice $L$ with $0$ and $1$ if and only if $y \leq x$ and there is a finite maximal chain joining $y$ with $x$. Assume that $L$ is equipped with an antitone mapping $\perp$ such that $0^\perp = 1$, $1^\perp = 0$ and $x \leq x^{\perp\perp}$ for all $x$. Thus $\perp\perp$ is a closure operator. Assume that it satisfies the axiom

$$x = x^{\perp\perp} \ \& \ \dim(x \vee y)/x < \infty \Rightarrow (x \vee y)^{\perp\perp} = x \vee y .$$

Let furthermore $K: \mathcal{V} \to \mathcal{V}$ be a map which satisfies (notation: $\mathrm{rad}\, x = x \wedge x^\perp$)

$$\mathrm{rad}\, x \leq \mathrm{rad}\, y \Rightarrow K(x) = \mathrm{rad}\, x \wedge K(y)$$

and the following finiteness assumptions

$$\dim(\mathrm{rad}\, x)^{\perp\perp}/K(x)^{\perp\perp} < \infty ,$$

$$\dim K(x)^\perp/(\mathrm{rad}\, x)^\perp < \infty ,$$

$$\dim \mathrm{rad}(x^\perp)/K(x^\perp) < \infty .$$

Then the sublattice of L which is $\perp$-stably and K-stably generated by a single element $v$ is given by the above diagram. This has been verified in [2]. The above finiteness conditions are responsible for the finiteness of our lattice; the precise values of the dimensions are irrelevant. This is the reason why the upper bound 52 for the cardinality of $\mathcal{V}(V)$, in Theorem 1, does not depend on $\dim S/T$. This upper bound does not depend on $\dim E$ either; $\dim E$ may be uncountable in Theorem 1.

## 3. Invariants of a subspace

We refer to the legend that goes with the diagram in the previous section. Let $(r/s)$ be the dimension of the quotient space $X/Y$ where $X$, $Y \in \mathcal{V}(V)$ have the numbers $r$ and $s$ respectively assigned to them.

We formally introduce the following "indices":

$$
\begin{array}{lllll}
a := (2/1)\,, & b := (3/2)\,, & c := (4/3)\,, & d := (5/4)\,, & e := (6/5) \\
f := (7/6)\,, & g := (8/2)\,, & h := (21/15)\,, & i := (27/14)\,, & k := (36/33 \\
\ell := (39/37)\,, & m := (42/39)\,, & n := (41/40)\,, & p := (45/44)\,. &
\end{array} \tag{1}
$$

One proves that

$$
\begin{array}{l}
(15/8) = (14/9) = (17/14) = (51/50) = c\,, \\
(46/45) = \ell\,, \quad (47/46) = e\,, \quad (48/46) = h+k\,, \\
(50/49) = d\,, \quad (52/51) = a\,.
\end{array} \tag{2}
$$

Countability of $\dim E$ - which we shall assume henceforth - is used in proving (2) only inasmuch as it guarantees the inequality "$\dim E/X^{\perp} \leq \dim X$". For arbitrary $\dim E$ one would have to introduce a few more indices such as $(52/51)$, etc.

By (2) all dimensions $\dim X/Y$ for nested elements $Y \subset X \in V(V)$ can be expressed in terms of the indices introduced in (1). Besides thes indices we get some more obvious invariants of the subspace $V$ by considering the map $Q$ restricted to the space $39 = \mathrm{rad}(V^{\perp})$. As $\Psi$ vanishes on this space we have for all $x \in 39$ that $0 = \Psi(x,x) = \kappa + \varepsilon\kappa$ $\kappa \in Q(x)$ ; therefore, $Q(x) \in S/T \subset k/T$ and the map $Q\colon 39 \to S/T$ is a homomorphism of k-vector spaces; its kernel is the space 21. Thus we have the following image in $L(S/T)$

(3)

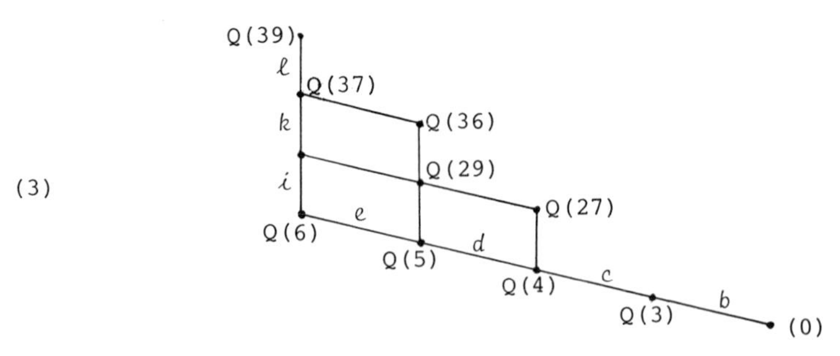


This lattice is an invariant attached to the orbit of $V$ .

For two subspaces $V$ , $\bar{V} \in E$ to belong to the same orbit it is necessary that $V$ and $\bar{V}$ are isometric. The isometry class of $V$ has, essentially, no bearing on that of $V^{\perp}$ . Therefore, we are going to assume in one lump that $V \cong \bar{V}$ and $V^{\perp} \cong \bar{V}^{\perp}$ . There is some overlapping with the lattice in (3). From $V \cong \bar{V}$ we obtain that $Q(6) = Q(\bar{6})$ ; from $V^{\perp} \cong \bar{V}^{\perp}$ we get $Q(39) = Q(\overline{39})$ . Thus it will be sufficient to make assumptions on $Q(3)$ , $Q(4)$ , $Q(5)$ , $Q(27)$ , $Q(36)$ . Except for one very degenerate case (Case II in Theorem 2) it will be seen that we have collected a complete set of invariants for the orbit of $V$ . This is remarkable. For there are many more obvious invariants: instead of merely considering the space $K(X) \subset \text{rad}\, X$ take, in each totally isotropic $X$ , the inverse image of the full lattice $L(S/T)$ under the map $Q\colon X \to S/T$ . This will, in general, furnish an infinitude of subspaces in each totally isotropic $X \in V(V)$ ; one may then pass to the $\perp$-stable lattice generated by all these spaces, etc. The reason why we do not have to penetrate into this thicket of invariants - it is fixed by the invariants which we have picked beforehand - is once more the assumption on the finiteness of $\dim S/T$ .

## 4. Characterization of the orbit of a subspace (Theorem 2)

We shall prove the following

Theorem 2 ([3]). Let $V$ and $\bar{V}$ be infinite dimensional subspaces of a nondegenerate quadratic space $(E,\Psi,\Omega)$ of dimension $\aleph_0$ over a division ring as specified in the introduction. Assume that $V \cong \bar{V}$ and $V^{\perp} \cong \bar{V}^{\perp}$ . We distinguish two cases.

Case I. $\dim V/\text{rad}\, V + \dim V^{\perp}/\text{rad}(V^{\perp}) = \aleph_0$ . In order that there is an isometry $\tilde{\tau}\colon E \to E$ with $\tilde{\tau}V = \bar{V}$ it is necessary and sufficient that the following hold.

(i) $Q(\text{rad}V \cap K(V)^{\perp\perp}) = Q(\text{rad}\bar{V} \cap K(\bar{V})^{\perp\perp})$ ("$Q(3)=Q(\bar{3})$")

(ii) $Q(\text{rad}V \cap K((\text{rad}V)^{\perp\perp})^{\perp\perp}) = Q(\text{rad}\bar{V} \cap K((\text{rad}\bar{V})^{\perp\perp})^{\perp\perp})$ ("$Q(4)=Q(\bar{4})$")

(iii) $Q(K(V)^{\perp\perp}) = Q(K(\bar{V})^{\perp\perp})$ ("$Q(27)=Q(\overline{27})$")

(iv) $Q(\text{rad}V \cap K(V^{\perp})^{\perp\perp}) = Q(\text{rad}\bar{V} \cap K(\bar{V}^{\perp})^{\perp\perp})$ ("$Q(5)=Q(\bar{5})$")

(4) (v) $Q(K(V^{\perp})^{\perp\perp}) = Q(K(\bar{V}^{\perp})^{\perp\perp})$ ("$Q(36)=Q(\overline{36})$")

(vi) $g(V)=g(\bar{V})$

(vii) $h(V)=h(\bar{V})$

(viii) $n(V)=n(\bar{V})$

(ix) $p(V)=p(\bar{V})$

Case II. $\dim V/\text{rad } V + \dim V^{\perp}/\text{rad}(V^{\perp}) < \aleph_0$ . In order that there is an isometry $\tilde{\tau}: E \to E$ with $\tilde{\tau}V = \bar{V}$ it is necessary and sufficient that (i), (ii), (iii), (vi) hold and, furthermore,

(x) $K(V^{\perp})^{\perp} \cong K(\bar{V}^{\perp})^{\perp}$ ("$47 \cong \overline{47}$") .

In Case II of the theorem we have

(5) $d = h = k = \ell = n = p = 0$ .

Since $\mathfrak{f} < \infty$ and $m < \infty$ by assumption we can chop off (finite dimensional) supplements of rad V in V and $\text{rad}(V^{\perp})$ in $V^{\perp}$ respectively. Since $V \cong \bar{V}$ , $V^{\perp} \cong \bar{V}^{\perp}$ we may, by the finite dimensional case of Arf's theorem (Chapter XV) assume without loss of generality that, in addition to (5),

(6) $\mathfrak{f} = m = 0$ .

If (5) and (6) hold then $\mathcal{V}(V)$ has the shape

(7)

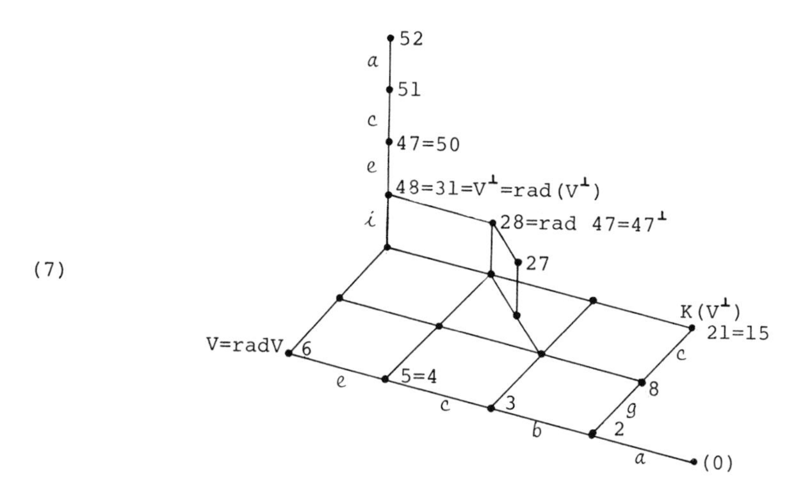


It is not difficult to see that, in general, the isometry class of 47 is not determined by the data (4) when (5) and (6) are assumed.

Let us look at (7) in the important special case where $\dim S/T = 1$. If $b + c + i \neq 0$ then the space $28 = \mathrm{rad}\ 47$ contains a vector $x$ with $1 \in Q(x)$. In that case we can find a totally singular supplement $P$ of $28$ in $\mathrm{rad}(V^{\perp})$, $\dim P = e$. If $P'$ is any supplement of $\mathrm{rad}(V^{\perp})$ in $47$ then $P \oplus P'$ is nondegenerate and a sum of hyperbolic planes $k(r_i, r_i')$ with singular $r_i$, $r_i'$. Thus the isometry class of $K(V^{\perp})^{\perp}$ is still determined by the data (4) in the theorem. On the other hand, if $b + c + i = 0$ in the lattice (7) then (4) does not determine the isometry type of $47$. Example: Take $E = k(r,r')$ a hyperbolic plane with $Q(r) = 1$ and $V = (r)$; here the index $e$ is $1$ and all other indices are zero. The isometry class of $k(r,r') = 47$ depends on $Q(r')$ which we have not specified. This legitimizes assumption (x) in Case II of Theorem 2. Furthermore, we have established the following

Corollary. Assume that in addition the division ring has $\dim S/T = 1$ (e.g. a perfect commutative field with $\varepsilon = 1$ and $\mathbb{1}$ as involution). Let $(E,\Psi,Q)$ be a $\aleph_0$-dimensional nondegenerate quadratic space. The orbit of a subspace $V \subset E$ under the action of the orthogonal group of $E$ is characterized by the collection of the following invariants.

α) The fourteen indices (cardinal numbers) $a, b, c, d, e, f, g, h, i, k, \ell, m, n, p$ defined by the lattice $V(V)$ (cf. (1), (2));

β) the isometry class of $V^{\perp}$ in case $0 < m < \infty$ ;

γ) the isometry class of $V$ in case $0 < f < \infty$ $(\Rightarrow d = h = k = \ell = n = p = 0)$ ;

δ) the isometry class of the space $47 = K(V^{\perp})^{\perp}$ in case $0 \leq m+f < \infty$ and $e \neq 0$ and $b = c = d = h = i = k = \ell = n = p = 0$ .

## 5. Proof of Theorem 2: Construction of the initial triple $(\tilde{\tau}_o , W_o , \bar{W}_o)$

It is presupposed that the reader is familiar with the proof of Theorem 1 in Chapter IV.

*

Assume that we are given two subspaces $V , \bar{V}$ in $E$ which satisfy (4) in Thm. 2. We first construct two finite dimensional isometric subspaces $W_o , \bar{W}_o$ in $E$ . It will be the germ of an isometry $E \cong E$ which maps $V$ on $\bar{V}$ .

We start out by having a look at (17; 10, 14, 16; 9) in the lattice $V(V)$ of Section 2. We claim that there are isometric supplements $C , \bar{C}$ with $3 \oplus C = 4$ and $\bar{3} \oplus \bar{C} = \bar{4}$ . Because $Q(4) = Q(\bar{4})$ and $Q(3) = Q(\bar{3})$ we can pick any $C$ and find a suitable $\bar{C}$ i.e. such that $Q(\bar{C}) = Q(C)$ .

We write $9 = 8 \oplus B$ , $15 = 8 \oplus C''$ . Hence $\dim C'' = c$ . We have $9 \oplus C = 10$ , $9 \oplus C'' = 16$ , therefore $17 = 8 \oplus B \oplus C \oplus C''$ . Let

$\{x_i \mid 1 \le i \le c\}$ be a basis of a supplement of 9 in 14 . Because $x_i \in 17$ we can decompose

$$x_i = y_i + b_i + c_i + c_i''$$

with $y_i \in 8$ , $b_i \in \mathcal{B}$ , $c_i \in \mathcal{C}$ , $c_i'' \in \mathcal{C}''$ . We set $c_i' := y_i + c_i''$ . $\{c_i' \mid 1 \le i \le c\}$ is a basis of a supplement $\mathcal{C}'$ of 8 in 15 , $15 = 8 \oplus \mathcal{C}'$ . Moreover, $c_i + c_i' = x_i - b_i \in 14$ . We see that $\{c_i + c_i' \mid 1 \le i \le c\}$ is the basis of a supplement of 9 in 14 .

In exactly the same manner we obtain a supplement $\bar{\mathcal{C}}'$ of $\bar{8}$ in $\overline{15}$ , and bases $(\bar{c}_i)$ , $(\bar{c}_i')$ of $\bar{\mathcal{C}}$ and $\bar{\mathcal{C}}'$ respectively, such that $(\bar{c}_i + \bar{c}_i')$ is a basis of a supplement of $\bar{9}$ in $\overline{14}$ . Because $\mathcal{C} \cong \bar{\mathcal{C}}$ by definition of $\bar{\mathcal{C}}$ and trivially $\mathcal{C}' \cong \bar{\mathcal{C}}'$ by total singularity we obtain an isometry $\gamma: \mathcal{C} \oplus \mathcal{C}' \cong \bar{\mathcal{C}} \oplus \bar{\mathcal{C}}'$ such that $\gamma(c_i) = \bar{c}_i$ , $\gamma(c_i') = \bar{c}_i'$ . Hence it is clear that

$$(9) \qquad \gamma(14 \cap (\mathcal{C} \oplus \mathcal{C}')) = \overline{14} \cap (\bar{\mathcal{C}} \oplus \bar{\mathcal{C}}') .$$

In a similar fashion (with less difficulties however) we deduce from the assumptions in Theorem 2 the existence of supplements $\mathcal{B}$ and $\bar{\mathcal{B}}$ , $\mathcal{D}$ and $\bar{\mathcal{D}}$ , $E$ and $\bar{E}$ , $I$ and $\bar{I}$ , $K$ and $\bar{K}$ , $L$ and $\bar{L}$ with

$$\begin{array}{llll}
2 \oplus \mathcal{B} = 3 , & \bar{2} \oplus \bar{\mathcal{B}} = \bar{3} , & \mathcal{B} \cong \bar{\mathcal{B}} \\
4 \oplus \mathcal{D} = 5 , & \bar{4} \oplus \bar{\mathcal{D}} = \bar{5} , & \mathcal{D} \cong \bar{\mathcal{D}} \\
5 \oplus E = 6 , & \bar{5} \oplus \bar{E} = \bar{6} & E \cong \bar{E} \\
14 \oplus I = 27 , & \overline{14} \oplus \bar{I} = \overline{27} , & I \cong \bar{I} \\
33 \oplus K = 36 , & 33 \oplus \bar{K} = \overline{36} , & K \cong \bar{K} \\
37 \oplus L = 39 , & 37 \oplus \bar{L} = \overline{39} , & L \cong \bar{L}
\end{array}$$

We set

$$W_o := \mathcal{B} \oplus \mathcal{C} \oplus \mathcal{C}' \oplus \mathcal{D} \oplus E \oplus I \oplus K \oplus L$$
$$\bar{W}_o := \bar{\mathcal{B}} \oplus \bar{\mathcal{C}} \oplus \bar{\mathcal{C}}' \oplus \bar{\mathcal{D}} \oplus \bar{E} \oplus \bar{I} \oplus \bar{K} \oplus \bar{L}$$

As all spaces in $W_o$ are pairwise perpendicular, and similarly in $\bar{W}_o$, we obtain an isometry

$$(10) \qquad \tilde{\tau}_o : W_o \cong \bar{W}_o$$

by joining the isometries $\mathcal{B} \cong \bar{\mathcal{B}}$, $\mathcal{C} \oplus \mathcal{C}' \cong \bar{\mathcal{C}} \oplus \bar{\mathcal{C}}'$, ... . Let $A \mapsto \bar{A}$ be the natural lattice isomorphism $\tau: \mathcal{V}(V) \to \mathcal{V}(\bar{V})$ which sends $V$ into $\bar{V}$. We assert that

$$(11) \qquad \tilde{\tau}_o(W_o \cap A) = \bar{W}_o \cap \bar{A} \qquad \text{for all } A \in \mathcal{V}(V)$$

$$(12) \qquad (W_o + A) \cap (W_o + B) = W_o + (A \cap B) \qquad \text{for all } A, B \in \mathcal{V}(V$$

$$(13) \qquad (\bar{W}_o + A) \cap (\bar{W}_o + B) = \bar{W}_o + (A \cap B) \qquad \text{for all } A, B \in \mathcal{V}(\bar{V}$$

As an illustration we shall verify (11).

Let $A = 14$. $14 \cap (\mathcal{B}+\mathcal{C}+\mathcal{C}'+\cdots+\mathcal{L}) = \mathcal{B} + 14 \cap (\mathcal{C}+\mathcal{C}'+\cdots+\mathcal{L}) = \mathcal{B} + 14 \cap (\mathcal{C}+\mathcal{C}')$. The assertion now follows from (9).

Let $A = 27$. $27 \cap (\mathcal{B}+\mathcal{C}+\mathcal{C}'+\cdots+\mathcal{L}) = 27 \cap (\mathcal{B}+\mathcal{C}+\mathcal{C}'+\mathcal{I}) = (\mathcal{B}+\mathcal{I}) + 27 \cap (\mathcal{C}+\mathcal{C}') = (\mathcal{B}+\mathcal{I}) + (27 \cap (\mathcal{C}+\mathcal{C}') \cap 17) = (\mathcal{B}+\mathcal{I}) + 14 \cap (\mathcal{C}+\mathcal{C}')$; hence the assertion by (9).

Let $A \neq 14, 27$. If $A$ is not one of $14, 27$ it is clear, by distributivity, that in $A \cap (\mathcal{B}+\mathcal{C}+\mathcal{C}'+\cdots+\mathcal{L})$ we may omit those summands of $W_o$ which lie "above" $A$ in the lattice. E.g. if $A = 18$ then we find by making use of modularity that $A \cap (\mathcal{B}+\mathcal{C}+\mathcal{C}'+\cdots+\mathcal{L}) = (\mathcal{B}+\mathcal{C}+\mathcal{C}'+\mathcal{D}) + 18 \cap (\mathcal{E}+\mathcal{I}+\mathcal{K}+\mathcal{L})$ and $18 \cap (\mathcal{E}+\mathcal{I}+\mathcal{K}+\mathcal{L}) = (0)$.

## 6. Proof of Theorem 2: The general step in Case I

We assume that we have constructed finite dimensional subspaces $W, \bar{W} \subset E$ and an isometry $\tilde{\tau}: W \to \bar{W}$ which extends (10) and which satisfies the three conditions analogous to (11), (12), (13). Given a prescribed vector $x \in E \setminus W$ we now show how to pick $\bar{x} \in E$ such that $\tilde{\tau}$

can be extended to an isometry $W \oplus (x) \to \bar{W} \oplus (\bar{x})$ which again satisfies the conditions (11), (12), (13) with $W \oplus (x)$ in lieu of $W$, etc.

Let then $\mathcal{M} = \mathcal{M}(W,x) := \{A \in \mathcal{V}(V) \mid x \in W+A\}$. $\mathcal{M}$ is a filter in view of (12) and, because $\mathcal{V}(V)$ is finite, it has a generator $D = D(W,x)$. In case $D$ is join reducible, say $D = D_1 + D_2$ with $D_1 \subsetneqq D$, $D_2 \subsetneqq D$ we express $x$ as $x = x_1 + x_2 + w$ $(w \in W, x_i \in D_i)$. By an induction on the height of $D$ we can adjoin $x_1$ to $W$ and then $x_2$ to $W + (x_1)$. The real task is of course with $D$ a join-irreducible element of $\mathcal{V}(V)$.

Since $W_o$, as defined in the previous section, is contained in $W$ the only candidates for join-irreducible $D$ as generators of $\mathcal{M}(W,x)$ are the following spaces (use the diagram in Section 2 and the legend that goes with it)

(14) 2, 8, 21, 7, 42, 41, 45, 46, 47, 48, 50, 51, 52

We may assume that $x \in D$. In order to find $\bar{x}$ we proceed as follows. As $\tilde{\tau}: W \to \bar{W}$ is an isometry for the quadratic form $(\Psi,Q)$ it certainly is an isometry for its associated trace-valued form $\Psi$. Hence by the constructions performed in IV.5 we can find a vector

$$\bar{x}_o \in \bar{D} \setminus (\bar{D}_o + \bar{W})$$

with $\Phi(\bar{x}_o, \tilde{\tau}w) = \Phi(x,w)$ for all $w \in W$ (here $\bar{D}$ is the element of $\mathcal{V}(V)$ that corresponds to $D$; $\bar{D}_o$ is its antecedent). We shall now look out for a vector $\bar{x}_1$ such that $\bar{x}_1 \in \bar{x}_o^{\perp} \cap \bar{W}^{\perp} \cap \bar{D}$ and $Q(\bar{x}_1) = Q(x) + Q(\bar{x}_o)$. Such an element $\bar{x}_1$ will enable us to extend $\tilde{\tau}$ to an isometry for $Q$ by sending $x$ into $\bar{x}_o + \bar{x}_1$. In order to save induction assumptions (12) and (13) we have to pick $\bar{x}_1$ such that

(15) $$\bar{x}_o + \bar{x}_1 \notin \bar{D}_o + \bar{W}.$$

Let us go through the list in (14). Spaces 2, 8, 21 are totally singular, hence we are done with $\bar{x}_1 = 0$ . If $0 < \mathfrak{f} < \infty$ we chop off a supplement of rad V in V and may assume $\mathfrak{f} \in \{0, \aleph_0\}$ . If space 7 shows up then $\mathfrak{f} = \aleph_0$ . Let $\bar{F}$ be a supplement of rad $\bar{V}$ in $\bar{V}$ . The radical of $\bar{F} \cap \bar{x}_o^{\perp} \cap \bar{W}^{\perp}$ is finite dimensional. Hence $\bar{F} \cap \bar{x}_o^{\perp} \cap \bar{W}^{\perp}$ contains an infinite dimensional nondegenerate space P . P contains hyperbolic planes $k(e_1, e_2)$ with $e_i$ totally singular because there is but one isometry class in dimension $\aleph_0$ (by our assumption on the fields in this chapter). Thus P contains vectors $\bar{x}_1$ of arbitrarily prescribed length $Q(\bar{x}_1) \in k/T$ . As $\dim P = \aleph_0$ we can certainly satisfy (15) by a suitable choice. This takes care of space $7 = V$ . Candidate $42 = V^{\perp}$ is treated alike. The same with $41 = V^{\perp\perp}$ (it does not show up when $\mathfrak{f} = 0$ for $\mathfrak{f} < \infty$ implies $n = 0$). The argument above with space 7 can be repeated for all the remaining spaces 45, ..., 52 in (14): as $\mathfrak{f}+m = \aleph_0$ each of these spaces X has $\dim X/\text{rad } X = \aleph_0$ .

From (15) it follows that $\bar{D}$ is the generator of the filter $M(\bar{W}, \bar{x}_o+\bar{x}_1)$ and this in turn makes the verification of the induction assumption (11) quite easy (refer to IV.5). The verification of the induction assumption (12) - with $W \oplus (x)$ in place of W - runs smooth because filters $M$ with generators D from the list (14) are prime (i.e. the elements in (14) are join-prime). This is exactly what our construction of $W_o$ in Section 5 was aimed at. We refer to IV.5 for details of the calculation. The verification of (13) is mutatis mutandis that of (12).

The proof of Theorem 2 in Case I is thus complete.

## 7. Proof of Theorem 2: Case II

The program is the same as that outlined at the beginning of the previous section. In Case II we assume that (5) and (6) hold. By looking at the diagram (7) we see that the join-irreducible $D$ which are candidates for generators of the filters $M(W,x)$ are the spaces 2, 8, 47, 51, 52 . The first two among them are totally singular and nothing remains to be done in these cases. As $\delta < \infty$ and $\dim V = \aleph_0$ we have $\alpha = \aleph_0$ ; hence we have again no problem with condition (15) when $D = 52$.
We are left with the two "hard" cases

$$D = 47 , 51 .$$

We shall not meet a new kind of problems here. The solution in Case II is simply more cumbersome because it is indispensable to closely look at the lattice. This is necessary because we lack the buffer spaces between $\mathrm{rad}\, V$ and $V$ (and between $\mathrm{rad}(V^{\perp})$ and $V^{\perp}$) which freely allowed for correction manoeuvers in Case I.

<u>Solution for $D = 51$</u> . Let $x \in 51 \setminus (50+W)$ be prescribed. We first find a vector $\bar{x}_o \in \overline{51} \setminus (\overline{50}+\bar{W})$ with $\Psi(\bar{x}_o,\tilde{\tau}w) = \Psi(x,w)$ for all $w \in W$ . (This by the method of IV.5.)

This means that $\bar{x}_o$ forms the correct "angles" $\Psi(\bar{x}_o,\bar{W})$ and has the right location vis-à-vis the lattice $V(\bar{V})$ . There remains to adjust the "length". If $Q(\bar{x}_o) = Q(x)$ we are done. Otherwise pick $t \in \overline{15}$ with $\Psi(\bar{x}_o,t) = 1$ (possible because $\bar{x}_o \notin \overline{50} = \overline{15}^{\perp}$). For $\kappa \in Q(x)$ , $\kappa_o \in Q(\bar{x}_o)$ we set $\bar{x}_1 := (\kappa+\kappa_o)t$ ; the vector $\bar{x}_o + \bar{x}_1$ is in $\overline{51} \setminus (\overline{50}+\bar{W})$ and has the right length $Q(\bar{x}_o+\bar{x}_1) = Q(x)$ . We may have altered the angles somewhat: decompose $W = (W\cap 15^{\perp}) \oplus V_1$ , $\bar{W} = (\bar{W}\cap\overline{15}^{\perp}) \oplus \tilde{\tau}V_1$ ; relative to $\bar{W} \cap \overline{15}^{\perp}$ the angles do not change if we switch from $\bar{x}_o$ to $\bar{x}_o + \bar{x}_1$ and the rest can be remedied as follows.

By the choice of $\bar{V}_1 := \tilde{\tau}V_1$ we have $\bar{V}_1 \cap \overline{15}^{\perp} = (0)$ ; we also have $(\bar{V}_1 \oplus (\bar{x}_o+\bar{x}_1)) \cap \overline{15}^{\perp} = (0)$ since otherwise $\bar{x}_o + \bar{x}_1 \in \overline{15}^{\perp} + \bar{V}_1 \subset \overline{50} + \bar{W}$ contradicts the choice of $\bar{x}_o$ . Hence we can find $\bar{x}_2 \in \overline{15}$ with prescribed angles on $V_1 \oplus (\bar{x}_o+\bar{x}_1)$ . The obvious choice is

$$(16) \qquad \begin{aligned} &\Psi(\bar{x}_2 , \tilde{\tau}w) = \Psi(x,w) + \Psi(\bar{x}_o+\bar{x}_1 , \tilde{\tau}w) , \quad w \in V_1 \\ &\Psi(\bar{x}_2 , \bar{x}_o+\bar{x}_1) = 0 . \end{aligned}$$

The vector $\bar{x} := \bar{x}_o + \bar{x}_1 + \bar{x}_2$ satisfies all requirements: $Q(\bar{x}) = Q(x)$ , $\Psi(\bar{x} , \tilde{\tau}w) = \Psi(x,w)$ for all $w \in W$ , $\bar{x} \in \overline{51} \setminus (\overline{50}+\bar{W})$ . This solves the case where $D = 51$ . We see that in Case II ($\delta+m < \infty$). the only critical case which remains is $D = K(V^{\perp})^{\perp} = 47$ . Hence we have the

<u>Corollary</u>. Let $e = \delta = m = 0$ (and hence $d = h = k = \ell = n = p = 0$ as well). In order that there is an isometry of $E$ which maps $V$ onto $\bar{V}$ the conditions (i), (ii), (iii), (vi) in (4) of Theorem 2 are necessary and sufficient.

*

<u>Solution for $D = 47$</u> . The plan is to reduce the problem to the situation of the Corollary by chopping off $2 \cdot e$-dimensional orthogonal summands of $E$ .

Because $47 \cong \overline{47}$ there are isometric supplements $K_o$ , $\bar{K}_o$ of $28$ in $47$ and $\overline{28}$ in $\overline{47}$ respectively. Then we have

$$(17) \qquad E = K_o \oplus K_o^{\perp} = \bar{K}_o \oplus \bar{K}_o^{\perp} .$$

By Arf's Theorem (for finite dimensions) we have $\bar{K}_o^{\perp} \cong K_o^{\perp}$ . It is obvious however that these supplements have to be chosen <u>suitably</u> if we are to arrive at an isometry $E \cong E$ which maps $V$ onto $\bar{V}$ . We should like to have that

$$(18)\quad \begin{array}{l} V = (V\cap K_o) \oplus (V\cap K_o^{\perp}) \quad , \quad \bar{V} = (\bar{V}\cap\bar{K}_o) \oplus (\bar{V}\cap\bar{K}_o^{\perp}) \\ V \cap K_o \cong \bar{V} \cap \bar{K}_o \end{array}$$

in order to make some use of (17). If (18) can be achieved then we extend the isometry between $V \cap K_o$ , $\bar{V} \cap \bar{K}_o$ to an isometry $K_o \cong \bar{K}_o$ (by Arf's theorem). Provided that there is an isometry $K_o^{\perp} \cong \bar{K}_o^{\perp}$ which maps $V \cap K_o^{\perp}$ onto $\bar{V} \cap \bar{K}_o^{\perp}$ we can simply join the two isometries to get an isometry of $E$ which, by (18), maps $V$ onto $\bar{V}$ . Our proviso is easily taken care of: if (18) can be achieved then $V_o := V \cap K_o^{\perp}$ in $E_o := K_o^{\perp}$ and $\bar{V}_o := \bar{V} \cap \bar{K}_o^{\perp}$ in $\bar{E}_o = \bar{K}_o^{\perp}$ $(\cong E_o)$ will qualify for the Corollary and we are done.

Thus we are left with the following _task_: Show that isometric $K_o$ and $\bar{K}_o$ can be chosen such that (18) holds. Let

$$\tau_1 : V \to \bar{V} \quad , \quad \tau_2 : K(V^{\perp})^{\perp} \to K(\bar{V}^{\perp})^{\perp}$$

be the isometries guaranteed by the assumptions of Theorem 2. Choose supplements $V_2$ of 5 in 6 and $K_1$ of 21 in 28 , say $K_1 = \mathcal{B} + \mathcal{C} + \mathcal{I}$ (Sec. 5). The assignement $\tau_3(x+y) := \tau_1(x) + \tau_2(y)$ for all $x \in V_2$ , $y \in K_1$ defines an injective isometry

$$\tau_3 : V_2 \oplus K_1 \to \overline{47} .$$

For, $\tau_1$ must map a supplement of 5 in 6 into a supplement of $\bar{5}$ in $\bar{6}$ as it respects $Q$ . $\tau_2$ preserves, of course, the radical of $K(V^{\perp})^{\perp}$ which is 28 . Hence $\tau_3(V_2) \cap \tau_3(K_1) = (0)$ . Since $\tau_3(V_2) \perp \tau_3(K_1)$ our assertion is clear. By Arf's Theorem as formulated in Chap. XV we can extend $\tau_3$ to all of a supplement $P$ of 21 in 47 : For, let $\bar{P}$ be a supplement of $\overline{21}$ in $\overline{47}$ that contains $\tau_3(V_2 \oplus K_1)$ ; $P \cong \bar{P}$ because $47 \cong \overline{47}$ . $K_1$ is the radical of $P$ and $\tau_3(K_1)$ is the radical of $\bar{P}$ . Since the domain of $\tau_3$ contains the radical of $P$ and $\tau_3$ maps it onto the radical of $\bar{P}$ it is now clear that $\tau_3$ can be extended. From the existence of this extension we learn that we can enlarge $V_2$ to a supplement $K_o$ of $28$ in $47$ which is mapped - under this extension of $\tau_3$ - onto an isometric supplement $\bar{K}_o$ of $\overline{28}$ in $\overline{47}$ , $\bar{K}_o \supset \tau_3(V_2) = \tau_1(V_2)$ . Now we have

$$E = K_o \oplus K_o^{\perp} = \bar{K}_o \oplus \bar{K}_o^{\perp} ,$$

$$\tau_3 : K_o \cong \bar{K}_o , \text{ hence } K_o^{\perp} \cong \bar{K}_o^{\perp} ,$$

$$V = (V \cap K_o) + (V \cap K_o^{\perp}) , \text{ to wit, } V = V_2 \oplus 5$$

$$(5 \subset 47^{\perp} \subset K_o^{\perp}) ; \text{ similarly}$$

$$\bar{V} = (\bar{V} \cap \bar{K}_o) + (\bar{V} \cap \bar{K}_o^{\perp}) , \text{ namely, } \bar{V} = \tau_3(V_2) \oplus \bar{5} .$$

$$V^{\perp} = (V \cap K_o) \oplus V^{\perp} \cap K_o^{\perp}$$

$$28 = K((\mathrm{rad}(V \cap K_o^{\perp}))^{\perp\perp})^{\perp\perp}$$

$$K(V^{\perp})^{\perp} = K_o \oplus 28$$

Since $V^{\perp} \subset K(V^{\perp})^{\perp}$ we read off that $V^{\perp} \cap K_o^{\perp} \subset 28$ and hence $V^{\perp} \cap K_o^{\perp} = 28$ . (Mutatis mutandis for $\bar{V} \cap \bar{K}_o^{\perp}$ in $\bar{K}_o^{\perp}$ .) This tells us in one stroke that the subspace $V_o := V \cap K_o^{\perp}$ in $E_o := K_o^{\perp}$ has "47 = 28" which is precisely the situation of the Corollary above. Our task is thus finished and so is the proof of Theorem 2 in Case II.

## 8. The irreducible objects

Let $(E_\iota, \Psi_\iota, Q_\iota)$ $(\iota \in I)$ be a family of nondegenerate quadratic spaces over the same division ring and $V_\iota$ a subspace of $E_\iota$ . If $E$ is the (external) orthogonal sum of the $E_\iota$ and $V$ the sum $\Sigma V_\iota$ in $E$ then we call $(V,E)$ <u>the sum of the pairs</u> $(V_\iota, E_\iota)$ , $(V,E) = \sum_{\iota \in I} (V_\iota, E_\iota)$ . A pair $(V,E)$ is called <u>reducible</u> if it is a sum $(V_1,E_1) + (V_2,E_2)$ such that $K(V^{\perp}) = K(V_1^{\perp}) + K(V_2^{\perp})$ and the dimensions of both $Q(\mathrm{rad}(V_1^{\perp}))$ and $Q(\mathrm{rad}(V_2^{\perp}))$ in the value space $S/T$ are nonzero.

One can give a complete collection of isometry classes of irreducible pairs $(V,E)$ by listing canonical representatives with dim $E \leq \aleph_0$ ; it can be found in [4]. As a result of the endeavor one obtains the following corollaries.

Corollary 1. Let $(E,Q)$ be a nondegenerate $\aleph_0$-dimensional quadratic space over a division ring as specified in the introduction and $V$ an infinite dimensional subspace. Then the pair $(V,E)$ is isometric to a finite sum of irreducible pairs from the list; the summands are not (and cannot be) uniquely determined in general.

Corollary 2. Let $V$ and $(E,Q)$ be as in Cor. 1 and $\Psi$ the associated sesquilinear form. The following are the only relations among the indices defined in (1) of Section 3.

1. $b + c + d + e + i + k + \ell \leq \dim S/T < \infty$
2. $a < \infty \Rightarrow b = c = g = i = 0$
3. $f < \infty \Rightarrow d = h = k = \ell = n = p = 0$
4. $h < \infty \Rightarrow d = k = 0$
5. $m < \infty \Rightarrow p = 0$
6. $a + f < \infty \Rightarrow m = \aleph_0$
7. $m < \infty$ & $\Psi$ is alternate $\Rightarrow$ $m$ is even
8. $f < \infty$ & $\Psi$ is alternate $\Rightarrow$ $f$ is even

*

## References to Chapter XVI

[1] C. Arf, Untersuchungen über quadratische Formen in Körpern der Charakteristik 2, J. reine angew. Math. 183 (1941), 148-167.

[2] F. Bolli, Verallgemeinerung des Verbands von Glauser, Master's Thesis, University of Zurich 1977.

[3] H.R. Glauser, Quadratische Formen in unendlichdimensionalen Vektorräumen im Falle von Charakteristik 2, Ph. D. Thesis, University of Zurich 1976.

[4] H. Gross, Untersuchungen über quadratische Formen in Körpern der Charakteristik 2, J. reine angew. Math. 297 (1978), 80-91.

APPENDIX I

# QUATERNIONS IN CHARACTERISTIC 2 AND A REMARK ON THE ARF INVARIANT A LA TITS

1. The multiplication table of quaternions is well known if the characteristic is not 2. To be sure, the table does remain meaningful if $2 = 0$ ; but the resulting algebra is commutative. On the other hand, if we think of the quaternion algebra as a crossed product of a separable quadratic extension of the center with its Galois group then this also makes sense for arbitrary characteristic and, this time, we obtain a noncommutative structure when the characteristic is 2. One is lead to an (associative) algebra of dimension 4 over its center $K$ which possesses a basis $e_0$ , $e_1$ , $e_2$ , $e_3$ with the following multiplication table $(\alpha, \beta \in K)$ :

(1)

| | $e_0$ | $e_1$ | $e_2$ | $e_3$ |
|---|---|---|---|---|
| $e_0$ | $e_0$ | $e_1$ | $e_2$ | $e_3$ |
| $e_1$ | $e_1$ | $\alpha e_0+e_1$ | $e_3$ | $\alpha e_2+e_3$ |
| $e_2$ | $e_2$ | $e_2+e_3$ | $\beta e_0$ | $\beta e_0+\beta e_1$ |
| $e_3$ | $e_3$ | $\alpha e_2$ | $\beta e_1$ | $\alpha\beta e_0$ |

Definition. Let $K$ be a commutative field of characteristic 2 and $\alpha$ , $\beta$ two elements of $K$ . An algebra $\left(\frac{\alpha,\beta}{K}\right)$ of quaternions is a 4-dimensional K-vector space with a basis $e_0$ , $e_1$ , $e_2$ , $e_3$ and equipped with a K-bilinear multiplication such that the elements $e_i$ multiply according to (1).

Given any $K$ and $\alpha$ , $\beta \in K$ the bilinear extension of (1) to the K-space of (formal) linear combinations $\lambda_0 e_0 + \lambda_1 e_1 + \lambda_2 e_2 + \lambda_3 e_3$

turns out associative. Hence there always is a quaternion algebra $\left(\frac{\alpha,\beta}{K}\right)$ .

If we replace $e_1$ by $e_1' := e_1 + e_0$ we obtain from (1) a table relative to the new basis $\{e_0, e_1', e_2, e_3\}$ and we see that this new table is just the "transpose" of the old one. In other words, the assignment $e_0, e_1, e_2, e_3 \mapsto e_0, e_1', e_2, e_3$ induces an isomorphism of $\left(\frac{\alpha,\beta}{K}\right)$ onto the opposite algebra $\left(\frac{\alpha,\beta}{K}\right)^{o}$ . In short,

(2) $$q = \lambda_0 e_0 + \lambda_1 e_1 + \lambda_2 e_2 + \lambda_3 e_3 \longmapsto q^* = (\lambda_0 + \lambda_1) e_0 + \lambda_1 e_1 + \lambda_2 e_2 + \lambda_3 e_3$$

is an involutory antiautomorphism of $\left(\frac{\alpha,\beta}{K}\right)$ . Calculation gives

(3) $$qq^* = q^*q = [\lambda_0^2 + \lambda_0\lambda_1 + \alpha\lambda_1^2 + \beta(\lambda_2^2 + \lambda_2\lambda_3 + \alpha\lambda_3^2)]e_0 ,$$
$$q \mapsto qq^* \text{ is multiplicative.}$$

<u>Example</u>. Let $K_0$ be any commutative field, $\operatorname{char} K_0 = 2$ , and $K = K_0(X,Y)$ where $X$ , $Y$ are algebraically independent over $K_0$ . The algebra $\left(\frac{X,Y}{K}\right)$ is a division ring because we find that $qq^* = 0$ if and only if $q = 0$ ; hence for $q \neq 0$ we have $q^{-1} = \mu q^*$ where $\mu = (qq^*)^{-1}$ .

<u>Remark 1</u>. Let $A$ be any noncommutative division ring of characteristic 2 with center $K$ and $\lambda \mapsto \lambda^*$ an antiautomorphism which leave $K$ pointwise fixed. Assume that $(A,*)$ is <u>reflexive</u> in the sense that $\lambda\lambda^*$ belongs to the center for all $\lambda \in A$ . From $(\lambda+1)(\lambda+1)^* \in K$ we conclude that $\lambda + \lambda^* \in K$ for all $\lambda \in A$ . Hence each $\lambda \in A$ is quadratic over $K$ , $\lambda^2 + (\lambda+\lambda^*)\lambda + \lambda\lambda^* = 0$ . One can prove that $A$ is a quaternion algebra and $*$ coincides with the "conjugation" defined in (2) ([1], p. 72 and p. 84).

<u>Remark 2</u>. In [2] (Theorem 7, p. 205) I. Kaplansky proved that if a commutative nonformally real field of characteristic not 2 admits a

unique quaternion division algebra over K (up to algebra isomorphism) then each nondegenerate form over K in dimension 4 is universal, i.e. has vectors of any length (see also [4] p. 152 and pp. 319-324 for the relevance of this topic). H.A. Keller proved the corresponding result for K of characteristic 2 ([3], Teorema 1, p. 79). A spandy new proof had to be devised for this purpose.

2. Let $k = \left(\frac{\alpha,\beta}{K}\right)$ be a division algebra, char $K = 2$. We shall consider 2-dimensional k-spaces $E$ equipped with quadratic forms $(\Psi, Q)$ in the sense of Definition 2 in XIV.3. We choose $\varepsilon = 1$ in the underlying structure $(k,*,\varepsilon)$ and let $*$ be as defined in (2). We see that the additive subgroup $P = \{\varepsilon\xi^* - \xi \mid \xi \in k\}$ coincides with the center $K$ of $k$. For $\lambda \in k$ we let $[\lambda]$ be its class in $k/P$.

Let $g_1$, $g_2$ be a basis of $E$ and $\alpha$, $\beta$ symmetric elements in $k$, $\alpha = \alpha^*$, $\beta = \beta^*$. We define a quadratic form $(\Psi, Q)$ on $E$ by setting

$$(4) \qquad \begin{array}{l} Q(g_1) = [\alpha], \quad Q(g_2) = [\beta], \\ \Psi(g_1,g_1) = \Psi(g_2,g_2) = 0, \quad \Psi(g_1,g_2) = 1. \end{array}$$

With each "symplectic" basis in $E$ of the kind (4) we can associate the element

$$T(\alpha\beta) := \alpha\beta + (\alpha\beta)^* .$$

Let $G$ be the additive subgroup $\{\xi^2+\xi \mid \xi \in K\}$ in $K$. It is not difficult to verify that the class of $T(\alpha\beta)$ in the factor group $K/G$ does not depend on the symplectic basis chosen; i.e. this class is an invariant attached to the plane $(E,\Psi,Q)$. (Use the fact that the group of symplectic $2\times 2$ matrices is generated by matrices of the type $\begin{pmatrix} 0 & 1 \\ 1 & 0 \end{pmatrix}$, $\begin{pmatrix} . & 0 \\ 0 & . \end{pmatrix}$, $\begin{pmatrix} 1 & 0 \\ . & 1 \end{pmatrix}$.) We call this class the <u>Arf invariant</u> of

$(E,\Psi,Q)$ . This definition is obtained by particularizing the definition by Tits (Corollaire 4, p. 37 in [5]) to the present situation. In the classical situation of a commutative field one has the fundamental fact that two planes $(E,\Psi,Q)$ , $(E',\Psi',Q')$ are isometric if and only if they have equal Arf invariants and contain a vector $e \in E$ , $e' \in E'$ of common length, $Q(e) = Q'(e')$ (for this reason the Arf invariant is also called pseudo-discriminant). We shall now show by an example due to H.A. Keller that in the nonclassical situation this fundamental property of the invariant is lost.

3. We keep $k$ as in the previous section and consider a second plane $(E',\Psi',Q')$ with $Q(g_1') = [\alpha]$ , $Q(g_2') = [\gamma]$ where $\{g_1',g_2'\}$ is a symplectic basis of $E'$ relative to $\Psi'$ . We shall arrange for $T(\alpha\beta) = T(\alpha\gamma)$ and $E$ , $E'$ being nonisometric.

Suppose that $(E,\Psi,Q)$ and $(E',\Psi',Q')$ are isometric. Then by Witt's Theorem (Thm. 1 in XV.2) there must exist an isometry

$$g_1 \mapsto g_1' , \quad g_2 \mapsto \omega g_1' + g_2' .$$

It follows that $\omega = \omega^*$ and $\alpha\omega\cdot\alpha\omega + \alpha\omega = \alpha(\beta+\gamma) + \alpha\upsilon$ for some $\upsilon \in K$. Thus, in order to produce our counter example it suffices to exhibit a quaternion algebra $k$ over $K$ and symmetric elements $\alpha , \beta , \gamma \in k$ with $T(\alpha\beta) = T(\alpha\gamma)$ such that

$$\xi^2 + \xi = \alpha(\beta+\gamma) + \alpha\upsilon \qquad (5)$$

admits no solution $\xi$ for any $\upsilon \in K$ .

We start out with a field $k_0$ sharing the following property: there is $d \in k_0$ such that the equations

$$(6) \qquad x^2 + x = 1$$

$$(7) \qquad x^2 + x = d$$

$$(8) \qquad x^2 + x = 1 + d$$

have no solutions in $k_0$ . An example would be $k_0 = \mathbb{Z}_2((X))$ with $d = X^{-1}$ (write $x$ as a powerseries and check the parity of the lowest exponent of $X$ in $x^2+x$ ). Set $K = k_0((t))$ . The equations (6), (7), (8) remain unsolvable over $K$ (if $x = at^n+\cdots$ solves one of the three equations we must have $n=0$ so $x^2+x = (a^2+a)+\cdots$ and we find a solution in $k_0$ , contradiction). In particular, as (7) is unsolvable over $K$ , we find that $k := \left(\frac{d,t}{K}\right)$ is a division algebra because $\lambda\lambda^* = 0$ $(\lambda \in k)$ if and only if $\lambda = 0$ . Let then $e_0 = 1$ , $e_1$ , $e_2$ , $e_3$ be a basis of $k$ as in (1). We choose

$$(9) \qquad \alpha = 1 + t^{-1}e_2 \ , \quad \beta = 1 \ , \quad \gamma = 1 + e_2 \ .$$

We have $T(\alpha\beta) = T(\alpha\gamma) = 0$ . We assert that (5) has no solution in $k$ for any $\upsilon \in K$ .

Indeed, taking traces in (5) we obtain $T(\xi)^2 + T(\xi) = 0$ and we shall therefore distinguish between two cases. If $T(\xi) = 1$ then $\xi^* = \xi + 1$ and so $\xi\xi^* = \xi^2 + \xi$ ; therefore

$$\xi\xi^* = \alpha(\beta+\gamma) + \alpha\upsilon = (1+t^{-1}e_2)(\upsilon+e_2) \ .$$

The right hand side is in the center if and only if $\upsilon = t$ so we obtain $\xi\xi^* = t(1+t^{-1}e_2)^2 = 1 + t$ . Because $T(\xi) = 1$ the element $\xi$ is of the form $\xi = \lambda_0 e_0 + 1\cdot e_1 + \lambda_2 e_2 + \lambda_3 e_3$ and thus

$$\xi\xi^* = (\lambda_0^2+\lambda_0+d) + t(\lambda_2^2+\lambda_2\lambda_3+d\lambda_3^2) = 1 + t \ .$$

But this is impossible because of the unsolvability in $k_0$ of (7) and (8). (We leave the caulking to the reader.) If, on the other hand, we should have that $T(\xi) = 0$ then $\xi^2 + \xi = \xi\xi^* + \xi$ . We put $\xi\xi^* = \mu$ and multiply the equation $\xi\xi^* + \xi = \alpha(\beta+\gamma) + \alpha\upsilon$ by its conjugate. We obtain $\mu^2 + \mu = \alpha^2(\beta+\gamma+\upsilon)^2$ , thus

$$\mu^2 + \mu = 1 + t + \upsilon^2 + \upsilon^2 t^{-1} .$$

Let $\upsilon = \gamma t^n + \cdots \in K$ , $\gamma \in k_0 \setminus \{0\}$ . If $n \leq 0$ then the right hand side in our last equation is a power series in $t$ that begins with a negative odd power of $t$ ; this yields a contradiction, obviously. Therefore $n > 0$ and the right hand side is of the form $1 + \cdots$ ; this time we arrive at a contradiction because of the unsolvability of (6) over $k_0$ . Q. E. D.

Summary. We have specified two quadratic planes $(E,\Psi,Q)$ , $(E',\Psi',Q')$ over a suitable quaternion division algebra such that $E$ and $E'$ contain vectors $e \in E$ , $e' \in E'$ with $Q(e) = [\alpha] = Q'(e')$ and have equal Arf invariants, $T(\alpha\beta) \equiv T(\alpha\gamma) \mod G$ , and such that $E$ , $E'$ are <u>not</u> isometric relative to the forms $Q$ , $Q'$ .

*

## References to Appendix I

[1] J. Dieudonné, Sur les groupes classiques. ASI 1040, Hermann Paris, 1958.

[2] I. Kaplansky, Quadratic forms. J. Math. Soc. Japan 5 (1953) 200-207.

[3] H.A. Keller, Algebras de cuaternios y formas cuadráticas sobre campos de característica 2. Notas matemáticas, Universidad Católica de Chile-Santiago, 8 (1978) 65-84.

[4] T.Y. Lam, The algebraic Theory of quadratic forms. W.A. Benjamin Inc., Reading Massachusetts, 1973.

[5] J. Tits, Formes quadratiques, groupes orthogonaux et algèbres de Clifford. Invent. math. 5 (1968) 19-41.

## Symbols and Notations

## Index of Names

Index